Best Practices Guide to Residential Construction

MATERIALS, FINISHES, AND DETAILS

Steven Bliss

WILEY

John Wiley & Sons, Inc.

Published by John Wiley & Sons, Inc., Hoboken, New Jersey

Published simultaneously in Canada

For general information about our other products and services, please contact our Customer Care Department within the United States at (800) 762-2974, outside the United States at (317) 572-3993 or fax (317) 572-4002.

Wiley also publishes its books in a variety of electronic formats. Some content that appears in print may not be available in electronic books. For more information about Wiley products, visit our web site at www.wiley.com.

Library of Congress Cataloging-in-Publication Data:

Bliss, Steven, 1951-
    Best practices guide to residential construction : materials, finishes, and details / Steven Bliss.
        p. cm.
    Includes bibliographical references and index.
    ISBN-10: 0-471-64836-1 (cloth)
    ISBN-13: 978-0-471-64836-9
    1. House construction. I. Title.
    TH4811.B583 2006
    690'.837—dc22   2005013964

Printed in the United States of America
10 9 8 7 6 5 4 3 2 1

To my daughters, Julia and Haley, who always inspire and delight me.

About the Author

Steven Bliss served as editorial director of *The Journal of Light Construction* for more than sixteen years and construction technology editor of *Progressive Builder* and *Solar Age* magazines. He worked in the building trades as a carpenter and design/build contractor for more than ten years and holds a masters degree from the Harvard Graduate School of Education.

Contents

Preface

The goal of this book is to help building professionals—builders, architects, designers, and tradespeople—make wise decisions about the materials they choose and how they are installed. When I started working in the building trades 35 years ago, the main choices for siding where I worked in southern New England were shingles and clapboards *or* vinyl or aluminum siding on low-end remodels. A low-slope roof got roll roofing or BUR (built-up roofing). Exterior trim was clear pine, or No. 2 pine if you were on a tight budget and did not mind shellacking the knots. Flashing was copper, lead, or aluminum. Ceramic tile went on mud, wood floors were solid oak and nailed in place, kitchen counters got "Formica," and, if the budget could swing it, picture windows were "thermopane."

So much has changed since then, it can make your head spin. Many of the exterior materials we now use have little or no wood in them at all, only some combination of plastics, polymers, cement, wood fibers, and fillers. Elastomeric membranes seal our windows, which have microscopic coatings on the glass that selectively reflect certain wavelengths of light, and exotic gases between the glass that add insulation value. Ceramic tile can go over an amazing range of substrates, ranging from waterproof gypsum (does such a thing really exist?) to a wide variety of cementitious boards, many with wood fiber or plastic aggregate. Many wood floors are "engineered" and use high-tech factory finishes with just about everything, including diamond dust, thrown in for extra durability. The range of composite and synthetic materials for kitchen counters seems endless.

What are these materials? How do they perform? What are their limitations? Their expected lifespan? You would have to be a chemist, physicist, and engineer with a lot of time on your hands to answer all these questions satisfactorily. Even then, you wouldn't really know the answers until you had wrestled with a material on a real job site and waited a few years to see how it held up over time. So, in practice, we read the promotional literature, ask our colleagues and the salespeople at the lumberyard, and make our best guess.

I've spent the last two decades at *The Journal of Light Construction* and, before that, at *Solar Age,* learning about new construction materials and technologies and trying to turn that knowledge into practical advice for people working in the field. Our formula for success at *The Journal* was to listen to the manufacturers, trade associations, and technical experts, but to base our final analysis largely on feedback from the people working in the field. In many ways, our readers were also our best source of information. They kept our information honest and relevant to the people who specify and install these materials and must stand behind them if they fail.

I've tried to stick to the same approach in this book, whenever possible, listening to people with hands-on experience to learn about practical solutions that have been tested on real job sites. I have also tried to give readers of this book enough understanding of the new materials that they can improvise workable solutions on the job site, as is so often required.

Working at construction magazines for 20 years, culminating in the writing of this book, combines two of my passions—for building and for writing. I hope you find the results a useful tool in your own work of designing and building durable and enjoyable living spaces.

Steven Bliss
October 2005

Acknowledgments

I owe thanks to many people for helping me complete this project. First, I would like to thank all the editors and authors at *The Journal of Light Construction* whom I have had the pleasure to work with over nearly two decades. While I received a good grounding in residential construction on the job site, it was my years at *JLC* that gave me the depth of understanding of the many topics that I have tried to convey in this book. In particular, I would like to thank the *JLC* authors who assisted with portions of this manuscript: Carl Hagstrom and Patricia McDaniel (for the section on flashings); Paul Fisette (sheathing wraps); Russell Kenney (EIFS); Ron Webber (stucco); Michael Poster (adhesives); Frank Woeste, P.E. (deck ledgers); and Michael Byrne (ceramic tile). Also, thanks to Danny Parker at the Florida Solar Energy Center, John Carmody at the Center for Sustainable Building at the University of Minnesota, and Christopher Gronbeck of Sustainable Design (www.susdesign.com) for its information on glazing systems; and to Frank Stewart of the Western Wood Products Association, Ted Bozzi of the National Kitchen and Bath Association, and Tony Benora of the Cedar Shingle and Shake Bureau. Many busy individuals from the manufacturing side of the industry also gave generously of their time. In particular, I would like to thank Randy Willard (Dupont); David Steiner and Andrew Visser (W. R. Grace); Brian LeVoguer (Bakor); David Olsen (Fortifiber); Jeff Pitcher (CP Industries); Rod Nichols (Pacific Wood Laminates); Mike Sundell (Basic Coatings); Pete Ragland (Gorilla Glue); Karla Krengle (Kitchens.com); Randy Blanchette (Lightolier); Richard Maurer (Noble Company); Joe Knife (Classic Products); Brian Liflie (Specialty Steel Industry of North America); Dwight Shackleford (American Aldes); David Robinson (Lifebreath); and David Hanson (Memphremagog Heat Exchangers). Special thanks to Julie Trelstad for believing in this project at the outset and to Clayton DeKorne for helping me refine the scope and organization of the book, to Josie Masterson-Glen for her invaluable help with copyediting, to Susan Bliss for her assistance with the manuscript, and to computer wizard Brad Bull, who bailed me out every time my aging laptop threatened to crash. Special thanks also to illustrator Peter Thompson for his beautifully rendered drawings and to Wiley illustrator Joni Burns for her many accurate and stylish drawings. Thanks also to Wiley editor John Czarnecki for maintaining high standards throughout the project and to production editor Leslie Anglin for her diligence and attention to detail in organizing the myriad pieces into a successful finished work. Finally, I wish to thank Fran, Julia, and Haley for their continued support and great patience during the long hours over many months devoted to this project.

CHAPTER ONE | Exterior Finish

Water leakage through building exteriors has been the source of numerous callbacks and lawsuits across the United States. In nearly every case, the problems have been traced back to missing or poorly designed flashings or to weather barriers that inadvertently directed large amounts of water into building cavities or interiors. Most of these leaks occur at window and door openings or at intersections between building components. In some cases, caulks and sealants forestalled leakage at these poorly designed joints for the first few years. But eventually most caulk joints fail, allowing water to enter.

All residential cladding systems are more or less porous to water, particularly during wind-driven rain when high air pressures on the windward side of a building force water to flow toward lower-pressure areas behind the siding. Under pressure, the water exploits butt joints, lap joints, nail holes, and other openings to flow inside (Figure 1-1). Even without wind, some water will migrate through tiny gaps to the back of siding through capillary action, the way water is siphoned up a stalk of celery. This is true of brick, wood, and stucco, as well as the newest composite materials.

In older construction, water that penetrated the outer cladding had ample opportunity to dry both to the interior and to the exterior as wind washed through the wall cavities, which were kept warm by heat leaking from the building's interior. In modern construction, however, with high levels of insulation, continuous air and vapor barriers, and low-perm sheathing panels, when water gets in, it is much slower to dry and more likely to cause damage.

While the exterior finish should be detailed to repel and shed water, a backup system is needed for the times when the primary system fails. The backup system needs to catch any water that penetrates the cladding and to drain it safely to daylight at the bottom of the wall. This backup layer, called a *water-resistive barrier* by the International

FIGURE 1-1 | Leakage Through Siding.

All exterior claddings allow some water to penetrate. The back of this beveled wood siding, installed directly to the studs, reveals significant water staining from leakage at nail holes and from splashback and snow buildup.
SOURCE: Photo by author.

Residential Code (IRC), typically consists of properly lapped building paper or plastic housewrap integrated with all flashings to safely drain water away. It is also called the *drainage layer* or *drainage plane*. In this approach, the outer cladding functions as a decorative "rain screen," slowing down wind and water, but it is not expected to be 100% waterproof.

RAIN-SCREEN PRINCIPLE

The optimal way to protect the structure, siding, and exterior finishes from moisture damage is to design the outer layer of the house as a decorative "rain screen" that is solid enough to shed rain, block wind, and protect the sheathing wrap, but porous enough to dry to the exterior when wet. This is accomplished by separating the outer cladding from the building's water-resistive barrier by using an air space. This system takes advantage of the fact that no siding system is entirely waterproof and relies, instead, on the drainage layer for waterproofing (see Figure 1-2).

The rain-screen system has four components: an exterior cladding, an air space, a drainage plane, and weep holes.

1. *Cladding.* While the main function of the exterior finish material in a rain-screen wall is aesthetic, its durability can have a big impact on the costs of home ownership. Frequent repair, repainting, or replacement can be very costly. The cladding also protects the sheathing wrap from wind and ultraviolet (UV) radiation, and sheds most of the water that strikes the side of the building. While some exterior claddings are more porous to water than others—for example, brick, vinyl, and vertical-wood sidings are particularly leak prone— all can function well with a proper drainage plane.

FIGURE 1-2 Rain-Screen Wall.

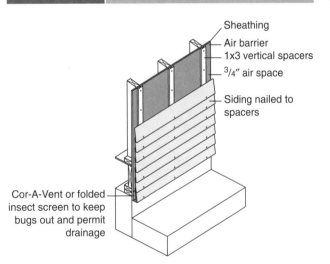

A rain-screen wall is the best guarantee against moisture problems with wood sidings, paints, and stains in harsh climates. The introduction of new draining housewraps promises to make the approach more practical and affordable.
SOURCE: Reprinted from *Fundamentals of Residential Construction*, Edward Allen and Rob Thalon, with permission of John Wiley & Sons, © 2001.

2. *Air space.* The air space serves several functions. First, it provides a space for any water that has penetrated the cladding to drain safely away. Second, it provides a capillary break between the cladding and the building paper. Wet wood siding or stucco has been shown to degrade both building paper and plastic housewrap if it is in direct contact with the wet cladding. Cedar and redwood sidings can leach out tannins that are particularly corrosive to building papers. Third, the air space helps promote drying from the back of wood siding or from the framing and sheathing in the event of a leak. With stained or painted wood sidings, the air space will significantly extend the life of the finish.

   Some siding materials, such as vinyl, aluminum, and wood shakes and shingles, are self-ventilating. For others, an air space can be created by installing vertical furring strips over the building paper. Although furring out the siding provides optimal protection for the siding and structure beneath, it also adds significant cost and complication to the job, so it is not commonly done. However, manufacturers are responding to this need with a variety of thin drainage materials that either install over the sheathing wrap or replace it (see "Draining Housewraps," page 5).

3. *Drainage plane.* The drainage plane typically consists of asphalt-impregnated building felt or a plastic housewrap that is fully integrated with all door, window, and wall flashings. The system must provide a clear drainage path out the bottom of the building. In general, the housewrap must be cut to lap over window and door cap flashings and under window and door sill flashings. In addition, the housewrap should lap over step flashings, the upper leg of abutting roof flashings, and deck ledger flashings. Upper courses of sheathing wrap should lap lower courses by at least 6 inches and vertical seams should lap 6 to 12 inches.

4. *Weep holes.* Any trapped water must freely drain to daylight at the bottom of the wall either through weep holes, as in brick veneer, through a weep screed in stucco, or out the bottom of vertical furring strips installed beneath wood siding. If furring strips are used, the openings at the bottom should be screened against insects. Short sections of corrugated plastic ridge vent material placed between furring strips work well to provide solid backing for the bottom course of siding.

Rigid Foam Sheathing

Although a rain-screen wall design will improve the longevity of any siding and finish, it is particularly critical when installing wood siding over foam sheathing. Research has shown that wood sidings installed directly over foam sheathings are more prone to cupping, cracking, and paint problems than when installed over wood sheathings. Wood sheathing acts as a reservoir for moisture that penetrates the siding. With foam, the moisture tends to build up on the back of the siding and cause problems. An air space, even a shallow space of $\frac{1}{4}$ to $\frac{1}{2}$ inch, between the siding and foam sheathing has been shown to reduce these problems.

SHEATHING WRAP

The primary goal of a sheathing wrap is to protect a building's structural components from water. At the same time, the sheathing wrap must be permeable enough to allow drying to the building's exterior if the framing or sheathing should get wet. While the permeance and water-resistance ratings of sheathing wraps vary significantly, how they are installed is far more important than the specific product used. The key is to always lap the sheathing wrap to shed water and to properly integrate the wrap with flashings so water is directed on top of the layer below.

All sheathing wraps fall into three basic types: asphalt felt, Grade D building paper, and synthetic housewrap. Grade D building paper is used primarily under stucco in the western United States and is essentially a lighter-weight version of asphalt felt. Comparing one material to another is difficult since there is no single standard for all products, and even where manufacturers follow the same test standard, test conditions may vary dramatically from one company to the next.

Code Requirements

The 2003 International Building Code calls for a "water-resistive barrier behind the exterior veneer" consisting of flashings and a "weather-resistant sheathing paper" lapped at least 2 inches horizontally and 6 inches vertically. It specifies asphalt-saturated felt that weighs at least 14 pounds per square and complies with ASTM D226, which means that most unrated No. 15 felt paper sold at lumberyards (which weigh closer to 7 pounds per square) do not comply.

Nearly all the plastic housewraps have been submitted to the model code authorities and accepted as substitutes for ASTM rated No. 15 building paper. If building in an area that follows the Model Energy Code (MEC), builders must either install a "vapor-permeable housewrap" on the exterior or seal all the penetrations in the building by using some combination of polyethylene, caulks, and gaskets on the interior.

Performance Measures

Permeance. Permeance ratings measure the rate at which water vapor passes through a material. One perm equals one grain of water vapor passing through one square foot of material per hour per inch of vapor-pressure difference. Under ASTM standard E96, manufacturers can use either test A (dry cup) or test B (wet cup), which yield somewhat different results. Grade D building paper uses yet another standard for measuring permeance, which is roughly equivalent to a permeance rating of about 5, similar to asphalt felt materials. Plastic housewraps range in permeance from around 5 to over 50.

In general, a sheathing wrap should have a permeance of at least 5 to enable wall assemblies to dry out should they get wet. Since common sheathing products like plywood and oriented-strand board (OSB) have permeance ratings of less than one, the sheathing is more likely to interfere with drying than the sheathing wrap.

Water Resistance. Several different tests are used to measure the ability of building papers to stop liquid water. Grade D building papers must have a 10-minute rating under ASTM D779, commonly called the "boat test," in which a piece of building paper is folded in the shape of a boat and floated in a dish of water until it soaks through and wets a powder on top. Some Grade D papers are rated as high as 60 minutes.

In general, products with very high vapor permeability, such as DuPont's Tyvek®, do poorly in the boat test since water vapor can pass through and wet the indicator powder. However, Tyvek® and other nonperforated plastic housewraps perform well in the alternate "hydro-head" test in which the material is placed under a 22-inch column of water and must not leak for five hours. More importantly, nonperforated plastic housewraps generally do a very good job of shedding liquid water in the field.

Air Infiltration. Many sheathing wrap suppliers tout their products' ability to block air infiltration, often citing proprietary test results. Some follow ASTM E283, in which an 8-foot-square wall section is tested before and after installation of the sheathing wrap. However, since the manufacturer is free to specify the type of wall assembly, one test is not comparable to another, and none simulates real job-site conditions with seams and holes in the sheathing wrap.

If a house already has a reasonably tight wall assembly, there is little evidence that a layer of housewrap will significantly tighten the building. In general, air-sealing efforts are better spent on the building's interior, using caulks and gaskets or a continuous polyethylene air/vapor barrier.

Sheathing-Wrap Materials

Installed carefully, any of the sheathing wraps can perform well and keep water out of walls. The three main choices are traditional asphalt felt, Grade D building paper, and the newer plastic housewraps. The optimal product will depend upon the siding choice, building details, and climate. With any sheathing wrap material, however, the key to good performance is to carefully lap the material to shed water. This job has been made easier by the introduction of a number of peel-and-stick membranes for use around windows, doors, and other trouble spots. General performance characteristics of sheathing wraps are summarized in Table 1-1.

TABLE 1-1	Sheathing Wrap Performance				
Material Type	Uses	Pros	Cons	Perms	Recommendations
Asphalt felt	Sheathing wrap under siding, roofing	Moderate resistance to liquid water. Absorbs and stores water. High permeability when wet.	Deteriorates under prolonged exposure to UV radiation. Prone to tear.	Approx. 5	Use min. 14 lb. material. Good option for all types of siding. Use with rain screen. Do not rely on for air infiltration barrier.
Grade D paper	Sheathing wrap under stucco	Resists water from 20 to 60 min. as per rating.	Deteriorates if saturated.	Approx. 5	Use two layers of rated material under stucco.
Plastic housewrap (perforated)	Sheathing wrap under siding	Good permeability.	Low resistance to liquid water	9 to 48	Use in relatively dry areas.
Plastic housewrap (nonperforated)	Sheathing wrap under siding	Most have high permeability and high resistance to liquid water. Creates air barrier if taped.	Could trap liquid water from leak.	6 to 59	Good option for all types of siding. Use with rain screen. Can also serve as air-infiltration barrier if taped at all seams and edges.
Draining housewraps	Sheathing wrap under siding. Drainage plane.	Provides drainage plane and air space for rain screen.	Relatively new and untested in field.	varies	Good option for all types of siding in areas subject to windblown rain.

Asphalt Felt. The old standby, asphalt felt, has a perm rating of around 5 and moderately good water resistance, making it suitable for use as a sheathing wrap. However, unlike plastic housewraps, asphalt felt will absorb water when wet. Once wet, its permeability jumps from around 5 to as high as 60. In the event of water leaking into the wall, asphalt felt may help store some of the water, and its high permeability when wet will promote drying to the exterior. Housewrap, in contrast, tends to trap any liquid water that gets behind it.

Some contractors find felt easier to install and weave into flashings because of its rigidity and narrow roll width. Felt, however, tends to get brittle and deteriorate under long-term exposure to UV radiation and is more prone to tear during installation than plastic housewraps. For situations where prolonged exposure is expected, plastic housewraps are better suited. Otherwise, asphalt building felt remains a valid choice for modern homes.

Although traditional 15-pound rag felt weighed 15 pounds per 100 square feet, the material sold today as No. 15 felt is made of recycled cardboard and sawdust and actually weighs only 7 to 8 pounds per square. Most of the lightweight building paper sold has no ASTM rating. ASTM-rated No. 15 felt is either a minimum of 7.6 pounds per square (ASTM D4689) or 11.5 pounds per square (ASTM D226). Similarly, the unrated variety of No. 30 felt typically weighs only 15 to 20 pounds per square versus 26 to 27 pounds for rated Type 2 felt (ASTM D226).

Grade D Building Paper. Grade D building paper is an asphalt-impregnated kraft-type paper, similar to the backing on fiberglass insulation. Unlike asphalt felt, it is made from new wood pulp, rather than recycled material.

Its most common use is under stucco in the western United States. The vapor permeance of Grade D paper is similar to asphalt felt. Its liquid water resistance ratings range from 20 to 60 minutes, as measured by using the boat test (see "Water Resistance," previous page).

Because Grade D paper tends to deteriorate under prolonged wetting, the trend in three-coat stucco is to use two layers of 30-minute paper. Because the paper tends to wrinkle, the two layers tend to form a small air space, creating a rain-screen effect.

Plastic Housewrap. There are a wide range of plastic housewraps on the market. Most are nonwoven fabrics made from either polyethylene or polypropylene. Some have perforations to let water vapor pass through and the others are designed to let water vapor diffuse through the fabric itself. Because there is no single testing standard for plastic housewrap performance, it is difficult to make apples-to-apples comparisons. However, published performance data and limited field studies suggest the following:

- *Permeance to water vapor.* The leading nonperforated products (Tyvek®, R-Wrap®, and Amowrap®) are significantly more permeable to water vapor, ranging from 48 to 59 perms, than nearly all the perforated materials.

- *Water resistance.* All sheathing wraps adequately shed water on vertical surfaces. Pooled water, however, will leak through most perforated plastic housewraps over time, while the nonperforated materials will contain liquid water indefinitely. No. 15 asphalt felt retains water moderately well, but it allows some penetration over time.

- *Damage from extractives.* Some researchers have noted that extractives leaching out of redwood and cedar siding can cause plastic housewrap to lose its water repellency and to deteriorate. Back-priming the wood siding or leaving an air space behind the siding will help prevent this. Stucco will also degrade plastic housewrap and is rarely installed over it.

- *Recommendations.* Given their high permeance ratings and excellent resistance to liquid water, the nonperforated housewraps are a good choice for most building applications. Traditional asphalt felt is also a good option. Many contractors find plastic sheathing wraps more convenient than asphalt felt in that they weigh only 1 to 2 pounds per square, and they are more flexible in the cold and more tear-resistant. Also they are available in 9- to 10-foot-wide rolls versus 3-foot rolls for asphalt felt.

Draining Housewraps. In the last few years, manufacturers have responded to the need for an air space and drainage plane with a variety of housewrap products that are either wrinkled or corrugated to provide an integrated air space. These include products intended primarily for stucco, such as DuPont's StuccoWrap®, and others developed for siding, such as Raindrop Housewrap, which is a plastic drainage mat from Pactiv, Inc. (see "Resources," page 47). The air space created by these products is minimal, ranging from 0.02 inch thick for StuccoWrap to 0.008 for RainDrop®. Although these materials may allow for some drainage, it is unlikely that they will provide any measurable airflow to promote drying.

A more promising approach is a $\frac{1}{4}$-inch nylon matrix, called HomeSlicker®, which has vertical drainage channels and installs between the sheathing wrap and siding. The material is rigid and thick enough to resist compression by the siding but thin enough that windows, doors, and trim can be installed without furring.

Sheathing Wrap Installation

The primary function of the sheathing wrap, whether building felt or plastic housewrap, is to protect against water leakage. It is critical, therefore, to cover the entire shell from roof to foundation, including gable ends and band joists, and always to lap upper layers over lower layers to shed water. It is also critical to integrate the sheathing wrap with all window, door, and other wall flashings if the weather barrier is to be successful.

The IRC requires asphalt felt to be minimum 14 pounds per square (ASTM D226), overlapped a minimum of 6 inches at vertical joints and 2 inches at horizontal laps. Plastic-housewrap manufacturers recommend 6 to 12 inches of overlap at vertical seams and 4 inches at horizontal laps, with all joints taped.

It is good practice to wrap corners at least 6 inches each way. If the walls are sheathed and wrapped before being raised, leave a 6- to 12-inch overlap at one side of each corner, and leave a 12-inch, unstapled flap at the bottom to cover the band joist area after the sheathing is nailed off. Wide staples with a minimum 1-inch crown are recommended every 12 to 18 inches for plastic housewraps.

FLASHING MEMBRANES

Peel-and-stick eaves membranes have been used for nearly 20 years to prevent roof leaks from ice dams and other roofing trouble spots. These are typically available in 36-inch widths and are used to protect eaves, shallow-pitch roofs, and other problem roof areas. Over the past few years, a new family of related products has been introduced to help seal walls against water intrusion.

Flashing Tapes

Typically ranging in width from 4 to 12 inches, these peel-and-stick membranes greatly simplify the task of creating a continuous barrier to water entry around doors, windows, decks, and other problem areas. Flashing tapes are faced with reinforced polyethylene or foil on the outer surface and a peel-away paper on the adhesive surface. The foil-faced products may be left exposed to the weather permanently, whereas the plastic-faced tapes should not be exposed to sunlight and weather for more than 30 days (longer for some brands) since UV radiation will degrade the facing.

Modified Bitumen vs. Butyl. Most flashing membranes are made from modified bitumen, the same rubberized asphalt used in eaves flashing. Some use a more expensive butyl rubber core, which stays more flexible in cold weather and is more stable at high temperatures. Butyl products also bond better to difficult substrates than modified bitumen and can be peeled off and adjusted during installation.

Moldable Flashing. A unique butyl-based flashing tape from DuPont, called Tyvek FlexWrap®, has a wrinkled facing that allows it to be molded easily to irregular shapes such as the head flashing of round-top windows. It can also be bent to create a pan flashing at window sills without any cutting and folding at the corners. Despite the higher material costs, labor savings make this product appealing for tricky applications.

Applications. These products offer several distinct advantages over metal flashings: They are easily bent or molded for an accurate fit, can accommodate settlement and shrinkage movement, are self-sealing around nail holes, and bond well to a variety of materials, including metal, wood, plywood, and vinyl window flanges. They provide long-lasting waterproof protection if installed correctly. Oriented-strand board (OSB), concrete, and other masonry materials, however, can be problematic for some of the rubberized-asphalt flashings and may require priming for a good bond. Consult with the product's specifications for compatible surfaces and priming requirements.

Installation of Flashing Membranes

To obtain the best results with these products and be protected by the manufacturer's warranty, it is advisable to follow the manufacturer's recommendations. These vary from product to product, but generally they address the same issues: application temperature, priming, installation techniques, and compatibility with surrounding materials.

Temperature. In general, the rubberized asphalt (modified-bitumen) products start to lose stickiness at around 50°F and will not bond much below 40°F. Unless you are working with a rubberized-asphalt product specifically formulated for low-temperature applications, a butyl-based product is a better choice in cold weather.

Very high temperatures can also be problematic for rubberized-asphalt membranes. When subjected to high temperatures and pressure, for example, when squeezed under a dark-colored metal flashing exposed to direct sun, the material will soften and begin to flow. Unless formulated for high temperatures and labeled "hi-temp," most modified bitumen will begin to soften between 185°F and 210°F. High-temperature formulations can tolerate up to around 240°F, but are generally not as sticky.

Substrates. Each manufacturer specifies which products are safe to bond to and which require priming. Solid wood, plywood, vinyl window flanges, and metal are usually fine as long as they are free of oil and dust. Some manufacturers of rubberized-asphalt tapes recommend that all materials be primed for best performance, particularly in cold weather. Most require that concrete and masonry be primed, and some require the priming of OSB and gypsum sheathing as well.

Many published details show asphalt-rubber flashing tapes bonded to asphalt felt and plastic housewraps. While these are rarely listed as suitable substrates in product literature, manufacturers of flashing tapes claim that their products will bond satisfactorily to both these materials as long as they are clean. Do not expect a good bond to dirty housewrap that has been exposed to the weather for a month or to any dirty job-site material. For that reason, it is always best to detail flashings and to layer materials so that they shed water even if the adhesive bond fails.

Compatibility. Rubberized-asphalt flashings should not be in direct contact with flexible vinyl flashings. The asphalt compound will draw the plasticizers out of the vinyl, causing the asphalt to soften and flow and the vinyl to become brittle. The rigid polyvinyl chloride (PVC) used in window flanges, however, is generally not a problem.

Rubberized-asphalt flashings should also not come into contact with any caulks or sealants unless specifically formulated for that use. Like soft vinyl, sealants may react with the asphalt, causing it to flow and stain the adjacent materials, such as window flanges.

Butyl-based flashings are compatible with most construction caulks and sealants, but they should never be installed in contact with any asphalt-based products such as roofing cement or bituminous flashing membranes. These may degrade the butyl and undermine its ability to seal. In these applications, rubberized-asphalt is a much better choice.

Applying Pressure. Flashing tapes must be pressed firmly into place to ensure full contact and a good bond. Some manufacturers recommend using a hard rubber roller for best results.

Splashback Protection and Other Uses. While most flashing tapes are used around doors and windows, they can be put to good use wherever water penetration is an issue. Other applications include band joists, deck ledgers, inside and outside corners, and any areas subject to frequent wetting. On wall areas adjacent to a deck or abutting a roof, for example, where splashback or snow buildup is likely to wet the siding, sections of membrane up to 36 inches wide can protect wall assemblies. Make sure to lap all layers of flashing, sheathing wrap, and adhesive membrane so that water is directed to the outside of the building, even if the adhesive bond fails.

Caution: Cold-Side Vapor Barrier. In cold climates, covering an entire wall section with waterproof membrane will create a cold-side vapor barrier, potentially leading to serious moisture problems and wood decay within the wall cavity. A section of membrane up to 3 feet wide, however, is unlikely to cause problems.

WALL FLASHING

Wall flashings are required at openings, corners, intersections, and wherever a roof terminates into a wall. While peel-and-stick tapes have replaced these flashings at many details, metal flashings are still preferred for many standard details and applications where the flashing is visible or needs to hold a shape or serve as a drip edge.

Flashing Materials

Choose metal flashings that are compatible with the adjoining building materials and are at least as durable as the siding and roofing materials where they are to be placed. (See "Galvanic Corrosion," page 83, for information on metal compatibility.)

Aluminum. Most residential wall flashing today is made from light-gauge aluminum coil stock. Aluminum is inexpensive, easy to bend, and holds paint well. However, it tends to oxidize and pit in salty or polluted air and, if unpainted, will corrode from contact with masonry due to the lime and acids. Aluminum cannot be soldered. If using aluminum, use at least .029-inch coil stock, preferably anodized or prefinished, which is much more resistant to corrosion.

Copper. When the budget allows, copper is a good choice. Copper flashings come in two types: soft and harder cold-rolled. Soft copper is very malleable and useful for molding into irregular shapes. The harder cold-rolled material is a better choice for most applications, because it is stronger and more durable.

Copper flashings solder easily and offer good corrosion resistance, even in polluted air and in contact with masonry. Over time, all unpainted copper will oxidize and develop a green patina that protects the underlying copper. While most people find the patina attractive, the runoff of the green oxidation can stain siding or trim.

Some experts also caution against using copper or lead-coated copper in contact with redwood or red cedar or its runoff. Over time, the copper surface will be etched by the acidic wood runoff. Although actual failures of copper flashings are rare, they have been reported in areas of the Northeast after 10 to 20 years of service. Acid rain, combined with exposure to runoff from red cedar or other corrosive materials, is suspected as the cause. (See also "Copper," page 83.)

Lead-Coated Copper. This is a sheet of copper with a lead coating on each side. Where staining of building components from runoff is a potential problem, lead-coated copper may be used, which has a less noticeable gray runoff. Also, without the lead coating, copper flashing will react with galvanized steel.

Lead. For special flashing applications where a high degree of malleability is required, lead is an option. In addition to being easily bent and molded, lead is very resistant to corrosion. Lead is relatively soft, however; so it should not be used where it will be bumped or walked on. Also, it is best if left unattached on one side; if rigidly fastened on all sides, it can tear from fatigue due to thermal movement.

Sidewall Flashing Details

Windows and Doors. Window and door flashings are discussed extensively in a later section (see page 119).

Termite Shields. Metal termite shields are widely used atop foundations in the southern United States and in tropical climates as a physical barrier to termites. They sit

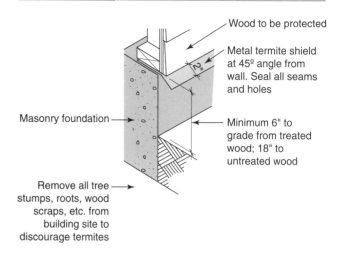

FIGURE 1-3 **Termite Shields.**

Wood to be protected

Metal termite shield at 45° angle from wall. Seal all seams and holes

Masonry foundation

Minimum 6" to grade from treated wood; 18" to untreated wood

Remove all tree stumps, roots, wood scraps, etc. from building site to discourage termites

Widely used in the southern states, termite shields do not stop termites, but they can slow down their progress and force them to build their tunnels in the open where they can be easily seen during inspections.

SOURCE: Adapted from *Architectural Graphic Standards, Residential Construction,* with permission of John Wiley & Sons, © 2003.

directly on top of foundation walls, piers, and other supports before the first piece of wood is installed (see Figure 1-3).

At one time termite shields were thought to block the entry of subterranean termites, the most widespread and destructive wood-boring insect in the United States. However, subterranean termites, which nest in the soil, will exploit the tiniest gaps in termite shields or other barriers to reach the wooden portions of a house and will build tunnels along exposed foundation walls and around termite shields if necessary. Although the shields do not stop termites, they slow down their progress and force them to build their tunnels in the open where they can be easily seen during inspections.

To work at all, the termite shield must have tightly sealed joints and be sealed around foundation bolts and other penetrations. Joints can be either soldered or mechanically interlocking. If the barrier is unsealed, termites will find any small gaps and render the effort worthless.

In general, termite shields should be a minimum of 6 inches above grade and extend out 2 inches on either side of the foundation at a 45 degree angle. In addition to making termite infestations visible, they also form a capillary break between the foundation and sill. Areas where a termite shield cannot be used, for example, where a concrete stairway abuts a foundation wall, are at high risk for termite entry.

In termite-prone regions, the only reliable way to prevent termite damage is to use treated wood in critical locations and treat the surrounding soil with termiticide.

Water Table. On many traditional homes, a wide board called the "water table" is installed along the foundation

FIGURE 1-4   Water Table.

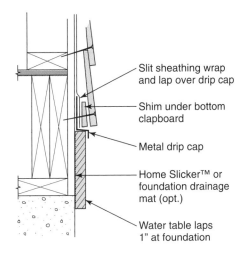

- Slit sheathing wrap and lap over drip cap
- Shim under bottom clapboard
- Metal drip cap
- Home Slicker™ or foundation drainage mat (opt.)
- Water table laps 1" at foundation

On many traditional homes, a water table along the foundation supports the first piece of siding. The water table should extend about an inch over the foundation and be protected by a metal drip cap installed under the sheathing wrap.

and supports the first piece of siding. The water table should extend about an inch over the foundation and be capped on top with either a preformed metal drip cap or a custom-bent flashing installed under the sheathing wrap. Cut a slit in the sheathing wrap along the entire length of the water table and slip the upper leg of the flashing under the wrap (see Figure 1-4).

Decks and Porches. It is critical to protect against leaks and water buildup at deck ledgers, since decay in this part of a building can lead to structural failure of the deck. At a minimum, install a cap flashing that tucks under the sheathing wrap and goes over the ledger (see Figure 4-8, page 145). Adding a second flashing, either peel-and-stick membrane or aluminum-coil stock, between the sheathing and ledger, as shown, is a worthwhile backup should any water get over, around, or through punctures in the cap flashing. Since pressure-treated wood can be corrosive to unfinished aluminum, use coated-aluminum or galvanized-steel flashing.

Corners. Corner boards are prone to leakage due to shrinkage of materials and wind exposure. For simple, effective backup protection, add a spline of asphalt felt paper at outside corners so that it extends 6 inches beyond the corner boards. Inside corners also benefit from a spline (Figure 1-5).

With this type of backup protection and with the end grain of the siding well sealed, it is unnecessary to caulk the siding joints at inside and outside corners. Leaving a small gap and not caulking these joints allows any water that penetrates to dry to the exterior. Eventually caulk joints will fail anyway, allowing water to leak in but inhibiting drying.

FIGURE 1-5   Corner Boards.

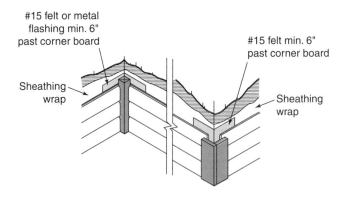

- #15 felt or metal flashing min. 6" past corner board
- Sheathing wrap
- #15 felt min. 6" past corner board
- Sheathing wrap

Protect building corners with felt splines that extend 6 inches beyond the corner boards. With this type of backup protection and the siding end-grain well sealed, it is not necessary to caulk the siding joints.

FIGURE 1-6   Step Flashings.

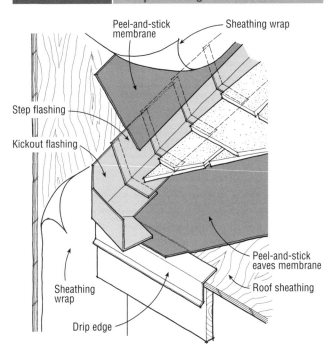

- Peel-and-stick membrane
- Sheathing wrap
- Step flashing
- Kickout flashing
- Sheathing wrap
- Drip edge
- Peel-and-stick eaves membrane
- Roof sheathing

Make sure the sheathing wrap overlaps all step flashings. Where snow buildup is anticipated, add a wide band of peel-and-stick membrane lapped over the step flashing but under the sheathing wrap.

Step Flashing. Integrate all step flashings with the sidewall-sheathing wrap by slipping the upper legs of the step flashing under the sheathing wrap (Figure 1-6).

Where snow buildup is anticipated, add a band of peel-and-stick membrane over the step flashing but under the sheathing wrap, as shown. Where the step flashing terminates along a sidewall, a preformed or custom-bent kick-out flashing is the best way to divert the water away from the siding.

Splashback. In wall areas subjected to splashback, snow buildup, or high moisture from other sources, rubberized asphalt membranes in widths up to 36 inches can be used to protect the wall sheathing and structure. Water damage from splashback is common in wall sections located under the eaves of a roof with no gutters. Walls above decks or flat roofs are also prone to moisture damage from splashback or snow buildup.

In all cases, make sure to detail the flashing membrane so that it tucks under the sheathing wrap above and over the step flashing or cap flashing below. If installed along the foundation, the membrane should cover the joint where the sill meets the foundation.

WOOD SIDINGS

Solid wood sidings remain popular in many sections of the United States despite their need for regular refinishing. In the Northeast, the most popular profile remains a simple bevel siding, or "clapboard." In the western states, heavier profiles such as channel rustic are more common.

Species

Red cedar remains the wood siding material of choice due to the natural decay resistance of the heartwood and its attractive appearance when stained or finished clear. Other decay-resistant woods are popular in the regions where they are produced: for example, redwood on the West Coast and cypress in the Southeast and Gulf Coast. On projects where premium wood species are not affordable, builders also use a wide variety of softwoods, including pine and spruce, which are not naturally resistant to decay. While most suppliers of wood siding now recommend back-priming and priming of cut ends, these details are even more critical with the less decay-resistant species.

Grading

Since wood siding is a nonstructural application, grading is generally for appearance only and is not governed by building codes. Most western species used for siding are graded according to one of the established grading agencies such as the Western Wood Products Association (WWPA). Still, manufacturers are free to name the grades as they choose for marketing purposes. So one company's

"Select" grade may be quite different from another's. For this reason, it is best to examine the material before specifying or purchasing.

Premium Grades. Western woods are generally labeled either *premium* or *knotty* grades. Premium grades have more heartwood and fewer defects and are typically kiln-dried. The highest grades of cedar are typically Clear VG (vertical grain) Heart and Clear Heart. Premium grades for other western woods include C Select, D Select, Superior, and Prime.

Knotty Grades. In general, "sound tight knots" or "select tight knots" (STK) indicates that there are no knots that will come loose or affect the performance of the siding. Other common designations are Select Knotty, Quality Knotty, 2 & Better Common, 3 & Better Common, and NPS (no prior selection). Since there are no uniform standards for these designations, an inspection of the material is important.

Moisture Content. Ideally, the siding should be installed at close to its equilibrium moisture content for the local climate (see Table 1-2). In general, unseasoned or green wood is shipped with a moisture content of greater than 19%. Air-dried or kiln-dried siding is shipped with a moisture content of 15 to 19%. In western woods, *dry* has a different meaning for premium and knotty grades. In premium grades, dry means that the siding has no more than 15% moisture content. In knotty grades, dry means that the moisture content does not exceed 19%.

Dry siding stored on the site (stickered if possible) will usually acclimate to local conditions in a week to 10 days. Unseasoned wood may need 30 days or longer to acclimate.

Siding Profiles. Because horizontal profiles naturally shed water, they resist water leakage better than vertical profiles. Also vertical wood siding is prone to wick up moisture from the bottoms of the boards, particularly where there is snow buildup or splashback. Diagonal siding is the most prone to leakage since water is conducted down the joints to window headers and other possible entry points. The most common profiles with typical installation details are shown in Figure 1-7.

TABLE 1-2	**Wood Moisture Content for Siding, Sheathing, and Exterior Trim**				
Recommended Moisture Content at Time of Installation					
Most of Continental U.S.		Dry Southwestern States		Warm, Humid Southeast Coastal Areas	
Average	Individual Pieces	Average	Individual Pieces	Average	Individual Pieces
12%	9%–14%	9%	7%–12%	12%	9%–14%

SOURCE: Adapted from the *Wood Handbook*, 1999, USDA Forest Products Laboratory.

FIGURE 1-7 Wood Siding Profiles.

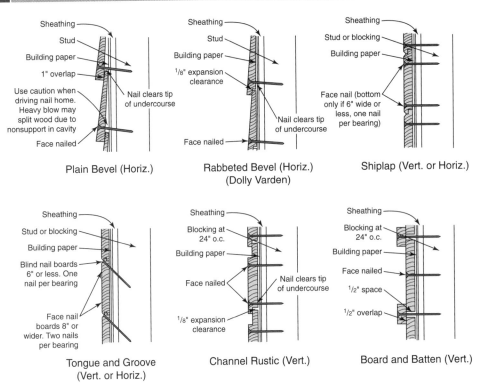

Plain Bevel (Horiz.)

Rabbeted Bevel (Horiz.)
(Dolly Varden)

Shiplap (Vert. or Horiz.)

Tongue and Groove
(Vert. or Horiz.)

Channel Rustic (Vert.)

Board and Batten (Vert.)

Horizontal profiles resist water leakage better than vertical profiles, which tend to wick up moisture from snow buildup or splashback. Diagonal siding is the most prone to leakage since water is conducted down the joints to window headers and other possible entry points.

SOURCE: Adapted from *Architectural Graphic Standards, Residential Construction,* with permission of John Wiley & Sons, © 2003.

Installation Details

While the premium grades of siding are more forgiving of installation and finishing problems than budget materials, all wood siding requires attention to detail to provide a durable and attractive exterior. Critical details are back-priming, air space, nailing, and finishing.

Drainage. An air space behind the siding, in addition to protecting the building shell (see "Rain-Screen Principle," page 2), also improves the performance of wood sidings. The siding material is less prone to moisture movement and paint is less likely to fail, even if the space is only $\frac{1}{4}$ inch wide. While the vast majority of wood siding is installed directly on the sheathing wrap, builders who have had problems with paint and siding have found that adding an air space is worth the additional cost. New products—such as wrinkled and corrugated sheathing wraps with an integral air space, and behind-the-wall drainage mats such as Benjamin Obdyke's Home Slicker®—are simplifying this step.

Back-Priming. The major trade associations representing siding manufacturers all recommend back-priming

and priming of cut ends. With cedar and redwood, back-priming will minimize the bleeding of extractives from the back of the siding, which can drip onto the face of the siding and stain the finish, and can also degrade sheathing wraps. With all sidings, back-priming will reduce the movement of moisture into and out of the siding, minimizing problems with cupping, warping, and checking.

Siding Over Foam. The need for back-priming and a ventilation air space is even greater when installing over foam sheathing. With no air space and no wood sheathing to temporarily store the moisture, any water that leaks through the siding or is driven in by the sun will tend to wet the back of the siding. The result, documented in a joint study conducted by wood siding and foam manufacturers, is increased cupping, cracking, and paint problems.

Plywood Siding. Plywood sidings are typically nailed directly to studs or through a layer of foam, and they provide a structural sheathing as well as an exterior finish. Most have vertical grooves to imitate vertical sidings. All plywood sidings should be painted or stained to protect the outer facing and prevent the panels from delaminating over time. Vertical joints are typically hidden by the vertical

FIGURE 1-8   **Plywood Siding.**

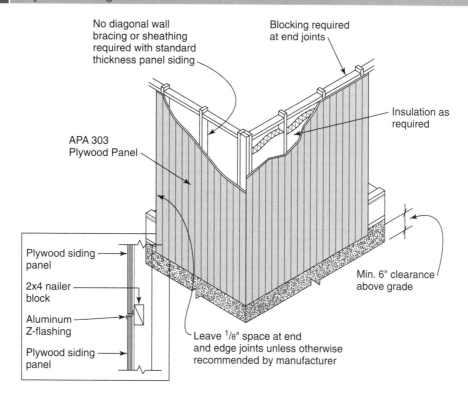

No diagonal wall
bracing or sheathing
required with standard
thickness panel siding

Blocking required
at end joints

Insulation as
required

APA 303
Plywood Panel

Plywood siding
panel

2x4 nailer
block

Aluminum
Z-flashing

Plywood siding
panel

Leave ⅛" space at end
and edge joints unless otherwise
recommended by manufacturer

Min. 6" clearance
above grade

*Plywood siding often provides a structural sheathing as well as an exterior finish, and
must be painted or stained to prevent delaminating. Vertical joints are hidden by grooves
in the pattern while horizontal joints require a Z-flashing (inset).*
SOURCE: Adapted from *Architectural Graphic Standards, Residential Construction,* with permission
of John Wiley & Sons, © 2003.

grooves in the pattern. Horizontal joints must be protected
by a Z flashing to shed water (Figure 1-8).

Fastener Types. Nailing requirements are shown in
Table 1-3. In general, nails should penetrate the sheathing
and studs or blocking by $1\frac{1}{2}$ inches, or $1\frac{1}{4}$ inches with ring-
shank or spiral-shank nails. Although specialized siding
nails with small heads and blunt tips are preferred, staples
are acceptable for some applications.

Since the cost of fasteners is a small percentage of a
siding job, it makes sense to use stainless steel, particularly with cedar and redwood, which can react with some
types of fasteners. The most common fastener choices are
as follows:

- *Stainless steel.* This is the best choice with all sidings,
  but it is particularly well suited to redwood and cedar,
  which react with some types of nails (galvanized and
  copper) and cause dark stains (see Figure 1-9).

  Ring-shanked or spiral-shanked siding nails can
  be set flush and painted over or countersunk and puttied before painting.

- *High-strength aluminum.* These are corrosion-
  resistant and can be used with all wood sidings.
  However, the aluminum can react with galvanized-
  steel flashing and cause corrosion.

- *Hot-dipped galvanized.* These can react with the
  tannins in cedar and redwood, causing black stains
  and streaking. Also the protective coating can
  chip when nailed, exposing the underlying steel to
  corrosion.

- *Electrogalvanized.* These are not recommended for
  any siding application since the coating is not thick
  enough and they are likely to corrode and stain the
  siding.

Nailing Schedule. Both the 2003 International Building Code (IBC) and the Western Wood Products Association require that solid wood siding products be nailed directly into studs or 2x blocking. Ring-shank nails should
penetrate $1\frac{1}{4}$ inches into wood (combined sheathing and
stud) and smooth-shank nails $1\frac{1}{2}$ inches.

With high-quality, dry, dimensionally stable siding
materials such as kiln-dried redwood and red cedar clapboards, some contractors nail siding directly to nominal
$\frac{1}{2}$-inch nail-base sheathings, such as OSB and plywood,
using ring-shank nails. Check with local codes before
taking this approach. To avoid problems, make sure joints
fall on studs or solid blocking (see Figure 1-10).

- *Horizontal sidings.* Nailing should be maximum
  24 inches on-center when over nail-based sheathing

| TABLE 1-3 | Nailing Recommendations for Wood Siding | | | | |

Siding Patterns		Description	Nominal Sizes (inches)	Nailing: Boards up to 6" Wide	Nailing: Boards 8" or Wider
Bevel or Bungalow		Bungalow has 3/4" butt vs. 1/2" for bevel. Lap 1" with either smooth or rough side out.	$\frac{1}{2}$ x 2 $\frac{3}{4}$ x 6 $\frac{1}{2}$ x 4 $\frac{3}{4}$ x 8 $\frac{1}{2}$ x 5 $\frac{3}{4}$ x 10 $\frac{1}{2}$ x 6 $\frac{5}{8}$ x 8 $\frac{5}{8}$ x 10	One siding or box nail per bearing, just above the 1" lap line. Do not nail through overlap.	One siding or box nail per bearing, just above the 1" lap line.
Rabbeted Bevel or Dolly Varden	1/8" gap	Thicker than bevel with rabbeted edge at overlap. Install with smooth or rough side out with $\frac{1}{2}$" overlap. Allow $\frac{1}{8}$" gap for wide boards or dry wood.	Standard $\frac{3}{4}$ x 6 $\frac{3}{4}$ x 8 $\frac{3}{4}$ x 10 Thick 1 x 6 1 x 8 1 x 10 1 x 12	One siding or box nail per bearing 1" up from bottom edge. Do not nail through overlap.	One siding or box nail per bearing, 1" up from bottom edge.
Shiplap	narrow wide 1/8" gap	Many patterns in smooth, rough, or saw-textured. Horizontal or vertical installation. Tongue edge up in horizontal applications. Allow $\frac{1}{2}$" overlap, with $\frac{1}{8}$" gap for wide boards or dry wood.	$\frac{3}{4}$ x 6 $\frac{3}{4}$ x 8 $\frac{3}{4}$ x 10	Face-nail with one siding or box nail, 1" up from bottom edge. Do not nail through overlap.	Face-nail with two siding or box nails, 3"–4" apart, 1" up from bottom edge.
Tongue and Groove	narrow wide	Available in many patterns. Horizontal or vertical installation. Tongue edge up in horizontal applications.	1 x 4 1 x 6 1 x 8 1 x 10 1 x 12 Note: Tongues may be $\frac{1}{4}$", $\frac{3}{8}$", or $\frac{7}{16}$"	One casing nail per bearing, blind-nailed at base of tongue. Do not nail through overlap.	Face-nail two siding or box nails 3"–4" apart per bearing.
Channel Rustic	narrow wide 1/8" gap	Available smooth, rough, or saw-textured for horizontal or vertical applications. 1" to 1$\frac{1}{4}$" exposed channel allows maximum dimensional change. Place wide rabbet up in horizontal applications. Allow $\frac{1}{2}$" overlap, with $\frac{1}{8}$" gap for wide boards or dry wood.	$\frac{3}{4}$ x 6 $\frac{3}{4}$ x 8 $\frac{3}{4}$ x 10	Face-nail with one siding or box nail per bearing, 1" up from bottom. Do not nail through overlap.	Face-nail with two siding or box nails 3"–4" apart per bearing.

and 16 inches on-center over nonstructural sheathings. Trade associations such as the Western Wood Products Association recommend against "double nailing" for most horizontal wood-siding profiles, including bevel siding. That is, nails should be driven above the overlap line of the siding board below to reduce the risk of cracking. Despite this recommendation, many contractors nail $\frac{1}{2}$-inch-thick clapboards just below the overlap line, catching the top edge of the piece below to avoid cracking the

siding during installation. While this approach may be acceptable with dry, premium-grade siding, it will likely lead to problems with lower quality materials (Figure 1-11).

- *Vertical sidings.* In general, vertical sidings are nailed to the top and bottom plates and to horizontal nailers installed every 36 inches for face-nailed siding and every 32 inches when blind-nailed. Because vertical sidings are vulnerable to leakage, they are not recommended for areas subject to wind-blown rain.

FIGURE 1-9 Nail Stains.

Corrosion from galvanized nails produced the striped pattern above on cedar siding. With redwood or cedar, stainless-steel nails are recommended.
SOURCE: Photo by author.

FIGURE 1-10 Siding Joints.

Locate all joints in bevel siding over studs, or the ends may cup and lift, as shown.

- **Plywood sidings.** Plywood siding is often nailed directly to studs or through an insulating sheathing and serves as a structural sheathing as well as the exterior finish. Use 6d box, siding, or casing nails for nominal $\frac{1}{2}$-inch plywood siding nailed directly to studs. For nail spacings, see Table 1-4.

FIGURE 1-11 Double Nailing.

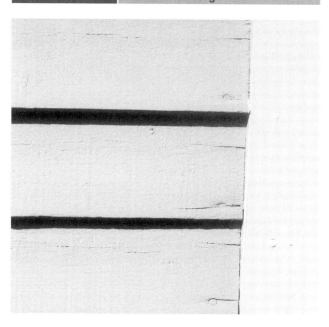

Nailing through the overlap can lead to shrinkage cracks in lower quality wood sidings.
SOURCE: Photo by author.

Wood Siding Details

Proper detailing at joints, corners, and openings makes for an attractive and durable job. Key details follow:

Lap Joints. The IRC requires that horizontal lap sidings have a minimum one-inch lap joint, or $\frac{1}{2}$ inch if the siding is rabbeted.

Butt Joints. In most climates, it is a good idea to slip a small spline of asphalt-felt paper behind each butt joint in horizontal sidings. Layer the spline so it overlaps the piece of siding below, directing any water out onto the siding (see Figure 1-12).

All end grain in the siding should be sealed after cutting with a water-repellent preservative (WRP) or primer.

Corners. Use overlapping 1x4s or 1x6s at outside corners or use $\frac{5}{4}$ stock for a heavier look. Use a felt paper spline, wrapped around the corner and extending 6 inches beyond the corner board, to protect the joints where the siding meets the corner boards (see Figure 1-5, page 8). Use a square length of $\frac{5}{4}$ stock at inside corners with a spline underneath. All end grain in the siding should be sealed after cutting with a water-repellent preservative (WRP) or primer. With the spline, there is no need to caulk the joint. With no caulk, the joint is free to dry out when wet.

Panel Siding Type	Nominal Thickness or Span Rating (in.)	Maximum Stud Spacing (in.)		Nail Size*	Nail Spacing (in.)	
		Face Grain Vertical	Face Grain Horizontal		Panel Edges	Intermediate
APA MDO EXT	$\frac{11}{32}$ and $\frac{3}{8}$	16	24	6d for panels $\frac{1}{2}$ inch or less. 8d for thicker panels.	6**	12
	$\frac{1}{2}$ and thicker	24	24			
APA-rated siding EXT	16 o.c.	16	24	"	6**	12
APA-rated siding EXT	24 o.c.	24	24	"	6**	12

TABLE 1-4 Nailing for Plywood Siding Direct to Studs or Over Foam Sheathing

*Use nonstaining box, siding, or casing nails.
**3 in. on-center (o.c.) for braced wall sections framed 24 in. on-center with $\frac{11}{32}$ or $\frac{3}{4}$ in. siding applied horizontally.
SOURCE: Adapted with permission from *Architectural Graphic Standards for Residential Construction* (2003).

FIGURE 1-12 Splines Under Bevel-Siding Joints.

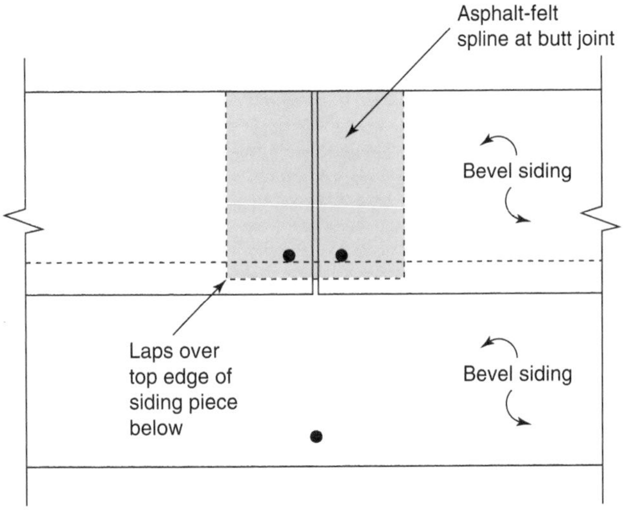

FIGURE 1-13 Joints in Vertical and Plywood Siding.

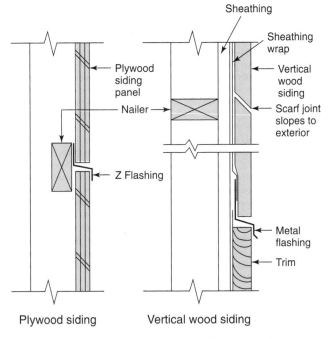

In areas subject to windblown rain and snow, protect joints in horizontal siding with felt-paper splines. Also, seal all end grain with a primer or water-repellant preservative.

Use a scarf joint to shed water to the building's exterior at horizontal joints in vertical wood siding. With plywood sidings, use a Z flashing.

Windows and Doors. If windows and doors are properly protected with splines of felt or flashing tape, there is no need to caulk the joints where siding meets the side casings. At the top of a door or window, always direct the sheathing wrap over the head flange or cap flashing. Never caulk the joint between the siding and the head casing or the sill, leaving these joints open to drain any trapped water.

At Step Flashings. Stop wood sidings at least 1 inch short of the bottom leg at step flashings and other roof flashings. Otherwise water will wick up into the

flashing leading to paint failures and decay (see Figure 1-6, page 8).

Vertical and Plywood Siding. Avoid horizontal butt joints in vertical siding. Where a butt joint is necessary, use a scarf joint sloped down toward the building's exterior. With plywood sidings, use a Z flashing at horizontal joints to shed water to the outside (see Figure 1-13).

TABLE 1-5	Maximum Exposures for Sidewall Shakes and Shingles		
Product Type	Grade	Single Course	Double Course
Number 1 Grade Products			
16" Shingles	1	$7\frac{1}{2}$"	12"
18" Shingles	1	$8\frac{1}{2}$"	14"
24" Shingles	1	$11\frac{1}{2}$"	16"
18" Shakes	1	$8\frac{1}{2}$"	14"
24" Shakes	1	$11\frac{1}{2}$"	18"
Number 2 Grade Products[1]			
16" Shingles	2	6"	10"
18" Shingles	2	7"	11"
24" Shingles	2	9"	14"
18" Shakes	2	7"	10"
24" Shakes	2	9"	14"

[1]No. 2 exposures based on Cedar Shake & Shingle Bureau recommendations. Others based on International Residential Code.

WOOD SHINGLES AND SHAKES

Cedar shingles and shakes are a popular choice for sidewall applications in coastal regions. Eastern white cedar shingles are often left unpainted in New England coastal areas. Red cedar shakes are often left unpainted on the West Coast. Red cedar shingles are sometimes left natural, but more often are painted or stained. Sidewall installation is similar for shakes and shingles with some variation in exposure (see Table 1-5).

Grades

Red cedar shingles come in four grades, but most sidewalls use grades No. 1 or No. 2. No. 1 is all heartwood and all edge-grain wood. They are available rebutted and rejointed (R&R) where a uniform appearance is desired and machine grooved for a textured surface. Red cedar shakes come either taper-split or untapered and are usually installed in Premium or No. 1 grade (see Table 1-6).

Eastern white cedar shingles are available in four grades. Most sidewall work uses Grade A (Extra), which is all clear heartwood, or Grade B (Clear), which has no knots on the exposed face (see Table 1-7).

Installation

The simplest and most common pattern for sidewall shingles and shakes is single coursing. For wider exposures and deeper shadow lines, shingles and shakes can also be installed in double courses. A rustic staggered pattern is also possible.

Underlayment. The Cedar Shake and Shingle Bureau recommends installation over Type 30 asphalt felt underlayment for red cedar shingles and shakes. Install the felt paper with minimum 6-inch overlaps on vertical joints, 2 inches on horizontal laps, and 4 inches wrapped each way at inside and outside corners. Creasing the felt at corners will help achieve a tight fitting corner.

For optimal performance, manufacturers of Eastern white cedar shingles now recommend installation over horizontal furring spaced equal to the shingle exposure or over a ventilating layer such as Benjamin Obdyke's Home Slicker®. They acknowledge that most sidewall installations still go directly over the wall sheathing covered with felt paper or plastic housewrap. Field experience suggests that an air space or drainage/ventilation layer is critical for longevity on roofs, but on sidewalls, good quality white cedar shingles perform adequately without these extra steps.

Single Coursing. The first course of shakes or shingles is doubled, with the outer course dropped $\frac{1}{2}$ inch lower to create a drip edge (see Figure 1-14). Tack up a length of 1x3 furring as a guide for the next course, moving up the wall with each successive course.

To create a weather-tight exterior, do not exceed the exposures shown in Table 1-5. Space No. 1 red cedar shingles $\frac{1}{8}$ to $\frac{1}{4}$ inch apart to prevent possible buckling. A $\frac{1}{4}$-inch space is recommended for No. 2 R&R red cedar shingles. White cedar shingles should be spaced from $\frac{1}{16}$ to

TABLE 1-6	Western Cedar Shingle and Shake Grades

Shingles

Grade	Length	Min. Clear Length	Characteristics
No. 1 Blue Label	16" 18" 24"	All clear	Premium grade for sidewalls and roofs. 100% clear, edge-grain heart-wood. Available rebutted and rejointed (R&R) with parallel edges and smooth butts. R&R available with one face sanded or machine grooved.
No. 2 Red Label	16" 18" 24"	(10") (11") (16")	Suitable for many applications. Flat grain and limited sapwood permitted. Available rebutted and rejointed.
No. 3 Black Label	16" 18" 24"	(6") (6") (10")	Utility grade for economy applications.
No. 4 Undercoursing	16" 18" 24"	(no min.)	Utility grade for undercoursing of double-coursed sidewalls. Not suitable for roofing starter course or other roofing applications.

Shakes

Grade	Length	Min. Clear Length	Characteristics
Certi-Split Handsplit Shakes	18" 24"	All clear	These have machine-split faces and sawn backs. Blanks are run diagonally through a bandsaw to produce two tapered shakes. Premium Grade is 100% edge-grain heartwood. No. 1 Grade allows up to 20% flat grain per bundle.
Certi-Sawn Tapersawn Shakes	18" 24"	All clear (except Grades 2 and 3)	Sawn both sides. Premium Grade is 100% clear edge-grain heartwood. No. 1 Grade allows up to 20% flat grain per bundle. No. 2 and 3 Grades also available.
Certi-Split Tapersplit Shakes	18" 24"	All clear	Produced by hand using a steel froe and mallet. The natural shinglelike taper is achieved by reversing the block end-for-end with each split. Premium Grade only.
Certi-Split Straight-Split Shakes	18" 24"	All clear	Produced by machine or in the same manner as taper split shakes except shakes are same thickness throughout. Premium Grade only.

NOTE: Based on grading criteria of the Cedar Shake & Shingle Bureau.

TABLE 1-7	White Cedar Shingle Grades

Grade	Characteristics	Recommended Uses
A. Extra (blue label)	Premium quality, pale color, all clear heartwood.	Siding or roofing in harsh climates. Highly resistant and durable.
B. Clear (red label)	Standard quality, rich color, no knots on exposed face (first 6").	Siding or steep-slope roofing.
C. 2nd Clear (green label)	Economical. Beige to brown color. Sound knots allowed on exposed face (first 6").	Siding of secondary buildings. Interiors.
D. Special (green label)	Width of 3" to 6". Variable color. Acceptable defects on full surface.	Starter course or undercourse for siding. Rustic interiors. Varied secondary uses.

$\frac{1}{4}$ inch apart depending on conditions—a $\frac{1}{16}$-inch gap would be appropriate for green shingles, which are prone to shrinkage, and a $\frac{1}{4}$-inch gap for kiln-dried shingles installed in a moist environment.

Also, offset joints in successive shake or shingle courses by at least $1\frac{1}{2}$ inches as shown in Figure 1-15.

Treat knots and other defects like an edge and offset adjacent courses at least $1\frac{1}{2}$ inches. With white cedar shingles, also make sure that two joints do not align if separated by only one course.

Double Coursing. For increased exposures and deeper shadow lines with red cedar shingles or shakes, apply in double courses, as shown (Figure 1-16).

Despite the greater exposures, considerably more material and labor are required. Installation starts at the

FIGURE 1-14   **Sidewall Shake and Shingle Installation.**

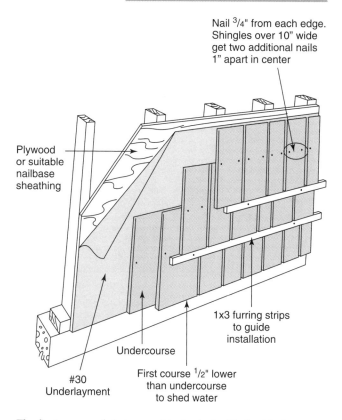

Nail $^3/_4$" from each edge. Shingles over 10" wide get two additional nails 1" apart in center

Plywood or suitable nailbase sheathing

Undercourse

#30 Underlayment

First course $^1/_2$" lower than undercourse to shed water

1x3 furring strips to guide installation

The first course of shakes or shingles is doubled, with the outer course dropped $\frac{1}{2}$ inch to create a drip edge. Tack up lengths of 1x3 furring as a guide for the next course, moving them up the wall with each successive course.

bottom with a triple layer and succeeding layers are doubled as shown.

Nailing. Use corrosion-resistant box or casing nails of either stainless steel, hot-dipped galvanized, brass, or aluminum. For concealed nails, hot-dipped galvanized are adequate. For exposed nails at corners and under eaves and windows, stainless steel, brass, or aluminum are less likely to stain the wood.

For red cedar shingles and shakes, nail 2 inches above the butt line and $\frac{3}{4}$ inch in from each end. Cedar shingles wider than 10 inches need two additional nails driven 1 inch apart near the center of the shingle. Nails should fully penetrate the sheathing (see Table 1-8). Aluminum or stainless-steel staples with $\frac{7}{16}$- to $\frac{3}{4}$-inch crowns are also an option for red cedar shingles if accepted by local codes.

White cedar shingles are nailed $1\frac{1}{2}$ inch above the butt line and $\frac{3}{4}$ inch in from each end. Manufacturers recommend a $1\frac{1}{4}$ inch (3d) box or shingle nail for new construction and a $1\frac{3}{4}$ inch (5d) nail when going over another siding material. Drive nails flush with the surface. Do not overdrive and set the nails or leave them projecting from the surface.

FIGURE 1-15   **Offsetting Shingle Courses.**

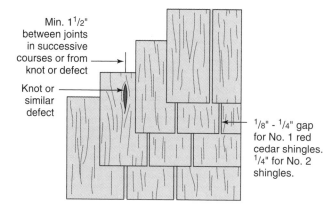

Min. 1$^1/_2$" between joints in successive courses or from knot or defect

Knot or similar defect

$^1/_8$" - $^1/_4$" gap for No. 1 red cedar shingles. $^1/_4$" for No. 2 shingles.

Offset joints in successive courses of shingles or shakes by at least $1\frac{1}{2}$ inches, as shown. Treat a knot or other defect like an edge, offsetting it from joints in adjacent courses.

FIGURE 1-16   **Double Coursing of Shakes and Shingles.**

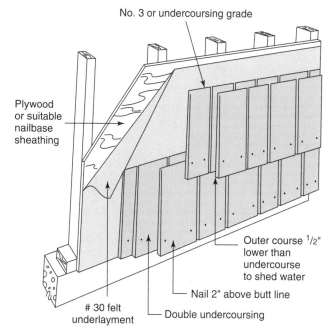

No. 3 or undercoursing grade

Plywood or suitable nailbase sheathing

30 felt underlayment

Double undercoursing

Nail 2" above butt line

Outer course $^1/_2$" lower than undercourse to shed water

Double coursing of shakes or shingles allows increased exposures and deeper shadow lines, but it requires more material and labor.

Clearance at Flashings. Keep all shingle bottoms a minimum of $\frac{1}{2}$ inch above the lower leg of any flashings to minimize the wicking of water, which can lead to staining and possible decay.

Corners. As extra protection, it is a good idea to add a layer of flashing or No. 30 felt paper at inside and outside corners. If felt is used, crease it at corners for a tighter fit. A simple, attractive inside corner can be achieved by butting the shingles to a $\frac{5}{4}$-inch square cedar strip nailed

TABLE 1-8	Fasteners for Red Cedar Shingles and Sidewalls	
Product Type	Nail Type and Min. Length for Single Coursing	Nail Type and Min. Length for Double Coursing
Certigrade, R&R and Sanded Shingles		
16″ and 18″ Shingles	3d box ($1\frac{1}{4}″$)	5d box ($1\frac{3}{4}″$) or same size casing nails
24″ Shingles	4d box ($1\frac{1}{2}″$)	5d box ($1\frac{3}{4}″$) or same size casing nails
Certigroove Singles		
16″ and 18″ Shingles	3d box ($1\frac{1}{4}″$)	5d box ($1\frac{3}{4}″$)
24″ Shingles	4d box ($1\frac{1}{2}″$)	5d box ($1\frac{3}{4}″$)
Certi-Split & Certi-Sawn Shakes		
18″ Straight-Split Shakes	5d box ($1\frac{3}{4}″$)	7d box ($2\frac{1}{4}″$) or 8d box ($2\frac{1}{2}″$)
18″ and 24″ Handsplit Shakes	6d box (2″)	7d box ($2\frac{1}{4}″$) or 8d box ($2\frac{1}{2}″$)
24″ Tapersplit Shakes	5d box ($1\frac{3}{4}″$)	7d box ($2\frac{1}{4}″$) or 8d box ($2\frac{1}{2}″$)
18″ and 24″ Tapersawn Shakes	6d box (2″)	7d box ($2\frac{1}{4}″$) or 8d box ($2\frac{1}{2}″$)

SOURCE: Courtesy of Cedar Shake & Shingle Bureau, © 2005 CSSB. All Rights Reserved.

FIGURE 1-17   Shingle Corner Details.

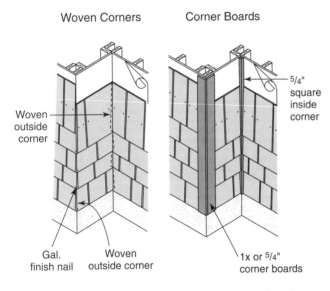

Woven Corners   Corner Boards

Woven outside corner

Woven outside corner

Gal. finish nail

5/4″ square inside corner

1x or 5/4″ corner boards

For fast, simple corners, butt the shingles to corner boards. Woven corners are attractive but labor-intensive. With either approach, flash behind the corner with a metal flashing or extra layer of felt paper.
SOURCE: Reprinted from *Fundamentals of Residential Construction,* Edward Allen and Rob Thalon, with permission of John Wiley & Sons, © 2001.

into the corner. For fast and simple outside corners, butt the shingles to corner boards made from 1x or $\frac{5}{4}$ stock (see Figure 1-17).

Another more labor-intensive approach is to "weave" inside and outside corners by alternating two shingles on one side with two on the other. On an outside corner shingled this way, the exposed edge alternates every course. To keep outside corners tight, nail through the butts with a

FIGURE 1-18   Prefab Shingles.

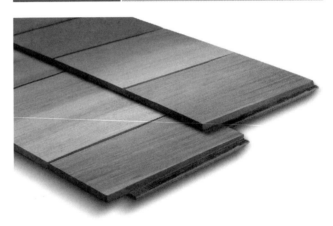

To speed up installation, several manufacturers offer sidewall shingle panels preassembled with either staples or adhesive. Panels can also simplify installation of outside corners, column wraps, and other labor-intensive details.
SOURCE: Photo courtesy of Shakertown 1992, Inc.

small hot-dipped galvanized finish nail. On woven inside corners, alternating courses keep the joints tight.

Panelized Installation. To simplify and speed up installation, several manufacturers offer sidewall shingles attached to panels with either staples or adhesive (see Figure 1-18).

The panels range from one-course panels 32 inches wide to three-, four-, and five-course panels 2 feet wide by 8 feet long, including panels with decorative patterns. Some manufacturers also offer prefabricated inside and outside corners, radiused panels for curved walls, column wraps, and other types of labor-intensive details.

FIGURE 1-19 The Natural Look.

Untreated white cedar shingles may weather unevenly despite salt air and a sunny exposure. In this case, the uneven weathering is due to splashback from an attached deck. For uniform weathering, a bleaching oil is recommended.
SOURCE: Photo by author.

Finishing Shingles. Eastern white cedar shingles are often left unfinished and tend to weather to an attractive silver gray—particularly with exposure to sun and salt air in coastal climates. However, splashback and other uneven weathering conditions can lead to dark streaks or splotches (see Figure 1-19).

To accelerate the weathering process and to guarantee uniformity of color, a bleaching oil is recommended. For a pigmented finish, use an oil-based, semitransparent stain. Prefinished white cedar shingles are available with a stained finish or pretreated with bleaching oil.

If left unfinished, red cedar shingles will tend to weather to a dark reddish-brown color. To guarantee uniformity of color, red cedar shingles should be finished with an oil-based clear finish, oil-based stain, or bleaching oil. If a painted finish is desired, use an oil-based primer with a 100% acrylic top coat for best results. Factory-finished shingles and shakes are available preprimed or prestained, ready for a top coat after installation. See more under "Exterior Wood Finishes" (page 42).

FIBER-CEMENT SIDING

Many synthetic alternatives to wood siding have fallen short either on aesthetics or durability. Fiber-cement, one of the newest entries into the field, holds great promise in that the material can be fashioned to resemble almost any exterior cladding, holds paint well, and is essentially impervious to decay, insects, UV radiation, and fire. It is also very dimensionally stable and resists shrinking and swelling, cupping, warping, and splitting. Warranties run from 30 to 50 years depending on the manufacturer and specific configuration. It is cost-competitive with vinyl and hardboard siding and significantly less expensive than premium wood sidings.

Fiber-cement is made up primarily of Portland cement, sand, and wood fibers. It is chemically similar to older asbestos sidings but contains no asbestos, glass fibers, or formaldehyde. It does, however, produce a very fine silica dust when cut with a saw or abrasive blade, which, if inhaled, can cause silicosis and other serious respiratory problems.

Fiber-cement boards are extremely straight and rigid when held edgewise, but they are much heavier than wood—about 20 pounds for a 12-foot length of $8\frac{1}{4}$-inch siding. They are flexible along the flat dimension, however, so any lumps in a wavy framing job will tend to telegraph through the siding. The material is fairly brittle and, if not handled carefully, can crack.

Styles and Sizes

Fiber-cement is available in a wide array of styles and finishes modeled after other materials ranging from horizontal wood siding to vertical sidings, wood shakes, bricks, and stones. The wood patterns are generally available either smooth or wood-grained and most are available factory-primed or finished as well as unfinished.

Fiber-cement horizontal siding planks are typically $5\frac{1}{4}$ to $12\frac{1}{4}$ inches wide by 12 feet long and are designed for a $1\frac{1}{2}$ inch overlap. Vertical siding panels measure 4x8, 4x9, or 4x10 feet, and shake and shingle panels are typically 16x48 inches. The thickness of most siding materials is $\frac{5}{16}$ inch. Smooth and textured soffit and trim boards are also available. Fiber-cement soffit material is typically $\frac{1}{4}$ inch and most trim stock is $\frac{7}{16}$ inch thick, but manufacturers have recently introduced thicker profiles (see section on fiber-cement trim, page 34).

Lap-Siding Installation

Fiber-cement siding products install similarly to the wood products they imitate. They can go over wood-based sheathings or rigid foam, but they must be nailed or screwed directly to studs or 2x blocking. Fasteners should penetrate solid wood by 1 to $1\frac{1}{4}$ inches, depending on the manufacturer's specifications.

The 12-foot-long fiber-cement planks can be held edgewise by a single person, but the boards may break in two or deform if picked up flat. One person can install a plank by driving a single nail near its center to hold it in place against guide nails driven into the sheathing to mark the upper edge.

Manufacturers recommend leaving $\frac{1}{8}$ inch between board ends and window casings and trim and caulking with a paintable 100% acrylic latex caulk. Butt joints between two planks can be either lightly butted and painted over or gapped $\frac{1}{8}$ inch and caulked. Manufacturers recommend priming cut ends on site if the joints are not being caulked. As with other siding products, leave at least $\frac{1}{2}$ inch clear at step and other flashings so the bottom edge does not soak up water.

Fastening. Fiber-cement siding should be nailed directly to studs with nail penetration into solid wood of 1 to $1\frac{1}{4}$ inches, depending on the manufacturer's specifications. Predrilling is required within $\frac{1}{4}$ inch of an edge or near sharp angles or other fragile shapes to avoid cracking. Predrilling may also be required when nailing through foam sheathing to avoid cracking the siding.

Manufacturers require a hot-dipped galvanized or stainless-steel siding nail (or roofing nail for blind nailing) that should be driven flush with the surface. Overdriving of nails can cause the material to shatter around the nail, weakening its holding power and, with some products, voiding the warranty. Staples and clip-head nails tend to penetrate too far, but coil nailers with adjustable depth-of-drive work well. Some contractors hand-nail the siding to avoid problems. Given the longevity of the siding, a long-lasting corrosion-resistant nail is recommended.

If fastening to metal studs, use corrosion-resistant pneumatic pins or self-tapping bugle-head screws.

Standard Nailing. In most installations, horizontal fiber-cement siding is nailed top and bottom into each stud, with the lower exposed nail going through both layers of siding (see Figure 1-20).

Butt joints should lie over studs. This is the most durable installation. Color-matched galvanized nails are available for the exposed nails on prefinished sidings.

Blind Nailing. Recommendations vary among manufacturers, but most permit "blind nailing" with siding planks less than $8\frac{1}{4}$ inches wide installed over 16-inch on-center framing. In this technique, the fasteners are hidden just above the lap line of the overlapping plank and put a slight curve in the siding, pulling it tight to the wall. Roofing nails work well because of their large heads (see Figure 1-21).

An occasional face nail may still be required to hold the lower edge tight to the wall where there is a bump or bulge in the framing. Since the lower edges remain unsecured, blind nailing is not suitable for high-wind areas.

Cutting. When cut with a diamond abrasive blade in a circular saw, fiber-cement creates a cloud of very fine silica dust which can cause silicosis and other serious

FIGURE 1-20   **Fiber-Cement Siding: Standard Nailing.**

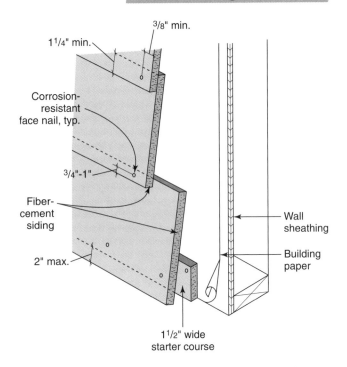

Horizontal fiber-cement siding is typically nailed top and bottom into each stud, with the lower exposed nail going through both layers of siding. Nails should be snug but not countersunk.
SOURCE: Adapted from *Architectural Graphic Standards, Residential Construction,* with permission of John Wiley & Sons, © 2003.

FIGURE 1-21   **Fiber-Cement Siding: Blind Nailing.**

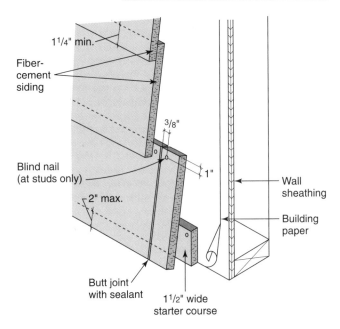

Most manufacturers permit blind nailing with narrow planks installed over 16-inch on-center framing. For best results, use a roofing nail just above the lap line. Occasional face nails are usually needed to keep the lower edge tight to the wall.
SOURCE: Adapted from *Architectural Graphic Standards, Residential Construction,* with permission of John Wiley & Sons, © 2003.

respiratory problems. Ordinary carbide-tipped blades produce less dust but wear out within a few hours compared to a few months for abrasive blades. In the last few years, manufacturers have responded with specialized diamond-tipped blades and tools, making the work easier and safer. The new fiber-cement blades cut smoother, create less dust, and outlast ordinary carbide blades. When used with the new dust-collecting saws designed for fiber-cement, cutting is safe and effective.

Many contractors also use electric shears, similar to a sheet-metal nibbler but specially adapted for fiber-cement. These make a clean cut with little dust, but are not as fast as a circular saw and cannot cut through multiple boards at once. Scoring and snapping, as for drywall, is also an option for quick cuts where a crisp edge is not needed.

Finishing. After installation, small dents or chips can be filled with any cementitious patching compound. Before priming or applying the top coat to preprimed material, wipe away any dust from cutting with a damp cloth or sponge or lightly hose down the siding and allow it to dry thoroughly. If the siding has been hosed down or power washed (unprimed siding only), allow at least two sunny days before priming. Painting should be completed within 90 days of installation to avoid deterioration of the surface from prolonged exposure to water.

For unprimed siding, manufacturers recommend an alkali-resistant, 100% acrylic primer specifically approved by the paint supplier for fiber-cement. Back-priming is not necessary; in fact, some manufacturers recommend against back-priming so any trapped moisture can dry from the back of the siding.

For the top coat, use a 100% acrylic latex paint. Because fiber-cement is dimensionally stable and largely inert, it holds paint well. Estimates range from 7 to 15 years for a quality paint job. Some of the prefinished products carry 15-year warranties on the finish.

VINYL SIDING

Vinyl siding is the leading choice for residential siding in the United States, accounting for an estimated 36% of the siding market. It owes its popularity to its low cost and low maintenance needs. When first introduced in the late 1950s, vinyl was criticized for fading rapidly, turning brittle in cold weather, and buckling (or "oil-canning") in hot weather. It is also vulnerable to blow-offs in high winds.

Through the use of additives to the resin and better installation techniques, however, manufacturers have addressed these concerns, and vinyl is finding its way onto more higher-end projects. Today's premium products typically carry a 50-year, or "lifetime," prorated warranty.

Materials

Vinyl siding is composed of the plastic polyvinyl chloride (PVC) blended with a number of additives for specific properties: plasticizers for flexibility; stabilizers to prevent oxidation; UV radiation absorbers, such as titanium dioxide, to prevent fading and degradation; and pigments to add color. Fillers are added to hold down costs, and the resin is extruded into a wide variety of the shapes that mimic natural siding materials. PVC is inherently fire-resistant and carries a Class 1(A) fire-rating.

Composition. While enhanced formulas have improved vinyl's performance over the years, it is not impervious to the elements. Oxidation still occurs and, over time, may cause a white dusting on the surface, particularly in wet, cloudy climates such as the Northeast or Northwest. In freezing weather, a stray baseball can still shatter a panel. Also sunlight tends to fade dark colors, and excessive heat will soften and potentially distort the vinyl. To minimize the effects of heat and sunlight, most vinyl colors are muted, although some darker colors are available with special additives to stabilize the vinyl.

Thickness. Nowadays most vinyl siding is extruded in a two-layer process that puts the more expensive weather-resistant resins only in the outer layer to save costs. While building codes allow vinyl siding as thin as .032 inch (32 mils), premium products range from about 40 to 50 mils, with the thicker products typically costing proportionately more. Some contractors prefer a heavier material for residing jobs to better smooth over the irregular substrate.

Profiles. The rigidity of the siding, however, is more a function of its profile and particularly the thickness of the butt edge, which typically ranges from $\frac{1}{2}$ to $\frac{3}{4}$ inch (Figure 1-22).

| FIGURE 1-22 | Vinyl Siding Profiles. |

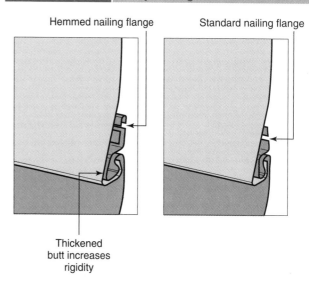

A thicker butt dimension (left) makes a more rigid siding with more pronounced shadow lines than a standard profile (right). The thicker profile at left also has a hemmed nailing flange to help reduce blow-offs in high winds.

In general, more rigid products are easier to install, but they have more pronounced shadow lines at joints. Siding panels come in several profiles, usually containing two to four courses of siding per panel. Panels range from 6 to 10 inches in width and are typically 12 feet long, although some manufacturers offer greater lengths. Finishes range from completely smooth to heavily textured. A lightly textured finish most closely mimics painted wood siding.

Lock and Nailing Flanges. All panels have a locking tab at the bottom of each panel that snaps over the top tab of the panel below. Because of problems with blow-offs in high winds, some of the premium panels feature reinforced nailing flanges, either with a thickened extrusion or a hem as shown in Figure 1-22.

Installation

Waterproofing. Vinyl siding is not waterproof. Since wind-driven rain will penetrate at lap joints, corner boards, and other penetrations, all new siding jobs should begin with the installation of a weather-resistant drainage plane consisting of building paper or plastic housewrap and integrated flashings. On residing jobs, any leaks should be repaired in the original flashing or cladding before installation begins.

Thermal Movement. Because of its high coefficient of expansion, the key to successful installation of vinyl siding is to allow it to move freely as temperatures change. A 12-foot length will vary in length up to $\frac{1}{2}$ inch over a 100°F temperature change. For that reason, manufacturers recommend leaving $\frac{1}{4}$ inch clearance at receiving trim located at corners, windows, mounting blocks, or other places where the siding terminates or is notched. Increase the clearance to $\frac{3}{8}$ inch when installing in temperatures below 40°F. Do not caulk the panels at overlap joints or at ends where they meet receiving trim.

Nailing. Nails can also restrict movement and cause buckling problems. To prevent this, do not nail the siding tight. Instead, "hang" the siding by driving nails in the center of the nailing slots and leaving $\frac{1}{32}$ to $\frac{1}{16}$ inch (the thickness of a dime) between the fastener head and the siding. Drive nails straight since the head of an angled nail can pinch and distort the siding. Use corrosion-resistant nails with heads at least $\frac{5}{16}$ inch in diameter, such as roofing nails, driven at least $\frac{3}{4}$ inch into solid wood (Figure 1-23).

Standard nailing is 16 inches on-center for horizontal panels, 12 inches for vertical panels. In high-wind areas, use extra nails and choose a product with a hemmed or reinforced nailing flange.

When locking the panels into position, do not force them up or pull them down to adjust the alignment. Too-tight panels can tear and too-loose panels can unlock and

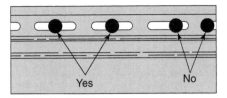

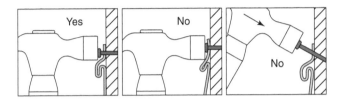

Drive nails in the center of the slot, with the nail heads slightly off the flange. Nails driven too tight or at an angle will pinch the siding, distorting the face and restricting thermal movement. SOURCE: Illustration courtesy of the Vinyl Siding Institute, Inc. © 2005 VSI.

come loose. One exception is at the band joint between the first and second floor where panels may come unlocked due to shrinkage of the framing. To compensate for this, some contractors pull the panels a little tight over the band joist area.

Overlaps. Where more than one panel is needed along a run, overlap the two panels by about an inch, with the overlapped edge facing away from high traffic areas so they will be less visible. Overlaps should be staggered at least 3 feet and in a random pattern to avoid creating a visual seam or step effect up the wall. Where possible, use a single piece of siding across the wall. The fewer joints, the more attractive and water-resistant the job will be.

Mounting Blocks. Exterior fixtures—such as light fixtures, electrical panels, and hose bibs—can also cause problems if they are fastened through the siding, restricting its free movement. Siding manufacturers sell mounting blocks with integral J-channel to hold panel ends and allow for movement. Or the contractor can install wood mounting blocks before installing the siding and trim them with J-channel or utility trim.

Trim for Vinyl Siding

The appearance of a vinyl siding job often has more to do with the trim details than with the siding itself. By using wider trim pieces and avoiding the overuse of J-channel, the installer can produce a more attractive finished product. Manufacturers sell a wide range of accessories in

PVC, aluminum, or vinyl-coated aluminum. Most contractors fabricate at least some of their trim pieces on site from either prefinished or vinyl-coated aluminum coil stock, using a sheet-metal break and other specialized tools.

Like vinyl siding, aluminum trim has a high coefficient of expansion so installation details need to accommodate movement. Avoid putting nails in the face of flat pieces of coil stock and allow $\frac{1}{4}$ inch at edges for expansion and contraction. Where possible, use a vinyl receiving channel, roofing drip cap, or another piece of trim to support long runs of flat aluminum trim, minimizing the use of nails. Where nails are required, use slotted nail holes, which can be made using a slot punch. Prepainted aluminum or stainless-steel nails are available to match siding and trim colors. A one-inch hem placed along one edge of flat trim, such as fascia, will help minimize buckling or oil-canning.

Soffit and Fascia.

Most vinyl siding jobs include aluminum fascia trim and vinyl or aluminum soffit panels. The fascia is typically secured at the top, either by the drip edge or a piece of vinyl utility trim, although it can also be fastened with a few face-nails (through slotted holes) if the nails will be hidden by a gutter. The bottom lip of the fascia should be nailed sparingly in slotted holes to allow movement (Figure 1-24).

The outside edge of the aluminum or vinyl soffit panels can be supported by the receiving channel (J-channel or F-channel) or by the L bend at the bottom of the fascia,

as shown in Figure 1-24. The back edge of the soffit is either supported by receiving channel or a wood or vinyl frieze board. These details allow the fascia and soffit panels to move freely to accommodate thermal expansion and avoid buckling.

Windows and Door.

Window and door trim is perhaps the most conspicuous part of a vinyl siding job. Good planning is important. If possible, plan the job so a full butt of the siding lands on top of the windows. At window bottoms avoid the use of $\frac{3}{4}$-inch J-channel, which lets the unrestricted siding buckle. Instead use utility trim or under-sill trim to hold the siding tight here and at other horizontal projections. Where the vinyl has been notched below a window, use a snap-lock punch to create raised lugs along the top edge of the siding, locking it into the under-sill trim.

At window side and head jambs, J-channel is the most common treatment, with end tabs on the top J-channel bent over the side channels to deflect water (Figure 1-25).

While this detail helps shed water, it is important to note that *J-channel does not serve as window head flashing.* The window head should be properly flashed with the window's top flange or drip cap lapped under the sheathing wrap and sealed to the sheathing with flashing tape.

To simplify installation and avoid the conspicuous look of J-channel around windows, one option is to use solid vinyl windows with an integral J-channel (Figure 1-26).

For the more traditional look of flat trim around the window, you can use vinyl widow casing, which is typically $2\frac{1}{2}$ inches wide with an integral J-channel.

FIGURE 1-24 **Vinyl and Aluminum Trim.**

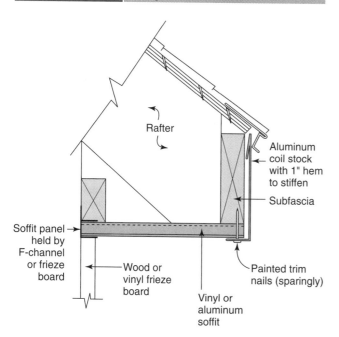

Most vinyl siding jobs include aluminum fascia and vinyl or aluminum soffits. A hem or bend in the aluminum-coil stock helps to stiffen the fascia and minimize oil-canning. The soffit panels should be free to float at one or, preferably, both edges.

FIGURE 1-25 **Window and Door Trim.**

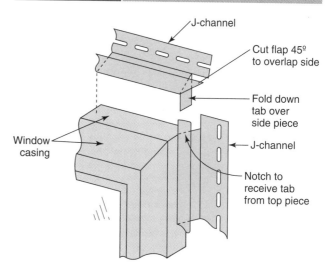

J-channel is the most common treatment at window side and head jambs, with the end tabs on the top J-channel bent over the side channels. While this helps shed water, the window should also have a proper head flashing sealed to the sheathing.

FIGURE 1-26   Integrated J-channel.

FIGURE 1-26   Integrated J-channel.

Solid vinyl windows with an integral receiving channel simplify siding installation and avoid the conspicuous look of J-channel around windows.
SOURCE: Photo by author.

Corners. Set the inside and outside vinyl corner posts about $\frac{1}{4}$ inch below the soffit or frieze above, and lock them in place with nails at the top of the uppermost nailing slots. Then nail in the center of the slots every 6 to 12 inches so that any movement is downward, not upward. Vinyl corner trim tends to be wavy, so following a snapped line is helpful (Figure 1-27).

Wood Trim. For those attracted to the low-maintenance appeal of vinyl siding but who want the look of traditional trim, builders can use wood or composite trim rabbeted or built out to create a receiving channel for the siding. For example, $\frac{5}{4}$-inch corner boards can be rabbeted to receive the siding, or standard $\frac{3}{4}$-inch stock can be furred out the thickness of the vinyl siding to create a similar effect (see Figure 1-28).

Window and door casings can be fashioned the same way. Either use a furring strip to raise the casing above the vinyl siding or use a thicker profile with a rabbet. At the bottom of the window, you can partially conceal the under-sill trim in the rabbet. To shed water, the head casing will

FIGURE 1-27   Vinyl Corner Posts.

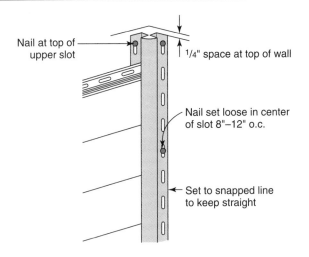

Align inside and outside vinyl corner posts to a snapped line, and drive nails tight at the top to lock the posts in place about $\frac{1}{4}$ inch below the soffit or frieze board. Center all other nails in their slots, driven loosely to allow movement.

still need conventional head flashing and J-channel, but these will be relatively inconspicuous.

STUCCO

No stucco system is impervious to water penetration, whether traditional three-coat stucco, modern one-coat systems, or exterior insulation and finish systems (EIFS). Since water may enter through cracks, penetrations, or through the stucco finish itself, all stucco exteriors rely on a backup waterproof drainage plane to protect the structure. The drainage plane under stucco is essentially the same as under other exterior claddings, with building paper layered to shed water and carefully integrated with all flashings at doors, windows, and other penetrations.

In addition, stucco systems need a weep screed or similar perforated flashing at the bottom of the wall to safely drain away any trapped water at the foundation. Without a continuous drainage plane, stucco systems are subject to serious water problems. While older, traditional stucco walls were designed to get wet and readily dry out, the newer synthetic systems are less permeable to moisture. If trapped water cannot readily drain away or dry to the exterior, the underlying structure is more vulnerable to moisture damage.

Drainage Plane

Traditionally, stucco contractors have used Grade D building paper rather than asphalt felt when applying stucco to wood-frame walls. Grade D building paper is an

FIGURE 1-28 | **Traditional Trim.**

To combine the look of traditional wood trim with vinyl siding, fur out $\frac{3}{4}$-inch stock to overlap the siding. No J-channel is required (left). Around light fixtures and other exterior details, use premanufactured mounting blocks with an integral receiving channel (right).
SOURCE: Photos by author.

asphalt-impregnated kraft-type paper, similar to the backing on fiberglass insulation. Unlike asphalt felt, it is made from new wood pulp, rather than recycled material. It has water-resistance ratings ranging from 20 to 60 minutes, depending on the thickness.

Although the IRC does not specify a required rating for stucco underlayment, the trend in the industry is to use two layers of 15- or 30-minute Grade D paper, isolating one layer from direct contact with the stucco and creating a secondary drainage space in the gaps between the two layers. Two layers are necessary, since the stucco tends to bond to the outer layer of building paper or plastic housewrap, compromising its water repellency. The wetter the climate, the heavier the paper should be. In coastal areas, some contractors use as much as two layers of 60-minute paper. The heavier papers provide better protection, but they are less flexible and more difficult to install.

Some contractors are starting to use plastic housewrap under stucco. How well it holds up in direct contact with stucco is in question. One option is to use plastic housewrap as the first layer and cover it with Grade D building paper, which has a longer track record in direct contact with stucco.

Other than the building paper, flashings are essentially the same as with any other cladding system. Metal or membrane pans are recommended at the bottoms of windows and doors. As with other cladding systems, it is critical that the building papers layer over window head flashings and that window pan flashings drain on top of the building paper. Do not caulk the horizontal joints at window head and pan flashings; this way, any trapped water can drain out.

To complete the system, the drainage plane behind stucco must have a perforated flashing called a "weep screed" at the foundation line. According to the IRC, this must be at least 4 inches above grade and must allow trapped water to drain to the outside of the building. Without a weep screed, the stucco tends to bond to the top of the foundation, creating a moisture dam.

Three-Coat Stucco

Three-coat stucco using Portland-cement plaster has been used successfully in the United States for nearly 200 years. It is applied about $\frac{7}{8}$ inch thick over metal lath, which creates a drainage space between the building paper and the stucco, allowing water to drain out through the weep screed at the foundation (see Figure 1-29).

Stucco relies on this drainage plane for waterproofing, since the stucco material itself is relatively porous. It tends

FIGURE 1-29 **Three-Coat Stucco.**

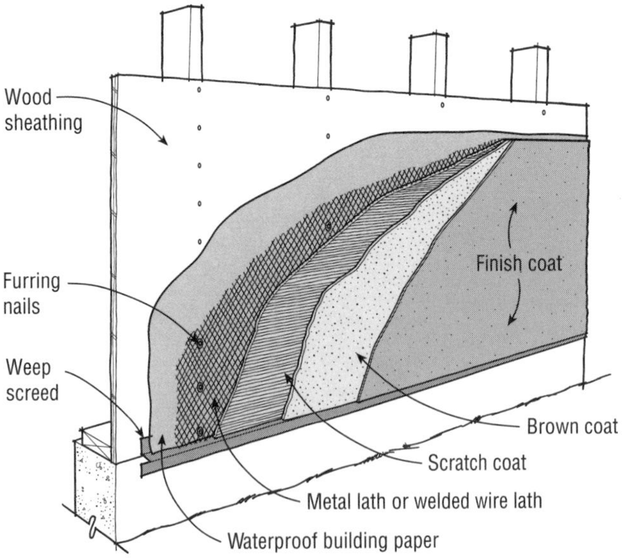

Traditional three-coat stucco is applied about $\frac{7}{8}$ inch thick over metal lath, which creates a space for water to drain out through the weep screed along the foundation. The drainage plane is essential since stucco itself is relatively porous.

to soak up water when it rains, but it dries out quickly since it is highly permeable to water vapor.

The Structure. Portland-cement stucco shrinks as it dries, which normally creates small hairline cracks in the finished surface. Larger cracks may form, however, if there is significant movement in the structure, since stucco is nonstructural and relatively rigid. A well-designed foundation and good-quality, dry framing lumber with adequate bracing will minimize this type of movement. On stucco jobs with no structural sheathing, still common in some western states, adequate bracing for racking strength and rigidity is particularly critical.

Cracking can also result from thin sections in the stucco finish. To avoid these problems, fur out or straighten any bowed or irregular walls before applying stucco.

Expansion Joints. Where stucco is applied over metal lath on wood framing, the Portland Cement Association recommends expansion joints every 10 feet, forming panels of no more than 150 square feet. Expansion joints are particularly critical at joints between dissimilar materials, such as where wood framing meets masonry, or wherever excessive movement is expected, such as the band joist area between two stories. Many residential projects are built without expansion joints, which can lead to cracking given the excessive movement associated with today's lower quality framing materials.

Metal Lath. Stucco will bond directly to most masonry surfaces, but on sheathed walls the stucco requires metal lath to form a mechanical bond to the wall. On residential projects, contractors use either expanded metal lath or "stucco netting," a 17- or 18-gauge galvanized wire woven into a hexagonal mesh that looks like chicken wire. Lath should always run perpendicular to the studs, and expanded metal lath must be installed with the correct side pointed up or the plaster will slip off when troweled on.

The lath is nailed or stapled approximately every 6 inches at studs and other framing members. Galvanized staples are now widely used to attach metal lath. However, unless the lath is the self-furring type, it should be installed with special furring nails that space the lath about $\frac{1}{4}$ inch from the wall, fully embedding it in the scratch coat, according to stucco expert Ron Webber, of Procoat Systems, in Orange, California.

Another common problem, according to Webber, is that lath installed too tightly at corners causes poor embedment of the mesh at the corners. This will cause cracks at the corners as the building undergoes normal movement with changes in temperature and humidity. To prevent corner cracks, installers should use a special corner bead called Cornerite™ or build up the corners with two layers of wire. Another option is to pull the lath away from the building at the corners to make sure it is properly embedded.

The corners at either side of window and door headers is another common location for stucco cracks. To reinforce these areas and reduce cracking, some contractors add a second layer of reinforcing at these corners using a rectangular section of metal lath placed diagonally at each corner.

The metal lath should form a continuous layer around the building with all laps wired together and vertical laps staggered. With large-mesh reinforcement, lap vertical and horizontal joints at least one full mesh and a minimum of 2 inches. For small-mesh reinforcement, the laps should be at least 1 inch.

Mixtures. Stucco is a mixture of Portland cement, sand, and water, with a little lime or a plasticizer added for workability. A proper mixture has good tensile strength and weather resistance and the ability to bond well to the mesh or substrate. It is also easy to trowel on and resists sagging. In cold climates, it must also have freeze-thaw durability, usually obtained by using air-entrained plaster.

The cement base can be masonry cement, plastic cement, or Portland cement, which may have air-entraining additives. Do not add lime or a plasticizer to masonry cement or plastic cement since these already contain plasticizers. While approximate proportions are well established, the right mix for a job depends on the weather exposure of the wall and weather conditions during application (see Table 1-9).

Other than the right proportions, the keys to a good stucco mix are clean, good quality sand and clean potable water. Since sand makes up about 97% of the stucco mixture by volume, it is critical to use good sand. The sand should be free of vegetable matter, loam, clay, silt, and soluble salts and should conform to ASTM C897, which

TABLE 1-9	Stucco Base-Coat Mixes (proportion by volume)					
	Cementitious Materials				Sand	
Plaster Type	Portland Cement (C)	Lime (L)	Masonry Cement (M)	Plastic Cement (P)	First Coat (Scratch)	Second Coat (Brown)
C	1	$0-\frac{1}{2}$	—	—	$2\frac{1}{2}-4$	3–5
CM	1		1		$2\frac{1}{2}-4$	3–5
L	1	$\frac{1}{2}-1\frac{1}{4}$	—	—	$2\frac{1}{2}-4$	3–5
M	—	—	1	—	$2\frac{1}{2}-4$	3–5
CP	1	—	—	1	$2\frac{1}{2}-4$	3–5
P	—	—	—	1	3–5	4–5

SOURCE: Courtesy of Portland Cement Association; adapted from ASTM C926.

designates the distribution of particle sizes (gradation). Impurities in the sand or water can affect the strength of the mix, and poor grading of the sand will hurt its workability. Salts can cause staining on the finished surface from efflorescence.

Application. Stucco can either be hand troweled or blown with a machine. Some stucco contractors use a pump for the base coats but apply the finish coat by hand. Although the mixes are slightly different for the two approaches, both can produce a high-quality finish. In three-coat stucco, the first and second coats are $\frac{3}{8}$ inch thick, and the finish coat is $\frac{1}{8}$ inch for a total thickness of $\frac{7}{8}$ to 1 inch.

Scratch Coat. The first, or "scratch," coat, which forms the base for the next two coats, should completely encase the reinforcement. While still wet, the plaster is scored horizontally with a special metal rake or trowel to create a good mechanical bond with the second coat (vertical scratching promotes cracking at studs). For proper curing, the scratch coat needs to be kept moist by misting or fogging with water for 48 hours. Except in very moist weather, misting should start as soon as the freshly applied stucco lightens in color and be repeated at the start and end of each day until the second coat goes on.

Brown Coat. The second, or "brown," coat should go on as soon as the first coat is hard enough to accept the second coat without cracking, but at least 48 hours later, according to the IRC. The second coat fills any cracks in the scratch coat, and the additional sand in the brown coat helps prevent new shrinkage cracks. Whether it is hand-troweled or machine-applied, it must be leveled with a straightedge ("rodded") and floated to produce an even surface for the final coat.

A short delay between the first and second coat helps to create a good bond between the two and strengthens the scratch coat by rewetting it for a more complete cure. Any cracks larger than $\frac{1}{16}$ inch in the brown coat should be patched before the top coat goes on. In the Southwest, where adobe is popular, the brown coat is often steel-troweled for an adobe look and serves as the final coat.

Finish Coat. After the second coat is allowed to cure for a minimum of 7 days (14 will allow a more complete cure), the top coat is applied to provide the finish color and texture. Many contractors now use premixed color coats, some with acrylic additives to increase water resistance and flexibility. Creating a uniform color and texture requires a skilled applicator, uniform mixing, favorable weather (avoid direct sun), and a uniform substrate without variations in texture or water absorption. Problems in the substrate will tend to show through the thin finish coat. It is best to do an entire side of the building in one batch with no cold joints. A modest amount of color variation is considered part of the character of traditional stucco, but too much is a sign of substandard work.

A certain amount of shrinkage cracking is also inevitable in stucco exteriors. Application over wood-frame construction results in more cracking than over concrete block or other more stable substrates. Coarse textures in the finish will tend to hide the cracks better than smooth finishes. Even under the best of conditions, small shrinkage cracks of less than $\frac{1}{16}$ inch will occur in the finished stucco and are to be expected. Generally these do not leak or indicate substandard work.

Weather. Temperature will speed up or slow down the hydration process that cures the cement in stucco. It is best to avoid application in extremely hot or cold temperatures. In hot, dry, and windy weather, frequent misting will be required on the scratch coat or the installer may need to tape polyethylene sheeting in place for proper curing. Direct sun tends to dry out the fresh stucco too fast, so installers should try to follow the shade around the building. Also, retardants are available that can be sprayed on the scratch or brown coat in hot weather to slow down the curing.

Cold weather also presents problems. Stucco should not be applied under 40°F, and it should not be allowed to freeze within 24 hours of appplication. Accelerators can be added to the stucco mix in cold weather, but these can weaken the material, and calcium-based accelerators can lead to efflorescence. Heating the materials and, if necessary, tenting the structure can permit work to proceed in cold, even freezing, weather.

Cool, moist weather is ideal. In humid weather, with relative humidity over 70% or heavy fog, misting is not usually required.

Thin-Coat Stucco

In an effort to speed up stucco application time and simplify the process, several manufacturers have introduced proprietary thin-coat stucco systems variously referred to as one-coat, two-coat, thin-coat, or fiberglass-reinforced stucco. All these systems apply a single base coat and a top coat with a total thickness of $\frac{3}{8}$ to $\frac{1}{2}$ inch, compared to $\frac{7}{8}$ to 1 inch for traditional three-coat stucco. The thinner finish weighs from 5 to 6 pounds per square foot, compared to 9 pounds for three-coat, and it is cost-competitive with traditional stucco.

Like traditional three-coat stucco, thin-coat is applied over wire mesh or expanded metal lath by hand or pump. It is backed up by a waterproof drainage plane consisting of Grade D building paper, integral flashings, and a weep screed along the top of the foundation to drain away any trapped water.

Some manufacturers, such as United States Gypsum, have introduced hybrid systems in which the stucco is applied to a cementitious board rather than to wire mesh. The advantage is that cement board is impervious to moisture. The drainage plane, and in some cases a layer of foam insulation, lies behind the cement board.

Application. The base coat in thin-coat systems has acrylic polymers and chopped fiberglass added to increase its strength and resistance to shrinkage cracking and to freeze-thaw cycles. The base coat is premixed with only sand and water added at the job site. Most contractors using these systems apply an elastomeric color coat, similar to a thick acrylic paint with fine aggregate, and formulated to bridge small gaps less than $\frac{1}{16}$ inch. This produces a smoother finish that is more water- and stain-resistant and less prone to cracking than a traditional stucco. The top coat can also be a traditional cement stucco finish.

Most of these systems require a 24- to 48-hour moist cure and a total of six or seven days of curing before the top coat is applied. Some require a primer for acrylic finishes.

Pros and Cons. To their credit, properly applied one-coat systems are more waterproof and less prone to shrinkage cracking than traditional stucco. It is easier to obtain a uniform color and texture with the synthetic color coat than with a traditional cementitious finish coat. Whether a customer prefers the uniform color of a synthetic finish or the more muted and variable color of cement stucco is a matter of taste.

On the downside, one-coat systems are less impact-resistant than traditional three-coat stucco. And with a thickness of only $\frac{3}{8}$ inch, one-coat systems are less able to hide irregularities in the framing and are more likely to have thin spots that are prone to problems. Also, one-coat systems are not completely waterproof. Over time, water will find its way in at joints, penetrations, or cracks, and the synthetic stucco will be slower to dry out than the more permeable traditional stucco.

Finally, each system is proprietary and must be installed according to the manufacturer's approved specs and details, which vary from system to system. Otherwise, warranties are voided and code approvals, which are based on building code evaluation reports, are invalid. For both reasons, contractors should avoid mixing and matching components from different thin-coat systems.

Exterior Insulation and Finish Systems (EIFS)

When originally imported from Europe to the United States in the 1970s, most exterior insulation and finish systems (EIFS) were "barrier" type systems. They were designed to create a waterproof exterior skin consisting of a thin layer of acrylic polymer-based synthetic stucco directly applied to foam insulation. The expanded polystyrene (EPS) foam was glued to the building's sheathing. A layer of fiberglass cloth embedded in the synthetic stucco provided reinforcement, and a thin acrylic finish coat added color and texture.

With the EPS glued directly to the sheathing, there was no place for building paper or conventional flashings at penetrations. Openings, joints, and penetrations relied on caulks and sealants for waterproofing. With no backup waterproofing or drainage layer, there was little margin for error.

While these systems performed adequately in Europe for nearly 25 years, the United States version had thinner base coats and lower polymer content, creating a weaker skin. Also, workmanship in the United States was often inferior due to lack of applicator training and quality-control programs. When water leaked into these systems through failed caulk joints, cracks in the stucco skin, or through the window frames themselves, it wet the foam insulation, sheathing, and sometimes the structural framework. This in turn led to more sealant failures and cracking of the surface and additional leakage. The EPS foam acted like a sponge, trapping water against the building, and the nonporous polymer coatings retarded drying. In many cases the leakage and resulting decay was extensive, resulting in widespread property damage and litigation.

Drainage EIFS. In response to these problems, most EIFS manufacturers have introduced new "drainage" or "water-managed" systems, which require the same type of waterproof drainage plane found behind traditional stucco systems (see Figure 1-30).

FIGURE 1-30 **Drainage EIFS.**

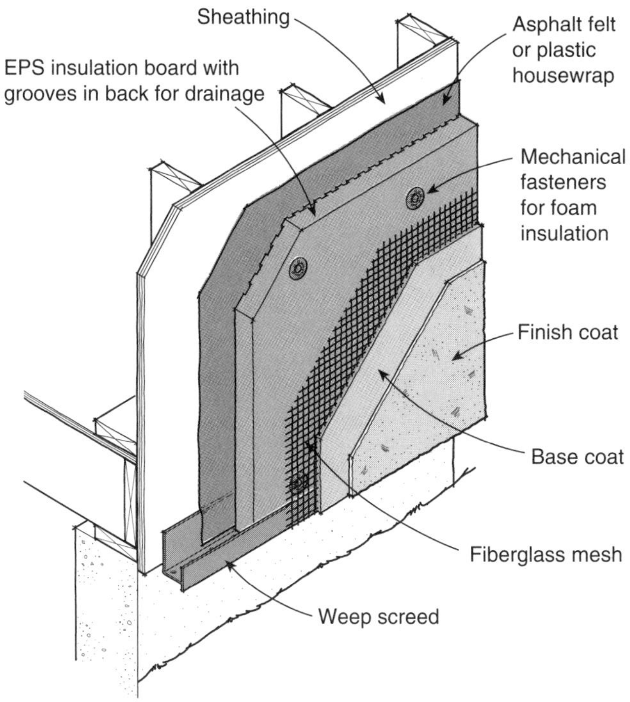

In response to widespread EIFS failures in the United States, manufacturers have introduced new systems that use a drainage plane similar to traditional stucco systems. Most of the new systems use mechanical fasteners, rather than adhesive, to attach the foam insulation to the sheathing. Grooves cut in the back of foam board create a capillary break and drainage space.

As with traditional stucco, layered building paper or plastic housewrap protects the framing and sheathing, and all exterior openings and penetrations are flashed to conduct any water to the outside of the sheathing wrap. Since window leakage was the single biggest contributor to EIFS failures, pan flashing is recommended at windows.

Rather than gluing the EPS foam to the sheathing, the new drainage EIFS typically use mechanical fasteners and are designed with a capillary break between the back of the EPS and the sheathing wrap to promote drainage. Some use special corrugated or wrinkled sheathing papers to create the drainage space, while others have vertical grooves cut into the back face of the foam insulation. In all cases, the drainage plane leads to a perforated weep flashing at the foundation to drain away any trapped water.

Workmanship. The backup drainage layer, however, should not provide an excuse for sloppy workmanship on the exterior skin. The new kinds of EIFS should still be made as waterproof as possible, since any water that leaks past the skin may be slow to dry out. EIFS consultant Russell Kenney, who has worked with these systems for nearly 20 years, recommends exceeding the minimum specs required by EIFS manufacturers. For example, Kenney recommends a higher-density EPS foam with only 2% water absorption by volume instead of the 4% allowed by

ASTM C584. In addition, Kenney recommends a heavier 6-ounce reinforcing mesh versus the typical 3-ounce cloth, as well as special high-impact mesh in high-traffic areas. He also recommends a $\frac{3}{32}$-inch base coat applied in two layers, with the first layer used to partially embed the fiberglass reinforcing and the second layer to fully cover and protect it. These steps will significantly improve the impact resistance of EIFS, but it is still less durable than traditional stucco or thin-coat stucco.

Apply Sealant to Base Coat. As with the original barrier EIFS, all penetrations require a high-quality elastomeric sealant. The sealant needs to be applied to the base coat since the finish coat tends to soften when wet, providing a poor substrate for sealant. For the caulk joints to last, they must be wide enough to tolerate the anticipated movement, typically $\frac{3}{8}$ to $\frac{1}{2}$ inch, and backed up by backer rod (see "Joint Design," page 37). While control joints are generally not needed along the length of the wall—unless it exceeds 75 feet and is in direct sun—they are required between floors on multistory buildings. Silicone sealant is recommended at all joints for its longevity and flexibility in cold temperatures.

In theory at least, drainage EIFS should function the same as any other exterior cladding systems. Any water that manages to penetrate the outer skin should be stopped by the drainage layer and safely drained away. However, given the low permeability of polymer-based coatings and the tendency of EPS foam to soak up and hold water, EIFS are best avoided in residential projects unless high-quality workmanship and regular maintenance of sealants can be assured.

WOOD AND COMPOSITE TRIM

As costs rise and quality levels fall for solid wood board stock, builders have become more receptive to a wide range of alternative products introduced over the past 10 to 15 years. Some of the products are variations on the material used in hardboard siding, a product that has been largely discontinued due to widespread problems with moisture absorption and buckling. Others are fiber-cement-based and offer the same durability and longevity as the siding. Still others make use of PVC, urethane, or other types of plastics, which promise longevity and low maintenance but may cost significantly more than the solid wood they replace (see Table 1-10).

Solid Wood

Solid wood is still the first choice of many builders for highly visible trim such as porch columns that require tight miters and smooth edges and need to tolerate a certain amount of wear and tear. Softwoods have served well in this capacity for many years, since they were traditionally inexpensive, dimensionally stable, and held paint well.

TABLE 1-10	Exterior Trim Options						

Material Type	Uses	Pros	Cons	Costs (1—lowest; 5—highest)	Dimensional Stability (1—best; 5—worst)	Recommendations
Solid wood	Boards, moldings; unfinished or preprimed	Easy to cut, shape, install, and finish. Decay-resistant species available.	Premium materials getting scarce and expensive. Inconsistent quality.	1–3 (depending on grade)	1 (with grain) 2 (across grain)	If budget permits, choose vertical-grain lumber of decay-resistant species. Prime all faces and cut edges.
Finger-jointed	Boards, moldings; unfinished or preprimed	Same as solid wood, but more consistent and dimensionally stable.	Joints may telegraph through finish.	2–3 (depending on grade)	1 (with grain) 2 (across grain)	Prime all faces and cut edges. Maintain finish.
Hardboard	Boards; preprimed or prefinished	Economical. Uniform consistency.	Swells from water intrusion at nail holes and exposed edges. May delaminate with persistent wetting. Nail heads left flush. Heavy.	1	3	Prime all faces and cut edges. Drive nails flush. Leave $\frac{1}{8}$-inch gap at butt joints and caulk. Avoid high-moisture applications.
LVL	Boards with textured wood or smooth MDO face; preprimed	Strong and stable. MDO facing takes paint well.	Plywood edges vulnerable to moisture.	2	1 (with face grain) 2 (across face grain)	Prime cut edges and avoid exposing edges to excessive moisture.
Fiber-cement	Boards, panels, soffits, exterior ceilings; smooth or wood-grained; unfinished, preprimed, or prefinished	Economical, impervious to insects, decay, fire. Paints well.	Requires special tools to cut. Hazardous dust. Heavy. Pneumatic nails only. Nail heads left flush (except with XLD)	1 (standard) 3 ($\frac{5}{4}''$ Harditrim XLD)	1	Use with fiber-cement siding for a durable exterior. Drive nails flush. Suitable for high-moisture applications.
Polymer	Moldings and decorative trim; preprimed	Easy to work. Installs with adhesive. Saves time with complex trim. Paints well.	Expensive. Expands and contracts with temperature changes.	5	5	Use for dentils, pediments brackets, and other complex trim. Labor savings offset high material costs.
Cellular PVC	Boards, moldings, Panels, soffits, exterior ceilings	Easy to work and install. Impervious to moisture. Can be left unpainted (white).	Expands and contracts with temperature change. Cannot paint dark colors.	3	4	Use with fiber-cement siding for a durable exterior. Suitable for high-moisture applications. Carefully detail to accommodate movement.

Decay Resistance. As smaller, faster-growing trees replace older virgin timber stands, high-quality wood has become more expensive and harder to find. Even when using decay-resistant species, the smaller trees harvested today have less heartwood, which is where the extractives are found in sufficient quantities to be effective against decay (Table 1-11). With any wood species, the sapwood is more prone to decay.

Paintability. Solid wood has virtually no shrinkage along the grain and, if finished on all sides, limited seasonal movement across the grain. In general, the denser

TABLE 1-11	Decay Resistance of Domestic Woods (heartwood only)	
Resistant or Very Resistant	**Moderately Resistant**	**Slightly or Nonresistant**
Catalpa	Cypress, tidewater red (young growth)[1]	Alder
Cedars	Douglas fir	Ashes
Cherry, black	Honey Locust	Aspens
Chestnut	Larch, western	Basswood
Cypress, Arizona	Oak, swamp chestnut	Beech
Cypress, tidewater red (old growth)[1]	Pine, eastern white[1]	Birches
Junipers	Southern pine (longleaf or slash)[1]	Cottonwood
Locust, black[2]	Tamarack	Elms
Mesquite		Hemlocks
Mulberry, red		Hickories
Oak		Maples
Osage orange[2]		Oak (red and black)
Redwood		Pines (other than longleaf, slash, and eastern white)
Sassafras		Poplars
Walnut, black		Spruces
Yew, Pacific[2]		True firs (western and eastern)
		Willows
		Yellow Poplar

[1]Southern and eastern pines and tidewater red cypress are now largely second growth with mostly sapwood. Little heartwood lumber in these species is available.
[2]Exceptionally high decay-resistance
SOURCE: *Prevention and Control of Decay in Homes,* USDA, 1980.

FIGURE 1-31	Cupping of Flat-Sawn Lumber.

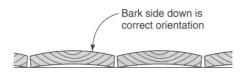

To keep corners tight and prevent edges from lifting, always install flat-grained trim "bark-side down," since the annual rings tend to straighten as the wood shrinks, causing the wood to cup as shown.
SOURCE: Reprinted from *Architectural Graphic Standards, Residential Construction,* with permission of John Wiley & Sons, © 2003.

a wood species is, the more it shrinks and swells with changes in moisture and the worse it is as a substrate for paint on a building's exterior (Table 1-12). Vertical-grain, or "edge-grain," softwoods, such as vertical-grain cedar or redwood, are the most stable and hold paint the best. The flat-grained woods more commonly used as trim are more prone to cupping and other moisture movement and do not hold paint as well.

To compensate for the cupping and to keep corners tight and edges from lifting, always install trim "bark-side down," since the annual rings try to straighten as the wood shrinks (see Figure 1-31). To improve paintability, it may be necessary to rough up the surface of flat-grained trim with 80 to 100 grit sandpaper before priming. As with siding, it is best to prime all surfaces and to prime cut ends to minimize water absorption through end grain. Also, hold trim pieces back at least $\frac{1}{2}$ inch short of flashings or other surfaces where water may collect and soak the end grain.

Finger-Jointed Wood

Many manufacturers now offer solid wood trim made up of short lengths of high-quality lumber that is finger-jointed and, in some cases, edge-glued to make boards as long as 24 feet. As with solid lumber, finger-jointed lumber is available in a number of grades and species.

Telegraphing at Joints. In general, finger-jointed stock is durable and dimensionally stable since short pieces of wood are less likely to warp and twist. The main concern is whether the glue joints will telegraph through the paint as the material swells and shrinks in response to changes in relative humidity. Because no two pieces of wood swell and shrink at exactly the same rate, the joints often do show through. The best protection is to keep excess moisture out of the wood by starting with quality preprimed stock or using a high-quality water-resistant prime coat and two top coats of paint.

TABLE 1-12	Finishing Characteristics of Selected Wood Species		
Wood Species	Paint Holding Ability (1—best; 5—worst)	Resistance to Cupping (1—best; 4—worst)	Conspicuousness of Checking (1—best; 2—worst)
Softwoods			
Cedar, western red	1	1	1
Cypress	1	1	1
Douglas fir	4	2	2
Eastern white pine	2	2	2
Ponderosa pine	3	2	2
Redwood	1	1	1
Southern yellow pine	4	2	2
Spruce	3	2	2
Western hemlock	3	2	2
Hardwoods			
Lauan (plywood)	4	2	2
Oak, white or northern red*	4–5	4	2
Yellow birch	4	4	2
Yellow poplar	3	2	1

*These woods have large pores that require filling for durable painting. With pores properly filled, paint-holding ability = 2.
SOURCE: Adapted with permission from *Finishes for Exterior Wood*, 1996, R. Sam Williams, Mark Knaebe, and William Feist. Courtesy of Forest Products Society.

When purchasing finger-jointed trim, look for long-term warranties against any delamination or glue lines telegraphing through the paint. As with any wood-based product, minimizing exposure to water and maintaining the finish are important for long-term performance.

Hardboard Trim

The leading alternative to solid wood trim, and the oldest in the marketplace, is hardboard, essentially the same material used in hardboard siding. Hardboard consists primarily of ground wood fibers and phenolic resin, the same adhesive used in exterior-grade plywood, along with additives to improve weather resistance. It is typically available with either a smooth finish or a wood-grain texture and is sold in 16-foot lengths. Hardboard weighs about 4 pounds per square foot for 1-inch stock, roughly twice as much as softwood.

Paintability and Dimensional Stability. To its credit, hardboard trim is very uniform in consistency and holds paint well. However, because it has no grain, it shrinks and swells equally in all dimensions—up to twice as much as wood along its length. In very dry or very humid climates this can lead to gaps or buckling over long sections. Most manufacturers recommend leaving an $\frac{1}{8}$-inch gap at butt joints and caulking with a high-quality paintable sealant to avoid problems.

Workability. While hardboard is relatively easy to nail, it does not hold nails well and is prone to split if edge-nailed. Drilling pilot holes will help. Compared to solid wood, it is more difficult to set nails and fill the holes in hardboard. Finish nails and pneumatic nails tend to pucker the surface, which must be sanded smooth before filling the holes. Round-headed nails driven straight in at a 90° angle leave a clean hole for filling. Most manufacturers recommend face nailing flush to the surface of the board to avoid these problems. If nails do penetrate the surface, sand the material smooth and fill the hole before painting.

Water Penetration. The biggest problem with hardboard is swelling and delamination where water has penetrated the material at unpainted cut edges, nail holes, or other penetrations (see Figure 1-32).

FIGURE 1-32A AND B | Water-damaged Hardboard.

This three-year old hardboard trim has swelled around the nail holes (left) and is crumbling at the door bottom (right). Even when primed and painted, hardboard trim is vulnerable to swelling and delamination in wet locations and at cut edges, nail holes, miters, and other penetrations.
SOURCE: Photos by author.

Even if primed, the material is vulnerable in wet locations, for example, at the bottom of an exterior door casing or in direct contact with a concrete slab or foundation. Sharp edges are also vulnerable to chipping, making this not the best choice where wear and tear is expected. To avoid problems, corners should be butted, not mitered.

Laminated-Veneer Lumber (LVL) Trim

Widely used in beams and headers, LVL has also been put to good use as a trim material with some minor modifications such as water-resistant edge sealing and adding a couple of cross-laminated layers to minimize cupping. Like LVL beams, LVL trim is dimensionally stable and is easy to cut, nail, and install, similar to a piece of plywood. Its weight falls in between solid wood and hardboard. It can be used for fascia, casings, corner boards, and most other exterior trim, and is available in lengths from 8 to 24 feet.

Facing Materials. One manufacturer, Pacific Wood Laminates, makes a preprimed Douglas-fir LVL trim faced either with textured wood veneer (Socomi Lam®) or medium-density overlay (Clear Lam®). Medium-density overlay (MDO) is a highly durable resin-impregnated paper that resists surface checking and holds paint well. It has a 20-year track record as a durable facing in concrete forms, outdoor signs, and other exterior applications. Pacific Wood's LVL trim is sealed on all edges with a water-based elastomeric coating that hides the end grain and resists moisture penetration. It is then primed on all faces, with a second prime coat applied to the finish face and edges.

Installation. As with other engineered trim products, all cut ends of LVL should be primed in the field and butt joints gapped $\frac{1}{8}$ inch and caulked. Miters and scarf joints can be used. Cut ends should be kept 6 inches off the ground, concrete, or other wet materials. Resistance to swelling and delaminating will be similar to plywood siding panels such as T-111. The manufacturer recommends that the nails be set below the surface either by hand or pneumatic nailer and caulked. Use a small-headed finish nail or siding nail so as not to make too large a hole in the MDO facing.

MDO Plywood. Conventional plywood with MDO facing also makes an attractive and economical smooth soffit material. Combined with LVL fascia, this should produce a durable and attractive eaves detail.

Fiber-Cement Trim

Contractors who install fiber-cement siding are looking for equally durable materials to trim out their houses. Some have found their answer in fiber-cement trim, which boasts the same dimensional stability, paint holding ability, and resistance to rot, insects, and fire as the siding. Standard $\frac{7}{16}$-inch fiber-cement trim is comparable in price to midgrade softwoods and is available with either a smooth face or a wood-grain texture.

Workability. Like the siding, fiber-cement trim is heavy, requires special tools and procedures to cut and machine, and presents a dust hazard to those cutting the material. Although special diamond-tipped tools make it easier to cut and form, standard fiber-cement trim is not easy to miter or to make complex cuts in, and it is too hard to hand-nail. Since it is not a nail base, all pieces need solid wood backing. Like the siding, fiber-cement trim must be face-nailed with the nails set flush to the surface.

Applications. Because of its durability and resistance to moisture, fiber-cement is a good choice for soffits, using $\frac{1}{4}$-inch-thick panels. For other trim, most manufacturers sell $\frac{7}{16}$-inch planks, which contractors typically build out with $\frac{1}{2}$-inch plywood or oriented-strand board (OSB) to create a thicker profile. In some markets $\frac{3}{4}$-inch fiber-cement board stock is also available, eliminating the need to fur out the trim but adding considerable weight.

Lightweight Option. To simplify installation, James Hardie Building Products has introduced a low-density fiber-cement trim board called HardiTrim XLD that handles and installs more like wood trim. The new material can be installed with pneumatic finish nails set below the surface and puttied. Unlike standard fiber-cement, XLD holds nails and can be mitered and edge-nailed, simplifying details like corner boards. The 1-inch-thick material weighs about $4\frac{1}{2}$ pounds per square foot versus about 2 pounds for an equal sized piece of white pine.

Polymer Moldings

Molded from high-density polyurethane, polymer moldings have been in use for over 20 years and have proven their durability in both interior and exterior applications.

FIGURE 1-33 | **Polymer Moldings.**

High-density polyurethane moldings are used primarily for complex profiles such as dentils, window pediments, and decorative brackets. These can often be handled by a single piece of molded polymer, saving time and labor costs.
SOURCE: Photo courtesy of Focal Point Architectural Products.

Because of its high cost, almost four times the price of premium-grade softwood, urethane is used almost exclusively for complex profiles, such as dentils, window pediments, and decorative brackets (see Figure 1-33).

Installation. Most material is sold preprimed and can be cut, planed, and sanded like wood, only more easily because of the lighter weight. Polymer moldings are installed with adhesive rather than nails, although a few finish nails are often used to hold it in place while the glue dries. Butt joints and miters are bonded with the same adhesive used to hold the molding to walls or soffits. Most manufacturers provide a proprietary adhesive for installation.

Because polyurethane foam expands and contracts with changes in temperature, installers should cut long runs of molding $\frac{1}{8}$ to $\frac{1}{4}$ inch long and "spring-fit" the material into place. If installed slightly compressed, the molding will not leave gaps when the temperature drops and the material shrinks. When joining two pieces of molding,

FIGURE 1-34 | Cellular PVC Trim.

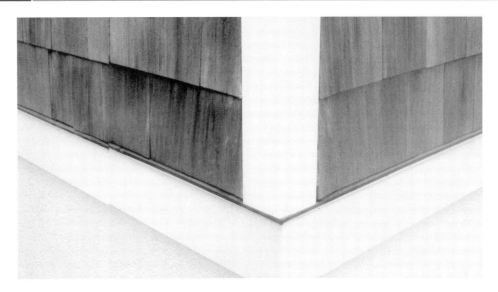

PVC boards and moldings are impervious to moisture and insects and typically carry a 25-year warranty. These one-inch-thick preformed corner boards have a welded seam guaranteed not to open.
SOURCE: Photo courtesy of AZEK Trimboards.

apply a generous amount of adhesive on both surfaces and clean the squeeze-out with a putty knife. Solvent may be needed to clean adhesive from the joint after it dries.

Painting. Before painting, fill any holes or dents with an exterior spackling compound and paint. Avoid leaving cut edges exposed, since without its hard skin, the material has a rough, irregular surface, even after painting. Since polyurethane will degrade from prolonged exposure to UV radiation, it should be painted soon after installation. If the surface is undamaged, it holds paint well.

Cellular PVC Moldings

Also called expanded PVC, or cellular vinyl, this relatively new material is a form of PVC that has been expanded with foaming agent and extruded into boards and a wide range of exterior molding profiles. Like other plastics and fiber-cement, cellular PVC is impervious to moisture and insects and is approved for contact with ground or masonry, making it well suited to moisture-prone applications such as garage-door trim. Warranties typically run to 25 years. Since cellular PVC is cost-competitive with premium wood, it is an attractive option for those seeking a more durable alternative.

Boards, Panels, and Moldings. Trim stock is available in thickness from $\frac{5}{8}$ to 1 inch and in lengths up to 20 feet. Sheet stock, which can be used for soffits and other panel applications, is available in $\frac{3}{8}$- to 1-inch thicknesses, with a smooth or beaded face. Manufacturer Marley Moldings makes a wide array of molding profiles, while Azek offers a wide range of trim materials, including a 1-inch-

thick prefabricated corner board for use with any type of siding (Figure 1-34).

Workability. Similar in density to pine, expanded PVC can be cut, drilled, sanded, and even routed like wood. It has moderately good nail holding ability and is installed similarly to wood trim, although with allowances for thermal expansion and contraction.

Thermal Expansion. Although it has less thermal expansion than polymer (polyurethane) foam, cellular PVC expands and contracts considerably more than wood, so it requires special detailing to avoid gaps, buckling, and other movement problems. Manufacturers recommend leaving a $\frac{1}{8}$-inch gap for movement for every 18 feet of length. This space is required where trim terminates into an inside corner or against an intersecting piece of trim, such as where corner boards meet the frieze board. A small back cut at the end of the board can help conceal the gap.

This amount of movement assumes that the PVC trim board is secured with two nails every 16 inches on-center to a solid substrate (three nails for boards 12 inches or wider). Nailing along the length of the board restricts its overall movement. You can further restrict movement in PVC trim by gluing it to a wood substrate with construction adhesive.

Installation. Because cellular PVC has less strength and stiffness than wood, it must be installed over a solid substrate, and material $\frac{1}{2}$ inch thick or less should be glued to a solid substrate. It will tend to conform to an uneven substrate, showing any waviness. Using PVC cement to weld one piece to the next, contractors can fashion corner

FIGURE 1-35 Installing PVC Trim.

Cellular PVC trim can be cut, drilled, and sanded like wood. Joints are usually glued with PVC cement and the assembly nailed in place.
SOURCE: Photo courtesy of AZEK Trimboards.

boards, window surrounds, or other trim assemblies, using screws or nails to hold the pieces in position while the glue dries (Figure 1-35). The assembled sections are then nailed or glued in place.

For long runs like fascia, scarf joints are best since they provide the greatest gluing area. Manufacturers recommend nailing on each side of the scarf joint to hold the trim in place while the glue dries. Construction adhesive behind the scarf joint—for example, between the PVC fascia and the wood subfascia—will also help reinforce the joint. As with other PVC joints, the plastic surfaces must be in direct contact to bond properly since the glue will not fill any voids.

Predrilling is generally not necessary as long as you use small diameter, blunt-head nails. Either galvanized or stainless-steel box nails are recommended, but contractors have used pneumatic nailers with ring-shank nails successfully. In general, the nailing pattern is the same as for wood trim except nails should be kept 2 inches from the ends of boards. Typically nails are set and puttied before painting.

Long Runs. Long runs of fascia require expansion space at each end or in the middle. Where a long run turns a corner, for example on a hip roof, one option is to glue the outside corners and leave an $\frac{1}{8}$-inch space along the center of each run where it will be less conspicuous than at the corner. To hide the joint, either leave an unglued scarf joint

or butt the two pieces of fascia and fill the gap with a polyurethane caulk. If building in hot weather, it is best to construct tight joints, since they will open in colder weather as the material contracts.

Painting. Like PVC windows and vinyl siding, PVC trim has UV inhibitors and does not need painting. However, if you want a color other than white, use 100% acrylic latex paint and avoid dark colors, which can cause the vinyl to overheat in direct sunlight. PVC trim requires no special preparation to paint and reportedly holds paint well.

TRIM DETAILS

The best exterior trim details are designed to keep water out but to provide easy drainage for any water that penetrates the exterior. This is particularly important when using trim materials that are vulnerable to decay or moisture damage, such as nondecay-resistant softwoods or hardboard. Caulking trim joints with sealants is a double-edged sword, since all caulk joints will eventually fail, and when they do, the remaining sealant will tend to keep the joint from drying.

Trim that is subject to frequent wettings from the weather, such as corner boards, water tables, or wrapped porch posts or balusters, is a good candidate for rain-screen

FIGURE 1-36 **Rain-Screen Trim.**

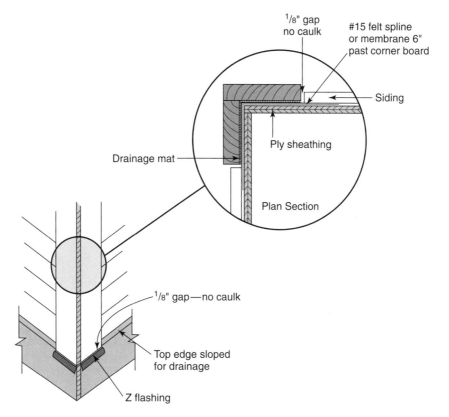

$^1/_8$" gap
no caulk

#15 felt spline
or membrane 6"
past corner board

Siding

Drainage mat

Ply sheathing

Plan Section

$^1/_8$" gap—no caulk

Top edge sloped
for drainage

Z flashing

Corner boards, porch posts, and other trim subject to frequent wetting will benefit from a ventilation and drainage space. Synthetic drainage materials developed for siding or foundations can work well.

installation, where a ventilation and drainage space is left behind the trim (Figure 1-36).

Make sure to leave $\frac{1}{8}$ to $\frac{1}{4}$ inch free at the bottom edge of the trim to drain to daylight. To create the drainage space, either use shims (do not block the drainage path) or synthetic drainage materials, such as Benjamin Obdyke's Home Slicker®. Foundation drainage materials can also work. If shims are used, add metal screening or drainage mutting at top and bottom to block insect entry.

Regardless of the specific detail, the following principles will help create long-lasting exterior trim:

- Wide roof overhangs, 8 inches minimum, at rake and eaves keeps water away from the side of the house, preserving siding and trim.

- Avoid wide horizontal wood surfaces exposed to water. Slope for drainage or cap with metal flashing. Cut drip groove under edge to shed water.

- Slope top edges of exterior railings and horizontal trim boards, such as water tables, to shed water.

- Avoid exposed end-grain facing upward in vertical trim boards. Heavily prime all end grain and exposed edges.

- Avoid exterior miter joints, which tend to open and absorb water.

- Use Z- or drip-cap flashing at horizontal joints, such as above windows or where corner boards meet the

water table. Leave $\frac{1}{8}$ inch clear above the flashing, and do not caulk the horizontal joint.

- Avoid caulk joints. Instead, flash well behind joints and leave a gap of $\frac{1}{8}$ inch for ventilation. Where caulking must be used, apply a properly shaped caulk bead (see "Joint Design," below).

CAULKS AND SEALANTS

While no residential exteriors should rely solely on caulks and sealants to keep water out, many details require caulk either to mask an expansion joint between materials or as the first line of defense against leakage. When choosing a caulk or sealant (another name for a high-performance caulk), look for a product that will bond well to the substrate materials and be sufficiently flexible to tolerate the anticipated movement (Table 1-13). Just as important is how the caulk bead is applied. The best quality caulk will fail if applied 1 inch thick and bonded on three sides of the joint.

Joint Design

The ideal caulk joint where movement is anticipated is an hourglass shape about twice as wide as it is deep (see Figure 1-37).

TABLE 1-13   **Caulk and Sealant Performance**

	Pros	Cons	Best Uses	Shrinkage	Joint Movement (typical)	Application Temperature °F (typical)	Service Temperature °F (typical)
Silicone	Good adhesion to most non-porous materials. Excellent flexibility, UV resistance, water resistance. Wide temperature range for application and service.	Not paintable, difficult to tool. May stain masonry and porous stone. Acid-cure incompatible with some metals and metallic coatings. May require primer. Acetone or special solvent needed for cleanup.	Sealing to glass, ceramics, porcelain, most plastics, painted wood. Kitchen and bath and wet locations.	<10%	+/−25% (up to 50% in extension)	−35 to 140	−80 to 400
Polyure-thane	Strong adhesion to a wide variety of materials. Easy to tool. Paintable, flexible, and good weather and UV resistance.	Few color choices. Less flexible in cold weather.	Sealing to wood, metal, glass, concrete, masonry and most construction materials. Metal roofs, flashing, masonry control joints.	<10%	+/−25% (up to 40% in extension)	30 to 140	−40 to 194
Butyl	Strong adhesion to a wide variety of materials. Very flexible. Highly resistant to water and UV radiation.	Sticky consistency makes tooling difficult. Incompatible with rubberized-asphalt flashing membranes. Mineral spirits for cleanup.	Wood, metal, glass, concrete, masonry and most construction materials. Metal roofs, gutters, foundations.	<20%	+/−25%	>35	−40 to 225
Acrylic Latex	Easy to apply, tool, and clean up. Paintable; colors available. New formulations improve performance.	Typically less flexible and durable than other sealants. Less flexible in cold weather.	Wood, metal, glass, ceramic tile, and most construction materials.	20%	+/−10%	40 to 100	−5 to 170

This shape allows a caulk bead to stretch without either failing in "adhesion" to the substrate materials or failing in "cohesion" by tearing itself. A good rule-of-thumb is that a caulk joint should be four times the width of the anticipated movement, limiting the sealant's stretching to 25%. For most residential building details, this requires at least a $\frac{1}{4}$-inch-wide joint.

In general, the sealant should be no more than $\frac{1}{2}$ inch deep. For deep joints, it is best to pack the joint with a backer rod, a flexible foam material that controls the depth of sealant and shapes it into the hourglass profile. Backer rod is made of either open-cell or closed-cell foam

and comes in diameters from $\frac{1}{4}$ inch to as much as 2 inches. In wet locations, such as concrete control joints, use closed-cell foam, since it will not absorb water. Use a backer rod a little bigger than the joint being sealed.

Bond Breakers. In addition to controlling the depth and shape of a caulk bead, the backer rod acts as a "bond breaker," preventing the caulk from sticking to the back side of the joint. A three-sided caulk joint tends to tear when the materials move. Corner joints subject to movement are also prone to fail. For corner joints, use a small-diameter backer rod or any other material that will not bond

FIGURE 1-37 **Correct Caulking.**

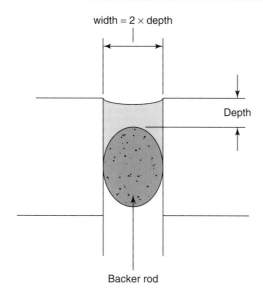

The ideal caulk bead forms an hourglass shape about twice as wide as it is deep, allowing the bead to stretch without tearing or pulling away from the substrate. In general, a caulk joint should be four time as wide as the anticipated movement—at least $\frac{1}{4}$ inch wide for most residential applications.

to the sealant. Plastic and foam tapes sold for weather-stripping can work in corners (see Figure 1-38).

Cleaning and Priming.
Since dirt, debris, and loose paint act as bond breakers, sealing to a dirty or flaking joint will fail when the joint moves. Also, the joint should be dry unless using a sealant approved for damp surfaces, such as some polyurethanes and some of the newer synthetic-rubber "Kraton" type sealants. Do not use compressed air to clean the joints unless a line filter is also used, since the oil from the compressor may coat the joint, interfering with the bond. Although priming is not required for most sealants used in residential construction, some metals may need priming with acid-cure silicones. Consult the sealant manufacturer's specifications.

Specifications.
For critical joints where movement is anticipated, choose a caulk that complies with ASTM C920 and is rated for +/−25% movement. ASTM C920 indicates that the sealant is highly weather-resistant, durable, and shrinks no more than 10%. For stationary joints, a +/−12.5% rating for joint movement is acceptable.

Silicones

Silicones bond well to nonporous surfaces, such as glass, tile, and metals, and they are the most flexible sealants made. A good silicone will stretch as much as 50% of its original width before tearing. Silicones are good in cold-temperature work and can be applied from well below 0°F to over 100°F. Once cured, they can also tolerate temperatures from well below 0°F to about 400°F, or higher for

FIGURE 1-38 **Caulked Corner Joint.**

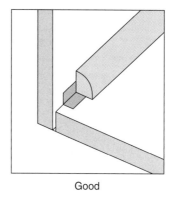

Good

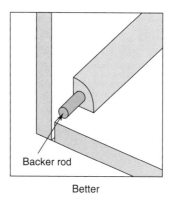

Backer rod

Better

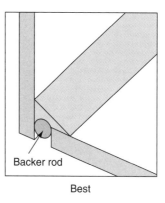

Backer rod

Best

Caulked corner joints subject to movement are prone to fail. Where movement is anticipated, use a bond-breaking tape or small-diameter backer rod.

special high-temperature formulations. Unlike most sealants, silicone stays flexible when cold. Silicones are also very resistant to UV radiation and water, making them a good choice for exteriors as well as kitchens and baths.

The main disadvantages of silicone are that it is messy to work with, difficult to tool, and does not hold paint well. Cleanup when wet requires acetone or special-order silicone solvent, and the residue is hard to remove when it is time to reapply. Because of the residue, once you've sealed a joint with silicone, it is best to reseal with silicone as well. Silicone does not bond well to unpainted wood and can stain or degrade porous stone and masonry materials.

Silicones come in two types: acid-cure (acetoxy) and neutral-cure (sometimes called "noncorrosive" silicone). The acid-cure type has a distinctive vinegarlike odor. Both types will stick well to glass, ceramics, and other nonporous surfaces. Acid-cure silicone, however, requires primer with most metals to bond well and to avoid corrosion. Neutral-cure silicones are compatible with most metals and metal finishes and bond somewhat better to wood.

Polyurethanes

Polyurethane is a versatile, water-resistant, high-performance sealant and has become the first choice of many contractors for exterior work. Polyurethanes provide excellent adhesion to a wide variety of materials from wood to masonry and remain flexible across a wide temperature range. Furthermore, they are relatively easy to tool, and some brands accept wet tooling with soapy water. Tooling time is adequate and shrinkage minimal. Polyurethanes are available in only a few colors, but the cured sealant holds paint well.

Although polyurethane is not naturally UV-resistant, UV inhibitors give it good durability in exterior applications. Because of its aggressive bond, polyurethanes are good for sealing between different materials. Polyurethanes are widely used on metal roofs, concrete and masonry control joints, flashing, and exterior trim.

Butyl

Butyl is a high-quality, tough, rubberlike sealant that is ideal for exterior jobs requiring a durable, watertight seal. Because of its longevity, temperature range, and high UV resistance, it has long been used as a glazing compound. Notable for its stickiness, butyl bonds very well to a wide range of materials, including wood, concrete, masonry, glass, and metal. Its stickiness, however, can also make its application messy and tooling difficult. Before curing, it can be cleaned with mineral spirits.

Because of its good adhesion and water resistance, butyl is often used to seal metal gutters, metal roofing, and around foundations. It is approved for use below grade. Butyl should not be used, however, in contact with modified-bitumen flashing tapes or roofing membranes, which can degrade it.

Acrylic Latex

The most economical and widely used caulking compound in residential work, acrylic latex caulks come in a wide variety of formulations and prices. To their credit, latex caulks are easy to apply, easy to tool, and can be cleaned before curing with water. They bond moderately well to a wide variety of materials and have a long tooling time. When cured they are highly paintable, making acrylic latex popular for caulking paintable trim in both the interior and exterior.

Lower-end acrylic latex caulks do not have the same flexibility, temperature range, and long-term durability as butyl, polyurethane, or silicone. Newer premium products, however, promise performance on par with some of the high-performance sealants. Added plasticizers make the material more flexible and other additives provide better UV and water resistance. For exterior work in joints subject to movement, look for an ASTM C920 rating and a rated joint movement of $+/-25\%$.

Most latex caulks cannot be applied under 40°F and should not be allowed to freeze in the tube or in place before cured. Also do not apply to wet surfaces or where rain is likely to fall before the caulk has a chance to fully cure.

EXTERIOR ADHESIVES

No exterior millwork should rely entirely on adhesives, since no glue is 100% waterproof, and any adhesive can fail with enough moisture cycling and movement in the wood. It is always wise to back up an exterior glue joint with mechanical fasteners, design the woodwork to shed water, and protect it with a good paint job. Still, there are several good options for gluing exterior work that should last indefinitely if well maintained (see Table 1-14).

There are several factors to consider in selecting a glue. For exterior woodwork, the biggest concerns are typically water resistance, strength, and cleanup. Working temperatures, clamping time, and gap filling abilities may also be important, depending on the specific job and conditions.

A glue's water resistance is classified as Type I or Type II. A Type I designation indicates that the glue bond can survive repeated submerging in boiling water. Type 1 glues are used for laminating structural timbers such as glulams. The most common Type 1 glue, resorcinol, has strict temperature and clamping requirements and is rarely used on residential job sites. Type II glues must maintain their bond after being soaked for four hours and then dried three successive times. These are suitable for all but the most punishing residential applications.

Type II Yellow Glue. Polyvinyl acetate (PVA) is the most common glue on residential job sites due to its low price, long shelf life, easy cleanup, and overall ease of use. The Type II version provides good water resistance and provides a very strong bond, and it is only slightly more expensive than the regular yellow glue. Similar to the older style white glue, yellow glue is formulated with a higher solids content to make it less runny and with other additives to make it set up quicker. Clamp time is about one hour. Any squeeze-out is simple to remove with a damp rag. Once the glue dries, however, it will resist paint and stain and needs to be scraped or sanded off.

In general, yellow glue should not be applied in temperatures below 50°F or allowed to freeze before it cures.

TABLE 1-14 Exterior Adhesives

	Pros	Cons	Cost	Bond Strength	Water Resistance	Application Temperature	Open Working Time	Clamping Time	Clean up	Shelf Life After Opening
Type II yellow glue	Easiest to use and to clean up. Excellent strength.	Dried glue resists finishes. Not gap filling.	Low	3,500 + psi	Moderate. ANSI type II	>55°F	5 minutes	1 hour	Cleans with water when wet. Scrape or sand dry.	One year in tightly closed container.
Polyurethane	Easy to use. Bonds to damp materials. Hardened glue sands easily. Very strong and water-resistant.	Cannot be cleaned from skin. Gap filling is not structural.	Medium	3,400 + psi 1,356 psi (hot-melt type)	Excellent. ANSI type II	40°F–130°F For best results: 68°F–130°F	15 minutes	1–4 hrs., depending on manufacturing.	Mineral spirits or acetone when wet. Scrape or sand dry.	Up to one year in tightly closed container (squeeze air out of container)
Epoxy	Bonds well to most materials. Fills gaps with little loss of bond strength. Minimal clamping required. Very strong and water-resistant.	Expensive. Requires mixing multiple components. Cleanup difficult. Dried glue blocks stain penetration. Can irritate skin.	High	2,500– 3,000 psi	Excellent, but bond to wood may fail with repeated wet-dry cycling.	Min. 40°F–60°F, depending on hardener.	10–60 minutes, depending on hardener and temperature.	Until epoxy cannot be dented with fingernail. Depends on hardener and temperature.	Alcohol, acetone, or lacquer thinner	Several years in tightly closed containers.

In freezing weather, store the glue indoors, since a couple of freeze-thaw cycles may ruin the glue. For exterior work subjected to moderate weather exposure, Type II yellow glue is a good option.

Polyurethane. One-part polyurethane glues have grown in popularity over the past few years due to their excellent strength and increased water resistance compared to yellow glue. Since polyurethane requires moisture to cure, it will bond to wood that has up to 25% moisture content. For wood that has less than 10% moisture content or appears dry, you should moisten one of the two surfaces being joined with a sprayer or damp cloth and apply a thin coating of glue to the other. Polyurethane bonds well to wood, stone, most metals (not stainless steel), and ceramics, as long as at least one of the surfaces being glued is porous.

Polyurethane foams up as it cures, expanding to three to four times its original size and filling any small gaps in the joint. But unlike epoxy, the filled gaps have no strength. Clamping time is one to four hours, depending on the specific formulation. For maximum strength, clamp for 24 hours.

Because of its tenacious grip, you should protect any materials or finished surfaces from drips and protect your hands with latex gloves, as the glue cannot be removed except by abrasive cleansers. If wet glue drips onto a finished surface, wipe with a dry cloth, since anything wet will activate curing. After the glue has dried, scrape away the squeeze-out with a sharp chisel and sand any residue. The glue dries to a brownish tan, which can be painted.

Where clamping is not practical, another option is hot-melt polyurethane. Hot-melt polyurethanes have been used in industrial settings for many years, but they have only recently been introduced for job-site use. Unlike its moisture-cured cousin, hot-melt polyurethane does not foam up and needs no clamping. It sets in about 30 seconds and provides the same level of water resistance as regular polyurethane but less than half the strength. Still, this is more than enough for many applications. Remove any squeeze-out with a putty knife or scraper as soon as it firms up, as it is difficult to sand clean.

Epoxy. Long the adhesive of choice for boat builders, epoxy has high adhesive strength and rigidity, low shrinkage, and good resistance to water and chemicals. It bonds well to wood, concrete, foam insulation, and other porous substrates—and to nonporous surfaces as long as they are lightly roughed up. It is comparable in strength and water resistance to polyurethanes, requires minimal clamping, and can fill gaps with little loss of strength, making it an ideal choice for less-than-perfect carpentry joints.

As a two-part system—with various hardeners to choose from and additives such as fillers to improve gap filling—epoxy is also the most complicated and costly approach. Once the resin and hardener are mixed, the working time ranges from about 10 minutes to an hour, de-

pending on the ambient temperature and whether a slow or fast hardener is used. Heat speeds up the curing, so a slow hardener is recommended in hot weather, a fast hardener in cold weather. It is important to mix the correct proportions and mix thoroughly or unreacted resin or hardener may remain in the cured epoxy.

For best results, use a disposable brush to coat both sides of the joint with liquid epoxy. After coating the joint, add fillers to the mix if required. Fillers change the viscosity of the mix and enable it to bridge gaps with minimal loss of strength (you can bridge small gaps up to about $\frac{1}{16}$ inch without fillers). A small amount of filler helps keep the mix from running. Once the fillers have been added, apply the thickened epoxy to one side of the joint and clamp just enough to squeeze out a little epoxy. A common mistake with epoxy is clamping too tightly. This will create a weak, "glue-starved" joint. Cleanup of the wet epoxy requires solvents such as acetone or lacquer thinner. Workers should use rubber gloves to protect their skin.

Any kind of clamping that holds the joint still is suitable, including staples, nails, or wood screws. Scrape off any squeeze-out with a putty knife or dry rag. Once the epoxy has cured to a solid state that cannot be dented with a fingernail, it has reached 90% of its final strength. Then the clamps can be removed and any excess sanded off. The epoxy continues to gain strength for several days and is paintable.

EXTERIOR WOOD FINISHES

The USDA Forest Service Forest Products Laboratory (FPL) has done extensive research on how to keep paints and stains on wood sidings and trim. In general, they recommend paint for the longest lasting finish and best protection of the underlying wood, followed by solid or semi-transparent stains. Clear finishes need the most frequent recoating and offer the least protection from water damage and UV radiation (see Table 1-15).

How long a finish will last depends on many variables, including the quality of the finish, type and texture of wood, application conditions, and exposure. South- and west-facing walls get the most sun and are, therefore, often the first to need recoating. Whether painting, staining, or finishing in any manner, the FPL makes the following recommendations:

Moisture Content. Never paint wood with a moisture content over 20%. Ideally, the wood should be painted at its average moisture content for that climate—about 12% for most of the United States, 9% for dry southwestern states (see Table 1-2, page 9).

Wood Surface. A rough-sawn wood surface will hold paint and stain much longer than a smooth, planed surface, which is why many contractors prefer to install siding rough side out. Also most lumber and siding today is

| TABLE 1-15 | Expected Life of Exterior Wood Finishes[1] | | | | | |

Type of Exterior Wood Surface	Water-Repellent Preservative and Oil		Semitransparent Stain[3]		Paint and Solid-Color Stain (One Prime Coat, One Top Coat)[4]		
	Suitability	Expected Life[2] (Years)	Suitability	Expected Life[2] (Years)	Suitability	Expected Life of Paint (Years)	Expected Life of Solid-Color Stain (Years)
Cedar/redwood siding (vertical grain)	High	1–2	Moderate	2–4	High	4–6	3–5
Cedar/redwood siding (rough-sawn)	High	2–3	High	5–8	High	5–7	4–6
Pine, fir, spruce siding (smooth, flat-grained)	High	1–2	Low	2–3	Moderate	3–5	3–4
Pine, fir, spruce siding (rough, flat-grained)	High	2–3	High	4–7	Moderate	4–6	4–5
Shingles	High	1–3	High	4–8	Moderate	3–5	3–4
Plywood (sanded or smooth-textured)	Low	1–2	Moderate	2–4	Moderate	2–4	2–3
Plywood (rough-sawn textured)	Low	2–3	High	4–8	Moderate	4–6	3–5
Plywood—MDO	Not suitable	—	Not suitable	—	Excellent	6–8	5–7
Hardboard (smooth or textured)	Not suitable	—	Not suitable	—	High	4–6	3–5
Millwork (usually pine)	High[5]	—	Moderate	2–3	High	3–6	3–4
Decking—new (smooth)	High	1–2	Moderate	2–3	Low	2–3	1–2
Decking—weathered (rough)	High	2–3	High	3–6	Low	2–3	1–2

[1]Expected life for average location in the continental United States. Will vary for extreme climates or exposure.
[2]Need for refinishing determined by presence of mildew on surface.
[3]One coat on smooth, unweathered surface; two coats on rough-sawn or weathered surface—the second coat applied while first is still wet.
[4]Applying a second top coat (three-coat job) will approximately double the life. Top quality, 100% acrylic latex paints have best durability.
[5]Exterior millwork, such as windows, should be factory-treated. Liberally treat other trim by brushing before painting.
SOURCE: Adapted from *Finishes for Exterior Wood*, 1996, by R. Sam Williams, Mark Knaebe, and William Feist. Courtesy of Forest Products Society.

flat-grained, which holds paint less well than vertical (or edge) grained. The combination of flat grain and planing can create a burnished surface called "mill glaze," which can cause problems with paint adhesion. To avoid problems, it is best to lightly sand with 50 to 80 grit sandpaper before painting smooth siding. The optimal approach is to first wet the lumber to raise its grain and then let it dry for two days before sanding.

Weathering Before Finishing. Some painters recommend letting smooth siding weather for a few weeks to open up the grain. However, research at FPL has shown that after two weeks of exposure, the wood surface begins to degrade and to loosen the wood fibers on the surface, which weakens the paint adhesion. The FPL therefore strongly recommends painting within two weeks of installation, whether the rough or smooth side is facing out. If you need to paint wood that is badly weathered, the wood should be sanded, power rinsed, and allowed to dry before priming. Once the primer is dry, the top coat should be applied as soon as possible.

Species and Grain. In general, less dense woods hold paint better than more dense woods (see Table 1.12, page 32). Also, within a single species, vertical-grain (also called edge-grain) wood holds paint much better than the more common flat-sawn lumber, primarily because flat-sawn wood shrinks and swells more from changes in relative humidity. Also vertical-grain wood has narrower bands of latewood, the denser and harder portion of each annual ring in a tree. When paint, particularly oil-based, becomes brittle with age, it tends to peel from the latewood.

Dense woods with wide, flat grain will present the greatest problems in holding paint. This is true for most hardwoods as well as dense softwoods with wide, flat grain, such as southern yellow pine and Douglas fir, especially if planed smooth.

Paints

Paints offer wood the greatest protection from the elements and can last from 7 to 10 years if properly applied with one prime coat and two top coats of quality paint. The longevity of a particular job will depend on a number of variables, including paint quality, surface preparation, climate and exposure, and the type of wood.

Latex vs. Oil-Based.

In addition to its easy cleanup, latex paint has always held certain advantages over oil. Perhaps most important, latex paints stay flexible over time while oil-based paints get brittle as they age. This is particularly true of 100% acrylics, which makes them less likely to crack due to seasonal movement of the substrate. Also, while oil is more resistant to liquid water, latex is more permeable to water vapor, making it less likely to blister in situations where moisture must pass through. Latex also fades less over time, is not prone to chalking, and is less likely to support mildew growth than oil-based paint. The best quality latex paints use 100% acrylic binders, offering increased flexibility and durability over latex-vinyl blends.

Oil-based paints, however, are still favored by many professional painters for their better appearance and better adhesion due to the oil penetrating the surface of the wood. Oil paint's flow characteristics help hide brush strokes and provide better coverage, particularly in high-gloss paints. Also, window sash and doors painted with oil paint dry to a harder finish that is less likely to stick to other painted surfaces.

In the past two decades, however, manufacturers have greatly improved the quality of latex paint, overcoming many of the problems associated with it in the past, while oil-based paints have suffered somewhat as manufacturers have had to adjust their formulas to comply with air-quality regulations that restrict the use of VOCs (volatile organic compounds) found in paint solvents. Since latex now dominates the market in residential paint sales, most development efforts now and in the future will focus on improving latex rather than oil-based paints.

Oil-Based Primers.

Many painters still prefer oil as a primer for woods with water-soluble extractives, such as redwood and red cedar, although specially formulated stain-blocking latex primers can also work for this application. Many painters also favor oil primer when re-painting over chalky or degraded surfaces because of its penetrating oils and strong adhesion. Painting over high-gloss surfaces also may be easier with oil-based paints.

Finally, oils offer greater temperature flexibility in both hot and cold weather. In hot weather, latex may dry too fast; while below about 50°F, latex should not be used without special additives. Oil-based paints can be safely

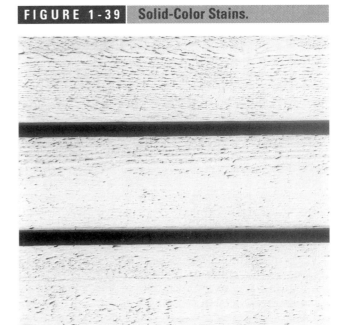

FIGURE 1-39   Solid-Color Stains.

Solid-color stains are not true penetrating stains since they form a paint-like film on surface of the wood. However, they allow the wood texture to show through, particularly on rough-sawn siding.

used to about 40°F. Newer formulations of latex paints, however, promise to extend their temperature range.

Solid-Color Stains

Solid-color, or "opaque," stains are not true penetrating stains, since they form a film on the surface of the wood as paint does. In fact, they are formulated the same as paints, only with fewer solids, leaving a thinner, less protective film. They may also contain water-repellents and preservatives. Like paints, they help protect wood from UV degradation; but also like paints, they can peel and blister if applied incorrectly. Most require a primer for best results.

The thinner coating of these products tends to hide the wood grain but allows the wood texture to show through, particularly on rough-sawn siding (see Figure 1-39).

Most solid-color stains sold today are latex-based, which makes them fast-drying and likely to show lap marks if not applied carefully. The most durable latex solid-color stains are 100% acrylic. Oil-based solid stains are sometimes used on redwood and cedar.

Two coats of top-quality latex solid stain over a primer on a solid-wood siding should provide 3 to 7 years of service versus as many as 10 years for an acrylic latex paint of equal quality.

Application of Paints and Solid Stains

The best paint in the world can fail within the first year if applied over a wet, dirty, or degraded substrate. So the first priority is to make sure that the material being painted is sufficiently dry and clean.

Sealers. For the best protection of the underlying wood and the longest lasting finishes, bare wood should be sealed with a water-repellent preservative (WRP) before priming and painting or staining. WRPs contain a small amount of wax or other water repellent and a mildewcide, fungicide, or both, usually in a solvent base. The preservatives help prevent mildew and decay in above-ground applications but are not meant for ground contact. Some WRPs contain UV blockers as well, which slow down the degradation of the outer wood fibers.

While sometimes formulated as a finish treatment for siding, some WRPs can be used as a pretreatment for painting and are recommended for that use by the USDA Forest Products Laboratory (FPL) and Western Wood Products Association (WWPA). Research shows that WRPs resist water entry better than acrylic primers. On bevel siding, they also reduce warping, splitting, and mildew growth. They can also improve paint performance on hard-to-paint woods, such as southern yellow pine and Douglas fir.

In new construction, the FPL recommends that siding and trim be coated on all sides with a paintable WRP such as DAP Woodlife® or Cuprinol's Clear Wood Preservative, preferably by dipping or with a brush, roller, or pad. If the siding or trim is already installed, they suggest treating all places vulnerable to water entry, including door bottoms, window sills, lap and butt joints, edges and ends of trim, and any end grain on panel products such as plywood sidings.

If used as a pretreatment for paint, apply to bare, dry wood when it is above 50°F, and use only a single coat or excess wax buildup on the surface could affect the paint adhesion. Allow two days of warm weather to dry, or up to a week if the material was dipped. If painted before the solvent has evaporated and the wax absorbed, the paint can be discolored and not bond well.

Priming. All paints and most solid stains require priming on new wood. Primers are formulated with a higher ratio of binder to pigment than paints. This forms a durable film that bonds well to the surface and blocks water. However, without much pigment, it offers limited UV protection.

For woods with water-soluble extractives, such as cedar and redwood, use an oil-based primer or a stain-blocking acrylic primer formulated to seal in the extractives. Also use a stain-blocking primer on any knots. Otherwise the extractives can bleed through the finish and stain the siding. For wood species relatively free of extractives, use a 100% acrylic latex primer. If sprayed or rolled on, back brushing is recommended for a good bond.

Many manufacturers now sell siding and trim preprimed. In addition to the convenience for the contractor, the factory-applied coating is applied uniformly without the risk of bad weather or other job-site variables. The only concern is the thickness of the primer. While most major manufacturers of preprimed siding do a good job, some third-party prefinishers may ship material with too thin a coating. In general, the primer should be 1.5 to 2 mils thick—thick enough that it hides the wood grain.

Back Priming. Most paint failures are related to moisture moving through the wood either from wind-driven rain that reaches the back of the siding or moisture escaping from the house. In some cases exposed end grain picks up moisture and causes localized peeling. Use of a water-repellent preservative or primer on the back of the siding and on all edges and cut ends, in addition to the visible face, will minimize these problems. Sealing the wood properly also helps prevent moisture from being driven through the siding by solar radiation.

Top Coats. For paints and solid stains, apply the top coat as soon as the primer is dry but *not* more than two weeks later. For best performance, apply two top coats. Latex paints can typically be recoated within a few hours. Oil must cure for one or two days between coats. Apply paint at the coverage recommended on the can. Too thin a coat will wear quickly and too thick a coat may crack.

While brushing provides the best adhesion, a properly done spray job can yield good results. When spraying or rolling, the best results are achieved by back-brushing the paint to help work it evenly into the wood, particularly on rough-sawn surfaces.

Temperature and Time of Day. Oil-based paints should be applied when it is over 40°F; for latex coatings the temperature should be at least 50°F during application and for 24 hours after. Also it is best not to apply paint too early or too late in the day. If the dew has not evaporated in the morning, both oil and latex may have adhesion problems. If applied within two hours of sunset and a heavy dew forms before the paint dries, latex paints may streak and oil-based paints may not cure properly.

Finishing Pressure-Treated Wood. Wood that is pressure-treated with waterborne preservatives, such as chromated copper arsenate (CCA), ammoniacal copper zinc arsenate (AZCA), and ammoniacal copper quaternary (ACQ), present special problems for painted finishes. First, pressure-treated lumber is often shipped to lumberyards with very high moisture contents. If painted while wet, the moisture may get trapped by the paint film and cause peeling. Also the species most commonly pressure-treated—flat-sawn southern yellow pine in the eastern United States and Douglas fir and Ponderosa pine in the West—do not hold paint well to begin with.

Whether or not you intend to paint the wood, pressure-treated exterior trim should be sealed with a water-repellent preservative as soon as the surface is sufficiently dry. This will protect cut ends and help keep the wood from checking, cupping, and warping as the wood dries out. If this is the only treatment, it will need recoating every one to two years. Factory-sealed treated lumber is now available that only requires treatment of cut ends when installed.

The most common treatment for pressure-treated wood is an oil-based, semitransparent stain. Since this type of finish is relatively permeable to moisture, for best results apply it over a sealer or over factory-sealed lumber. While the sealer can be applied to wood that is still wet inside, it is best to air dry the wood before staining. This will take from a few days to a few weeks, depending on conditions, with two weeks on average. Two coats of an oil-based, semitransparent stain over a sealer should last several years. The second coat should be applied before the first coat dries completely, or the second coat cannot penetrate the wood.

If a painted finish is desired, you will need to seal the wood first and allow it to dry for two to three weeks before applying a compatible primer and two coats of 100% acrylic top coat. The longer the wood dries, however, the greater the risk that UV radiation will damage the wood surface, interfering with the paint's adhesion. To avoid these problems and the long delays, consider using kiln-dried treated lumber that can be finished immediately.

Discoloration Problems

Extractive bleeding and mildew can discolor either bare wood or finished surfaces. They should be removed before finishing or refinishing. After washing, it is important to allow the surface to dry before applying the new finish.

Extractive Bleeding. Excess moisture in wood species such as cedar, redwood, Douglas fir, and mahogany can dissolve the natural tannins in the wood and cause them to migrate to the surface, leaving a reddish-brown stain on the finish. Sealers and stain-blocking primers help to minimize this problem but do not always eliminate it. If staining occurs, the first step is to eliminate the moisture problem. Then, if the extractive bleeding is mild, remove the stains with a mild detergent and water. More severe cases will require cleaning with an oxalic acid solution.

Carefully follow the manufacturer's instructions when using oxalic acid, as the bleaching solution will harm plants and may bleach existing finishes on siding, trim, and other woodwork. After washing, the oxalic acid must be thoroughly rinsed with clean water and the wood dried before finishing or refinishing. If the extractive bleeding has been allowed to bake in the sun, it may have hardened and be difficult to remove. In this case, you will need to apply a stain-blocking primer before refinishing.

Mildew. This common fungus grows on just about any surface with sufficient moisture and heat. In new construction, it can be minimized by storing wood off the ground and providing adequate ventilation. Although sealers and stains contain a mildewcide, any mildew should be removed before finishing or refinishing, or it will continue to grow through the new finish.

To remove mildew, use a sodium hypochlorite solution, which can be made with household chlorine bleach. Depending on the severity of the problem, the solution should range from 1 to 8 parts bleach to 1 part water. Spray the solution onto the siding (avoid sprayers with aluminum parts), starting at the top and working down. If two applications do not remove the stains, you may need to scrub in the solution with a brush. Thoroughly rinse everything with water.

Bleach can harm plants, discolor the finishes on trim, and corrode aluminum, brass, and copper. It is best to cover plants with tarps and protect any stained or painted surfaces.

Semitransparent Penetrating Stains

Most semitransparent stains are oil-based, and they penetrate the surface of the wood. They have a moderate level of pigment that offers some UV protection and provides some color without hiding the wood grain. Because these stains do not form a film on the surface, they are not subject to blistering and peeling like paints and solid-based stains.

Penetrating stains last longer on rough than on smooth siding materials. One coat of oil-based penetrating stain on rough-sawn siding or plywood will last two to five years, depending on exposure and other variables; two coats may last as long as seven or eight years. In general, subsequent coats last longer than the first coat because the weathered wood will accept more stain. For decks, steps, or other wood subject to foot traffic, use a special deck stain formulated with better abrasion resistance (see "Finishes for Decking," page 154).

Like paints, penetrating stains can be applied by brush, spray, or roller. If sprayed or rolled on, back-brushing will improve the penetration and performance. Also spraying without back-brushing can cause a splotchy appearance. If two coats are desired, apply the second coat before the first has fully dried or the second coat will not be able to penetrate the surface.

Because oil-based stains are thin and dry quickly, lap marks may form if the applicator is not careful to maintain a wet edge. It is best to work on a small area at a time and, if possible, to work in the shade to extend the drying time.

Clear and Lightly Tinted Finishes

Some customers want to retain the look of "natural" wood siding, particularly with the warm-toned hues of premium red cedar or redwood. Unfortunately, there is no finish that will magically preserve the look of new wood.

Wood turns gray as UV radiation degrades the outer surface and as mildew spores develop. Clear water-repellent preservatives (described under "Sealers," previous page) with UV blockers can slow down this natural aging process, but will need to be reapplied every year or two to keep the wood from turning a weathered gray.

To retain the tone of new wood, the best approach is to use a WRP or penetrating oil with UV blockers and a tint added to match the redwood or cedar. Amteco's Total Wood Protectant (TWP®), Flood's Clear Wood Finish (CWF®), and Penofin® (Performance Coatings Inc.) are proprietary formulations designed to maintain a natural wood appearance. A similar product called Sikkens Translucent Cetol® (Akzo Coatings) darkens the wood somewhat and creates a thin film, but it does not peel like paint or varnish. Apply one to three coats, according to the manufacturer's recommendations. Even with "one-coat" finishes, a second coat may be worthwhile on south or southwest sides of the building due to increased UV exposure.

If applied correctly, a high-quality tinted finish can keep redwood or cedar siding looking close to new for three to five years. Before recoating, you may need to clean the siding with a bleach solution to remove any mildew and dirt that has started to discolor the siding. After cleaning, another coat of the original finish should restore the new wood look for another three to five years.

Bleaching Oils

In some regions, homeowners like the silver-gray, weathered look of unfinished cedar shingles, but they do not want the splotchy, uneven coloring that sometimes results from uneven wetting and sun exposure. Bleaching oils solve this problem by combining a lightly pigmented semi-transparent stain with a bleaching agent. Initially, the pigment colors the wood a silver-gray color, and over time, the bleach lightens the underlying wood to a uniform color.

The uniform weathered look can last for a number of years, but the oil and pigments in the original finish protect the wood for only two or three years. Beyond that, a clear water-repellent preservative can be used periodically to protect the wood from UV degradation and decay. If, after several years, the siding begins to darken or lose its uniform appearance, another coat of bleaching oil should restore the original look.

TABLE 1-16	**Relative Decay Resistance of Untreated Heartwood**	
Resistant or Very Resistant	Moderately Resistant	Slightly or Nonresistant
Bald cypress (old growth), cedar, white oak, redwood	Bald cypress (new growth), Douglas fir, western larch, eastern white pine, southern yellow pine (longleaf, slash), tamarack	Pines other than longleaf, slash, and eastern white; spruces, true firs

SOURCE: U.S. Forest Products Laboratory *Wood Handbook*, 1999.

Unfinished Siding and Trim

Due to their high level of extractives, the heartwood of some species is naturally resistant to decay and insects and can be used on the exterior unfinished. The woods most commonly used this way are western red cedar, northern white cedar, redwood, and bald cypress (see Table 1-16).

In salty coastal air with good exposure to sunshine, untreated wood tends to weather to an attractive silver gray. In other regions, uneven staining from mildew is likely. Even in coastal regions, areas of the house that get frequent wetting from splashback, snow, or other types of weather exposure may become darkened from mildew (see Figure 1-19, page 19).

Also, the wood extractives do nothing to prevent cupping, warping, or cracking from uneven absorption of moisture. For a uniform appearance without leaving the results to chance, it is best treat the wood with a WRP or bleaching agent.

RESOURCES

Suppliers

Grade D Building Paper

Fortifiber Co.
www.fortifiber.com

Hal Industries
www.halind.com

Plastic Housewrap

Benjamin Obdyke
www.benjaminobdyke.com
Homeslicker drainage mat and Homeslicker Plus Typar

DuPont
www.construction.tyvek.com
Tyvek, HomeWrap, and StuccoWrap

FirstLine
www.firstlinecorp.com
FirstWrap

Johns Manville
www.jm.com
ProWrap perforated housewrap

Ludlow Coated Products
www.ludlowpc.com
Barricade housewrap, Weathertrek draining housewrap

Owens Corning
www.owenscorning.com
PinkWrap and PinkWrap Plus

Pactiv
www.pactiv.com
Classic wrap, Ultra wrap, and Raindrop Housewrap

Reemay
www.typarhousewrap.com
Typar housewrap

Flashing Tapes and Membranes

Avenco
www.avenco.com
Butyl flashing tape

Bakor, Inc.
www.bakor.com
*Blueskin self-adhesive, rubberized-asphalt
flashing tape*

Carlisle Coatings and Waterproofing
www.carlisle-ccw.com
*Self-adhesive, rubberized-asphalt flashing tapes
and membranes*

Dupont
www.construction.tyvek.com
*StraightFlash and moldable FlexWrap Butyl flashing
tapes*

Fortifiber
www.fortifiber.com
*Moistop and FortiFlash self-adhesive and nonstick
rubberized-asphalt flashing tape*

Grace Construction Products
www.graceconstruction.com
*Vycor self-adhesive, rubberized-asphalt flashing tapes
and membranes*

Illbruck Sealant Systems
www.willseal.com/usa
Self-adhesive butyl and foil-faced butyl flashing tapes

MFM Building Products Corp.
www.mfmbp.com
*FlexWrap (foil-faced) and FutureFlash self-adhesive,
rubberized-asphalt flashing tapes and membranes*

Polyguard Products
www.polyguardproducts.com
*Windowseal self-adhesive, rubberized-asphalt flashing
tapes and membranes*

Protecto Wrap Co.
www.protectowrap.com
*Standard and moldable (Protecto Flex) self-adhesive,
rubberized-asphalt flashing tapes*

Sandell Manufacturing Co.
www.sandellmfg.com
Rubberized-asphalt, PVC, and EPDM flashing tapes

Prefabricated Cedar Shingle Panels

Cedar Valley
www.cedar-valley.com

Maibec Industries
www.maibec.com

Shakertown
www.shakertown.com

Fiber-Cement Siding and Trim

Cemplank
www.cemplank.com
Cemplank fiber-cement siding and trim

James Hardie
www.jameshardie.com
Hardiplank, Hardipanel, fiber-cement shingles

Nichiha Wall Systems
www.n-usa.com
*Fiber-cement lap siding and simulated shakes, brick, and
stone*

Certainteed
www.certainteed.com
*Fiber-cement lap, vertical, and shingle sidings, and soffits
and trim*

GAF
www.gaf.com
Weatherside fiber-cement siding

Exterior Insulation and Finish Systems (EIFS)

Dryvit Systems
www.dryvit.com

Parex
www.parex.com

Senergy
www.senergyeifs.com

Sto Corp.
www.stocorp.com

Hardboard Trim

ABTco
www.abtco.com

The Collins Companies
www.collinswood.com

Georgia-Pacific
www.gp.com

Masonite Corp.
www.masonite.co

Temple-Inland Forest Products
www.templeinland.com

Polyurethane Trim

Custom Decorative Mouldings (CDM)
www.custom-moulding.com

Focal Point Architectural Products
www.focalpointap.com

Flex Trim
www.flextrim.com
Flexible polymer composite moldings

Fypon
www.fypon.com

Mid-America Building Products
www.midamericabuilding.com

Nu-Wood Decorative Millwork
www.nu-wood.com

Outwater Plastics Industries, Inc.
www.outwater.com

Ras Industries
www.rasindustries.com

Resin Art
www.resinart.com
Duraflex flexible moldings

Cellular Polyvinyl Chloride (PVC) Trim

AZEK Trimboards
www.azek.com

Edge Building Products
www.permatrimboard.com

Gossen Corp.
www.gossencorp.com

Marley Moldings
www.marleymoldings.com

LVL Trim

Pacific Wood Laminates
www.pwlonline.com
Clear Lam (textured wood facing) and Socomi Lam (MDO facing)

Caulks and Sealants

Bostik
www.bostikfindley.com
Construction sealants

Chemrex
www.chemrex.com
Polyurethanes and other high-performance sealants

DAP
www.dap.com
Acrylic latex caulks

Dow Corning Sealants
www.dowcorningsealants.com
Silicone sealants

GE Silicones
www.gesilicones.com
Silicone sealant

Geocel Corp
www.geocelusa.com
Acrylic latex, tripolymer, copolymer, Kraton, and clear sealants

Macklanburg-Duncan
www.mdteam.com
Acrylic latex sealants

OSI Sealants Inc
www.osisealants.com
Polyseamseal PVA-based caulk. Pro Series includes latex, polyurethane, and Kraton sealants.

Phenoseal
www.phenoseal.com
Phenoseal vinyl adhesive caulk

Red Devil
www.reddevil.com
Acrylic, silicone, and butyl sealants

Sashco Sealants
www.sashco.com
Big Stretch and Mor-Flexx water-based sealants, Lexel Kraton sealant

Sika Corp
www.sikaconstruction.com
Complete line of Sikaflex polyurethane-based sealants, butyl sealant

Tremco Inc.
www.tremcosealants.com
High-performance, architectural-grade sealants,

UGL
www.ugl.com
Acrylic latex caulks

White Lightning
www.wlcaulk.com
Tripolymer, butyl, polyurethane, silicone, elastomeric, and other high-performance sealants

Exterior Adhesives

Abatron
www.abatron.com
Epoxy, BestBond polyurethane glue

Ambel
www.excelglue.com
Excel polyurethane glue

Elmer's Products
www.elmers.com
Yellow glues, ProBond polyurethane glue

Custom-Pak Adhesives
www.custompak.com
Resorcinol and yellow glues

DAP Inc.
www.dap.com
Weldwood contact cement, resorcinol, and construction adhesives

Franklin International
www.titebond.com
Titebond yellow glue, Liquid Hide Glue, and construction adhesives

Gloucester Co. Inc.
www.phenoseal.com
Phenoseal adhesive caulk

Gougeon Brothers
www.westsystem.com
West System epoxy

Gorilla Group
www.gorillaglue.com
Gorilla polyurethane glue

MACCO Adhesives
www.liquidnails.com
Liquid Nails construction adhesive

OSI Sealants
www.osisealants.com
PL400 construction adhesive

System Three Resins
www.systemthree.com
Quick Cure epoxy

Water-Repellent Preservatives (WRPs)

Cuprinol
www.cuprinol.com
Cuprinol Clear Wood Preservative

Dap
www.dap.com
DAP Woodlife

Wolman
www.wolman.com
Premium Water-Repellent Sealer

Clear Wood Finishes

Amteco
www.amteco.com
Total Wood Protectant (TWP)

The Flood Company
www.floodco.com
Clear Wood Finish (CWF)

Performance Coatings Inc.
www.penofin.com
Penofin wood finishes

Sikkens/Akzo Nobel
www.nam.sikkens.com
Sikkens Cetol finishes

For More Information

California Redwood Association
www.calredwood.org

Cedar Shake and Shingle Bureau
www.cedarbureau.org

USDA Forest Products Laboratory (FPL)
www.fpl.fs.fed.us

Vinyl Siding Institute
www.vinylsiding.org

Western Wood Products Association (WWPA)
www.wwpa.org

CHAPTER TWO | Roofing

ASPHALT SHINGLES

Asphalt shingles, which cover 80 to 90% of residential roofs, have undergone much change in the last 20 to 30 years. Until the late 1970s, all asphalt shingles were manufactured from a heavy organic felt mat that had established a reputation for both strength and flexibility and generally outlasted their 15- to 20-year life expectancy. Since their introduction in the late 1970s, fiberglass shingles have come to dominate the market, accounting for over 90% of shingles sold today. However, premature failure of some fiberglass shingles in the 1980s and 1990s tarnished the product's reputation and spawned a number of lawsuits and resulted in a toughening of standards and a general improvement in fiberglass shingle quality.

Shingle styles have changed as well. The common three-tab shingles of the 1950s and 1960s are now joined by no-cutout shingles, multitab shingles, and laminated "architectural" shingles (Table 2-1). Laminated shingles provide deep shadow lines and a heavily textured appearance, some simulating wood or slate. These now account for over half the shingles sold.

Shingle Quality

Shingle quality is often difficult to determine visually since it is based largely on hidden factors such as the strength of the reinforcing mat (organic felt or fiberglass), the strength and flexibility of the asphalt, and the amount and type of fillers used. In most cases, however, the guide-lines outlined below can help to select shingles that perform as promised.

Organic Felt vs. Fiberglass. Organic shingles are built around a thick inner mat made from wood fibers or recycled paper saturated with soft asphalt. Fiberglass shingles, on the other hand, use a lightweight nonwoven fiberglass held together with phenolic resin. Both shingles are then coated on top with a layer of harder asphalt and fillers and topped with colored stone to create a decorative surface and protect against ultraviolet light. A thin layer of asphalt on the bottom is coated with a nonsticking dusting that keeps the shingles from sticking in the bundle. Each type has its pros and cons (Table 2-2).

Organic. In general, organic shingles have better tear resistance and resistance to nail pull-through than fiberglass shingles, making them less likely to blow away during a cold weather installation when they have not yet had a chance to seal. Also, some roofers find that organic shingles are more pliable and easier to work with in cold weather. On the downside, the organic mat is neither fireproof nor waterproof. Organic shingles therefore typically carry only a Class C fire rating.

Although uncommon, manufacturing defects that allow water penetration into the mat can lead to premature curling and cupping of organic shingles. Blistering and curling in warm climates has also been occasionally reported. Organic shingles cost more than comparable fiberglass shingles, but remain popular in colder regions and

TABLE 2-1   Asphalt Shingle Types

	Weight per square	Length	Width	Exposure	ASTM Fire Ratings	ASTM Wind Ratings
Three-tab	200–300 lb.	36–40 in.	12–13 1/4 in.	5–5 5/8 in.	A or C	Wind resistant
Multi-tab	240–300 lb.	36–40 in.	12–17 in.	4–7 1/4 in.	A or C	Many wind resistant
No-cutout	240–360 lb.	36–40 in.	11 1/2–14 1/4 in.	4–6 1/8 in.	A or C	Many wind resistant
Laminated	200–300 lb.	36–40 in.	12–13 1/4 in.	5–5 5/8 in.	A or C	Wind resistant

Reprinted by permission of the Asphalt Roofing Manufacturers Association. © 1993

TABLE 2-2   Organic vs. Fiberglass Shingles

	Pros	Cons	Recommendations
Organic Shingles	Better resistance to tearing and nail pull-through. Easier to install in cold weather with fewer blow-offs.	More expensive. Occasional blistering or curling from moisture penetration or excessive heat. Only Class C–fire rating.	Good for cold climates or cold-weather installations. Look for ASTM D255 certification.
Fiberglass Shingles	Less expensive. Less prone to cupping or curling from heat or moisture. Class A–fire rating. Lighter weight.	Less tear-resistant. More prone to blow-offs in cold weather. Some premature failures reported, primarily with lightweight, low-end products.	Best in moderate and warm climates. Look for ASTM D3462 certification from an independent lab such as UL.

throughout Canada. With organic shingles, shingle weight tends to be a good predictor of performance and longevity since the added weight usually indicates a thicker mat saturated with more soft asphalt.

Fiberglass. Fiberglass shingles, built on a thin nonwoven fiberglass core, were first introduced in the late 1970s and now account for over 90% of the shingles sold. Because they use less asphalt, they are lighter and generally less expensive than organic shingles. Because fiberglass mats are more fire-resistant and moisture-resistant than felt, most fiberglass shingles carry a Class A (severe exposure) fire rating and are less prone to cupping and curling from moisture damage. On the downside, fiberglass shingles are generally not as tear-resistant as organic shingles, making them more prone to blow-offs in cold weather when the shingles have not properly sealed. After they have sealed, they can still tear from movement in the sheathing, since fiberglass shingles have little give, unlike organic shingles. In this situation, if the bond strength of the adhesive strip exceeds the tear strength on a lightweight shingle, the shingles can crack.

Premature failure of some fiberglass shingles due to splitting or cracking led to a number of class-action lawsuits in the 1980s and 1990s. The problems were primarily with lower-end shingles with lightweight mats, types that have been largely eliminated from the market. But it still pays to buy ASTM-rated products from a reputable company that provides a good warranty.

Laminated Shingles. Also called "architectural" or "dimensional" shingles, these have two layers laminated together at the lower half of the shingle, giving the roof a thicker textured appearance with deeper shadow lines. Depending on the shape and size of the cutouts, half or more of the exposed shingle area is triple thickness and the rest double. With the added thickness and without the tabs, which typically wear out first in three-tab shingles,

FIGURE 2-1 **Laminated Asphalt Shingles.**

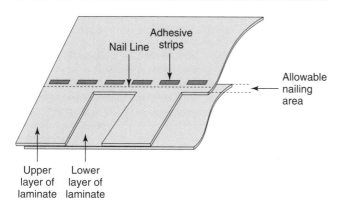

In general, laminated or "architectural" shingles perform well due to their heavy profile and lack of cutouts. Nails placed too high, however, may miss the lower layer of laminate, allowing it to come loose over time.

most laminated shingles carry longer warranties as well as higher wind ratings, some as high as 120 mph.

While not immune to the problems of other shingles, such as premature cracking, it is reasonable to expect good performance from a reputable brand. One problem unique to laminated shingles is the loosening of the bottommost piece of the shingle caused, in part, by nailing above the line where the double thickness ends (Figure 2-1).

On many laminated shingles, nails must be precisely placed so they are high enough to stay hidden while still penetrating both layers.

Wind Resistance. Most shingles carry a wind-resistance rating of 60 miles per hour as tested under ASTM 3161 or UL 997, while specialty shingles may be rated to as much as 130 miles per hour. While laboratory tests may not predict actual performance in a storm, a higher rated shingle will likely perform better than a lower rated one. Shingles rated at over 100 mph are often special-order items and typically require six rather than the usual four nails per shingle. Adding two extra nails and extra dabs of plastic roofing cement to a regular shingle can also increase its performance in high-wind conditions (see "Fastening," page 58, and "Manual Sealing," page 57).

A wind-resistance rating is not the same as a warranty. Shingles that carry a wind-resistance warranty generally require that the shingle tabs have been adequately sealed to the adhesive strip and most limit wind coverage to five or ten years from installation. In cold, cloudy weather or on a steep north-facing slope, manual sealing with roofing cement may be necessary.

Algae Resistance. Black streaks on shingles caused by algae or fungal growth used to be limited to warm, humid climates, but now this can be seen on houses as far north as Canada. Some experts attribute the spread to the increased use of crushed limestone as a filler material in asphalt shingles. Limestone is economical and makes a durable shingle, but the calcium carbonate in the limestone supports algae growth. In algae-resistant (AR) shingles, zinc or copper granules are mixed in with the colored stone topping. When the shingles get wet, the zinc or copper is released, inhibiting algae growth. Warranties for algae resistance are usually for less than 10 years since the protection ends when the mineral washes away. Some shingles have longer lasting protection than others due to a higher percentage of AR granules.

Manufacturing Standards. Fiberglass shingles are covered by ASTM D3462, which includes a tear test as well as a new nail-pull-through test added after fiberglass shingle failures started occurring in the late 1980s. A new pliability test was also added in recent years. Organic asphalt shingles are covered under their own standard, ASTM D255. In the past, most companies did their own testing, but under pressure from contractors' associations and others, most now use independent certifiers such as UL. With fiberglass shingles, look for the UL label next to the ASTM D3462 certification. This is not the same as a UL listing for a fire rating, which is printed on most fiberglass shingle packages. More and more jurisdictions are requiring compliance with ASTM standards, but discount shingles are still available with no certification.

As with many consensus standards, the ASTM D3462 requirement for tear strength of fiberglass shingles is considered by many experts to be a bare minimum rather than a guarantee of high quality. Also, once installed the shingles' strength will likely diminish. So finding products that exceed the minimum is recommended for demanding applications.

Warranties. Shingle warranties run from 20 to over 50 years. Although products with longer warranties are usually of higher quality, in some cases, the longer warranties are more of a marketing strategy than an accurate predictor of shingle life. While the specific terms of the warranty are important, more important is the manufacturer's reputation for warranty service in the local area. All manufacturers retain the right to void the warranty if installation instructions are not closely followed, and they can often find a way to avoid honoring a claim if so inclined. Key issues to consider in a warranty are as follows:

- Is the warranty prorated from the date of installation, or is there an introductory term of 5 to 10 years when the full value can be recovered?

- How long are warranties valid against wind damage, algae growth, or other types of damage?

- Does the warranty cover a portion of the labor costs of tear-off, disposal, and installation, or does it cover materials only?

- Is the warranty transferable?

- Perhaps most importantly, does the manufacturer have a strong reputation for warranty service in the local area?

Underlayment

The roof deck should be sound and level before laying the underlayment. Fifteen-pound or heavier felt underlayment is required by code in some areas. Whether or not it is required, underlayment is cheap insurance against problems. There are several good reasons to install underlayment:

- It protects the roof deck from rain before the roofing is installed.

- It provides an extra weather barrier in case of blow-offs or water penetration through the roofing or flashings.

- It protects the roofing from any resins that bleed out of the sheathing.

- It helps prevent unevenness in the roof sheathing from telegraphing through the shingles.

- It is usually required for the UL fire rating to apply (since shingles are usually tested with underlayment).

Standard Slopes. On roofs with a slope of 4:12 or greater, use a single layer of 15 lb. asphalt-saturated felt, starting at the eaves and lapping upper courses over lower by a minimum of 2 inches. Vertical joints should lap a minimum of 4 inches and be offset by at least 6 feet in successive rows (see Figure 2-2).

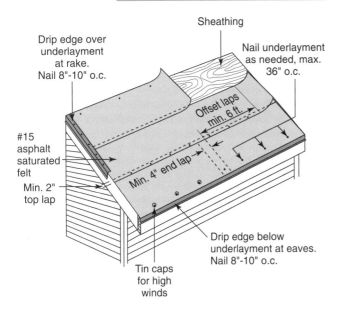

| FIGURE 2-2 | **Underlayment for Asphalt Shingles.** |

On roofs with a slope of 4:12 or greater, use a single layer of minimum No. 15 asphalt-saturated felt, starting at the eaves and lapping upper courses over lower. Run the felt 6 inches over ridges and hips from each direction, and 6 inches up any adjoining walls.

Secure each course along seams and edges with enough corrosion-resistant nails to hold it in place until the roofing is installed. In high-wind areas, apply fasteners a maximum of 36 inches on-center along overlaps.

For best protection against leaks, run felt 6 inches over ridges and hips, from each direction, and 6 inches up any adjoining walls. Valleys should be lined with a full width of roofing felt (or bituminous membrane) pushed tight into the valley so there is no slack. Side courses of underlayment should run over the valley lining and extend 6 inches past its edge. (See "Valley Flashing" page 59.)

Low Slopes. Asphalt shingles can be used on roofs with a slope of 2:12 to 4:12 if double-coverage underlayment is used. Start with a 19-inch strip of 15 lb. asphalt-saturated felt along the eaves, and lap succeeding courses by 19 inches as shown in Figure 2-3.

Wherever there is a possibility of ice or snow buildup or the backup of water from leaves or pine needles, install a self-adhering bituminous membrane along the eaves that extends up the roof to a point at least 36 inches inside the interior wall line. An alternative approach, not widely used anymore, is to seal all laps in the lower courses of roofing felt with lap cement or asphalt plastic cement.

In areas with extensive snowfall or windblown rain, the best protection against leakage is to cover the entire low-slope roof area with a bituminous membrane, as shown in Figure 2-4.

Vertical end laps should be at least 3 inches and horizontal laps 6 inches. If the roof changes to a steeper slope, for example, where a shed dormer joins the main roof, extend the membrane 12 to 18 inches up the main roof slope. Bituminous membranes are self-healing around nail holes, and because they bond fully to the sheathing, any leaks that occur cannot spread. As a safeguard against expensive callbacks, many roofers now apply membrane to the entire surface of any roof with a slope of 4:12 or less.

Eaves Flashing. The best defense against ice dams in cold climates is a so-called "cold roof," consisting of high levels of ceiling insulation separated from the roof surface by a free-flowing vent space (see "Preventing Ice Dams," page 97). Where a cold roof cannot be achieved due to complex roof shapes, unvented roofs, or retrofit constraints, ice dams may form during severe winters, in some cases, causing pooled water to wet wall cavities and interior finishes.

Where adequate insulation and ventilation cannot be assured, self-adhering bituminous eaves flashing should be installed. The membrane should go from the lower edge of the roof to a point at least 24 inches inside the interior wall line (Figure 2-5).

Where two lengths of eaves flashing meet at a valley, run each across the valley, starting with the length from the roof with the lower slope or lesser height. The valley flashing should later lap over the eaves flashing.

FIGURE 2-3 | **Low-Slope Underlayment.**

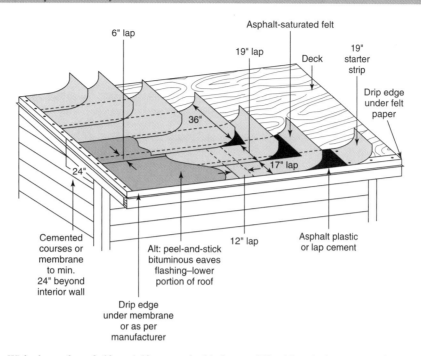

With slopes from 2:12 to 4:12, use a double layer of No. 15 underlayment as shown. Where water may back up from ice or debris from trees, protect the lower portion of the roof with a bituminous eaves flashing or fully cemented felt, as shown.

FIGURE 2-4 | **Low-Slope Underlayment in Cold Climates.**

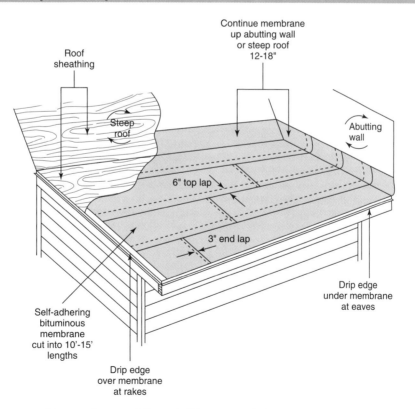

In areas with extensive snowfall or windblown rain, the best protection against leakage is to cover the entire low-slope roof area with a bituminous membrane. Extend the membrane to the top of skylight wells and up any adjacent walls or roof slopes by 12 to 18 inches.

FIGURE 2-5   **Eaves Flashing.**

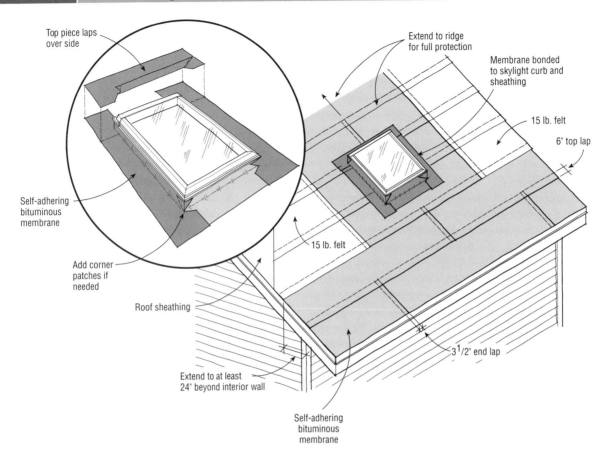

Where adequate roof insulation and ventilation cannot be assured, install a bituminous eaves flashing to a point at least 24 inches beyond the inside surface of the wall. Also, continue the membrane up to and around skylights where heat leaks can melt snow and contribute to ice dams.

Skylights. With deep snow, melting water from above and around the skylight can lead to ice dams below. For full protection, some contractors extend the eaves membrane up to the bottom of any skylights and continue it around the sides and top of the skylight. By wrapping the skylight curb with membrane as well, any potential flashing leaks are also eliminated as shown in Figure 2-5 (see also "Skylight Flashing," page 127.)

If it is impractical to install membrane all the way from eaves to skylight, install a 3-foot-wide band of membrane below the skylight, lapping the bottom edge of the membrane over the row of shingles where the membrane terminates.

Drip Edge. Drip edge should always be used along the eaves to kick water away from the fascia, and it is a good idea along rakes as well. Drip edge should lap over the underlayment at the rakes and under it at the eaves (as shown in Figure 2-6). Overlap joints in the drip edge by 2 inches. Shingles can be set even the with the drip edge or overlap by up to $\frac{3}{4}$ inch.

Some manufacturers of eaves membranes specify that the drip edge be installed on top of the membrane along the eaves, violating the principle that upper layers of flashing

FIGURE 2-6   **Drip Edge.**

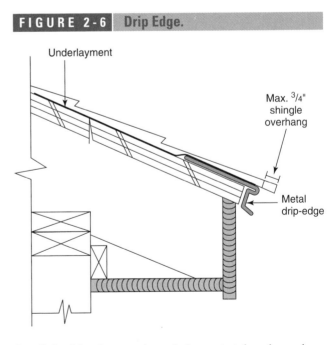

Install the drip edge over the underlayment at the rakes and under it at the eaves. Shingles may lie flush with the metal edge or overhang by up to $\frac{3}{4}$ inch.

should overlap lower layers. To remedy the problem, the manufacturers suggest using a second strip of membrane to seal the top of the drip edge to the eaves membrane. In practice, however, most installers place the drip edge first and lap the eaves membrane over it, consistent with good building practice.

Installation

Installation Temperature. Ideally, shingles should be installed at temperatures ranging from 40°F to 85°F. Below 40°F, shingles are brittle and crack easily when hammered or bent. Above 85°F, it is easy to tear the shingles or mar the granular coating. In hot temperatures, roofers often start very early in the morning and break at midday. In cold temperatures, it is best to store the shingles in a heated enclosure until they are installed.

Manual Sealing. In cold climates, the sealant strip may not set up properly and may require manual sealing. For three-tab shingles, place two quarter-size spots of plastic roof cement under the lower corners of each tab (as shown in Figure 2-7). With laminated shingles, place four to six quarter-sized dots, spaced evenly, about one inch above the bottom of the overlapping shingle.

FIGURE 2-7   **Manual Sealing of Asphalt Shingles.**

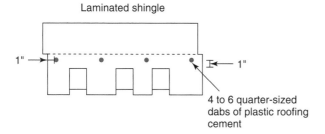

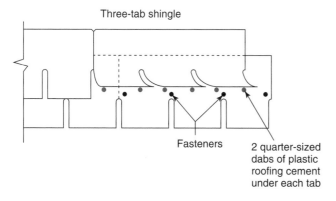

Manual sealing of shingles may be required in cold weather or on slopes over 21:12. Three-tab shingles (bottom) require two spots of plastic roof cement under the lower corners of each tab. Laminated shingles (top) require four to six spots spaced evenly about one inch above the lap line.

FIGURE 2-8   **Asphalt-Shingle Offsets.**

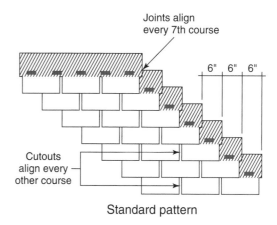

Standard pattern

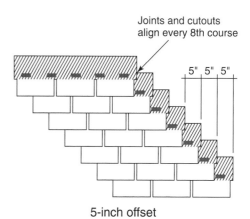

5-inch offset

In the standard installation pattern (top), cutouts line up every other course and generally wear out first from channeled water. A 5-inch offset (bottom) creates a more attractive random appearance with less wear in the cutouts.

Starter Course. After the underlayment and drip edge are installed, a starter course of asphalt shingles, with the tabs removed, is nailed along the eaves so its sealant strip seals down the first course.

Offsets. Successive courses are typically offset 6 inches (half a tab) on a 36-inch shingle in a stepped fashion, making cutouts align every other course and butt joints align every seventh course (Figure 2-8). For a more random pattern where cutouts align only every eighth course, offset shingles only five inches. Both of these patterns effectively resist leakage, but the 5-inch offset may provide longer wear since water will not be channeled down the cutouts thereby eroding the stone topping.

For ease of installation some roofers install shingles straight up the roof, staggering shingles 6 inches or 18 inches back and forth (Figure 2-9). Since this lines up butt joints every other course, this is considered a less watertight roof and may leak under extreme situations, such as windblown rain on a low pitch. It is not recommended by any roofing manufacturers. Manufacturers also claim that shingle color patterns may create splotches or stripes if laid this way.

Fastening. The preferred fastener is galvanized roofing nails with a minimum 12-gauge shank and head diameter of at least $\frac{3}{8}$ inch. Although staples are allowed in some jurisdictions, they do not provide the same holding power. Both nails and staples should be long enough to penetrate the roof sheathing by $\frac{3}{4}$ inch or penetrate $\frac{1}{4}$ inch through the sheathing if it is less than $\frac{3}{4}$ inch thick. Fasteners should

be driven straight and flush with the shingle surface (Figure 2-10). Overdriven nails or staples can cut into the shingle or crack it in cold weather.

Fastener Location. Standard nailing for three-tab shingles is four nails per shingle, about 1 inch in from either end and one over each slot. Placement should follow manufacturers specs, which typically require nailing and stapling just below the sealant strip (Figure 2-11).

Nailing too high can allow wind to get under the shingles. Nailing too low will expose nails to the weather and to view from below. Nailing through the sealant strip can interfere with sealing.

- **High winds.** For areas subjected to high winds, use six nails as shown in Figure 2-11 or add two dabs of sealant at the bottom of each tab (as shown in Figure 2-7). Also special wind-rated shingles with heavier sealing strips are available by special order and may be required in some jurisdictions.

- **Laminated shingles.** With laminated shingles, standard nailing is four fasteners spaced equidistant as shown in Figure 2-12, or six fasteners equidistant for heavy-duty installations. It is important that fasteners go in the designated nail area where they will penetrate both laminations. Nailing too high will leave the bottom lamination loose and subject to slipping out of place.

FIGURE 2-9   **Straight-Up Installation of Asphalt Shingles.**

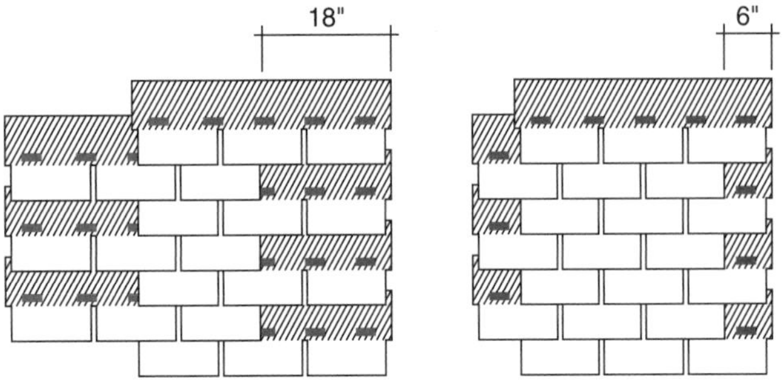

Half-shingle offset   Half-tab offset

While some roofers install shingles straight up the roof, staggered by either a half tab or half shingle as shown, manufacturers recommend against this approach. With butt joints lined up every other course, leaks could occur on shallow slopes or with windblown rain.

FIGURE 2-10   **Asphalt-Shingle Fasteners.**

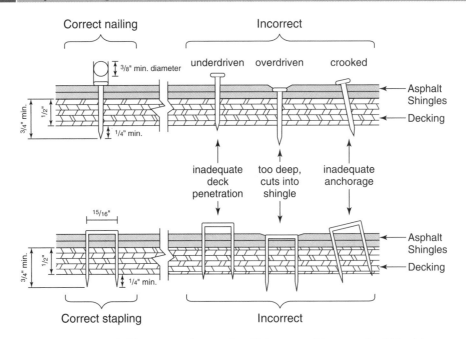

Nails provide better holding power than staples. Either type of fastener should be driven straight and flush with the surface. Overdriving or under-driving weakens the fastener's holding power.

SOURCE: Reprinted by permission of the Asphalt Roofing Manufacturers Association, © 1993.

FIGURE 2-11 **Fastening Three-Tab Shingles.**

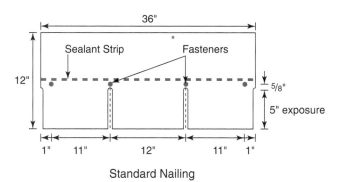

Standard Nailing

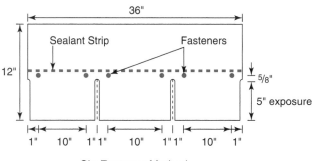

Six-Fastener Method
(For high winds or steep slopes)

Standard nailing for three-tab shingles is four nails per shingle just below the sealant strip. For more demanding conditions, such as high winds or steep pitches over 21:12, use six fasteners, as shown.

FIGURE 2-12 **Fastening Laminated Shingles.**

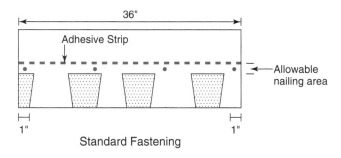

Standard Fastening

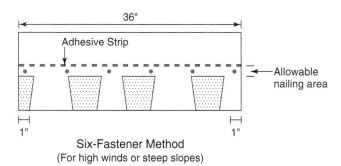

Six-Fastener Method
(For high winds or steep slopes)

Standard nailing for laminated shingles is four nails per shingle; six per shingle for high-wind conditions or slopes over 21:12. Nails must penetrate both laminations or the bottom layer may come loose.

Low Slopes

Asphalt shingles can be installed on roof slopes of 2:12 to 4:12 if special procedures are followed for underlayment (see "Low Slopes," page 54). Eaves flashing to a point at least 24 inches inside the interior wall is recommended if there is any possibility of ice dams or water backup from leaves or pine needles. A conservative approach is to run self-adhering bituminous membrane over the entire low-slope area. Once the underlayment is complete, shingles are installed in the standard fashion. In cold weather, manual sealing may be required as wind uplift will be greater on shallow roofs (see "Manual Sealing," page 57).

Steep Slopes

Asphalt shingles should not be installed on vertical walls, but they can be used on steep slopes, such as mansard-style roofs. For slopes greater than 21:12, apply underlayment in the normal fashion. However, shingle sealing may be a problem, particularly on shaded portions of the roof. For best performance, use the six-fasteners-per-shingle method (Figure 2-11) and manually seal the shingles with plastic roofing cement (see "Manual Sealing," page 57).

Flashings

Flashings for asphalt shingles should be corrosion-resistant metal with a minimum thickness of 0.019 inch. A cricket or saddle should be installed on any chimney greater than 30 inches wide and can be covered with flashing or the same materials used as a roof covering.

Valley Flashing

Because valleys catch water rushing down two roof planes, they are likely places for roof leaks. Leaks can be caused by water rushing up the opposite side of the valley or from wear and tear caused by the channeled water, snow and ice buildup, or traffic on the roof. For that reason all valleys should start with a leakproof underlayment system to back up the shingle or metal valley detail.

Valley Underlayment. Start by cleaning any loose nails or other debris and nailing down any sheathing nails that are sticking up. If eaves flashing is used, it should cross the valley centerline each way and be installed before the valley underlayment (see "Eaves Flashing," page 54). Next install a 36-inch-wide strip of self-adhering bituminous membrane in 10- to 15-foot lengths up the valley. Keep the membrane tight to the sheathing at the valley center, since any hollow sections could be easily punctured. Next install the 15-pound felt underlayment across the roof, lapping over the valley flashing by at least 6 inches. Roll roofing is also an acceptable underlayment for asphalt shingle valleys, although it is more prone

TABLE 2-3	**Asphalt Shingle Valley Types**		
	Pros	Cons	Recommendations
Woven	Minimal cutting and no sealing required. Very weather resistant due to double coverage across valley.	Both sides must be shingled together. Hollows left under weave may be torn or punctured. Heavy laminated shingles create uneven appearance.	Make sure shingles are pushed tight to sheathing. Use extra nail in top corner of end shingles.
Closed Cut	Each side can be shingled separately. Provides clean appearance with any shingle type.	Single coverage at valley center. Shingles on overlapping side must be clipped at corners and sealed with roofing cement.	Push lower shingles tight to sheathing and use extra nail in top corner of end shingle. Seal shingles carefully. Use bituminous underlayment as backup.
Open Metal	Most durable if appropriate metal is used. Attractive appearance with heavy laminated shingles.	Most costly. Metal lining requires prefabrication. Shingles on both sides must be clipped at corners and sealed with roofing cement.	Use with heavy laminated shingles where decorative appearance is desired. Avoid uncoated aluminum. Best with copper, lead-coated copper, or steel (enameled, galvanized, or stainless)

to crack and is not self-healing around nails. After the underlayment is complete, the valley can be completed in any of the following ways (Table 2-3).

Woven Valley. On the first course across the valley, the shingle from the larger or steeper roof plane overlaps the shingle from the smaller or shallower plane. Extend the end of each shingle at least 12 inches beyond the valley centerline and avoid placing any butt joints near the valley center. Press the shingles tight into the valley when nailing and place no fasteners within 6 inches of the valley center. Add an extra nail at the end of each shingle that crosses the valley (see Figure 2-13).

Continue to the top of the valley. Done correctly, woven valleys are very weather-resistant and best for high wind regions, but they are somewhat slow to install. They work better with three-tab shingles than with heavy laminated shingles, which do not conform well to a crisp valley line.

Closed-Cut Valley. This starts the same way as a woven valley, with the first course of shingles run across the valley from both roof planes, lapping the shingle from the larger or steeper roof plane over the shingle from the smaller/shallower plane. Then continue roofing the smaller or lower-slope roof plane, running each course at least 12 inches past the valley centerline. Press the shingles tight into the valley and nail in place, locating no fasteners within 6 inches of the valley center and adding an extra nail at the end of each shingle that crosses the valley (see Figure 2-14). Do not allow any butt joints to fall in the valley.

Next, snap a chalk line 2 inches out from the valley centerline on the opposite slope and shingle up the other side of the valley, holding nails back 6 inches from the valley center. Trim each shingle to the guide line as you go, or run them long and trim them later. In either case,

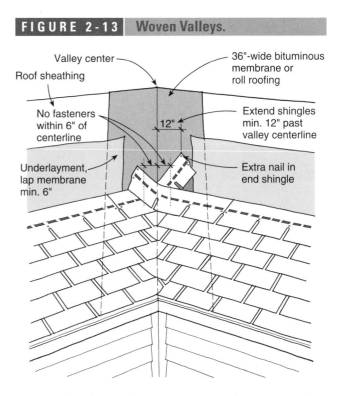

FIGURE 2-13   **Woven Valleys.**

Woven valleys do not rely on roofing cement for a water seal and are very weather-resistant. However, thick laminated shingles may not conform well to a crisp valley line.

clip about 1 inch off the uphill corner of each shingle to help direct rushing water into the valley. Finally seal each shingle to the valley and to the overlapping shingle with a 3-inch-wide bead of plastic roofing cement.

Closed valleys go up quickly and provide a clean appearance with either standard or laminated shingles. If sealed well, they provide adequate protection.

FIGURE 2-14 **Closed-Cut Valleys.**

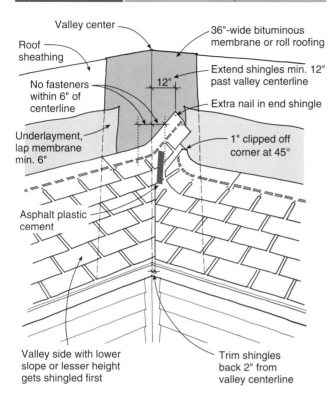

Closed valleys are popular since they go up fast and allow one roof plane to be shingled at a time. However, they only provide single coverage at the valley center and rely on a good sealant job for weatherproofing.

FIGURE 2-15 **Open Valleys.**

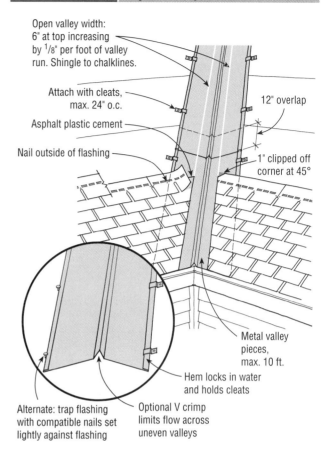

Open valleys should be wider at the bottom to handle the increased water flow along the valley. To avoid wrinkling in metal valleys, limit sections to 10 feet in length, and fasten with cleats that allow movement. A V crimp in the valley center keeps water from flowing across the valley where the two roof slopes are uneven in size or slope.

Open Valleys. With a heavy-gauge, noncorrosive metal lining, open valleys are the most durable valley and the most costly (see "Flashing Materials," page 6). An economical version uses two layers of roll roofing for the lining, which should last as long as an asphalt roof. The bottom layer of roll roofing goes on with the gravel facing downward; the top layer with the gravel facing upward. Nail along the edges every 12 to 18 inches, keeping the material tight against the roof sheathing.

The valley lining, whether asphalt or metal, should have 6 inches open at the top (3 inches on either side of the valley centerline) and increase by $\frac{1}{8}$ inch for each foot of valley length to accommodate the greater flow further down the valley. So a 16-foot valley would have 6 inches open at the top and 8 inches at the bottom (see Figure 2-15).

Metal valley linings should be 2 to 3 feet wide and no more than 8 or 10 feet in length to prevent wrinkling from lengthwise expansion. Overlap valley sections by 12 inches, and seal the lap with a flexible sealant, such as polyurethane or butyl, on roofs shallower than 5:12. Where two valleys meet, for example above a gable dormer, a soldered joint is likely to break from the movement. A lead cap overlaid 6 inches onto each valley is an effective way to seal the top.

Where the roof slopes are uneven or one roof is larger than the other, a 1- to $1\frac{1}{2}$-inch-high V crimp in the middle of the metal valley will prevent the uneven flow from running up one side of the valley. The crimp also stiffens the valley. A hem is also desirable, both to stop any overflow water and to provide a place to attach nailing clips, which hold the flashing securely while allowing movement. Nails wedged against the edge of the flashing and driven lightly against the flashing may also be used. Clips and nails should be the same metal as the valley or a compatible metal that will not cause galvanic corrosion (see "Galvanic Corrosion," page 83).

Shingles should overlap the valley lining by at least 6 inches. With a roll roofing valley, keep the nails at least 6 inches from the valley centerline. With a metal liner, nail $\frac{1}{2}$ inch outside the liner. Seal each shingle to the liner and overlapping shingle with a 3-inch-wide bead of plastic roofing cement.

Reroofing

Reroofing saves the cost, trouble, and risks (water damage while the roof is exposed) associated with a tear-off. If the roof is structurally sound, most building codes allow for two layers of asphalt shingles and some allow for a third on roofs with a 5:12 or steeper pitch. If the original shingles are not badly curled and the sheathing is sound (check for bouncy areas), then a reroof is a good alternative.

Shingle Type. The heavier the shingle on the new layer, the less likely it is that irregularities in the surface below will telegraph through. Laminated or other heavy-textured shingles work well, as they do not need to be carefully fitted to the existing shingles, and the irregular texture will conceal any small bumps or dips from the original roof.

Prep Work. Clip any curled shingle corners and remove any curled tabs, replacing them with new shingle scraps as shims. Install new drip edge on rakes and eaves. Specialty drip edge profiles designed for retrofitting wrap around the exposed roof edge, leaving a neat protected edge. If the roof had no eaves flashing and one is needed, use a retrofit membrane such as AC Evenseal (NEI, Brentwood, New Hampshire).

Starter Course. If laying three-tab shingles over three-tab shingles, it is important to nest the new shingles against the old to create a flat surface. This process starts with a 5-inch starter strip fit along the eaves and set against the second course of existing shingles (see Figure 2-16).

Next install a course of shingles cut down to 10 inches wide, so they fit against the bottom edge of the existing third course (this creates a new 3-inch first course). After that, shingling should proceed normally, fitting each course up against the bottom of an existing course.

Fastening. Use galvanized roofing nails long enough to fully penetrate the sheathing, typically $1\frac{1}{2}$ inches for a second roof and $1\frac{3}{4}$ inches for a third. Nesting each new row below an existing one keeps the new nails 2 inches below the existing, which will help minimize any splitting of the sheathing.

Flashings. Depending on their condition and accessibility, some flashings can be reused. New shingles may be able to tuck under existing step flashing, chimney flashings, and front-wall flashings. If they are deteriorated, they must be replaced along with vent boots.

Valleys. Any type of valley flashing will work and simply lays over the existing flashing (except in a tear-off, where all flashings should be replaced). Unless a metal valley flashing is used, the first step is to line the existing

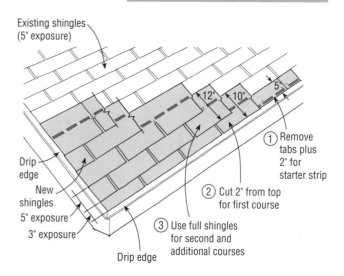

FIGURE 2-16 Reroofing with Asphalt Shingles.

Existing shingles (5" exposure)

Drip edge

New shingles
5" exposure
3" exposure

Drip edge

① Remove tabs plus 2" for starter strip

② Cut 2" from top for first course

③ Use full shingles for second and additional courses

When laying new three-tap shingles over existing ones, it is important to nest the new shingles against the old to create a flat surface. Begin with a 5-inch starter course, followed by a 10-inch first course—each set against the bottom edge of the existing shingles. Proceed with full-size shingles, each nested under the existing course above.

valley with a new underlayment consisting of either 90-pound roll roofing or a more durable modified bitumen membrane. Then install either a closed or woven valley as described above.

CLAY, CONCRETE, AND COMPOSITE TILE

Tile roofing accounts for about 8% of new residential roofs in the United States, primarily in the Southeast, Southwest, and on the West Coast. In addition to its durability and natural beauty, tile is impervious to fire, insects, and rot, and it can be formulated to withstand freeze-thaw cycles. When colored white, tile roofing has been shown to reduce cooling costs by up to 22% for barrel or flat tile (compared to black asphalt shingles in tests conducted by the Florida Solar Energy Center). Since most tile roofs carry a 50-year warranty and a Class A fire rating, they are a popular choice for high-end projects, particularly in warm climates.

Nearly all roofing tiles in the United States were traditional clay until the 1960s when concrete tile first gained acceptance. Concrete tile now dominates most tile roofing markets, primarily due to its lower cost (see Table 2-4). Where weight is a concern, options include lightweight concrete tiles or fiber-cement shingles, which typically weigh even less. Fiber-cement roofing typically simulates

TABLE 2-4	Roof Tile Materials			
	Material Costs (typical)	Advantages	Disadvantages	Recommendations
Clay Tile	$400–600 per square	Traditional material. Colorfast, lasts 50+ years, and is virtually maintenance-free. Impervious to fire, insects, and decay. Resists high winds. Can be formulated to resist freeze-thaw cycles. Class A fire rating.	More expensive than concrete tiles. Typically weighs 800–1,200 lb per square. Roof structure may need extra support. High shipping costs unless manufacturer nearby. Very susceptible to breakage from foot traffic.	Confirm structural support is adequate. In cold climates, use vitrified tile with low water absorption. Choose larger tiles for best prices.
Concrete tile	$100–$300 per square	Less expensive than clay tile. Lasts 40–50 years in Southwest; 20–30 years in Southeast. Impervious to fire, insects, and decay. Resists high winds. Can be formulated to resist freeze-thaw cycles. Can simulate shakes or slates. Lightweight version available (550–800 lb per square). Class A fire rating.	Installed costs about three times asphalt shingles. Typically weighs 900–1,200 lb per square. Roof structure may need extra support. High shipping costs unless manufacturer nearby. Colors tend to fade. Lightweight type are very susceptible to breakage from foot traffic.	Confirm structural support is adequate. In cold climates, check warranty for freeze-thaw resistance. Use through-color tiles in cold climates.

slate or wood shakes and provides a Class A fire rating at a cost comparable to wood shakes.

Tile Shapes

All tiles can be classified as high-profile, low-profile, or flat (see Figure 2-17). Common high-profile tiles include two-piece pan-and-cover Mission tile and one-piece Spanish S-tiles. Low profile styles include a wide variety, many with a double-S shape that creates multiple water courses. Many flat tiles are shaped and colored to simulate slate or wood shakes. In general, patterns using smaller tiles cost more per square for both materials and labor than patterns using larger tiles.

Clay Tile

To make tiles, moist clay is extruded through a die or cast in a mold and then fired in a kiln until the clay "vitrifies," fusing the particles together. Complete vitrification will create a strong tile with very low water absorption, which protects tile from freeze-thaw damage in cold climates or damage from salt air in coastal areas. Where regular freeze-thaw cycling is expected, roof tiles should comply with ASTM C1167 Grade 1, which allows minimal water absorption. Grade II tile provides moderate resistance to frost action, and Grade III tile is porous and should not be used in freeze-thaw areas.

When buying clay tile, look for at least a 50-year warranty on both durability and fading. Costs vary widely, depending on quality, style, and the shipping distance required. In general, patterns using smaller tiles will cost more for both materials and labor.

FIGURE 2-17	Tile Shapes.

Flat profile: surface rises up to 1/2"

Low profile: height is 1/5 width or less

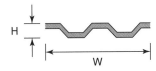

High profile: height is greater than 1/5 width

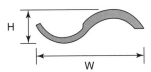

The industry generally classifies tiles as high-, low-, or flat-profile depending on the ratio of width to height. Flat tiles often mimic slates or wood shakes.

Color. Clay tiles come in a wide range of colors. Color-through tile takes the natural color of the clay, ranging from light tan to pink and red. Other colors can be added to the tile surface as a clay slurry before firing, but slurry coatings are only suitable for warm climates, as they cannot withstand freeze-thaw cycles. Clay tile can also be colored with ceramic glazes to create a durable, glass-like surface in just about any color. In general, clay tiles do not fade in the sun.

Blended Patterns. Some jobs require the installer to mix two or three different colors in a random pattern. The best way to achieve this is to premix bundles on the ground with the correct proportion of each color, then send them up to the roof for installation. Periodically inspect the roof from the ground for hot spots or streaking.

Clay Tile Styles. Clay roof tiles are available in traditional two-piece styles, one-piece profiles, and flat profiles (Figure 2-18).

Designs are either overlapping or interlocking, with protruding lips that lock the tiles together and form a weather seal. Many flat clay tiles interlock. Interlocking designs are recommended for regions with heavy rain or snow. Manufacturers provide special trim tiles to seal the voids formed at ridges, rakes, and hips.

- *Pan and cover.* This traditional two-piece style, also called barrel- or Mission-style, is installed in pairs with the cover tile overlapping the pan tile. It provides an attractive high-profile look but is labor-intensive and expensive to install. Variations include Roman and Greek profiles, which have flat, rather than

curved, pan tiles. Tiles typically range from 8 to 12 inches in width and from 16 to 19 inches long.

- *Spanish S-tile.* These one-piece tiles provide the high-profile look of traditional pan-and-cover tile but with simpler installation. The most popular S-tiles measure about 13 inches wide by $16\frac{1}{2}$ inches long. Other common sizes are $8\frac{3}{4}$x11 and 9x14 inches.

- *Flat shingle tile.* These are laid in a double thickness, like slate. Widths range from 6 to 8 inches, lengths from 12 to 18 inches.

- *Interlocking tile.* These are either flat or low profile and are laid in a single thickness with a 3-inch overlap. They have interlocks on the sides with channels or ribs, and butts may also lock into the tops of the underlying shingles. Contours and ribs add strength to the tiles. Widths typically range from 9 to 13 inches and lengths from 11 to 16 inches.

Concrete Tile

Concrete tiles were introduced to the United States in the early 1900s, but they did not catch on until the 1960s. They now account for more than half the tiles sold in the United States. In Europe, over 90% of new houses have concrete tile roofs. Concrete tiles cost as little as half as much as clay and offer both traditional and flat styles that simulate slate roofing and wood shakes.

High-quality concrete tiles should last up to 50 years in arid climates and up to 30 years in hot, humid climates. While some early products faced problems with freeze-thaw cycling, most newer formulations are made to withstand winter weather. In cold climates, make sure the product is warranted for freeze-thaw durability.

Special lightweight concrete tiles weighing under 600 lb per square are gaining in popularity. Although they cost more than standard concrete tiles and are more prone to breakage, they are easier to handle and suitable for applications where the roof structure cannot support the weight of standard tiles. Lightweight tiles cannot support foot traffic without adding walking pads to distribute weight or filling the space under the tiles with polyurethane foam. They are also not recommended for high-snow regions.

Color. Concrete tiles can be surface colored with a slurry of iron-oxide pigments applied to the surface or have the color added to the concrete mix for a more durable, and expensive, through-color. Through-color choices are more limited, and the colors are more subdued. Either type of tile is also sealed with a clear acrylic spray to help with curing and efflorescence. While the color-through tile will hold its color better than the slurry type, particularly under freeze-thaw cycling, all concrete tile coloring can be expected to fade and soften over time. Surface textures can also be added to flat concrete tiles to simulate wood shakes or shingles.

FIGURE 2-18 **Typical Clay-Tile Profiles.**

High-Profile

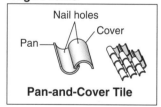

Pan-and-Cover Tile

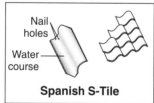

Spanish S-Tile

Flat Profile

Flat Shingle Tile

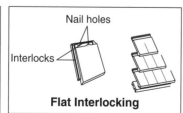

Flat Interlocking

Low Profile

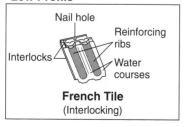

French Tile
(Interlocking)

FIGURE 2-19 **Typical Concrete-Tile Profiles.**

Spanish S-Tile

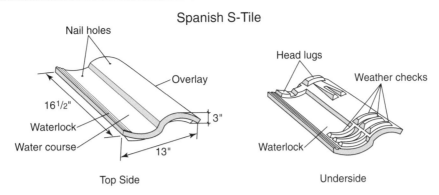

Low-Profile Tile (Double Roman)

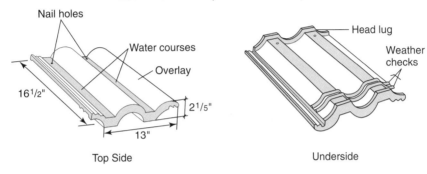

Flat Tile (Shake)

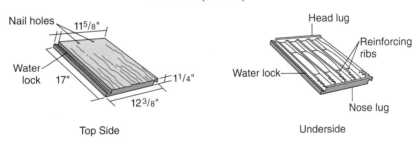

Concrete Tile Styles. Concrete tiles are available in shapes that simulate traditional clay styles as well as flat profiles that simulate wood or slate (Figure 2-19).

Most are designed with an interlocking channel on the left edge that is lapped by the next tile. Underneath each tile is a head lug at the top and series of ridges at the bottom. The head lug fits over the top of a horizontal 1x batten, if these are used. Otherwise it sits directly on the roof deck. The ridges at the bottom (called nose lugs or weather checks) match the profile of the tile below, creating a barrier against windblown rain and snow. Manufacturers provide special trim tiles to fill in the large voids that profile tiles leave at ridges, rakes, and hips. While many sizes are available, the most common concrete tiles measure 12 to 13 inches wide by $16\frac{1}{2}$ or 17 inches long.

- **Spanish S-tile.** These provide the look of traditional two-piece Mission tiles but with simpler installation. Nearly all have interlocking side channels.

- **Interlocking low-profile tile.** These have a less pronounced double-S shape and interlocking joints and side channels. Heads and butts may also interlock or simply overlap.

- **Interlocking flat tile.** These simulate clay roof tiles, wood shakes, and slate. Ridges, hips, and rakes are easier to seal than with curved tiles.

Fiber-Cement Tile

Early generations of fiber-cement roofing products using asbestos fibers were used successfully in the United States for over 50 years. Newer formulations introduced in the 1980s and 1990s used wood fibers instead of asbestos and were marketed widely in the western United States as a fire-resistant alternative to wood shakes. Made from a mixture of Portland cement and wood fibers, they weighed 400 to 600 pounds per square and were designed to imitate slates or wood shakes. They promised excellent resistance to insects, fungus, fire, and weathering and carried warranties ranging from 25 to 50 years.

Performance Problems. Within five years of installation, however, many of the fiber-cement shakes began to deteriorate. Problems included surface crazing, cracking, delamination, and softening and resulted in a number of lawsuits against key manufacturers and several companies abandoning the product. The problems were generally linked to high water absorption, which created an alkaline solution that was corrosive to the wood fibers.

Some products have fared better than others. In general, products that are steam-cured in an autoclave will have lower water absorption, but they tend to be more brittle. Many products are represented as complying with ASTM C1225, a standard for nonasbestos fiber-cement roofing shingles; but in its current form, this standard does not guarantee long-term durability. Only a product with a proven long-term track record in a specific climate zone should be considered.

Roof Slope

Most manufacturers recommend minimum slope requirements for their tiles as well as special underlayment and fastening techniques for low-slope installations. Typical minimums are shown in Table 2-5. Some manufacturers allow specific tile types to be installed on roofs as shallow as $2\frac{1}{2}$:12 if a full waterproofing layer, such as a built-up roof or single-ply membrane, is installed. Reduced exposure and special fastening techniques may also be required for low slopes. On slopes less than $3\frac{1}{2}$:12, roofing tile is considered decorative only. The underlying roof provides all the necessary waterproofing.

In general, there is no maximum slope for tile roofs. However, on extremely steep roofs above 19:12 or on vertical applications, wind currents may cause tiles to rattle. To avoid this, use wind clips on each tile along with a

construction grade silicone sealant or other approved sealant.

Roof Sheathing

While spaced sheathing is allowed under the codes, most installations today are done on solid wood sheathing with or without battens. The sheathing must be strong enough to support the required loads between rafters. Minimum requirements are nominal 1 inch for board sheathing or $\frac{15}{32}$ for plywood and other approved panel products.

Underlayment

Because of the long service life of tile, a long-lasting underlayment should be used as well. Underlayments play a key role in tile roofing, since most tile roofs are not completely waterproof. At a minimum, use a Type II No. 30 or No. 43 felt, lapped 2 inches on horizontal joints and 6 inches at end laps. The underlayment should lap over hips and ridges 12 inches in each direction and turn up vertical surfaces a minimum of 4 inches (Figure 2-20).

At tricky areas, such as around roof vents, chimneys, and skylights, self-adhesive bituminous membrane can help achieve a watertight seal. In windy areas, use tin caps or round cap nails to hold the underlayment securely. The fastening schedule for the underlayment will depend on local wind conditions.

For harsher conditions or shallower slopes, use mineral-surface roll roofing, self-adhering bituminous membrane, or other durable waterproofing systems. For slopes below $3\frac{1}{2}$:12, the underlayment must provide complete weather protection, and the tiles are considered merely decorative. Underlayment recommendations for different types of tiles and climate conditions are shown in Tables 2-6 to 2-8.

TABLE 2-5	Minimum Slope Recommendations (Typical)	
	Type	Minimum Slope
Clay	Flat shingle tile	5:12*
	Interlocking flat tile	3:12
	Interlocking low-profile (French) tile	3:12
	Pan-and-cover tile	4:12–5:12*
	S-tile	4:12*
Concrete	Interlocking flat tile	4:12
	Interlocking low-profile tile	4:12
	Interlocking S-tile	4:12

*May be reduced to $2\frac{1}{2}$:12 or 3:12 by using full waterproofing underlayment as per manufacturer and code.

FIGURE 2-20 | Underlayment for Roofing Tile.

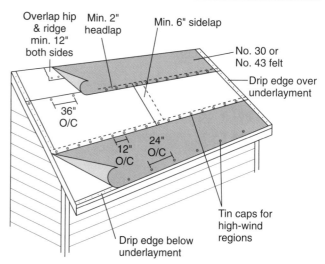

Use minimum No. 30-felt underlayment for moderate climates and No. 43-felt or mineral-surface roll roofing for high-wind and coastal regions. Tin caps or round cap nails are recommended in windy regions.

TABLE 2-6 Underlayments for Mechanically Fastened Tile Roofing Without Head Lugs			
	Roof Slopes from 2½:12, to Less Than 3:12*	Roof Slopes from 3:12 to Less Than 4:12	Roof Slopes 4:12 and Greater
Moderate climates	• Three-ply built-up roofing membrane. • An approved single-ply membrane.	• Two layers of type II No. 30 felt, lapped 19 inches on horizontal (36-inch roll), and 6 inches on vertical. • One layer Type 90 granular surface roll roofing. • An approved single-ply membrane.	Type II No. 30 felt, lapped 2 inches on horizontal, 6 inches on vertical.
Subject to windblown snow, ice dams, or high winds	• Three-ply built-up roofing membrane. • An approved single-ply membrane.	Two layers of type II No. 30 felt, lapped 19 inches on horizontal (36-inch roll), and 6 inches on vertical. Also eaves flashing of self-adhered membrane, or two layers No. 30 felt fully cemented.	Type II No. 30 felt, lapped 2 inches on horizontal, 6 inches on vertical. Also eaves flashing of self-adhered membrane, or two layers No. 30 felt fully cemented.

*Tile considered decorative only at this slope.
SOURCE: Based on recommendations of the Tile Roofing Institute and Western States Roofing Contractors Association.
All recommendations subject to local code.

TABLE 2-7 Underlayments for Tile Roofing with Projecting Head Lugs		
	Roof Slopes Less Than 4:12	Roof Slopes 4:12 and Greater
Moderate climates	Consult manufacturer and local codes.	Type II No. 30 felt, lapped 2 inches on horizontal, 6 inches on vertical.
Subject to windblown snow, ice dams, or high winds	Consult manufacturer and local codes.	Type II No. 30 felt, lapped 2 inches on horizontal, 6 inches on vertical. Also eaves flashing of self-adhered membrane or two layers No. 30 felt fully cemented.

SOURCE: Based on recommendations of the Tile Roofing Institute and Western States Roofing Contractors Association. All recommendations subject to local code.

Prep Work

Battens. Tiles with projecting head lugs can be installed either directly on the deck or with the lugs fitting over pressure-treated wooden battens nailed horizontally across the roof. Battens are typically nominal 1x2 or 1x4 lumber, but they may be larger to accommodate snow loads or unsupported spans over counterbattens. Battens should be made from pressure-treated lumber except in very dry climates. They are nailed at minimum 24 inches on-center with spaces for drainage every 48 inches. Lay out battens to provide equal courses with a minimum 3-inch head-lap, unless the tile profile is designed for a specific head-lap. Fasten with 8d galvanized nails or corrosion-resistant $1\frac{1}{2}$-inch 16-gauge staples with $\frac{7}{16}$-inch crowns.

Battens are recommended on roof slopes greater than 7:12 to provide solid anchoring and on slopes below 3:12 to minimize penetration of the underlayment. On low slopes and in areas subject to ice damming, counterbattens

TABLE 2-8 Underlayments for Tiles in High Wind and Coastal Conditions		
	Roof Slope 2:12 and Greater	Roof Slope 4:12 and Greater
Mechanically fastened tile (battens or direct deck)**	• Hot-mop or cold process: No. 30 or No. 43 felt base ply plus 90 lb roll roofing or modified cap sheet.* • Self-adhered underlayment applied to wood deck. • No. 30 felt plus self-adhered underlayment.	• Single-ply No. 43 or 90 lb organic or modified cap sheet.* • Two-ply No. 30 or No. 43 underlayment (battens only)
Mortar set tile (direct deck)	Hot-mop or cold process: No. 30 or No. 43 base ply sheet plus 90 lb roll roofing or modified cap sheet*	
Adhesive set tile (battens or direct deck)	• Single-ply 90 lb roll roofing or modified cap sheet. • Hot mop or cold process: No. 30 or No. 43 felt base ply plus 90 lb roll roofing or modified cap sheet. • Self-adhered underlayment to wood deck. • No. 30 felt plus self-adhered underlayment.	

*Plastic cement or approved sealant at fasteners penetrating underlayments.
** All applications below 3:12 must have both vertical and horizontal battens. Above 7:12, battens required for tiles with head lugs.
SOURCE: Based on recommendations of the Florida Roofing, Sheet Metal, and Air Conditioning Contractors, Inc. and the Tile Roofing Institute.
All recommendations subject to local code.

FIGURE 2-21 Tile Installation Over Battens.

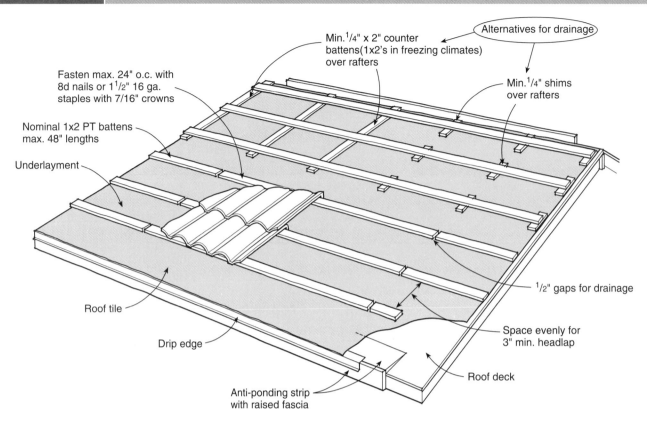

Fasten max. 24" o.c. with 8d nails or 1¹/₂" 16 ga. staples with 7/16" crowns

Min.¹/₄" x 2" counter battens(1x2's in freezing climates) over rafters

Alternatives for drainage

Min.¹/₄" shims over rafters

Nominal 1x2 PT battens max. 48" lengths

Underlayment

Roof tile

Drip edge

Anti-ponding strip with raised fascia

¹/₂" gaps for drainage

Space evenly for 3" min. headlap

Roof deck

Horizontal battens are recommended for tiles with projecting head lugs on roof slopes greater than 7:12. Battens are also used on low slopes to minimize penetration of the underlayment. Vertical counter-battens are recommended to promote drainage on low slopes and in regions with high snowfall.

nailed vertically up the roof slope are also recommended to promote drainage. Counterbattens should be minimum $\frac{1}{4}$x2 inches thick in moderate climates, $\frac{3}{4}$ inch thick in areas subject to ice damming. When battens are nailed directly to the deck, allow a $\frac{1}{2}$-inch gap every 4 feet or set the battens on minimum $\frac{1}{4}$-inch shims placed at each nail (see Figure 2-21).

Layout and Stacking. Lay out the courses so that tile exposures are equal with a head-lap of at least 3 inches (unless the tile specifies a different lap). Snap lines on the underlayment along the top of each course or along each batten. One or more vertical lines can also be helpful in keeping the tiles aligned. Accurate layout is critical with most tile patterns.

Next, carry tiles up to the roof and distribute the weight equally across the roof, as tiles weigh as much as 10 pounds each. Depending on the tile, stacks of about 6 to 10 tiles is workable. If mixing different colored tiles, arrange bundles with the correct proportions on the ground before stacking them on the roof.

Fastening Tile

The preferred method of attachment depends on the type of tile, climate conditions, and slope of the roof.

Loose Laid. For standard concrete tiles with lugs set on battens, building codes still allow tiles to be laid loose at slopes less than 5:12 (except for one nail per tile within 36 inches of hips, ridges, eaves, or rakes). Loose-laid tiles are not allowed, however, in snow regions, areas subject to high winds, or with tiles weighing less than 9 pounds per square foot installed.

Nail on. Nails are the least expensive and most common method for attaching concrete and clay tiles. Tiles can be nailed either directly into the roof sheathing or tiles with lugs can be nailed to battens. Corrosion-resistant nails must be minimum 11 gauge, with $\frac{5}{16}$-inch heads, and long enough to penetrate the sheathing by $\frac{3}{4}$ inch—typically 8d nails. Ring-shank nails or hot-dipped galvanized nails hold better than smooth-shank nails in areas subject to heavy winds. Whether driven by hand or pneumatic nailers, nails should be driven so heads lightly touch the tile but not so tight as to risk cracking tiles. Because of the longevity of a tile roof, some contractors use copper or stainless-steel roofing nails. No. 8 or 9 stainless-steel or brass screws also work well and are sometimes used in high-wind regions.

Most tiles have two prepunched nail holes. On curved tiles, use the hole closest to the deck surface unless a nail there would penetrate a critical flashing. The other hole is

TABLE 2-9	Minimum Nailing for Concrete and Clay Tile in Moderate Climates[1]			
Roof Slope	Spaced or Solid Sheathing With Battens	Solid Sheathing Without Battens	Spaced Sheathing Without Battens	Perimeter Tile All Sheathing Types[2]
Less than 5:12	None required[3]	One per tile	As per manufacturer	One per tile
5:12 to less than 12:12	One per tile every other row	One per tile	One per tile every other row	One per tile
12:12 to 24:12 [4]	One per tile	One per tile	One per tile	One per tile

[1]For roofs 40 ft high or less in areas without significant snowfall or repeated winds of over 80 mph. In snow regions, min. two nails per tile.
[2]First three tile courses, but not less than 36 in. wide, at any ridge, hip, eave, or rake. Also any roof overhang.
[3]Tiles less than 9 psf installed require one nail.
[4]On roofs over 24:12, the nose end of all tiles should also be securely fastened with wire clips or roofing cement.

also used for cut tiles or applications requiring two nails. For example, all flat, noninterlocking tiles require two nails. And in snow regions, codes require two nails per tile for all types and slopes. Otherwise follow the guidelines in Table 2-9, or the manufacturer's guidelines if they are more stringent.

High-Wind and Seismic Installations

In areas prone to high winds, such as Florida, setting the tiles in mortar was once considered the strongest system. However, newer anchoring systems using wires, special clips, and, in some cases, specialized adhesives have proven more reliable and have replaced mortar-set systems as the preferred approach. Wire and clip systems also perform better than rigid attachment systems in seismic zones, as the flexible systems tend to absorb the shockwaves of an earthquake and protect the tiles from cracking.

Codes vary in their requirements for high-wind and seismic areas but most permit one or more of the anchoring systems described below. Model specifications for high-wind installations are available in the *Concrete and Clay Roof Tile Installation Manual,* jointly published by the Florida Roofing, Sheet Metal and Air Conditioning Contractors Association and the Tile Roofing Institute. General guidelines for high-wind installations or roofs over 40 feet above grade include:

- Fasten the head of every tile.
- Fasten the nose of every tile with clips or other approved methods.
- Secure all rake tiles with two fasteners.
- Set the noses of all ridge, hip, and rake tiles in a bead of approved roofer's mastic.

Twisted Wires. This approach is used on roofs ranging from 2:12 to 24:12 in seismic zones and areas with moderate winds. Rather than nail the tiles to the roof, each tile is wired to a length of twisted 12-gauge wire (galvanized, copper, or stainless steel) running from eaves to ridge under each vertical course of tiles. The twisted wire has a loop to tie into every 6 inches and is attached every 10 feet with special anchors, making relatively few holes in the underlayment (see Figure 2-22).

Because wire systems allow some movement, seismic forces do not tend to break the tiles. Also, damaged tiles are easy to replace by snipping the tie wire and wiring in a new tile. Installation is labor-intensive, however, compared to nailing.

Hurricane Clips. A hurricane clip, also known as a storm clip or side clip, is a concealed L-shaped metal strap designed to lock down the water-channel side of a roofing tile near the nose (Figure 2-23).

Clips are well-suited to concrete tile and are used in conjunction with nails, screws, or other systems that secure the head of the tile. They are approved for use in some hurricane areas, but they should be combined with a nose clip or similar device for maximum protection. Used alone, they may deform or loosen after several storms.

Nose Clips. Also known as nose hooks, butt hooks, or wind locks, these simple metal clips hold down the bottom (nose) end of a roofing tile to prevent strong winds from lifting and breaking the tiles (Figure 2-24).

Nose clips are nailed in place through the underlying tile or attached to the tie wires in wire systems. They are compatible with all methods of tile attachment and are recommended for high-wind areas and slopes greater than 7:12. The main drawback to nose clips is that they are visible at the nose of each tile, which some homeowners find objectionable.

Tile Nails. This innovative fastener, used mostly with S-tile or two-piece Mission tile, functions as both a nail and a nose clip. Because the nail is driven about 6 inches above the tile, there is no risk of breakage and the nail hole can be easily sealed with mastic (Figure 2-25).

Tile nails are approved for all slopes and are especially useful in high-wind areas and on very steep pitches such as mansards. They are also useful for securing the first course of two-piece Mission tile. Examples include the Tyle Tye® tile nail from Newport Tool & Fastener Co. and the Hook Nail from Wire Works, Inc.

Tile Adhesives. Another way to prevent uplift in windy conditions and to keep tiles from rattling on steep slopes is

FIGURE 2-22 Twisted-Wire Systems.

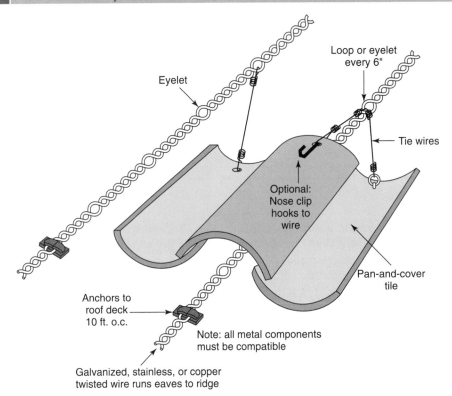

Eyelet

Loop or eyelet every 6"

Tie wires

Optional: Nose clip hooks to wire

Pan-and-cover tile

Anchors to roof deck 10 ft. o.c.

Note: all metal components must be compatible

Galvanized, stainless, or copper twisted wire runs eaves to ridge

Popular in high-end construction and commercial jobs, twisted wire systems create a nonrigid attachment that performs well in seismic zones. Wires are secured from eaves to ridge with special anchors. High-wind performance can be further enhanced by adding nose hooks, hurricane clips, or adhesives. Avoid mixing components of different metals.

FIGURE 2-24 Nose Clips.

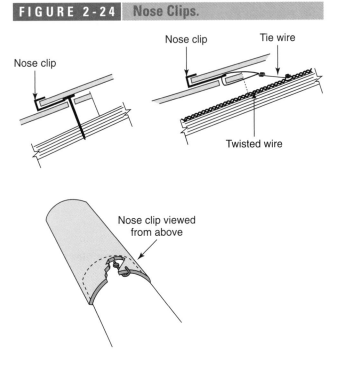

Nose clip

Nose clip

Tie wire

Nose clip

Twisted wire

Nose clip viewed from above

FIGURE 2-23 Hurricane Clips.

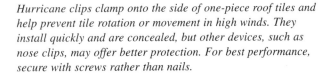

Hurricane clip nails or screws to deck at water channel

Hurricane clips clamp onto the side of one-piece roof tiles and help prevent tile rotation or movement in high winds. They install quickly and are concealed, but other devices, such as nose clips, may offer better protection. For best performance, secure with screws rather than nails.

Nose clips are recommended for steep slopes and high-wind regions to prevent tiles from lifting in the wind. They are nailed in place through the underlying tile (left) or attached to the tie wires in wire systems (right). Their main drawback is their visibility from the ground, particularly on low roofs.

FIGURE 2-25 Tile Nails.

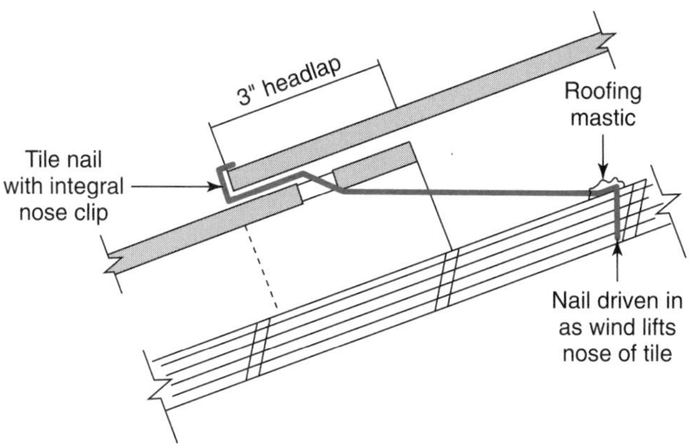

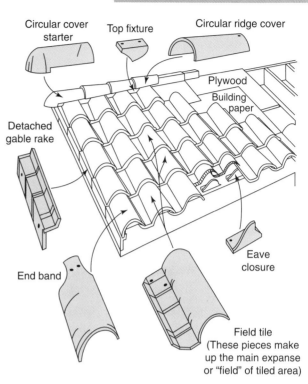

This innovative fastener functions as both a nail and nose clip. Because it is driven about 6 inches above the tile, there is no risk of breakage. Also, uplift pressure on the nose of the tile tends to drive the nail deeper rather than pull it out. Tile nails are useful in high-wind areas and on very steep slopes such as mansards.

to set the butt edge of each tile in a dab of roofing cement. Over time, however, roofing cement may become brittle and fail. New proprietary tile adhesives promise to last longer and stay flexible over time. In hurricane-prone areas, some contractors are applying adhesive to every tile—in some cases combined with other fastening methods, such as twisted wires. While long-term performance has not been well-established, testing by manufacturers has demonstrated that adhesives can outperform mortar systems in hurricane-force winds.

Installation Details

A number of specialized flashings, tiles, and fittings simplify modern tile installations. Key details for interlocking flat and profile tiles are shown in Figures 2-26 and 2-27.

Eaves Closure. Both profile and flat tile need special treatment at the eaves to raise the bottom edge of the first tile to the correct height and to close off any openings to birds and insects. For profile tile, many contractors use a metal *birdstop,* a preformed L-shaped strip with the vertical leg cut to match the underside of the first tile and fit snugly between the weather checks (see Figure 2-28).

With some high-profile tiles, a special eaves-closure tile achieves the same effect as shown in Figure 2-26.

With flat tiles, the first course may be raised with a special starter tile, as shown in Figure 2-27, or by a metal eaves closure, raised fascia, or wood cant strip. With a cant strip or raised fascia, a beveled wood or foam *antiponding strip* is required to prevent ponding of water along the eaves (Figure 2-29).

Ridges and Hips. Unless hip and ridge tiles are going to be set into a continuous bed of mortar, special nailers are required to install them. The hip and ridge boards

FIGURE 2-26 Spanish S-Tile—Typical Installation.

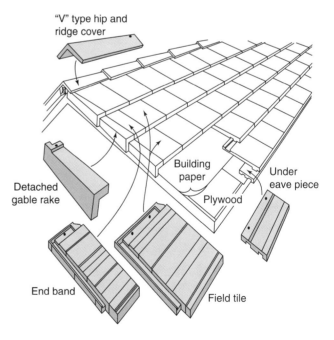

SOURCE: Adapted from *Architectural Graphic Standards, Residential Construction,* with permission of John Wiley & Sons, © 2003.

FIGURE 2-27 Flat Interlocking-Tile—Typical Installation.

SOURCE: Adapted from *Architectural Graphic Standards, Residential Construction,* with permission of John Wiley & Sons, © 2003.

are typically 2x3s to 2x6s set on edge to hold the trim tiles in an even plane. They are toenailed in place and individually wrapped with felt (Figure 2-30).

Hip and ridge tiles are later nailed on with a 2-inch head-lap, and the lower ends are sealed at the overlap with roofing cement or an approved tile adhesive. Finally, mortar, special trim tiles, or other weatherblocking is applied to fill in gaps between the ridge and hip tiles and the field tile.

Rakes. Rakes may be finished with detached gable-rake tiles (as shown in Figures 2-26 and 2-27) or with high-profile tiles, trimmed simply with half-round trim tiles as shown in Figure 2-31.

Flashings

Because of the longevity of a tile roof, high-quality flashing materials should be used. The International Residential Code calls for a minimum 26-gauge metal. Galvanized steel should have a minimum of 0.90 ounces of zinc per

square foot (G90 sheet metal). More expensive options include prepainted galvanized steel or 16-ounce sheet copper.

At Openings and Walls. At walls, dormers, chimneys, and other vertical surfaces, extend the flashing up at least 6 inches and counterflash. Extend flashing under the tile a minimum of 6 inches or as specified by the tile manufacturer. With flat shingles, use step flashing with a minimum 6-inch vertical leg and 5-inch horizontal leg with a hemmed edge. Profile tile along a wall should receive channel flashing turned up at least one inch on the lower flange (Figure 2-32).

Pipe Flashing. Pipe flashings generally get both a primary flashing when the underlayment is installed and a secondary soft-metal underlayment that conforms to the tile. For profile tile, this can be $2\frac{1}{2}$-pound lead or dead-soft aluminum with an 18-inch-wide skirt (Figure 2-33).

Valleys

According to the International Residential Code (IRC), valley flashing in tile roofs should extend at least 11 inches each way from the valley centerline, and the flashing should have a formed splash diverter at the center at least one inch high. The code requires a minimum underlayment at the valley of 36-inch-wide Type I No. 30 felt in addition to the underlayment for the general roof areas. In cold climates (average January temperature of 25°F or less), a self-adhering bituminous underlayment is recommended. Battens, if used, should stop short of the valley metal.

Tiles along the valley edge may be laid first and cut in place along a chalked line. Cut pieces are attached by roofing cement or a code-approved adhesive, or they may use wire ties, tile clips, or batten extenders.

Open Valleys. Open valleys permit free drainage and are recommended in areas where leaves, pine needles, and other debris are likely to fall on the roof. They are also recommended in areas subject to snow and ice buildup.

| FIGURE 2-28 | Eaves Closures for Profile Tile. |

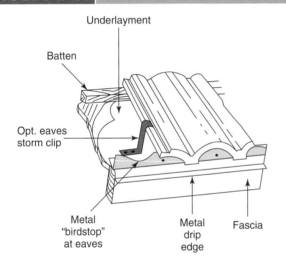

With profile tile, many contractors use preformed metal birdstop, which is contoured to fit under the first course of tile and lock into the tiles' weather checks.

| FIGURE 2-29 | Eaves Closures for Flat Tiles. |

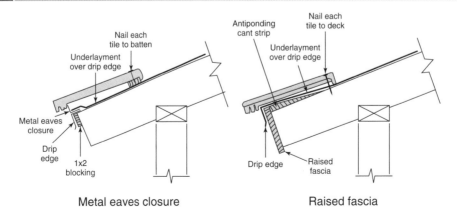

Metal eaves closure   Raised fascia

With flat tiles, use either a metal eaves closure (left) or a raised fascia with a beveled cant strip (right), to prevent ponding at the eaves. With a metal closure, no additional antiponding detail is required.

FIGURE 2-30   **Tile Ridge and Hip Boards.**

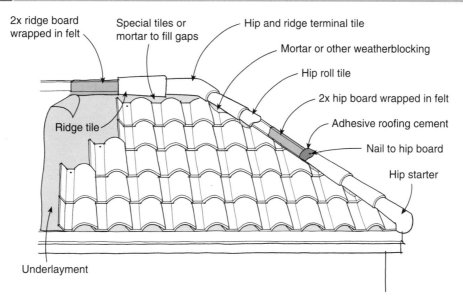

2x ridge board wrapped in felt

Special tiles or mortar to fill gaps

Hip and ridge terminal tile

Mortar or other weatherblocking

Hip roll tile

2x hip board wrapped in felt

Adhesive roofing cement

Nail to hip board

Hip starter

Ridge tile

Underlayment

Unless hip and ridge tiles are set in continuous mortar, 2x nailers are required to hold them in place. Hip and ridge tiles are later nailed at their heads and cemented where they overlap. Gaps between the hip and ridge tiles and field tiles are sealed with mortar, mastic, or special trim tiles.

FIGURE 2-31   **Tile Rake Trim.**

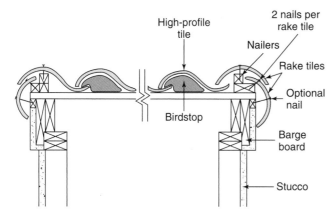

High-profile tile

2 nails per rake tile

Nailers

Rake tiles

Optional nail

Birdstop

Barge board

Stucco

With profile tiles, rakes may be trimmed with simple rolled tiles nailed into a nailer or barge board.

The valley flashing should have hemmed edges and be installed with cleats that allow individual sections to expand and contract (Figure 2-34).

Closed Valleys. In this type of valley, the flashing carries the runoff and the tile in the valley is only decorative. These are not recommended where debris from trees may fall on the roof or where the two roof planes joining at the valley have different pitches or length, causing uneven flows.

Foot Traffic

To prevent breakage, walk on tiles with extreme caution. Profile tile and lightweight tile are the most vulnerable,

FIGURE 2-32   **Channel Flashing for Tile Roofs.**

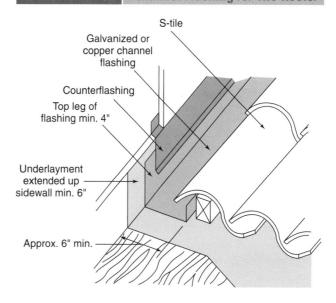

S-tile

Galvanized or copper channel flashing

Counterflashing

Top leg of flashing min. 4"

Underlayment extended up sidewall min. 6"

Approx. 6" min.

Where profile tile runs along a chimney, wall, or other vertical surface, place a channel flashing of 26-gauge galvanized steel or 16-ounce copper.

and concrete tiles are more fragile when they are freshly manufactured or "green." If possible, place antennas and other roof-mounted equipment where it is easy to access without crossing many tiles. When it is necessary to walk on tiles, step only on the head-lap (lower 3 inches) of each tile. With Mission- or S-tiles, it is best to step across two tiles at once to distribute the weight. When significant rooftop work is required, place plywood over the tile to distribute the load.

FIGURE 2-33 **Pipe Flashings for Tile Roofs.**

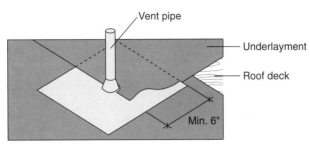

Step 1   Primary Flashing

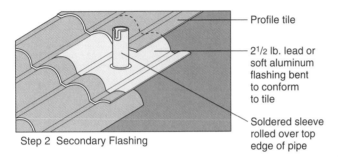

Step 2   Secondary Flashing

Vent pipes in tile roofs generally get a primary flashing, when the underlayment is installed, and a secondary soft-metal flashing that conforms to the tile surface.

Replacing Broken Tiles. If a tile is cracked, gently lift the overlapping tile and wiggle loose the damaged tile. Remove the tile nail, screw, or clip with a slate ripper or hacksaw blade. Seal any nail holes with roofing cement and slip a new tile into place, securing the butt end with an L-hook or bent copper wire (as shown in Figure 2-35).

METAL ROOFING

Residential installations of metal roofing have more than doubled in the past several years, and they are now estimated to account for over 10% of residential roofs. Originally associated with agricultural and commercial buildings, new metal roofing products aimed at the residential market are designed with simplified installation systems and offer more choices in materials, finishes, and design. The installed cost of premium metal roofing is three to four times more than asphalt shingles, but metal roofing offers a variety of attractive benefits:

- *Fire resistance:* Many metal roofs carry a Class A fire rating.
- *Low weight:* Most metal roofing products range from 125 to 175 pounds per square. Some lightweight aluminum shingles weigh as little as 40 pounds per square.
- *Wind resistance:* Many systems have earned a Class 90 wind-uplift rating, UL's highest rating.
- *Impact resistance:* Metal roofing systems offer moderate to excellent resistance to impact from hail, some earning UL's Class 4 rating.

- *Mold-resistant:* Metal roofing resists the type of algae and mildew growth that attacks asphalt and wood roofs.
- *Energy efficiency:* In a test conducted by the Florida Solar Energy Center, white metal roofing showed the greatest reductions in cooling loads of all roofing types, with 23 to 30% savings (compared to a control home with dark asphalt shingles).
- *Recycled content:* Many metal roofing products use recycled material, ranging from 25% with some steel products to over 90% with some aluminum modular shingles.
- *Longevity:* Metal roofs typically carry a 30-to 50-year warranty.

Noise Transmission. One frequently cited disadvantage of metal roofing is that it generates a noticeable noise when struck by rain, hail, or even dropping acorns. If installed directly to purlins with no roof sheathing, the noise might be heard in the building interior. However, when installed over a solid substrate, with normal levels of insulation, the noise should not be noticeably different than with other roofing types.

Walkability. Panels laid flat on solid decking are generally walkable. However, if panels are installed on battens, workers should be careful to step directly over battens or to use planking that spans multiple battens. Modular shingle panels generally use fairly light-gauge material, but it is stiffened somewhat by the stamped textures. In general, modular steel panels are walkable, but aluminum ones should be reinforced by foam inserts in sections expected to see a lot of foot traffic.

Minimum Slopes. Most metal roofing systems can be installed on slopes of 3:12 and greater and standing-seam systems from 2:12 and greater. Special standing-seam systems designed for slopes as shallow as $\frac{1}{2}$:12 require field crimping machinery and have sealant in all seams. The height of the ribs at seams and whether they are protected with a sealant affect how weathertight a roof will be under extreme weather.

There are three general types of residential metal roofing: Exposed-fastener panels, standing-seam, and modular panels.

Exposed-Fastener Panels

Steel and aluminum panel roofing with exposed fasteners has been a popular choice on agricultural buildings for decades. In recent years, these "ag panels" have grown increasingly popular for rural homes as well, since they can provide a long-lasting roof at a cost comparable to asphalt shingles. The products installed on homes, while essentially the same material as the agricultural panels, generally use better metal coatings, and installers pay more attention to sealing and watertight detailing.

FIGURE 2-34 **Open Valleys in Tile Roofs.**

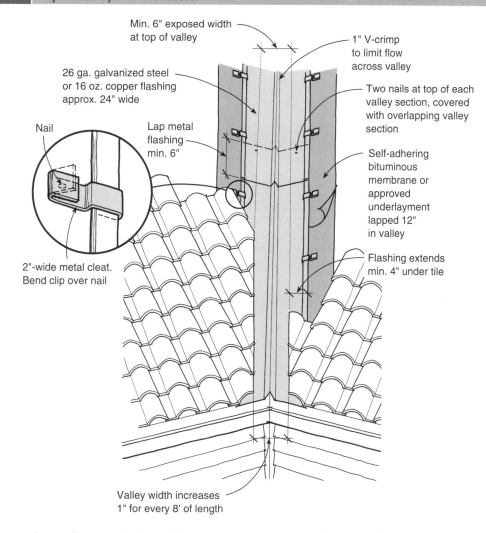

Min. 6" exposed width at top of valley

26 ga. galvanized steel or 16 oz. copper flashing approx. 24" wide

Nail

Lap metal flashing min. 6"

2"-wide metal cleat. Bend clip over nail

1" V-crimp to limit flow across valley

Two nails at top of each valley section, covered with overlapping valley section

Self-adhering bituminous membrane or approved underlayment lapped 12" in valley

Flashing extends min. 4" under tile

Valley width increases 1" for every 8' of length

Open valleys provide clear drainage and are recommended for areas subject to tree debris. Code requires fully waterproof underlayment and a flow area free of nail holes. Secure cut tile pieces with adhesive, clips, or wires.

FIGURE 2-35 **Replacing Damaged Tiles.**

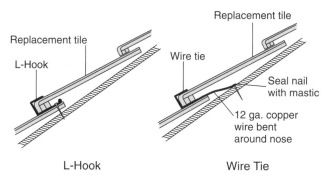

Replacement tile

Replacement tile

L-Hook

Wire tie

Seal nail with mastic

12 ga. copper wire bent around nose

L-Hook

Wire Tie

Gently lift the overlapping tile and twist loose the damaged tile. After filling any nail holes, slip in a new tile and secure with an L-hook (left) or bent copper wire (right).

While a carefully installed exposed-fastener roof should be free of leaks upon completion, small installation errors can result in leakage later as the metal panels undergo normal thermal movement that places stress on the fasteners. With so many exposed holes in the panels, periodic inspections are recommended. Also, the exposed fastener heads, in addition to lending a rural look to the building, tend to catch leaf debris and restrain sliding snow.

Materials. Exposed-fastener panels are typically 26 to 29 gauge, compared to the heavier 22 to 26 gauge used in standing-seam roofing. The ribs in exposed fastener roofing are also lower and closer together than in standing-seam roofing and may be squared, rounded, or v-shaped (see Figure 2-36). Most panels are 2 to 3 feet wide and formed with galvanized steel, Galvalume®, or aluminum.

• *Panel length.* While some stock sizes are available, ordering panels factory-cut to exact lengths simplifies installation and reduces corrosion at field cuts. Panels can be ordered in any shippable length, although excessive thermal movement can be a problem for steel panels longer than 40 feet or for aluminum panels

FIGURE 2-36 Exposed-Fastener Metal Panels.

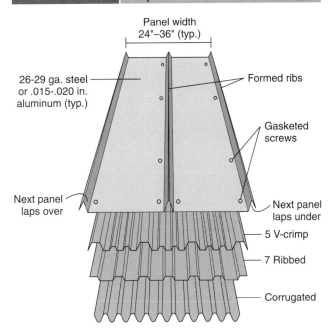

Panel width
24"–36" (typ.)

26-29 ga. steel
or .015-.020 in.
aluminum (typ.)

Formed ribs

Gasketed
screws

Next panel
laps over

Next panel
laps under

5 V-crimp

7 Ribbed

Corrugated

Exposed panel roofing typically uses light-gauge stock and stiffens it with square, round, or V-shaped ribs. Some patterns imitate standing-seam roofing, but most have more frequent ribs to provide greater rigidity.

longer than 16 feet. In regions with very wide temperature swings, contractors should use shorter lengths (see "Thermal Expansion" in Table 2-10, page 83).

Installation. While traditionally installed over battens, most panels in residential installations are now installed over a solid plywood deck with minimum No. 30 felt underlayment. Metal roofing manufacturers recommend plywood over oriented-strand board (OSB) due to plywood's better screw-holding ability. Roofing felt should be installed with plastic cap nails rather than metal buttons, which can deteriorate the metal roofing by galvanic action (see "The Galvanic Scale," page 83).

• *Align to eaves.* After installing drip edges and valley flashing, the first panel is fit along one rake, square to the bottom edge of the roof. If the roof is not square, the first panel may need to be cut at a bevel along the rake. Start at the downwind end of the roof, so the edge of each overlapping panel faces away from the prevailing winds.

• *Cutting panels.* Where panels need to be cut, use snips or shears rather than an abrasive blade, which overheats the steel coatings and leaves a rough edge prone to rust. Abrasive blades also produce hot metal filings that can embed in the paint and cause rust on the face of the panels.

• *Side and end laps.* After the first panel is screwed down, the next panel is set in place, lapping over the first. Side laps are typically sealed with butyl tape and

FIGURE 2-37 Gasketed Screws.

Too Tight   Too Loose   Correct

Exposed-fastener panels are vulnerable to leaks if screws are installed poorly or if excessive thermal movement loosens the screws or stretches the screw holes. To minimize leaks, place screws in the flat sections between ribs, and drive screws straight and snug but not overtight. Also, do not exceed the maximum panel length recommended by the manufacturer.

held together with gasketed sheet-metal screws. Where more than one panel is used up the run of the roof, the upper panel laps the lower by 6 inches and is sealed with butyl tape.

• *Fasteners.* Fasteners are typically special wood screws with integral EPDM or neoprene gaskets that compress under the screw head to seal the hole. Fasteners should be driven at a right angle to the roof plane and should be snug but not so tight as to deform the washer (see Figure 2-37). Nearly all manufacturers recommend placing screws in the flat sections between ribs. Although making holes in the flat section may seem unwise, placing screws in the ribs is discouraged for two reasons. First, the long exposed screw shaft passing through the rib is prone to snap over time due to thermal movement of the panels. Second, it is easy to overdrive the screws and crush the panels. Higher-cost EPDM washers are less likely to leak than neoprene.

Reroofing. Panels can go directly over a single layer of asphalt shingles in good condition. If the shingles are curled or uneven, install 2x horizontal purlins at 16 inches on-center. In either case, put down a new layer of No. 30 underlayment before installing the panels.

Flashings and Accessories. Most manufacturers supply preformed flashings, drip edges, rake moldings, and ridge caps color-matched to their roofing panels, as well as color-matched coil stock for fabricating custom pieces on-site. They also provide rubber closure strips or expandable foam tapes to seal panel ends against water and insect intrusion at eaves, valleys, ridges, and other terminations.

Pay particular attention to panel ends at valleys. Some manufacturers supply special closures for the angled cuts through ribs, but closures may need to be fashioned by cutting up standard closure strips. Some manufacturers also provide an expandable foam sealant tape that conforms to the rib pattern for a tight seal up the valley. Depending on the panel profile, the end treatment will vary, but ends should be fully sealed. Remember to place screws in flat sections and to use extra screws up the valley (Figure 2-38). For a vented ridge, place short sections of a matrix-type ridge vent between the ribs and secure with a preformed metal cap (Figure 2-39).

FIGURE 2-38 | **Valley Flashing for Exposed-Fastener Panels.**

Section View

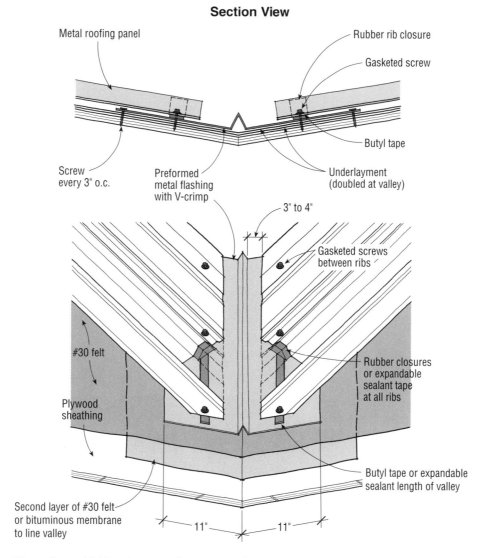

Metal roofing panel

Rubber rib closure

Gasketed screw

Butyl tape

Screw every 3" o.c.

Preformed metal flashing with V-crimp

Underlayment (doubled at valley)

3" to 4"

Gasketed screws between ribs

#30 felt

Plywood sheathing

Rubber closures or expandable sealant tape at all ribs

Butyl tape or expandable sealant length of valley

Second layer of #30 felt or bituminous membrane to line valley

11"   11"

Line valleys with bituminous membrane or two layers of No. 30 roofing felt. To seal the open ribs facing into the valley, manufacturers provide self-adhesive neoprene closures or an expandable asphalt-impregnated tape that conforms to the profile of the roofing.

FIGURE 2-39 | **Vented Ridge for Exposed-Fastener Panels.**

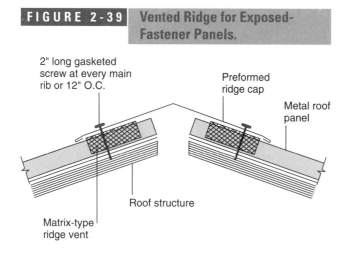

2" long gasketed screw at every main rib or 12" O.C.

Preformed ridge cap

Metal roof panel

Roof structure

Matrix-type ridge vent

For a vented ridge, use a preformed ridge cap with short sections of matrix-type ridge vent installed between the ribs.

For plumbing vents, most manufacturers recommend a moldable aluminum jack bent to conform to the profile of the roofing (Figure 2-40).

Rectangular openings, such as skylights and chimneys, typically require both base and counterflashing so roof panels are free to move with changes in temperature. Depending on the panel profile, either use a pan flashing or an L-flashing sealed to the top surface of the roofing panel with sheet metal screws and butyl tape. On large openings, a cricket is needed on the upslope to divert water around the penetration. Custom-made, one-piece curbs with built-in diverters simplify this type of installation. All flashing joints should be sealed with butyl tape or a manufacturer-recommended sealant.

Sealing. For the watertight performance required on homes (as opposed to barns), metal roofs need careful

FIGURE 2-40 Plumbing Vents for Metal Roofing.

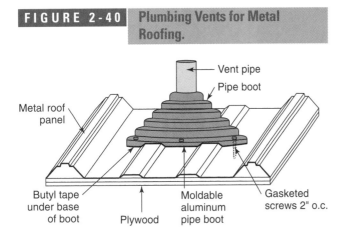

Vent pipe
Pipe boot
Metal roof panel
Butyl tape under base of boot
Plywood
Moldable aluminum pipe boot
Gasketed screws 2" o.c.

At plumbing vents, most exposed-panel manufacturers recommend a surface-mounted soft aluminum jack that molds to the profile of the roofing.

sealing around all penetrations, side laps, and end laps. On side seams and lap joints, the sealant should always go on the uphill, or "dry," side of any fasteners (Figure 2-41). Sealant should also be used at ridge caps, valleys, and wherever flashings lap over or under the metal roofing.

The preferred sealant for most concealed seams in roof panels is butyl tape, which absorbs movement and will not shrink. Gunnable terpolymer butyl or urethane caulk can also be used, as specified by the manufacturer. But never use acid-cure silicone caulking (the common type with vinegar odor) or asphalt roofing cement, as they will damage most metal coatings.

Panel Movement. Metal panels were originally designed for installation on purlins that can absorb the normal movement as the panels expand and contract from temperature changes. The thermal movement of a long panel installed over solid plywood, however, can cause problems. Typically, either the hole in the roofing elongates—creating a potential leak—or the screw becomes loosened, making the roof vulnerable to blow-off. The problems are greatest with aluminum, which has 70% more thermal movement than steel and less tensile strength. To avoid problems, experts recommend the following:

- With exposed-fastener panels, avoid lengths over 40 feet for steel or 16 feet for aluminum—less for climates subject to wide temperature swings. Break the run into two panels.

- On long runs of painted roofing, choose lighter shades, preferably white.

- Use screws in the flat part of the panel, not on the ribs. Screws should penetrate the sheathing fully, plus $\frac{1}{4}$ to $\frac{1}{2}$ inch.

- Where leak-free performance is critical, fasten the roofing to Z-shaped metal purlins screwed horizontally across the plywood sheathing. Or switch to a concealed fastener system.

Oil-Canning. Thermal expansion in light-gauge metal panels can cause a wavy appearance called "oil-canning" in the flat areas. In general, this does not signal a performance problem, but it may be visually objectionable. Oil-canning tends to be most visible in bright light from a close distance, and it is generally more noticeable on shiny metals, such as Galvalume®, than on colored metal panels. It is primarily a problem in profiles with few ridges to stiffen the panels. To reduce the effect, some manufacturers provide self-adhesive foam strips that are attached lengthwise to the bottom of metal panels.

Standing-Seam Roofing

Standing-seam roofing consists of individual panels that run the length of the roof with a high rib up each side of the panels. The ribs overlap and lock together, concealing the fasteners and giving the roofing its name. The hidden fasteners allow thermal movement in the panels and are less likely to leak than exposed fasteners. However, some trim pieces are still fastened with exposed screws.

The smooth surface of a standing-seam roof provides a cleaner appearance and is easier to keep clear of leaf debris than tile, wood, or other textured roofing surfaces. Also, it can be walked on when necessary. Snow slides off easily as well, making this a popular choice in high snow regions. The cost is generally 25% to 50% more than an exposed-fastener roof of similar materials.

Materials. Standing-seam panels are 8 to 24 inches wide and available in steel, copper, and aluminum with a wide array of finishes (discussed below). Stiffening ribs may be added to wider panels to reduce waviness (oil-canning). Thicknesses for quality residential applications are typically 24 or 26 gauge, but lighter and heavier stock is also available. Installers can form panels on-site from coil stock with portable roll-forming equipment, or they can order factory-made panels from a growing number of metal roofing manufacturers. Most factory-made panels have snap-together seams, eliminating the need for special crimping equipment used by site fabricators. In most cases, panels are fabricated to run from eaves to ridge, eliminated the need for end lap joints.

- *Clips vs. flange.* Standing-seam panels either have an integral screwing flange (through-fastener panels) or are installed with clips placed 20 to 24 inches on-center (Figure 2-42). Clip systems are more costly to manufacture and to install, but they have better wind resistance and a higher water-lock at the seams. Also, because the clips allow unlimited panel movement, panels can be fabricated to any length. The flange type should be limited to 40 feet for steel and 20 feet for aluminum for normal climate conditions.

- *Site vs. factory fabrication.* For those with the equipment, site fabrication provides flexibility and saves on shipping costs, which can be high. Site fabricators can

FIGURE 2-41 **Exposed-Fastener Metal Roofing—Typical Details.**

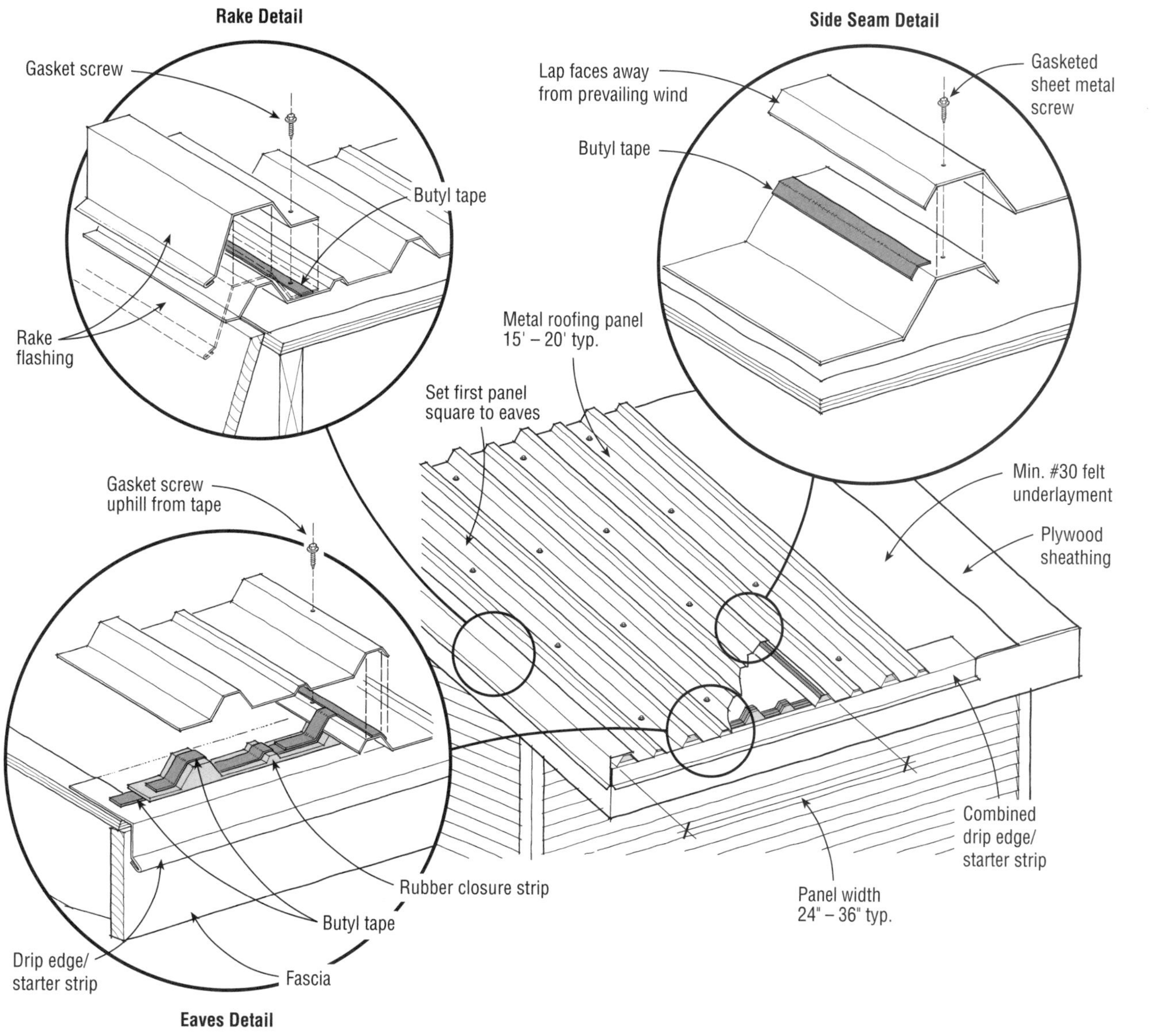

Rake Detail

Gasket screw

Butyl tape

Rake flashing

Side Seam Detail

Lap faces away from prevailing wind

Gasketed sheet metal screw

Butyl tape

Metal roofing panel 15' – 20' typ.

Set first panel square to eaves

Min. #30 felt underlayment

Plywood sheathing

Gasket screw uphill from tape

Rubber closure strip

Butyl tape

Drip edge/ starter strip

Fascia

Eaves Detail

Combined drip edge/ starter strip

Panel width 24" – 36" typ.

For profiles with deep ribs, manufacturers provide rubber closure strips to seal ends at eaves, valleys, and other terminations. Butyl tape is the preferred sealant for panel laps, joints, and flashings. Position panels carefully on the tape, as they are difficult to adjust afterward. Note that screws go either through the sealant or on the "dry side" of the butyl tape.

also produce matching flashings and accessories to match the specific needs of the job. Factory-made panels, on the other hand, offer consistent quality, as well as preformed flashings and fittings that simplify installation. Using factory-produced panels, however, requires detailed planning since every piece of roofing must be preordered to length.

Installation. On new homes, most panels are installed over a solid plywood deck with minimum No. 30 felt underlayment. Metal roofing manufacturers recommend plywood rather than OSB due to plywood's better screw-holding ability. Install the felt with plastic cap nails rather than metal buttons, which can cause corrosion when in contact with the roofing panels (see "Galvanic Corrosion," page 83).

After installing the drip edge, install the first panel, making sure it is square to the bottom edge of the roof. If the roof is not square, pull the panel away from the rake so the first rib does not overhang the rake edge. Later, the rake trim piece will cover any small discrepancies. If the panels have an integral screw flange, keep the screws just snug so the panels can move with temperature changes. The clips are designed to allow thermal movement.

The next panel fits over the first with an overlapping rib. Fit each panel to a line snapped up the roof, marking the edge of each panel. Without layout lines, the panels can build up an incremental error, throwing off the layout.

FIGURE 2-42 Standing-Seam Fasteners.

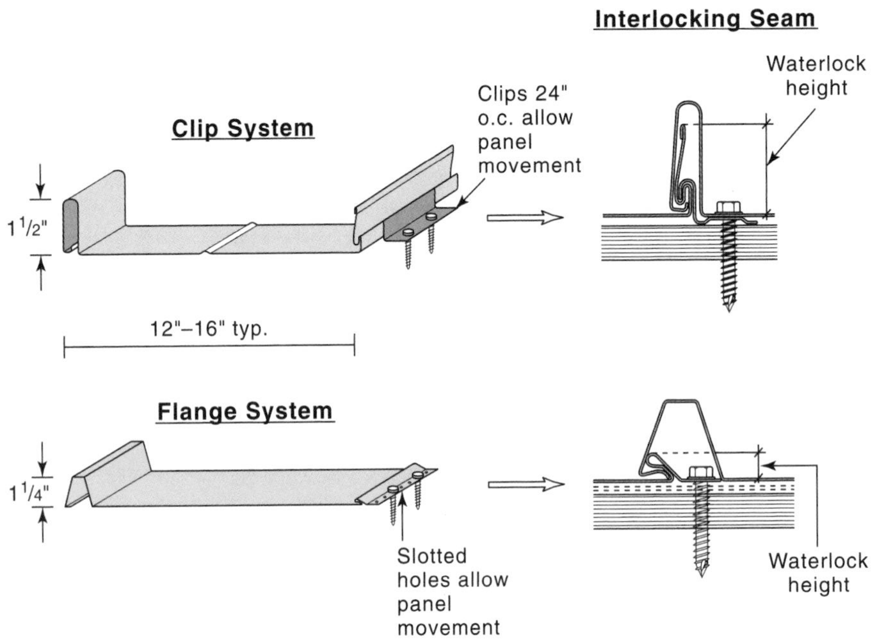

Factory-made standing-seam panels are secured with either clips (top) or an integral nailing flange (bottom). Both systems interlock at seams, using hand pressure. Clip systems are more costly but have better wind resistance and a higher water lock. Also they can better accommodate thermal movement, allowing longer panel lengths.

As panels are installed and secured, the joints are easily locked together with hand pressure. Traditional standing-seam roofing required special motorized crimpers to lock the seams. While these are still used on some low-slope systems, most residential installations now use snap-together panels. Unless the layout works perfectly, the last panel will need to be cut along the opposite rake and bent with a hand seamer to form the end rib.

Reroofing. Many installers will not install standing-seam roofing over existing asphalt shingles since the rough surface will tend to bind the panels and cause "oil-canning," as the panels move with temperature changes. One option is to install the new metal roofing over 2x4 purlins nailed through the old roofing and shimmed to form an even plane. Follow manufacturer's recommendations for spacing of purlins, typically no more than 24 inches on-center.

Flashing and Sealing. Manufacturers of preformed roofing panels provide eaves and rake flashings, ridge caps, and sidewall flashings in matching finishes, as well as coil stock for site fabrication. Many flashings are designed with hidden fasteners; others require exposed gasketed screws. Typical details are similar to those found in Figure 2-41. Follow manufacturers' recommendations regarding which sealants to use for compatibility with the roofing (typically butyl tape, or gunnable terpolymer butyl or urethane sealant). In general, avoid acid-cure silicone

(the type that smells like vinegar) as it can be corrosive to many metal finishes.

MODULAR SHINGLES

Modular metal shingles comprise the fastest growing segment of the metal roofing industry. Using light-gauge steel, copper, or aluminum, panels are stamped to imitate slates, shakes, asphalt shingles, or tiles. Some have aggregate stone finishes that closely resemble asphalt shingles. Most carry warranties from 20 to 30 years against fading and from 50-year to "lifetime" warranties against cracking or delamination of the shingle itself.

Modular shingles carry a Class A or B fire rating, depending on the material and installation details, and are highly resistant to wind uplift and damage from hail. Installed prices range from two to three times the cost of premium asphalt shingles. Installers accustomed to asphalt shingles or tile should have little trouble adjusting to metal shingles.

Materials. Modular shingles are typically stamped from lightweight .0165-inch metal, which is thinner than other types of metal roofing but stiffened by the textured patterns. Typical rectangular panel sizes range from 24 to 48 inches long by 12 to 16 inches wide, but they also include tile and diamond shapes and other specialty patterns. Weights range from 40 pounds per square for aluminum shingles to 140 pounds per square for steel shingles with a heavy stone aggregate. The lightweight patterns are well-suited to reroofing where weight is a concern. Most panels can be walked on, if done with care, but areas with heavy foot traffic should be reinforced with foam backers provided by the manufacturer.

Installation. Modular shingles are either nailed directly to the wood deck or attached to 2x2-inch battens installed at the exposed panel width, usually about 15 inches. Installation on battens allows more deeply etched patterns, such as simulated tiles. Either type can be installed with pneumatic nailers.

Underlayment is minimum No. 30 asphalt felt held with plastic caps to avoid contact between incompatible metals. Many manufacturers recommend proprietary laminated underlayments, such as VersaShield (Elk Premium Building Products, Inc.), which are tougher and less slippery than felt and provide better fire ratings. Aluminum shingles require fire-resistant underlayments to achieve an A or B fire rating.

- *Direct to deck:* Shingle panels installed directly to the deck are attached with concealed nails, either through clips or a nailing flange along the top, and have interlocking edges along all four sides (Figure 2-43). As they are installed, each panel locks to the panel below and to the left.

- *Over battens:* Modular panels designed for installation on battens have a nailing flange along the bottom of each shingle panel with nails going horizontally into the batten (Figure 2-44). Battens are useful for retrofits where the surface is irregular. Also, the air space boosts energy savings, especially when using shingles with solar-reflective surfaces.

FIGURE 2-43 **Modular Metal Shingles Installed Direct-to-Deck.**

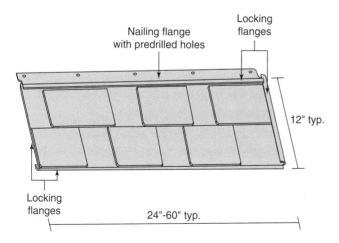

Shingle panels installed directly to the deck are attached with concealed nails either through clips or an integral nailing flange. Panels interlock along all four sides, providing excellent wind resistance.
SOURCE: Photo courtesy *of Accel Roofing Products.*

FIGURE 2-44 **Modular Metal Shingles to Battens.**

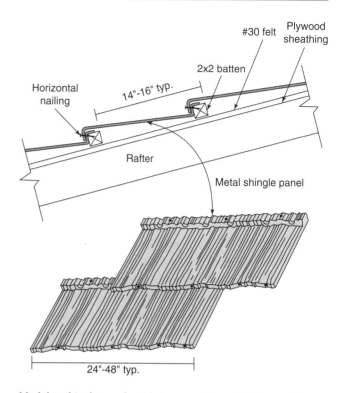

Modular shingle panels with deep profiles typically install over battens. Nails are driven horizontally into the batten through predrilled holes at the bottom lip of each shingle. The air space created boosts energy savings, especially when combined with new solar-reflective finishes.

Both systems begin with the installation of a drip edge and gable trim designed for the specific system. Working from left to right, the first shingle panel hooks into the drip edge, which also serves as a starter strip. Successive courses are staggered as specified by the manufacturer.

Reroofing. In general, most modular shingles can be installed over existing asphalt shingles if they are in good condition without excessive curling and deformation. Shingles designed to go over battens have more flexibility, since the battens can be shimmed to create a level surface.

Flashing and Sealing. Manufacturers provide standard flashings similar to those for standing-seam products. Eaves and rake flashings typically have concealed fasteners and lock the shingles in place. Ridge and headwall flashings often require exposed fasteners. Depending on the shingle profile, sidewall, chimney, and skylight flashings are either pan or step flashings. Typical details are shown in Figure 2-45.

Metal Choices

While some companies offer roofing products in copper, zinc, and stainless steel, the vast majority are coated steel

FIGURE 2-45 **Modular Metal Shingles— Typical Details.**

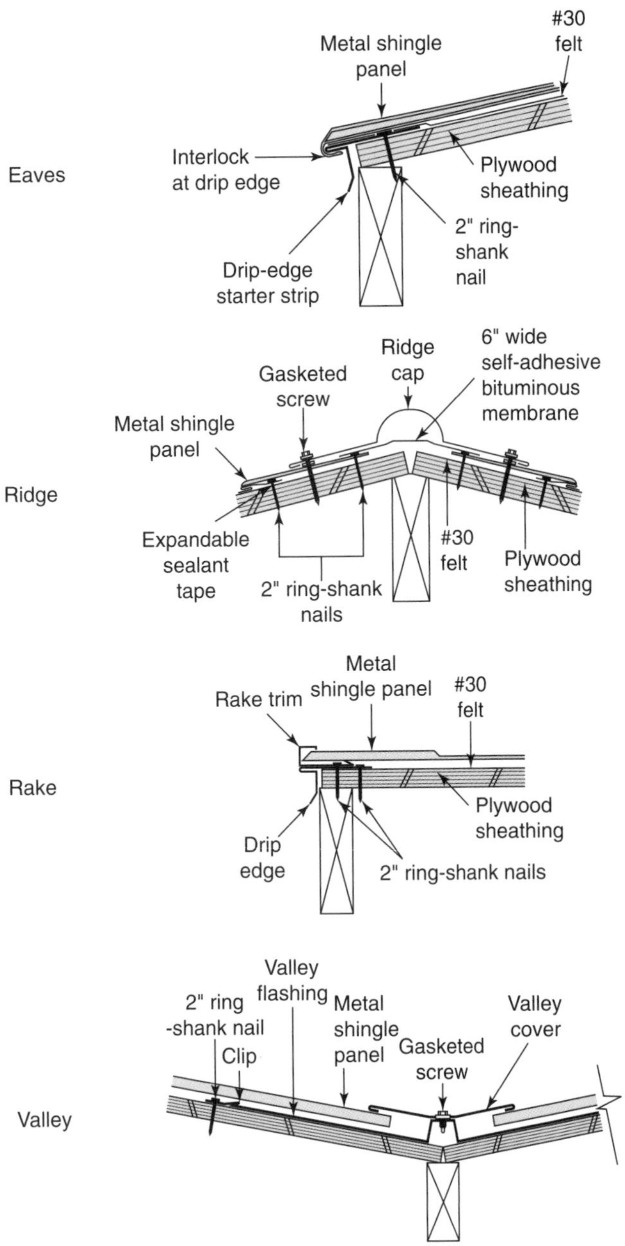

Modular metal shingle systems include prefinished flashings, sealing tapes, and accessories to handle most common details. For some situations, installers may need to bend prefinished coil stock on site.

and aluminum. Coated steel products are the most common and least expensive. In its favor, steel moves relatively little with temperature changes, has good structural characteristics, and resists denting. Its high melting point gives it a Class A fire rating. All coated steel materials, however, are vulnerable to corrosion at field-cut edges—although Galvalume® is the least affected (Table 2-10).

Galvanized Steel. To protect against corrosion, the steel is bonded to a layer of zinc, which works as a sacrificial coating on the surface and also offers some protection to cut edges and nicks by flowing to these areas. The heavier the zinc coating, the longer the protection. The Metal Roofing Alliance recommends G-90 galvanized steel for roofing, which has 90 ounces of zinc per square foot. Unpainted G-90 galvanized steel is typically warranted against corrosion for 20 years under normal conditions. It often lasts longer, but it may show visible corrosion in as few five years under harsh conditions, such as salt spray, significant air pollution, or low-slope applications in wet climates. Field cuts made with an abrasive saw are prone to corrosion.

Aluminized Steel. Developed in the 1950s, this is similar to galvanized steel, but it uses aluminum as the coating instead of zinc. The aluminum provides a physical barrier against corrosion and creates a reflective surface that helps reduce heat transfer to attics. However, aluminum does not have the self-healing properties of zinc, so exposed edges and scratches are more susceptible to rust. Aluminized steel generally outlasts galvanized steel but has largely been replaced in the market by Galvalume®.

Galvalume. Also sold under the tradenames Zincalume® and Galval®, Galvalume® was developed in the early 1970s. The underlying steel is coated with a zinc-aluminum alloy that combines the long-lasting protection of aluminum with the self-healing properties of zinc. It also has the reflective qualities of aluminum, reducing attic temperatures and cooling loads. The most common application weight is AZ 55, which has about a 1-mil-thick coating on each side. Unpainted Galvalume® is warranted against corrosion for 20 years, but it has stood up well in weathering tests for 30 years and is projected to last up to 40 years under normal conditions. Cut edges hold up very well, but cutting the material with an abrasive blade is discouraged as the filings will mar the surface. Galvalume® costs about 10% more than standard galvanized steel.

Aluminum. Aluminum that is anodized or painted is highly resistant to corrosion, making it well-suited to coastal environments (although lightweight aluminum flashings tend to pit and oxidize in salty air). Its light weight is an advantage in reroofing. Because of its high coefficient of expansion, however, attachment systems must be designed to accommodate the movement of long panels. And since it has a lower tensile strength than steel, more fasteners may be required to achieve wind ratings comparable to a steel roof. Also, aluminum has a low melting point so it relies on two layers of fire-resistant underlayment, such as VersaShield, to get a Class A fire rating. Most aluminum used in roofing has a baked-on paint finish rather than an anodized finish. Although anodized aluminum is less costly, new paint technologies such as Kynar® and Hylar® carry better warranties and are available with a low-gloss finish generally favored on roofs. Some coated aluminum products come with transferable lifetime warranties.

TABLE 2-10 | Metal Roofing Characteristics

Material	Advantages	Drawbacks	Incompatible Materials	Longevity*	Thermal expansion** (10^{-6} in/in/°F)
Galvanized steel	Least expensive. Strong and dent-resistant. Zinc coating heals small cuts and scratches.	Rusts after zinc wears away from oxidation. Field-cut edges vulnerable to corrosion.	Brass, bronze, untreated iron and steel, redwood, cedar, pressure-treated (PT) lumber.	Unpainted: 15 to 30 years. Exposed to salt spray: 5 to 10 years.	7.5
Aluminized steel	Provides a true barrier to corrosion rather than sacrificial coating.	Cuts and nicks not self-healing and prone to corrosion.	Brass, bronze, lead, copper, wet mortar, redwood, cedar, PT lumber, graphite (e.g., pencil marks).	Unpainted: 20 to 40 years.	7.5
Galvalume	Combines barrier protection of aluminum with healing characteristics of zinc. Reflects solar radiation.	Field-cut edges vulnerable to corrosion in coastal areas.	Lead , copper, unprotected steel, wet mortar, PT lumber, and graphite.	Unpainted: 30 to 40 years.	7.5
Aluminum	Superior corrosion resistance. Lightweight. Good for coastal areas.	Expensive. High level of thermal expansion. Relatively soft. Low melting point.	Brass, bronze, lead, copper, unprotected iron and steel, wet mortar, redwood, cedar, or PT lumber, and graphite.	Unpainted: 30 to 40 years	12.7
Copper	Easily roll formed. Superior corrosion resistance. Attractive green patina.	Very expensive. Greenish runoff can stain building. Avoid contact or runoff from cedar shingles.	Aluminum, stainless steel, zinc, unprotected iron and steel, galvanized steel, lead, brass, bronze.	60+ years	8.8
Zinc	Easily formed into intricate patterns. Superior corrosion resistance. Bluish-white patina.	Very expensive. Runoff can stain building.	Brass, bronze, copper, untreated iron and steel, stainless steel, redwood and cedar.	60+ years	15.1

* Longevity is affected by many variables, including slope of roof, wet vs. dry climate, air pollution, and exposure to salt spray.

**Average values. To find predicted expansion in inches, multiply the length of the metal (inches) times the change in temperature (°F) times the number in the chart. Divide the answer by 1,000,000.

Copper. This high-end material is highly resistant to corrosion and easily formed into panels. Copper roofs have been known to last for over a century and are a common sight on churches and historic buildings. Left unfinished, the material will oxidize to the familiar green patina that protects the underlying metal. In arid areas, the color may be more reddish-purple. Special clear acrylic coatings can be applied that will help copper retain its original color. One concern is that runoff from a copper roof can stain building components below if not managed with gutters. Also, premature failure of copper flashing and roofing has been linked to acid rain and runoff from cedar shingles (see "Flashings" under "Wood Shakes and Shingles," page 92).

Clients interested in copper should consider a newly developed proprietary sheet metal called Suscop™, which has copper plating over a stainless-steel core. The material combines the strength and durability of steel with the natural patina of real copper . Because of its greater strength, a lighter-weight sheet (0.4mm) can be used in place of 16-ounce copper, significantly reducing material costs.

Zinc. Zinc roofs are similar to copper in their durability but weather to a bluish-white color rather than green. The material is very malleable and can be formed into intricate patterns for metal shingles.

Galvanic Corrosion

With metal roofing or any metal building components, the safest strategy is not to mix metals that come in direct contact with one another. Use aluminum flashing and fasteners with aluminum roofing, copper flashing and copper nails with copper roofing, etc. When this is not possible, choose a second metal that is not likely to lead to galvanic corrosion or use a physical barrier to separate the two metals.

The Galvanic Scale. The galvanic scale (see Table 2-11) ranks a metal's tendency to react in contact with another metal in the presence of an electrolyte, such as water or even moisture from the air. Metals at the top of the chart are called *anodic*, or active, and are prone to corrode;

metals at the bottom are *cathodic,* or passive, and rarely corrode. The farther apart two metals are on the chart, the greater their tendency to react and cause corrosion in the more active metal. Metals close to each other on the scale are usually safe to use together.

The Area Effect. The rate of corrosion is controlled by the area of the more passive metal. For example, a galvanized steel nail (active) will corrode quickly if surrounded by a large area of copper flashing (passive). If a copper nail is used in galvanized steel flashing, however, the corrosion of the steel will be slow and spread over a large area, so it may not be noticeable. In each case, the active metal corrodes, and the passive metal is protected.

Galvanic Corrosion of Roofing. Because they are made from active metals, aluminum and zinc roofing panels, as well as steel roofing with aluminum and zinc coatings (galvanized steel, Galvalume®, etc.), are vulnerable to galvanic corrosion if allowed to come in contact with more passive metals. For example, never use copper or lead flashings with aluminum, zinc, or galvanized roofing materials. Even water dripping from a copper pipe,

flashing, or gutter can lead to corrosion of coated-steel or aluminum roofing materials. How common flashing materials react with metal roofing and other metal building materials is shown in Table 2-12.

Where incompatible metals must be used in close proximity, use the following precautions:

- Separate the two dissimilar metals with building paper, bituminous membrane, durable tapes, or sealants so they are not in direct contact.
- Coat the cathodic (less active) metal with a nonconductive paint or bituminous coating.
- Avoid runoff from a cathodic metal (e.g., copper gutters) onto an anodic metals (such as galvanized steel).

Other Incompatible Materials

In addition to galvanic corrosion, a number of other common building materials can harm the finishes on metal roofing or lead to etching or corrosion of the material itself:

Wet Mortar. Aluminum roofing materials and aluminum-based coatings can be damaged by alkali solutions such as wet mortar. Where contact with wet mortar cannot be avoided, one option is to spray the metal with lacquer or a clear acrylic coating to protect it until the mortar is dry.

Pressure-Treated Wood. Roof panels treated with aluminum and zinc coatings should not come into direct contact with pressure-treated (PT) wood, which can damage the finish and accelerate corrosion.

Sealants. Use only sealants recommended by the manufacturer. Never use acid-cure silicones (the most common type, with a vinegar smell) or asphalt roofing cement with coated-steel roofing, as these will mar the finish. Commonly recommended products include butyl tape and gunnable terpolymer butyl or urethane sealant.

Salt Spray. Saltwater spray is very hard on metallic-coated–steel products and may lead to corrosion within

TABLE 2-11	The Galvanic Scale
Most anodic or active (likely to corrode)	Zinc
	Aluminum
	Galvanized steel
	Mild steel, cast iron
	Lead
	Tin
	Brass, bronze
	Copper
	Silver solder
	Stainless steel (passive)*
	Silver
Most cathodic or passive (protected from corrosion)	Graphite
	Gold

*Most stainless steel used in light construction is passive, typically Type 304. Type 316 is recommended for exposure to salts or saltwater.
Note: Avoid placing dissimilar metals in direct contact unless they are close together on the galvanic scale.

TABLE 2-12	Galvanic Corrosion Potential Between Common Metals						
	Zinc	Alum.	Galvanized Steel	Iron/Steel	Lead	Brass, Bronze	Copper	Stainless Steel (passive)
Zinc	—	low	low	high	low	high	high	high
Aluminum	low	—	low	medium	medium	high	high	low
Galvanized steel	low	low	—	medium	low	medium	medium	medium
Lead	low	medium	low	low	—	medium	medium	medium
Copper	high	high	medium	high	medium	medium	—	high
Stainless steel	high	low	medium	medium	medium	high	high	—

Low: No significant galvanic action is likely to occur.
Medium: Galvanic corrosion may occur under certain conditions or over a long period of time.
High: Galvanic corrosion is likely so avoid direct contact.

5 to 7 years. In these areas, the best choices are copper, stainless steel, or painted aluminum. Hylar/Kynar® finishes hold up best.

Paints and Coatings

While unpainted metal roofs are common on utility buildings and some rustic homes, most homeowners prefer a painted surface. In addition to improving the appearance, a high-quality factory finish can significantly extend the life of metal roofing. In general, factory finishes are durable and flexible enough to tolerate factory roll-forming and bending on-site. The best finishes carry decades-long warranties against cracking and peeling, and "excessive" chalking and fading (as defined by the manufacturer). The quality of the finish is determined by the type of resin and the stability of the pigments.

Polyester. Polyester-resin paints are the least expensive and are commonly used on exposed-fastener panels. These have a medium to high gloss when applied, but they will fade significantly within 5 to 7 years on surfaces exposed to direct sun. Bright red, for example, may fade to pink. Fading will be less noticeable on light colors, making them a better choice. Warranties are typically for 3 to 5 years and rarely cover fading or chalking.

Silicone-Modified Polyesters. SMPs (silicone-modified polyesters) use polyester resins blended with silicone additives to improve performance. In general, the higher the silicone content, the more durable the finish. These are available in medium- and high-gloss colors, and they resist fading and chalking much better than standard polyester paints. Warranties against excessive fading and chalking typically run from 10 to 20 years, depending on the formulation.

Fluoropolymers. Based on a fluorocarbon-based resin called PVDF, these are the most technically advanced and most expensive finishes. Sold under the trade names Kynar 500® and Hylar 5000®, fluorocarbon-based paints provide a smooth and dense medium-gloss finish that offers excellent durability and long-lasting resistance to fading and chalking, even under intense sun exposures. The Teflon-like coating also resists dirt retention and holds up better in coastal environments than other finishes. The finish is softer than SMPs, however, and can be damaged by the roofing installers, if they are not careful. Typical warranties run 20 years or greater, with 10- to 20-year protection against excessive fading.

Reflective Finishes. White metal roofs can reduce cooling loads by as much as 30%, according to tests conducted by the Florida Solar Energy Center. More modest savings are now available with dark colors as well by using metal shingles coated with special paints formulated to selectively reflect the sun's infrared and ultraviolet radiation. These "Hi-R" paints are now standard options with Hylar/Kynar® finishes. Tests indicate that aluminum shakes with a reflective brown finish reject 30% to 40% of the total solar radiation compared to 67% for a white metal roof.

Granular Coatings. Some metal shingles are available with a textured finish consisting of crushed stone or ceramic granules blended into an acrylic resin. These are applied over a special primer and sealed with a clear acrylic sealer. The multicolored granules give the appearance of an asphalt shingle and protect against scratching from foot traffic. The finishes also help protect against denting from hail and help conceal any small dents.

WOOD SHAKES AND SHINGLES

Wood shakes and shingles are traditional American roof coverings dating back to Colonial times. They remain popular in many coastal areas and are common or even mandated in certain historic districts. Traditionally, wood roofs were laid on spaced sheathing, which provided good ventilation around the shingles and contributed to a service life of 30 years or more. New wood roofs set on solid sheathing have been known to fail in 10 years or less unless the installer takes adequate precautions to allow for good drainage and drying of the wood roofing materials. With installed costs of over $600 per square for premium materials, it is important to design a roof that will last.

Materials

Wood shakes and shingles soak up water through their end grain, dry unevenly in the sun, and slowly erode on the surface from a combination of ultraviolet radiation, wind, and precipitation. In humid conditions, wood shingles may become a breeding ground for moss, lichen, and decay fungi. To survive those harsh conditions, wood roofing should be made from a durable wood species that is either naturally decay-resistant or pressure-treated.

Wood Species. The most commonly used wood on roofs today is western red cedar. The heartwood of red cedar is rich in extractives that provide natural decay resistance. Eastern white cedar also has good decay resistance and is commonly used on the East Coast. However, white cedar is typically flat-sawn and has a mix of heartwood and sapwood, making it less durable on a roof and more prone to cupping and splitting. Other less common species with good track records are Northern white cedar, Alaskan yellow cedar (actually a cypress), and white oak.

Whatever species is selected, use the best grade available. With red cedar and other decay-resistant species, the heartwood is far more decay-resistant than the sapwood.

Edge-grain wood is more stable and less prone to cupping and splitting than less expensive flat-grain wood. The best choice for wood roofing is all-heart, edge-grained shakes or shingles.

Grades. Make sure the lumber to be purchased has been graded under the authority of a recognized grading agency such as the Cedar Shake and Shingle Bureau for red cedar or the Southern Pine Inspection Bureau for yellow pine. A blue label on the packaging, for example, may simply be a marketing tactic and does not necessary indicate that the shakes or shingles are certified as Grade 1.

Warranties. If installed in accordance with the Cedar Shake and Shingle Bureau's specifications by a certified installer, the CSSB will guarantee wood roofing for 20 to 25 years, depending on the thickness of the shake or shingle. Some pressure-treated shakes and shingles carry warranties of up to 50 years.

Preservative Treatment. If premium red or white cedar is too expensive, consider pressure-treated southern yellow pine shakes and shingles. In its favor, yellow pine is a tougher and stronger wood, and although not as pretty as red cedar when new, over time they will both weather to a similar silver gray. Because penetration of the treatment is nearly 100%, pressure-treated pine shingles carry guarantees against decay for up to 50 years, making them well-suited to high-moisture environments, shallow slopes, and shady wooded sites where organic matter may collect on the roof. The preservatives should not leach out over time.

One drawback to yellow pine shingles and shakes is that many are flat-grained, so most come pretreated with a water repellent to help them resist cupping and splitting. However, retreatment with a water repellent at some point may be required for optimal performance. Western red cedar shingles are also available pressure-treated for severe applications where standard cedar shingles are prone to decay.

Shingles. Shingles are sawn from blocks of wood, which gives them two smooth faces. They are relatively thin and cut to a taper. Red cedar shingles come in four grades, but most roofs use No. 1 or No. 2, which are all edge-grain heartwood (Table 1-6, page 16). They are available rebutted and rejointed (R&R), where a uniform appearance is desired, or machine-grooved for a textured surface.

Eastern white cedar shingles are also available in four grades. Most roofing work uses Grade A (Extra), which is all-clear, all-heartwood, or Grade B (Clear), which has no knots on the exposed face (see Table 1-7, page 16).

Shakes. Shakes are split from large blocks of wood and may be resawn to create a taper. They are heavier than shingles, less uniform in thickness, and are generally rough-textured on one or both sides creating a more rustic appearance. Grades and characteristics for red cedar shakes and shingles are found in Table 1-6, page 16. Red

cedar shakes come either tapered or untapered and are usually installed on roofs in Premium or No. 1 grade.

Fire-Retardant Treatment. Once popular on the West Coast, wood roofs have been banned in many residential areas by fire regulations designed to slow the spread of wildfires. Fire-retardant treated (FRT) shingles and shakes have been developed to address these issues and can obtain a Class B or C rating when combined with other components in a fire-resistant roof system. With pretreated shingles, consult with the treating company regarding fastener requirements and any special application instructions.

Slope and Exposure

Recommended exposures for shakes and shingles are shown in Tables 2-13 and 2-14.

TABLE 2-13	Wood Shingle Roofing Weather Exposure		
Shingle Length (in.)	Grade (label)	Maximum Exposure (in.)	
		3:12 to < 4:12 Slope	4:12 or Steeper Slope
16	1	$3\frac{3}{4}$	5
	2	$3\frac{1}{2}$	4
	3	3	$3\frac{1}{2}$
18	1	$4\frac{1}{4}$	$5\frac{1}{2}$
	2	4	$4\frac{1}{2}$
	3	$3\frac{1}{2}$	4
24	1	$5\frac{3}{4}$	$7\frac{1}{2}$
	2	$5\frac{1}{2}$	$6\frac{1}{2}$
	3	5	$5\frac{1}{2}$

SOURCE: Based on the 2003 International Residential Code and recommendations of the Shake and Shingle Bureau.

TABLE 2-14	Wood Shake Roofing Weather Exposure (4:12 and steeper)		
Type of Shake	Size (in.)	Grade	Max. Exposure (in.)
Hand-split shakes of naturally durable wood	$24 \times \frac{3}{8}$	1	$7\frac{1}{2}$
All other naturally durable wood shakes or pressure-treated taper-sawn shakes	18	1	$7\frac{1}{2}$
	18	2	$5\frac{1}{2}$
	24	1	10
	24	2	$7\frac{1}{2}$

SOURCE: Based on the 2003 International Residential Code and recommendations of the Shake and Shingle Bureau.

- *Minimum slopes.* The minimum recommended slope for standard installation of shingles is 3:12, and 4:12 for shakes.

- *Low slopes.* On lower slopes, shingles or shakes may be installed over a fully waterproof built-up roof (BUR) or membrane roof. Over the membrane, install vertical 2x4 battens lined up with the rafters, then spaced sheathing as described below.

- *Climate factors.* In warm, high-moisture climates, low-slope wood roofs need extra maintenance, particularly in areas with overhanging trees. If pine needles, leaves, or other organic debris is allowed to accumulate on a shaded section of the roof, moss, lichen, and algae will grow and retain moisture. This, in turn, will lead to premature curling, splitting, and decay of the shakes or shingles. Periodic cleaning, as well as chemical treatment, helps to avoid these problems (see "Maintenance," page 93). Pressure-treated shakes or shingles are recommended in these conditions.

Sheathing and Underlayment

Other than selecting a durable wood, the most important factor in determining a wood roof's longevity is its ability to dry out from both top and bottom when wet. While this was a natural feature of traditional installations over spaced sheathing, new methods and products are required for installation over solid sheathing. The two main approaches are:

- Create a system of spaced sheathing above the solid sheathing using vertical and horizontal battens; or

- Use a breathable underlayment applied over the sheathing.

Spaced Sheathing. The traditional way to lay wood shakes and shingles on spaced sheathing was ideal for wood roof longevity, but it has largely fallen by the wayside. Spaced sheathing is especially beneficial in warm, high-moisture climates, since the gaps in the substrate allow the shakes or shingles to dry out from both sides. It is not recommended in areas of windblown snow and not always permitted structurally. Where allowed, spaced sheathing typically uses nominal 1x4s for shingles or 1x6s for shakes. Code requires a minimum 1x4, and the spaces between battens should not exceed $3\frac{1}{2}$ inches (Figures 2-46 and 2-47).

The boards are spaced on centers equal to the weather exposure of the shakes or shingles, and they are lined up so the nailing falls in the center of each board. In areas where the average daily temperature in January is 25°F or less, solid sheathing is required on the lower section of the roof to support an eaves membrane. The eaves membrane

---

FIGURE 2-46   **Wood Shingles Over Spaced Sheathing.**

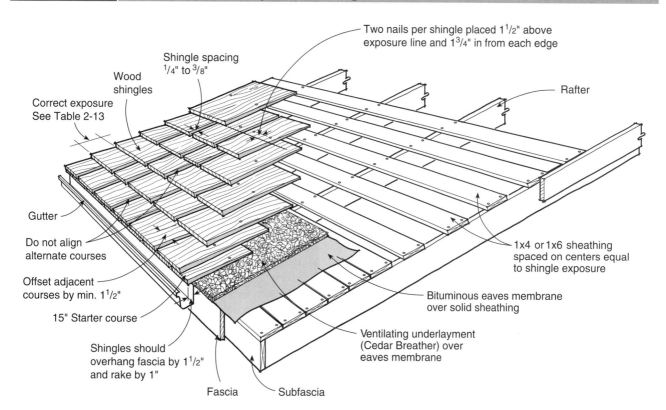

Where permitted by code, spaced sheathing provides the longest life to cedar shingles, but it is not recommended in areas of wind-blown rain or snow. Where an eaves membrane is required, use a ventilating underlayment to promote drying over the area of solid sheathing.

FIGURE 2-47   Shakes Over Spaced Sheathing.

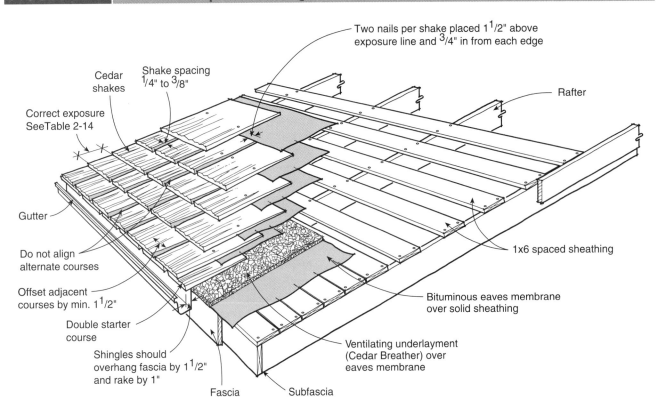

Two nails per shake placed 1$^1/2$" above exposure line and $^3/4$" in from each edge

Cedar shakes

Shake spacing $^1/4$" to $^3/8$"

Rafter

Correct exposure See Table 2-14

Gutter

1x6 spaced sheathing

Do not align alternate courses

Offset adjacent courses by min. 1$^1/2$"

Double starter course

Bituminous eaves membrane over solid sheathing

Shingles should overhang fascia by 1$^1/2$" and rake by 1"

Ventilating underlayment (Cedar Breather) over eaves membrane

Fascia   Subfascia

Because their irregular surface provides some self-ventilation, shakes can be installed over solid sheathing. However, in warm, humid climates, spaced sheathing is recommended for best performance. Interlaid strips of roofing felt are required in all installations to keep out windblown snow and rain.

should extend into the house 24 inches past the interior face of the outside wall.

Solid Sheathing.   This is required in areas of high wind or seismic activity and wherever else a solid roof diaphragm is required by code. Solid sheathing is also recommended in areas subject to windblown snow. Because of their irregular surface, rustic-style shakes are partially self-ventilating and may perform adequately on solid sheathing in relatively dry climates. Pressure-treated shingles or shakes can also be installed over solid sheathing. Shingles or smooth-surface (taper-sawn) shakes, however, are more prone to moisture buildup over solid sheathing, so a batten system or a ventilating underlayment is recommended, as described below.

Battens Over Solid Sheathing.   This provides the full benefit of spaced sheathing on top of a solid roof deck. After laying down No. 30 felt underlayment, install vertical 2x battens lined up with the rafters beneath for solid nailing. Next, place horizontal 1x4 or 1x6 battens (see "Spaced Sheathing," above) and nail into the vertical battens (Figure 2-48).

At the upper and lower edges of the roof, use insect screening or matrix-style roof vent material to block the entry of insects and other pests. Shake and shingle installation proceeds as for spaced sheathing.

Underlayment

• **Shingles:** Over solid sheathing, use minimum No. 30 felt lapped at least 3 inches horizontally and 6 inches at end laps. Over spaced sheathing, no underlayment is used except at the eaves if eaves flashing is required.

• **Shakes:** Over solid or spaced sheathing, use 18-inch-wide "interlayment" strips of No. 30 felt installed between shakes, as described below (Shake Installation, next page).

Ventilating Underlayments.   Many installers are shifting to a ventilating underlayment such as Cedar Breather (Benjamin Obdyke), which is easy to install and only adds about 10% to the cost of a wood roof. Cedar Breather is three-dimensional nylon matrix with dimples on the bottom and a smooth top surface. It lays over the felt paper and is tacked in place. It creates a continuous air space below the roofing, helping the shingles to dry out more rapidly and evenly. Although the air space is only about $\frac{1}{4}$ inch, contractors report that it reduces cupping and splitting. And by speeding up drying time, the air space should also help reduce the growth of decay fungi. However, ventilating underlayments are too new to draw conclusions about long-term performance. Installation details are shown in Figure 2-49.

FIGURE 2-48 | Wood Shingles Over Solid Sheathing with Battens.

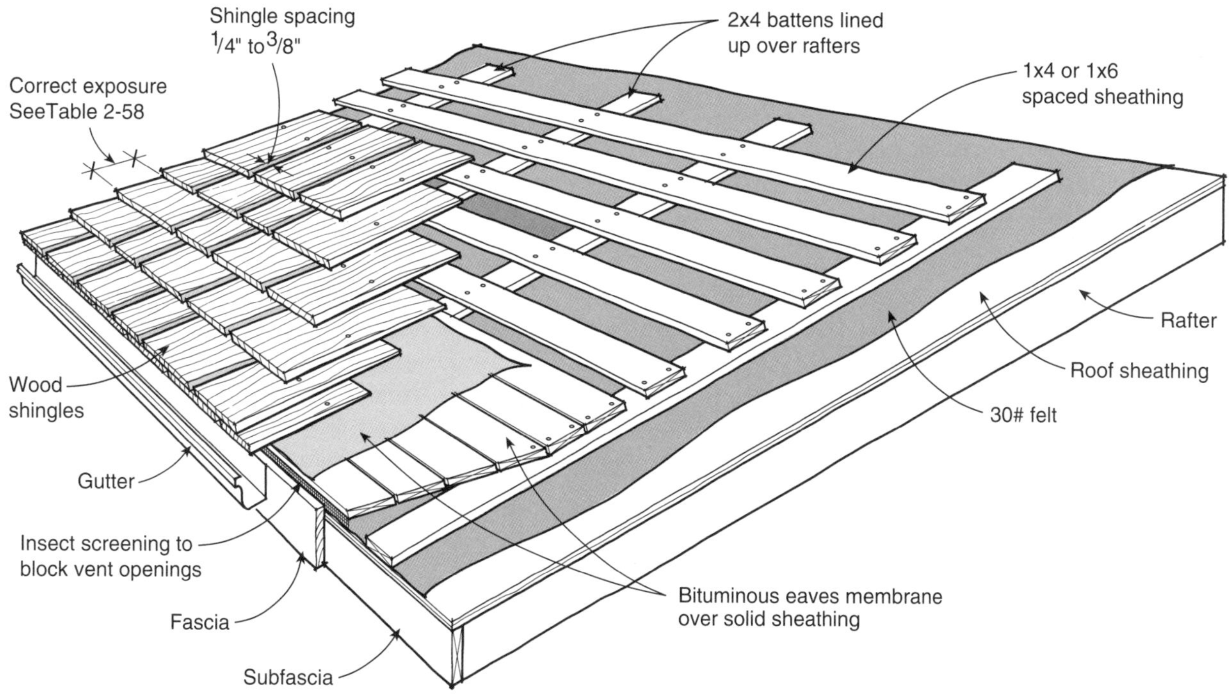

Where solid roof decking is required, it is best to provide ventilation under the shingles or use pressure-treated shingles. A system of spaced sheathing laid over 2x4 vertical battens provides optimal ventilation for either shingles or shakes.

Eaves Flashing. Apply eaves flashing to either spaced or solid sheathing in regions with an average daily temperature of less than 25°F (under the IRC) or in other areas prone to ice and snow buildup. The eaves flashing should extend up the roof to a point 24 inches inside the building. Where eaves flashing is required with spaced sheathing, install solid sheathing along the bottom section of the roof to support the eaves flashing.

Fasteners

All nails should be either stainless steel (type 304 or 316), hot-dipped galvanized, or aluminum. Staples should be either stainless steel or aluminum. Galvanized staples will not last the life of the roof. Treated shingles may require stainless steel or other special fasteners. Consult with the treatment company for recommendations. Stainless steel is also the first choice in coastal environments.

- *Nails* should be box type and penetrate the sheathing by $\frac{3}{4}$ inch (Table 2-15, page 91).
- *Staples* should have crowns between $\frac{7}{16}$ and $\frac{3}{4}$ inch wide and penetrate the sheathing by $\frac{3}{4}$ inch.
- *Drive flush.* Do not drive nail heads or staple crowns below the surface of the shingle. Underdriving or overdriving weakens the shingle attachment.
- *Placement.* Each shake or shingle should receive only two nails. Place one fastener $\frac{3}{4}$ inch in from each edge and about $1\frac{1}{2}$ inches above the exposure line (Figure 2-50).

Shingle Installation

Whether installed over solid sheathing or spaced sheathing, follow these guidelines:

- For the starter course, double or triple the shingles in the first row.
- Each shingle gets two nails about $\frac{3}{4}$-inch in from each end, and $1\frac{1}{2}$ inches above the butt line of the overlaying shingle.
- The first course should overhang the fascia by $1\frac{1}{2}$ inches. All courses should overhang the rake by about 1 inch.
- Leave a gap of $\frac{1}{4}$ to $\frac{3}{8}$ inch between adjacent shingles for expansion when wet.
- Offset joints in successive courses by at least $1\frac{1}{2}$ inches (Figure 2-50). Also, no more than 10% of joints should line up with joints in alternate courses (two courses away).
- Flat-grain shingles wider than 8 inches should be split into two shingles before installing.
- Treat knots, similar defects, and centerline of heart as if they were joints between shingles, and locate the defect $1\frac{1}{2}$ inches from joints in the row above or below.

Shake Installation

Whether installed over spaced or solid sheathing, shakes should always be interlaid with 18-inch-wide strips of No. 30 roofing felt. The felt strips acts as baffles to keep

FIGURE 2-49 | **Wood Shingles with Ventilating Underlayment.**

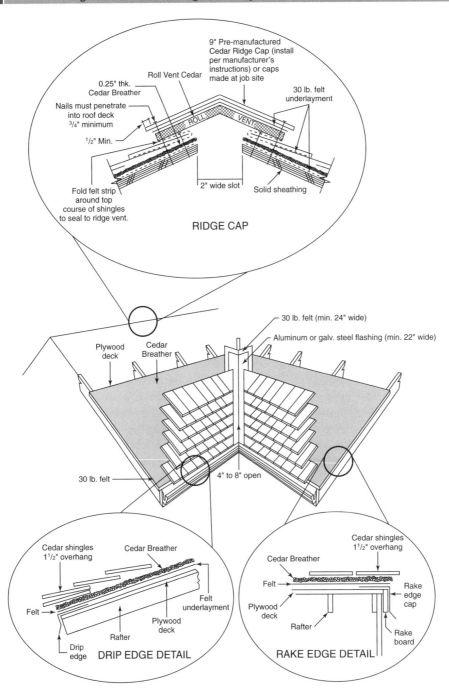

RIDGE CAP

9" Pre-manufactured Cedar Ridge Cap (install per manufacturer's instructions) or caps made at job site

Roll Vent Cedar

0.25" thk. Cedar Breather

Nails must penetrate into roof deck ³/₄" minimum

½" Min.

30 lb. felt underlayment

Fold felt strip around top course of shingles to seal to ridge vent.

2" wide slot

Solid sheathing

Plywood deck

Cedar Breather

30 lb. felt (min. 24" wide)

Aluminum or galv. steel flashing (min. 22" wide)

30 lb. felt

4" to 8" open

DRIP EDGE DETAIL

Cedar shingles 1½" overhang

Cedar Breather

Felt

Felt underlayment

Plywood deck

Rafter

Drip edge

RAKE EDGE DETAIL

Cedar shingles 1½" overhang

Cedar Breather

Felt

Plywood deck

Rafter

Rake edge cap

Rake board

New ventilating underlayments have simplified the job of creating a vent space below wood shingles and shakes. Cedar Breather, shown above, is a three-dimensional nylon matrix that creates a $\frac{1}{4}$-inch air space, helping to reduce cupping, splitting, and premature failure of shakes and shingles. Increase roofing nail lengths by $\frac{1}{4}$ inch.
SOURCE: Drawings courtesy of Benjamin Obdyke Inc.

windblown snow and other debris from penetrating the roof system during extreme weather. The felt "interlayment" also helps shed water to the surface of the roof. It is important to locate each felt strip above the butt of the shake it is placed on by a distance equal to twice the weather exposure (Figure 2-51).

Placed higher, the felt strips will be ineffective. Placed too low, they will be visible in the keyways and will wick

up water, leading to premature failure of the shakes. In addition, follow these guidelines:

- For the starter course, use either a single layer of shakes or two layers separated by a strip of felt interlayment (installed up from the eaves by a distance equal to the weather exposure). Fifteen-inch shakes are available for the bottom layer of a double starter course.

- Each shake gets two nails about $\frac{3}{4}$ inch in from each end and $1\frac{1}{2}$ inches above the butt line of the overlaying shake.
- The first course should overhang the fascia by $1\frac{1}{2}$ inches.
- All courses should overhang the rake trim by about 1 inch.
- Leave a gap between adjacent shakes of $\frac{3}{8}$ to $\frac{5}{8}$ inch for expansion when wet.
- Offset joints in successive courses by at least $1\frac{1}{2}$ inches.

TABLE 2-15	Fasteners for Red Cedar Shakes and Shingles on Roofs
Type of Certi-label Shake or Shingle	Nail Type and Minimum Length
Certi-Split & Certi-Sawn Shakes	
18 in. Straight-split	5d box ($1\frac{3}{4}$ in.)
18 in. and 24 in. Handsplit-and-Resawn	6d box (2 in.)
24 in. Tapersplit	5d box ($1\frac{3}{4}$ in.)
18 in. and 24 in. Tapersawn	6d box (2 in.)
Certigrade Shingles	
16 in. or 18 in. Shingles	3d box ($1\frac{1}{4}$ in.)
24 in. Shingles	4d box ($1\frac{1}{2}$ in.)

Courtesy of Cedar Shake & Shingle Bureau © 2005 CSSB. All Rights Reserved.

Reroofing

Under some conditions, shakes and shingles can be installed over existing roofing, as follows:

Existing Asphalt Shingles. If the existing asphalt shingles are not overly cupped or deteriorated, split or rough-sawn shakes can be installed over the shingles using interlaid strips of felt, as described above. Installing wood shingles over asphalt, however, requires a ventilating underlayment such as Cedar Breather or a system of battens (as shown in Figures 2-47 and 2-48).

Existing Wood Shingles. If the shingles are not badly curled or deteriorated, they can form an adequate surface for new shingles or shakes. Do not place building felt under the new shingles as that could inhibit drying, but if there is a high risk of decay (moist environment, low slope, overhanging trees), a layer of Cedar Breather is recommended. Shakes should be installed in the normal fashion with interlaid felt. Use nails long enough to penetrate the sheathing.

Existing Shakes. In most cases, these will need to be removed before reroofing, as the surface is too irregular, and nailing through the shakes into solid sheathing is impractical.

Hip and Ridge Details

The traditional treatment at hips and ridges is a labor-intensive "woven" cap, consisting of alternating sets of

---

FIGURE 2-50 Shake and Shingle Alignment and Nailing.

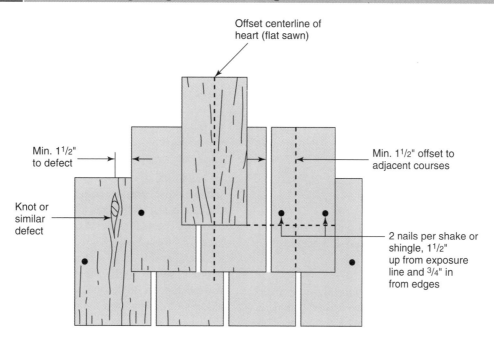

Joints in successive rows should be at least $1\frac{1}{2}$ inches apart and the same distance from knots and other defects. In lower grade shingles with flat grain, do not align joints with the centerline of heart. Only use two nails per shingle, as shown.

FIGURE 2-51 Shake Installation.

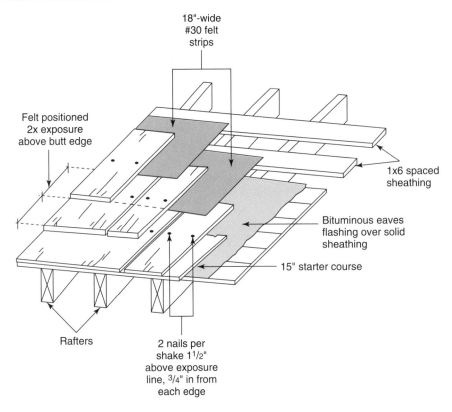

Whether installed over spaced or solid sheathing, shakes are always interlaid with strips of roofing felt. The felt interlay helps shed water to the roof surface, and keeps windblown rain and snow from penetrating the roof system during extreme weather.

two beveled shingles. Many installers now use factory-assembled cap pieces that speed up the process.

Hips. Lap the underlayment over the hip before installing the shingles. Then install a strip of roofing felt or metal flashing up the hip on top of the shingles before nailing the caps in place. Use nails or staples long enough to penetrate the sheathing by $\frac{3}{4}$ inch.

Ridge. For a vented ridge, use a plastic, matrix-type ridge vent. Cover the ridge vent with a strip of roofing felt and install factory-assembled ridge cap pieces. To prevent splitting of ridge-cap shingles, it is best to install them with a pneumatic nailer or stapler.

Flashings

Roof flashings should be at least 26-gauge, corrosion-resistant sheet metal, preferably painted galvanized steel or painted aluminum.

Copper and Cedar. Copper is a popular flashing material with wood roofs, although some experts caution against using copper in direct contact with red cedar or its runoff, since the soluble tannins in cedar can etch copper and, in extreme cases, lead to perforation of the flashing within 10 to 20 years (see also "Copper," pages 7, 83).

Premature failures have been documented in areas of the eastern United States that are subject to acid rain, leading the Cedar Shake and Shingle Bureau to advise against using copper flashing in areas east of the Great Lakes that are exposed to acid rain. Another approach endorsed by the Copper Development Association is to design flashing joints with a cant or hem that holds the edge of the cedar shingle slightly away from the flashing. The gap prevents water from being wicked into the joint, bathing the copper in the acidic solution.

Valleys. Wood roofs typically use open valley designs. While the International Residential Code (IRC) only requires the valley flashing to extend a minimum of 10 inches up each side of the valley for shingles and 11 inches for shakes, most contractors install 24- to 36-inch-wide valley flashing based on the area and pitch of the roof planes being drained. The valley metal should be protected by an extra layer of 36-inch-wide No. 30 felt installed directly under the metal or a layer of self-adhesive bituminous membrane applied directly to the sheathing. It is best to set aside the widest shingles or shakes for use in the valley to keep nails at least 12 inches from the valley centerline (Figure 2-52).

Chimneys and Skylights. These are flashed conventionally, using step flashing on the sides in accordance with Table 2-16. Use a soldered apron flashing below the

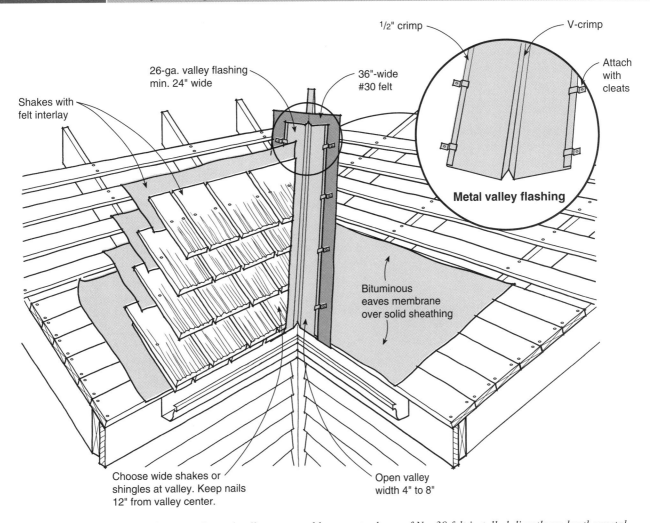

| **FIGURE 2-52** | **Valley Flashing for Shakes and Shingles.** |

Use a minimum 24-inch-wide crimped metal valley protected by an extra layer of No. 30 felt installed directly under the metal. Choose the widest shingles for use in the valley and keep nails at least 12 inches from the valley centerline.

TABLE 2-16	**Step Flashing Dimensions**	
	Horizontal Leg	Vertical Leg
Shakes	4 in.	3 in.
Shingles	$2\frac{1}{5}$ in.	$2\frac{1}{5}$ in.

chimney and a soldered head flashing at the top. Larger chimneys with significant water flow behind them should have a cricket above.

Maintenance

A number of factors affect the longevity of a wood roof. Key factors include the durability of the wood, local humidity and precipitation levels, and whether the roofing was installed with adequate ventilation. Other factors include the slope of the roof (steeper slopes shed water faster) and the presence of overhanging trees that shade the roof and drop organic debris onto the roof, trapping moisture on the surface. Some of these factors can be controlled by the contractor; some managed by the homeowner. Others, like the weather or the reduced durability of second-growth cedar, are beyond our control.

Some simple steps that a homeowner can take to prolong the life of a wood roof include:

- Trim overhanging branches that drop pine needles or leaves on the roof.

- Clean debris out of gutters and off the roof, both the surface areas and the keyways between shakes or shingles. A garden hose can do an adequate job.

- Ensure adequate year-round ventilation of the attic or roof assembly.

- Install strips of zinc or copper at the ridge (can serve also as a ridge cap) and midway across the roof on long slopes. Runoff from these strips forms a mild solution that reduces the growth of moss, mold, and mildew. This is effective for up to 15 feet downslope from the metal.

- If moss or lichen begin to grow, scrape it away and scrub the area with a solution of 1 quart household bleach, 1 ounce detergent, and 3 quarts warm water.

Over time, the natural extractives in cedar and other decay-resistant species will leach out, making the wood vulnerable to decay. Also, as the shingles dry out, they are prone to cupping, checking, and splitting. At some point, it may make sense to wash and treat the entire roof.

Washing. Cleaning wood roofs with high-pressure equipment is controversial and, in untrained hands, can cause significant damage. It is best to use normal garden hose pressure along with a brush or pump sprayer. To remove dirt, mildew, and weathered gray residue, a consortium of wood technology and coatings experts, including the U.S. Forest Products Laboratory (FPL), recommend a solution of sodium percarbonate (disodium peroxydicabonate) and water. With redwood and cedar, a second wash with a solution of oxalic acid may be needed to remove brown and black discoloration caused by tannins that leached out of the wood. Concentrated oxalic acid is toxic and should be handled with care.

Preservative Treatment. There are a number of commercial treatments available to restore decay-resistance to an aging wood roof. One effective and relatively benign (to plants) treatment consists of a copper-naphthenate compound called Cunapsol 5, which is diluted 1:4 with water and can be applied with a garden sprayer. The treatment needs to be repeated approximately every five years.

Oil-Borne Preservatives. Although Cunapsol 5 and similar waterborne treatments offer good protection against mold, mildew, and decay fungi, they will not do anything to slow down the cupping and splitting caused by weathering. For that, an oil-borne treatment is required. Effective treatments include copper naphthenate with a 3 to 4% metal content and copper octoate with a 1 to 2% metal content. These can be brushed on or dipped (before installation) or professionally applied with spray equipment.

Semitransparent Oil-Based Preservative Stains. Semitransparent oil-based preservative stains work well on rough-textured wood, such as shakes and shingles. They provide some pigmentation and protect the roof from decay for several years. Look for a product with both a wood preservative and a water repellent. Stains with a high percentage of pigment provide the best protection against UV degradation. While preservative stains are best applied before installing the shingles, a surface application can significantly extend the life of a wood roof.

Treatments to Use and to Avoid. According to the Shingle and Shake Bureau, one should use only products that are marketed and labeled as a cedar roof treatment, that have an MSDS available, and that contain one or more of the following: a water repellent, UV inhibitor, or U.S. EPA-registered wood preservative.

The following treatments should never be used:

- Film-forming finishes, including paints, solid stains, waterproofers, sealants, and plasticizers

- Any product with more than 40% solvents

- Any products that contains unfortified linseed oil or diesel fuel

- Any topical treatment marketed with fire-retardant claims

LOW-SLOPE ROOFING

Most roof coverings can be applied on roofs as shallow as 2:12 as long as a fully waterproof membrane is installed over the decking. In this case, the finish roofing material, whether asphalt shingles, wood, or tile, functions mainly as a decorative element but also helps protect the underlying membrane from UV radiation and physical damage.

At slopes lower than 2:12 on residential structures, the primary roofing options are built-up roofing (BUR), often called "tar and gravel," modified bitumen, and EPDM (see Table 2-17). In addition, a handful of proprietary single-ply membranes designed for easy application to small jobs have entered the market and offer a few new choices. While some of these products look promising, how long a new product will perform over 20-plus years is uncertain.

Minimum Slope. With any roofing material, a slope of at least $\frac{1}{4}$ inch per foot is recommended to promote drainage and minimize ponding. Where deflection from snow or other live loads is a concern, a greater slope will be needed to prevent any ponding. Most manufacturers of low-slope roofing products specify a minimum slope of between $\frac{1}{4}$ and $\frac{1}{2}$ inch per foot in their warranties.

While membranes, such as vinyl or EPDM, are unaffected by standing water, it will shorten the life of asphalt-based materials, such as BUR and modified bitumen. With any roofing material, ponding of water increases the likelihood of leakage, increases deflection in the roof framing, and contributes to rooftop growth of mosses, algae, and other plant life. Also, the freezing and thawing of ponded water can harm most roof surfaces.

Roll Roofing

The simplest product to install on a small section of low-slope roof is 90-pound roll roofing. This consists of a heavy, asphalt-saturated organic or fiberglass felt with a granular surface. Rolls are 36 inches wide and weigh 90 pounds. Single-coverage roll roofing typically has a 2-inch lap with exposed nails and is used mainly on utility structures.

Double-coverage roll roofing is installed with a full 19-inch lap joint, leaving a 17-inch exposure, with a 2-inch head-lap. Nails are concealed under the lap joints that are sealed with asphalt lap cement. With two layers of protection, double-coverage roll roofing is acceptable for small roof areas and can be used on roofs as shallow as 1:12.

TABLE 2-17	Low-Slope Roofing Options		
Material Type	**Pros**	**Cons**	**Avg. Longevity**
Roll roofing (double-coverage)	Inexpensive. Easily installed by carpenters. Concealed fasteners.	Short life span. Not suitable for cold weather installation.	Approx. 10 years
BUR	Long track record. Forgiving of installation errors due to multiple plies.	Expensive for small jobs. Heavy equipment, odors, potential spills during installation. Ponding water can cause deterioration. Leaks are hard to detect and repair.	15 to 20 years
Modified bitumen (torch-applied)	Durable heat-fused seams do not rely on adhesives. Self-flashing at openings. Easy to inspect and repair. Compatible with asphalt materials. Low temperature installations possible. Self-adhesive and cold-process versions available.	Requires careful installation for proper seaming. Fire risk during installation. Ponding water can reduce longevity. Different chemical formulations difficult to evaluate.	20+ years
EPDM	Relatively easy to install. Proven track record. Lightweight and UV-resistant. Self-flashing. Tolerates building movement and extreme heat and cold. Easy to inspect and repair. Self-adhesive version available for small jobs.	Requires careful installation for proper seaming. Can be damaged by petroleum products, solvents, and grease.	20+ years

BUR

Built-up roofing (BUR) systems dominated the commercial and residential low-slope roofing markets until the 1980s, when single-ply membranes became widely accepted. BUR roofs consist of layers of asphalt-impregnated felt bonded with hot asphalt, or in some parts of the country, hot coal tar. The average life span of a hot-mopped BUR roof is 15 to 20 years, although this can be extended by applying an aluminum coating every three to five years to reduce UV degradation and alligatoring.

BUR roofs can have either a smooth coated surface or a stone surface created by spreading crushed stone or gravel into a thick flood coat of hot asphalt or tar. Aggregate-faced roofs are typically more durable due to the heavier flood coat and the protection offered by the stone from UV radiation, hail, and other environmental wear and tear. However, the stone coating makes leaks harder to find and repair.

Proper detailing of metal flashings at openings, parapet walls, and roof edges is critical, and these areas need regular inspection and maintenance. The most likely place for leaks is flashings, particularly metal edge flashings due to their thermal movement. Asphaltic or rubber flashings may also become brittle and crack.

Pros and Cons. BUR roofs are reliable if properly installed, and their multiple layers provide some protection against small installation errors. However, the long set-up time makes BUR expensive for small residential jobs. Also the heavy equipment, odors, and potential spills associated with a hot-mop job are not welcome on many residential job sites.

Modified Bitumen

Most modified-bitumen roofs are torch-applied, although there are also self-adhesive and cold-process systems. The waterproofing membrane, sometimes called "single-ply modified," consists of asphalt bitumen reinforced with a polyester or fiberglass fabric and modified with polymers to give it greater strength, flexibility, resistance to UV degradation, and resistance to heat and cold. A variety of different chemical formulations have been tried over the years. It is best to stick to a product with an established track record. In general, modified-bitumen roofs can be applied to slopes as shallow as $\frac{1}{4}$ inch per foot.

Installation. A torch-applied, or *torchdown*, roof starts with a nonflammable base sheet made of asphalt-saturated felt or fiberglass that is mechanically attached to the roofing deck. In residential construction, the base sheet is usually attached with roofing nails driven through metal caps. The second layer is the waterproofing membrane, or *cap sheet*. This is heated with a torch as it unrolls, fusing it to the base sheet, to itself at seams, and to penetrations such as skylights. Installers must learn to heat the membrane so it is hot enough to fuse but not so hot as to burn through. Membranes may be either smooth or have a granular surface like roll roofing. Smooth-faced membranes need a third coating, which has colored or reflective pigments to protect against UV radiation. The smooth type is preferable where foot traffic is expected or where decking is going over the roofing.

Torchdown roofing is self-flashing and uses no adhesives or solvents to seal around openings. The material can be run up parapets and abutting wall, and patches are used to seal around metal skylight curbs and similar openings. A special patching compound is used to seal to PVC stacks. If applied correctly, the torchdown membrane is essentially seamless.

Pros and Cons. Modified bitumen is easily repaired without solvents or adhesives. It is compatible with asphalt shingles and asphalt compounds, although patching with roofing cement is not recommended. The reinforced fabric layer isolates the membrane above from building movement and gives the material enough strength to support occasional foot traffic.

The main drawback is the risk of fire during installation. While the risk of fire is low in the hands of trained installers, care must be taken when using torchdown on a wood-frame structure. A number of fires have started with sawdust that has accumulated in empty cavities, such as crickets and parapets. Inspection of the roof for sawdust pockets while it is being framed is advised.

EPDM

While a variety of single-ply roofing membranes are used on commercial jobs, only EPDM has become widely used on residential sites. EPDM, a form of synthetic rubber, owes its popularity to its relative ease of installation combined with exceptional durability. If installed correctly, roofs often exceed 20 years of service and callbacks are exceedingly rare.

While some commercial EPDM systems are loose-laid or ballasted, residential applications are typically fully adhered. Rolls typically vary from 10 to 50 feet in width and from 50 to 200 feet in length, but many distributors will cut a piece to size for smaller jobs. If possible, use a single piece with no seams for the field of the roof. EPDM membranes are available in two thicknesses: .045 inch and .060 inch. For fully adhered applications or any application where foot traffic or decking is planned, the thicker membrane is recommended.

Substrates. EPDM can be bonded to a wide variety of substrates, including plywood, OSB, fiberboard, and urethane insulation board. The substrate should be smooth, even, and free of debris. Fasteners should be driven flush except in the case of insulation fastening caps, which project their shape though the membrane. If the surface is uneven or deteriorated, a layer of fiberboard or thin plywood should be installed first.

Installation. After cutting the material to fit, installers use a roller to apply a proprietary contact cement to both the membrane and the substrate. Typically, a length of roofing is set in place and folded in half lengthwise so one-half can be glued at a time. The adhesive should be fully dry on both surfaces before bonding, or bubbles may develop. Also, care must be taken to smooth out wrinkles and air pockets as the two surfaces are mated. Where seams are required, the material is lapped 4 to 6 inches and sealed with either double-faced seam tape or a special adhesive used for bonding rubber to rubber.

At openings, inside corners, outside corners, and other irregular shapes where the membrane has been cut, patches of *uncured* EPDM are applied using the rubber-to-rubber adhesive. The uncured form of EPDM is highly elastic and can be stretched to conform to irregular shapes.

The material is lapped up abutting walls and serves as its own flashing. Other terminations are usually sealed with an aluminum termination bar or an aluminum flashing covered with a strip of EPDM. Finally all exposed edges of EPDM at laps, patches, and terminations are sealed with a bead of proprietary caulking that protects the edge and acts as an extra water stop.

Self-Adhesive. For small jobs, a few manufacturers offer a peel-and-stick version of EPDM. Installation is similar to standard EPDM but may require a primer on plywood and OSB substrates. Seams generally require a proprietary adhesive with special caulking on exposed edges. Although the square foot cost is greater than with site-glued EPDM, on small jobs labor savings offset the higher material costs.

Pros and Cons. While not intended as a walkway, EPDM works well as a substrate under rooftop decks. Leftover strips of membrane should be used to cushion the roofing from wood sleepers. Leaks are rare and usually can be traced to sloppy sealing of joints. Leaks are also relatively easy to identify and fix. One caution is that EPDM can be damaged by grease and petroleum-based products, a potential problem with outdoor grills and spillage of oil-base finishes used on siding or wood decking.

WALKABLE ROOFING MEMBRANES

For rooftops that will also serve as decks (see "Rooftop Decks," page 150), one option is to use a roofing material designed for foot traffic. Duradek (Duradek U.S. Inc.) is a sheet vinyl membrane similar to Hypalon but with a non-skid wear surface. It was developed over 25 years ago for waterproofing decks, balconies, and outdoor living spaces. For use over a living space, the manufacturer recommends its 60-mil Ultra series, which is warranted against leakage for 10 years.

Duradek is made of reinforced PVC sheet with heat stabilizers and additives for resistance to fire, UV degradation, and mildew. The wear surface is textured for slip resistance and available in a variety of colors.

Installation. Installation is similar to other single plies and must be done by factory-certified contractors. The membrane glues to almost any clean substrate with either a proprietary contact cement or a special water-based adhesive applied with a notched trowel. Seams are heat welded with a heat gun, the most critical step. Like other single-ply membranes, the material is self-flashing at abutting walls and penetrations.

Pros and Cons. Duradek creates an attractive and durable no-skid deck surface that can withstand normal wear and tear, direct sun exposure, high winds, and freeze-thaw cycles. However, because the membrane is also the wear surface, it can be damaged by cigarette burns, punctures, and heavy abrasion.

ROOF VENTILATION

All residential building codes require some form of roof ventilation. These rules were first developed in the 1940s, when attic spaces first started to develop problems with mold and mildew due to excess moisture. With the growing use of plywood, asphalt shingles, insulation, and better doors and windows, houses were being built tighter. The tighter spaces retained more of the normal household moisture generated by cooking, bathing, household plants, crawlspaces, and exposed basement slabs. As the stack effect drove this moisture up into attic spaces, problems ensued.

Code Requirements

The rules of ventilation developed by researchers in the 1940s were adopted first by the Federal Housing Administration (FHA) and later by all the major residential building codes, including the 2003 IRC, with few changes. Most asphalt shingle manufacturers will void their warranties if these rules are not followed. They require:

- 1 square foot of *net free vent area* (NFVA) per 150 feet of attic floor.

- 1 square foot of NFVA per 300 square feet of attic floor if a vapor barrier is installed on the ceiling below.

- The IRC adds that the NFVA ratio can also be reduced to 1:300 if 50% to 80% of the required ventilation is located in the upper portion of the attic (or cathedral ceiling) and the rest is located at the eaves, with the upper vents at least 3 feet above the lower.

Tight Ceiling

Although the code-mandated ventilation rate has proven adequate under normal conditions, homes with high-moisture levels and air leaks in ceilings may still experience problems such as moldy sheathing. Cathedral ceilings are at the greatest risk due to the limited ventilation path. The best defense against problems is to create a continuous air and vapor barrier between the living space and attic or roof cavity by carefully sealing all air leaks. The ceiling air barrier may consist of foam insulation with taped seams, taped polyethylene sheeting, or finished drywall that is sealed at corners and top plates with gaskets or sealants.

Penetrations. Pay special attention to penetrations in the ceiling plane, particularly in cathedral ceilings. Chimneys, recessed lights, plumbing chases, and holes drilled through top plates for plumbing or wiring should all be sealed (Figure 2-53).

FIGURE 2-53 | **Typical Ceiling Air Leaks.**

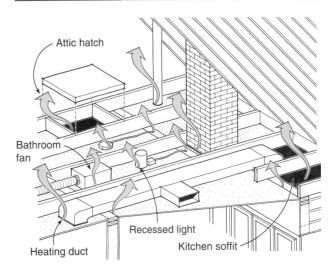

Standard attic ventilation will prevent moisture problems and ice dams as long as the ceiling plane is properly sealed, controlling air and heat leakage from the living space below. Common leakage paths are shown above.

Plug holes with durable materials, such as expandable urethane foam, foam backer rod, EPDM, or sheet metal, and use long-lasting sealants such as high-quality urethanes, silicones, and butyls.

With a tightly sealed ceiling, attic moisture is no longer a significant problem. Attic ventilation is still recommended for three other reasons:

- Preventing ice dams in cold climates

- Reducing cooling loads in hot climates

- Extending shingle life

- Allowing roof components to dry out in the event of a leak

Preventing Ice Dams

Ice dams form when heat leaking into attics or roof cavities from the building below, or from attic ductwork, melts the bottom layer of snow on the roof. The melt water runs down the length of the roof to the eaves, where it refreezes, forming a dam and icicles. In the worst cases, liquid water pools behind the dam and flows under the shingles and into the building (Figure 2-54).

Research has indicated that the ice-dam risk is greatest when temperatures range between 15°F and 20°F—when it is warm enough for snow to melt but cold enough for it to refreeze at the eaves. Also, the greater the depth of snow on the roof, the greater the risk of ice dams due to the insulating value of the snow itself.

Cold Roofs. Ventilation helps prevent ice dams by keeping the roof surface cold enough to limit uneven melting. Tests conducted in 1996 at the U.S. Army Corps of Engineers Cold Regions Research and Engineering Laboratory (CRREL), showed that the traditional 1:150 ventilation

FIGURE 2-54 Ice-Dam Formation.

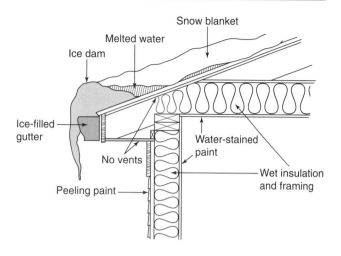

Ice dams form when melted snow, caused by excessive heat loss through the roof, runs down and refreezes at the eaves. Pooled water behind the dam can flow into ceiling and wall cavities causing extensive damage. The best protection is a well-ventilated "cold roof."

TABLE 2-18 Roofing Color and Cooling Loads*

Roofing Type	Reduction in Peak Load	Annual Savings
Dark gray asphalt shingle (control)	—	—
White asphalt shingle	4%	17%
Terra-cotta tile	3%	13%
White S-tile	20%	32%
White flat tile	17%	34%
White galvanized steel	23%	28%

*Based on an average-sized south Florida home with R-19 ceiling insulation and cooling ducts located in the attic.

SOURCE: Adapted from the FSEC report *Comparative Evaluation of the Impact on Roofing Systems on Residential Cooling Energy Demand in Florida*, 2000, by Danny Parker, Jeffrey Sonne, John Sherwin, and Neil Moyer. Courtesy of Florida Solar Energy Center.

rule was sufficient to prevent ice dams on roofs with R-25 or greater ceiling insulation. The 1:300 rule proved adequate for roofs with R-38 or greater insulation. Since most standard eave and ridge vents sold today meet the higher ventilation rates, most new homes are protected as long as there are no large heat leaks into the attic, or tricky sections of the roof with inadequate ventilation.

Reducing Cooling Loads

Experts recommend using attic ventilation in hot climates as part of an overall strategy to reduce cooling loads. Ventilation helps even more when used in combination with radiant barriers.

Ventilation Alone. Researchers at the Florida Solar Energy Center (FSEC) have found that adequate attic ventilation can modestly lower sheathing and shingle temperatures, and reduce an average home's cooling load by about 5%.

Ventilation and Radiant Barriers. For greater savings on cooling, consider adding a radiant barrier to the underside of the roof sheathing or draped between the rafters. This can reduce peak cooling loads by 14 to 15% and seasonal loads by an average of 9%. By doubling the roof ventilation from 1/300 to 1/150, the annual savings from radiant barriers rises to 12%. These numbers assume R-19 ceiling insulation and cooling ducts located in the attic, which are typical in Florida. With R-30 ceiling insulation, the cooling benefits of radiant barriers are less dramatic.

Roofing Color. Tests at FSEC also indicate that simply switching from dark to white asphalt shingles in a cooling

climate can reduce peak cooling loads by 17% and seasonal loads by 4%. The greatest savings resulted from using white metal roofing (see Table 2-18.)

Unvented "Hot" Roofs

In cathedral ceiling configurations where it is difficult to provide ventilation, some builders have eliminated the vent space, relying instead on careful sealing of the ceiling plane to prevent moisture problems. While experts concede that this should work in theory, most caution that it is difficult to build a truly airtight ceiling assembly. Also, cathedral ceilings are slow to dry out if moisture problems do occur, whether from condensation or roofing leaks. If a hot roof is the only option for a section of roof, take the following precautions:

- Install a continuous air and vapor retarder, such as 6-mil poly, carefully sealed at all junctures.

- Do not use recessed lights or other details that penetrate the ceiling plane.

- Carefully seal all penetrations in the ceiling assembly, including top plates of partitions, with durable materials.

- Use a nonfibrous insulation, such as plastic foam, and install it without voids where moisture could collect.

- In regions prone to ice dams, use enough insulation to maintain a cold roof—preferably R-38 or greater.

- Eliminate all sources of excess moisture in the home (wet basements, uncovered crawlspaces, unvented bathrooms).

Attic Ventilation Details

Soffit and Ridge Vents. For both attics and cathedral ceilings, roof ventilation works best when it is balanced

FIGURE 2-55 · Balanced Roof Ventilation.

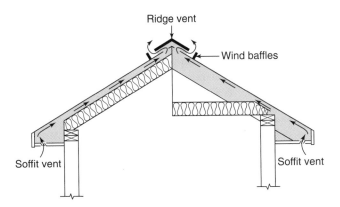

The soffit vent area should be equal to or a bit larger than the ridge vent area, so there is always sufficient makeup air. Also, ridge vents should have either external or internal baffles to limit the infiltration of windblown rain and snow.

between high and low to take advantage of natural convection (Figure 2-55).

This configuration also tends to evenly wash the underside of the roof with ventilation air. The soffit-vent area should be equal to or slightly larger than the ridge-vent area. Ridge vents should either have external or internal baffles to minimize infiltration of windblown rain and snow. Use insulation baffles or modified framing to make sure that the ceiling insulation does not block airflow at the eaves (Figure 2-56.)

Alternatives. Where ridge vents are not an option, combine any type of upper vent such as gable-end vents, roof vents, or turbines, with soffit vents. Where soffit vents are not possible, use gable-end vents on both ends of the roof, which will ventilate adequately under wind pressure.

Avoid High Vents Alone. Do not use ridge vents or other rooftop vents without low vents to provide makeup air. The suction created could help pull moist household air into the attic.

Cathedral Ceiling Ventilation Details

Cathedral ceilings require the same continuous air barriers, and balanced soffit and ridge vents, as attics. Both air sealing and ventilation are more critical, however, since any trapped moisture in the roof cavity will remain longer and potentially cause greater damage than in an open attic. Also, since there is little or no communication from bay to bay, an effective ventilation system must reach every bay (Figure 2-57).

Hips and Valleys. Ventilating hips and valleys can be challenging with a cathedral ceiling. One approach is to use a double or triple hip or valley rafter one size smaller than the common or jack rafters. This will create a vent space along the top of the hip or valley rafter that can be used to supply ventilation air to the jack rafters (Figure 2-58).

FIGURE 2-56 · Soffit Vents.

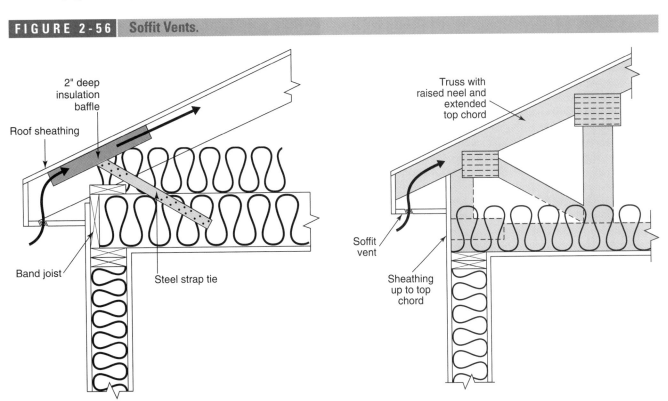

Place soffit vents close to the fascia for best performance. With high levels of insulation, a raised-top plate (left) or raised-heel truss (right) are recommended to allow full insulation at the plate area. With a raised top plate, use steel strap ties or similar connectors to securely anchor the rafters to the joists.

Skylights. Localized hot spots such as skylights can also lead to ice dams below, due to blocked ventilation as well as melt water from skylight heat loss. Notching the rafters on either side of the skylight will help maintain airflow above the skylight (Figure 2-59).

If icing is still a problem, add an interior storm window to reduce heat loss through the glass in cold weather. As a backup, it is always a good idea to seal the skylight curb and surrounding roof area with a bituminous membrane (see Figure 2-5, page 57).

FIGURE 2-57 Cathedral Ceiling Ventilation.

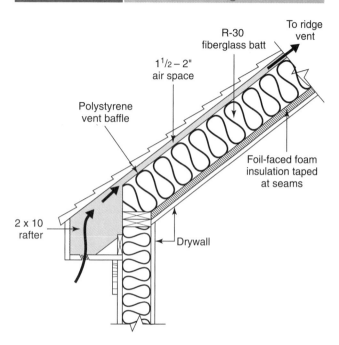

Builders have devised many methods to effectively ventilate cathedral ceilings. The key elements for success are an airtight ceiling plane and a minimum $1\frac{1}{2}$ inch free vent space from soffit to ridge.

FIGURE 2-59 Venting Around Skylights in Cathedral Ceilings.

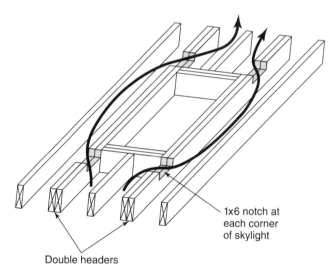

Notching the tops of the rafters on either side of a skylight will help maintain airflow to the roof area above the skylight.

FIGURE 2-58 Ventilating Cathedral Ceiling Hips and Valleys.

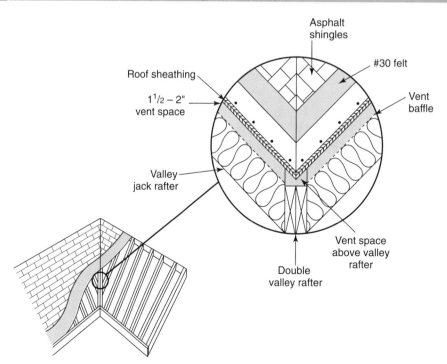

To ventilate the rafter bays between hip or valley jacks, use a double or triple hip or valley rafter one size smaller than the common or jack rafters. The space left above the hip or valley rafter provides an air inlet for hips or an air inlet for valleys.

RESOURCES

Manufacturers

Asphalt Shingles

Atlas Roofing Corp.
www.atlasroofing.com
Fiberglass and organic felt shingles

Certainteed Roofing
www.certainteed.com
Fiberglass shingles

Elk Premium Building Products
www.elkcorp.com
Fiberglass shingles

GAF Materials Corp.
www.gaf.com
Fiberglass shingles

Georgia-Pacific Corp.
www.gp.com/build
Fiberglass and organic felt shingles

IKO
www.iko.com
Fiberglass and organic felt shingles

Owens Corning
www.owenscorning.com
Fiberglass shingles

Tamko Roofing Products
www.tamko.com
Fiberglass and organic felt shingles

Concrete Roof Tiles

Bartile Roofs
www.bartile.com

Eagle Roofing Products
www.eagleroofing.com

Entegra Roof Tile
www.entegra.com

MonierLifetile
www.monierlifetile.com

Vande Hey-Raleigh
www.vhr-roof-tile.com

Westile
www.westile.com

Clay Roof Tiles

Altusa, Clay Forever LLC
www.altusa.com

Ludowici Roof Tile
www.ludowici.com

MCA Clay Tile
www.mca-tile.com

U.S. Tile Co.
www.ustile.com

Tile Fasteners and Adhesives

Dow Building Products
www.dow.com/buildingproducts
Tile Bond polyurethane foam tile adhesive

Fomo Products
www.fomo.com
Handi-Stick polyurethane foam tile adhesive

Newport Fastener
www.newportfastener.com
Twisted wire systems, hurricane clips, nose clips, and the Tyle-Tye TileNail

OSI Sealants
www.osisealants.com
RT 600 synthetic rubber tile adhesive

Polyfoam Products
www.polyfoam.cc
Polyset and Polyset One polyurethane foam tile adhesives

Wire works, Inc.
www.wireworks-inc.com
Tile hooks, hook nails, copper and stainless-steel nails

Metal Roofing

Classic Products
www.classicroof.com
Modular metal shingle panels and standing seam panels

Decra Roofing Systems
www.decra.com
Modular metal shingle, tile, and shake panels

Dura-Lok Roofing Systems
www.duraloc.com
Modular metal roofing shingles with granular coating

Fabral
www.fabral.com
Exposed fastener and concealed clip metal roofing panels

Gerard Roofing Technologies
www.gerardusa.com
Modular metal shake and tile panels with granular coating

Met-Tile
www.met-tile.com
Modular metal roof-tile panels

Atas International
www.atas.com
Modular metal shingle, tile, and standing-seam panels

Custom-Bilt Metals
www.custombiltmetals.com
Modular metal shakes and standing seam panels

Low-Slope Roofing Membranes

Duradek
www.duradek.com
Vinyl roofing and walkable deck membrane

Firestone
www.firestonebpe.com
RubberGard EPDM residential roofing system

GenFlex Roofing Systems
www.genflex.com
Peel-and-stick TPO membrane

Hyload, Inc.
www.hyload.com
Kwik-Ply self-adhering polyester and coal-tar roofing membrane

Ridge Vents

Air Vent/A Gibraltar Company
www.airvent.com
A complete line of roof ventilation products, including shingle-over and exposed-ridge vents with exterior wind baffles and internal weather filters. Also soffit and drip edge vents and passive and powered attic turbine-type vents.

Benjamin Obdyke
www.benjaminobdyke.com
Shingle-over ridge vents. Low-profile Roll Vent uses nylon-matrix. Extractor vent is molded polypropylene with internal and external baffles.

Cor-A-Vent
www.cor-a-vent.com
Shingle-over low-profile ridge vents, including Cor-a-vent, Fold-a-vent, and X-5 ridge vent, designed for extreme weather. Corrugated core.

GAF Materials Corp.
www.gaf.com
Cobra vent: roll-out shingle-over ridge vent with a polyester-matrix core

Mid-America Building Products
www.midamericabuilding.com
Ridge Master and Hip Master shingle-over molded plastic ridge vents with internal baffles and foam filter

Owens Corning
www.owenscorning.com
VentSure corrugated polypropylene ridge vents; also passive roof vents and soffit vents

Trimline Building Products
www.trimline-products.com
Shingle-over low-profile ridge vents, Flow-Thru battens for tile roofs

Elk Premium Building Products
www.elkcorp.com
Highpoint polypropylene shingle-over ridge vents

Tamko Roofing Products
www.tamko.com
Shingle-over ridge matrix–type Roll Vent and Rapid Ridge (nail gun version) and Coolridge, which is molded polypropylene with external and internal baffles

Venting Underlayments

Benjamin Obdyke
www.benjaminobdyke.com
Cedar Breather, a $\frac{3}{8}$-in.-thick matrix-type underlayment designed to provide ventilation and drainage space under wood roofing

For More Information

Asphalt Roofing Manufacturers Association (ARMA)
www.asphaltroofing.org

Cedar Shake and Shingle Bureau
www.cedarbureau.org

Metal Roofing Alliance
www.metalroofing.com

Tile Roofing Institute
www.tileroofing.org

CHAPTER THREE

Windows and Doors

WINDOW TYPES

Windows have a bigger impact on the quality of life in a home than almost any other building component. They affect heating and cooling costs, natural lighting levels, ventilation quality, and the comfort of occupants year-round. Subjected to high ultraviolet (UV) exposure, extremes of weather, and hundreds of operations over their service lives, windows must be well-engineered from durable materials if they are to provide satisfactory service. It is well worth investing the time and money to select the right windows for the job and to install them properly.

The most common types of operable windows are double-hung, casement, and sliding. Awning-style windows are commonly used either near grade to let light into basements or high on a wall for privacy. Tilt-turn windows, European imports that swing inward like a door on hinges, are often used in large sizes as emergency exits. Windows with compression-type seals are the tightest, and windows that swing open provide the best ventilation (see Figure 3-1).

Common window types and their characteristics are summarized in Table 3-1.

Single- and Double-Hung

The most common windows in the United States, double-hungs have upper and lower sash offset so both can slide up and down. Only the lower sash moves in the less common single-hung window. In older homes, the operable sash were connected by rope to heavy iron counterweights to hold the upper sash in place and to assist with raising the lower sash. Modern double-hungs ride up and down in metal or plastic tracks called jamb liners and use hidden springs in place of sash weights. On many of the newer models, the sash are designed to tilt in for easy cleaning.

FIGURE 3-1   Operable Sash Types.

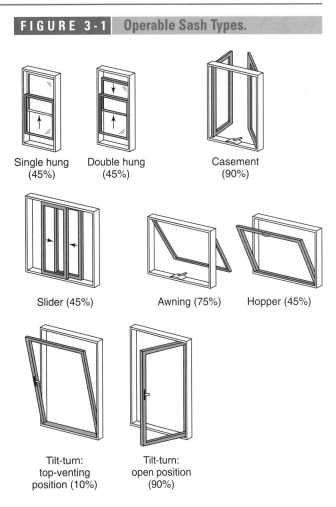

Single hung (45%) Double hung (45%) Casement (90%)

Slider (45%) Awning (75%) Hopper (45%)

Tilt-turn: top-venting position (10%) Tilt-turn: open position (90%)

Casement-style windows have the greatest effective open area (numbers shown in parentheses). And, if oriented to catch prevailing winds, will act as wind scoops, directing breezes into the house. Sliding windows provide about half the effective ventilation area.

TABLE 3-1 Window Types

Window Type	Effective Ventilation	Infiltration	Rain Penetration	Convenience	Durability/ Maintenance	Recommendations
Single- and double-hung	45% of frame area	Moderately tight but seals prone to wear. Strong winds may penetrate.	Overlapping top sash helps unit shed water. Lower sash can be open in light rain.	Exterior screens out of way. Units with tilt-in sash easy to clean. Sash easy to operate.	In general, very durable with low maintenance. Weather-stripping may eventually need replacement.	Practical, versatile choice, but meeting rails may block views. In windy areas, look for models with low air leakage.
Horizontal Sliders	45% of frame area	Average tightness. Seals prone to wear.	Water tends to collect in bottom track and can cause problems. OK in light rain.	Exterior screens out of way. Outside glass hard to reach for cleaning. Sideways operation of sash can be awkward.	Weep holes in sill track can clog and trap water leading to damage. Track can also collect grit and impair sliding action.	Low-cost option for wide, open views. However, sliding operation may be difficult on low-quality products.
Casement	90% of frame area. Can orient open sash to catch prevailing winds.	Compression seals very tight.	Rain bounces off top sash and into window when open. Top of sash must be protected from decay. Also vulnerable to strong gusts.	Crank operation convenient for hard-to-reach places. Interior screens easily damaged and may collect leaf debris and dust. Open sash may interfere with screen doors or deck space. Hard to reach for cleaning.	Hardware needs occasional lubrication. Large, heavy sash may sag over time affecting the fit. Cranks or linkage occasionally break.	Good over sink, tub, or other hard-to-reach spaces. Excellent ventilation. Look for quality hardware and sturdy build, particularly for oversized units. Cladding should protect top of sash.
Awning	75% of frame area, but sash can deflect breeze.	Compression seals very tight.	Can be left open in moderate wind and rain.	Crank operation easy in hard-to-reach and high places. Outside glass hard to reach for cleaning.	Durable with low maintenance. Occasional lubrication of crank mechanism helpful.	Useful for privacy windows high on wall. Also can be combined with fixed glazing to allow ventilation at top or bottom.
Hopper	45% of frame area, but sash can deflect breeze.	Compression seals very tight.	Best if used in sheltered location.	Crank operation easy in hard-to-reach and high places. Outside glass hard to reach for cleaning. Inward opening sash may interfere with furniture, living space.	Durable with low maintenance. Occasional lubrication of crank mechanism helpful.	Same applications as awning window where in-swinging window preferred.
Tilt-Turn	90% of frame area	Among the tightest windows available due to 4- to 6-point locking mechanism.	Acts like either an in-swinging casement or a hopper (that opens only a few inches). Resists leakage in hopper position.	Easy to operate and versatile. Can be left in tilt mode for ventilation with house locked. In-swinging sash may interfere with living space. Large sizes can function as emergency exits.	Durable with low maintenance. Occasional lubrication of tilt-turn hardware helpful.	Good where ventilation is desired in locked house. In-swinging function good for emergency exits, but may interfere with interior space.

Because the top sash overhangs the bottom, and both are recessed in the frame, double-hung windows shed rain well and can be left open at the bottom in a light rain. They use exterior screens that are out of the way and inconspicuous. However, ventilation is limited to half the area of the frame, and visibility is somewhat hampered by the meeting rails, which often sit near eye level.

Also, because they rely on slide-by rather than compression-type weather-seals, many double-hungs have air leakage rates nearly twice that of casements. With improvements in materials and designs in high quality units, however, the performance gap has narrowed, at least when the windows are tested (with brand-new weather-seals).

Although probably not the tightest windows in a high-wind location, in general, double-hungs offer a versatile, moderately priced, and trouble-free option.

Sliding Windows

Sliders can offer large horizontal expanses of glass and operable sash that do not interfere with interior or exterior space. They are more common in western states, while double-hungs and casements prevail on the East Coast.

Designing a window that slides sideways presents a few challenges. First, the lower track must rely on weep holes to drain away water, and second, any grit that collects in the lower track tends to impair the sliding action. If the weeps clog up, water may find its way into the sill or framing over time. Also, pushing a stuck window sideways is an awkward motion that tends to strain the back. As with double-hung windows, ventilation is limited to 50% of the framed opening, and slide-by weather-seals are less effective than compression seals.

To avoid problems with sliders, look for high-quality windows that slide freely. Also, consider alternatives such as a picture window with a casement along one side for ventilation.

Casements

These provide a more contemporary look than traditional double-hungs and can provide large uninterrupted views. If oriented to open into the prevailing winds, the sash acts like a big wind scoop, directing breezes into the house. And when closed, the compression seals create a tight fit that only gets tighter with oncoming winds.

Casements are not without problems, however. The outward swinging sash is vulnerable to water damage if the top is not fully clad, and the sash can clash with screen doors or encroach on deck spaces. Also, hardware problems such as stripped crank handles or broken linkages are not uncommon, particularly on lower-end units or large units with heavy sash. An occasional squirt of lubricant on the crank mechanism and hinges can go a long way toward preventing problems.

Large, heavy sash can also rack slightly out of square over time, requiring a push from the outside to fully close. Sometimes this can be fixed by tweaking the hardware, but a new sash may be required.

For best results with casements, choose units with sturdy sash construction, heavy-duty hardware, and sash that are protected on top from the elements.

Awning and Hopper Windows

Awning windows swing outward from top hinges, and hopper windows swing inward from bottom hinges. Awnings are useful for privacy windows and other high-up locations like clerestories, while hoppers are often used in basements. Both awnings and hoppers can be combined with fixed glazing to add ventilation below a picture window. Because they rely on cranking mechanisms, these windows have some of the same problems as casements. But without the heavy vertical sash, they are less prone to malfunction. For hard-to-reach locations, a pole or motorized operators can be used.

Tilt-Turn Windows

Developed in Europe, tilt turn windows rely on intricate hardware controlled by a single lever that allows the windows to either swing in like a door or tilt in at the top like a hopper. In tilt mode, the sash are open only a few inches, allowing the windows to provide ventilation in a locked house (see Figure 3-2).

When closed, the lever locks the window tightly in four to six locations, providing the tightest fitting windows available as well as excellent security against would-be thieves. And with their easy-to-operate in-swinging mode, large-sized tilt-turn windows often do double duty as emergency exits.

While tilt-turn windows offer several useful features, they are not commonly seen in the United States, perhaps because of the in-swinging sash and relatively high cost. Also, they do not typically come with screens, making them problematic in areas with summer pests. While not heavily marketed in the United States, they can be found here in wood, aluminum, aluminum-clad, and solid vinyl frames, the least expensive option.

WINDOW MATERIALS AND CONSTRUCTION

For many years, the material choices for new residential windows were limited to wood, clad wood, and aluminum. Wood and clad wood remain the leading materials, accounting for almost 50% of the new and replacement window market. Wood use has been declining, however, with the rapid growth of solid vinyl windows.

Solid vinyl windows made inroads into the replacement window market in the mid-1980s; but they were not widely accepted in new homes until the 1990s, when their use skyrocketed. They now account for an estimated 30% of the new-home market and 60% of the replacement market. Aluminum windows account for about 15% of window sales, with the remaining share of the market spread among fiberglass windows and a variety of hybrids and composites that have entered the fray, making window selection today anything but simple.

Wood and Clad-Wood Windows

Wood is the traditional material of choice for residential windows. It can be milled into highly detailed designs,

FIGURE 3-2   Tilt-Turn Windows.

Tilt-turn windows, popular in Europe, have intricate hardware controlled by a single lever that allows the window to swing in like a door or tilt in at the top for ventilation.
SOURCE: Photo courtesy of Marvin Windows and Doors.

such as true divided lites, and easily fashioned into just about any custom configuration. In addition to its aesthetic appeal, wood has excellent insulation value, and if well-maintained can last indefinitely.

On the downside, wood must be stained or painted and well-maintained or, over time, it will be vulnerable to peeling paint and decay. Windows with wood exposed directly to the weather, such as open casements, are particularly vulnerable. Although wood is dimensionally stable with temperature changes, it does swell and shrink with changes in relative humidity, causing tight-fitting windows to stick in humid summer weather.

When selecting a wood window, look for materials that are factory-treated with a water-repellant preservative, which will help prevent decay and also improve paint

retention and dimensional stability. The vast majority of wood windows made today have a vinyl- or aluminum-clad exterior. A clad exterior is the most practical choice, providing a maintenance-free exterior with the look of a wood window on the interior. The only downside is a limited color choice.

Construction.   Some very high-end wood windows are manufactured from rot-resistant species, such as mahogany, but most use clear pine for the sash and visible portions of the frame. The hidden portions of most wood windows use some combination of solid wood, finger-jointed lumber, and various types of engineered lumber, including laminated-strand lumber (similar to oriented-strand board, or OSB). Composites are beginning to be used as well, such as Fibrex, a proprietary wood-vinyl composite used in the subsills of some of Andersen's replacement windows and patio doors.

In general, wood windows are sufficiently strong and rigid for most residential applications. However, it is always best to use windows approved by one of the three agencies that certify windows and doors (see "Window Certification," page 110). In coastal areas or other areas prone to high winds or hurricanes, look for products with a suitable pressure rating from the American Architectural Manufacturers Association (AAMA) or the Window and Door Manufacturer's Association (WDMA).

Vinyl Cladding. While vinyl cladding is only offered by a few manufacturers, one of them is Andersen Windows, by far the largest supplier of residential windows in the United States. Andersen's clad windows' strong record of durability, reliability, and moderate cost has helped make vinyl-clad windows one the most popular options today. Good quality vinyl-clad windows have a heavy-gauge covering, and heat-welded corners to provide a durable seal against water entry. Vinyl cladding is also more energy-efficient than aluminum and is preferred by some in coastal environments subject to salt spray.

On the downside, vinyl comes in only a few colors, typically white and beige, and cannot be painted. Some manufacturers, such as Andersen, now offer a limited number of dark tones as well, using newer technologies that resist the fading and heat problems characteristic of dark-colored vinyl. Vinyl is also vulnerable to cracking in cold weather if struck by an errant baseball or hammer.

While all manufacturers cover the exterior frame and exterior face of the sash, some also protect the top of the sash, which is important in casement windows. Andersen wraps the entire sash inside and out, providing excellent protection (see Figure 3-3).

But for those seeking the look of wood on the interior of the sash, consider windows with exterior-only cladding, such as those from Weathershield and MW Manufacturing.

Aluminum Cladding. A number of manufacturers offer aluminum cladding, using either relatively thin roll-formed aluminum or heavier-gauge extruded aluminum. The advantage of extruded aluminum is that it adds strength and rigidity to the window and resists denting better than thinner stock. Also, aluminum can be formed to crisper profiles than vinyl, creating a less bulky appearance. Other advantages of aluminum include a wider choice of colors and the ability to be painted if desired. One downside to aluminum cladding is a slight reduction in energy-efficiency compared to vinyl-clad windows. On average, a vinyl-clad unit has an R-value about 10% higher than for a comparable aluminum-clad window.

Fiberglass Cladding. A few manufacturers, including Marvin and Milgard, offer wood windows clad on the ex-

FIGURE 3-3 **Vinyl Cladding.**

Andersen wraps the entire sash, inside and out, in its vinyl-clad casement windows, providing excellent protection from the elements. Other manufacturers leave natural wood on the interior face of the sash for a more traditional look.
SOURCE: Photo courtesy of Andersen Windows.

terior with a tough fiberglass composite manufactured in a process called *pultrusion* (see "Fiberglass Windows," page 109). Pultruded fiberglass is an ideal cladding material due to its durability, energy efficiency, and very low rate of thermal expansion. Unaffected by heat, cold, and moisture, manufacturers claim that pultruded fiberglass will never crack, peel, or warp. Fiberglass-clad windows come prepainted with a factory finish and can be repainted on site if desired.

Vinyl Windows

Solid vinyl windows can attribute their surge in popularity over the past decade to the fact that they have delivered a high-quality, maintenance-free product for about a third less cost than a clad window of comparable quality. Contractors like the fact that the interior sash do not need painting, and homeowners like the notion that the frame and sill cannot rot (although vinyl windows do not live forever).

Solid vinyl windows have shed their image as cheap plastic replacement windows by continually improving to the point where premium quality vinyl windows are stronger and more dimensionally stable, more fade resistant, and better looking with crisper extrusions and better hardware (Figure 3-4). As an added benefit, vinyl window manufacturers claim that their products can match the energy performance of wood windows.

While improvements in extrusions have made vinyl windows less bulky looking, they still do not have the crisp lines of a wood window, particularly on the interior,

and vinyl window sash are unmistakably white plastic. Still, vinyl windows have shed their stigma as a low-end product and are finding their way into more and more new homes across the spectrum from spec to upper-end custom.

Construction. Vinyl window quality starts with the chemistry. A number of additives are blended into the raw vinyl to make it more dimensionally stable and more resistant to UV radiation, which otherwise can cause the vinyl to fade, chalk, and become brittle over time. The vinyl is then extruded into long sections with multiple internal chambers that give the material its rigidity and insulation value from the trapped air (see Figure 3-5).

The thin-walled plastic ranges in thickness from about .065 to .085 inch. All other things being equal, the thicker

| FIGURE 3-4 | Solid Vinyl Windows. |

Solid vinyl windows have largely shed their one-time image as cheap plastic replacement windows. New models have greater strength, stability, and fade resistance and are better looking, with crisp extrusions. The one shown above mimics brickmold but with an integral channel for vinyl siding.
SOURCE: Photo by author.

| FIGURE 3-5 | Vinyl Extrusions. |

Vinyl is extruded into long sections with multiple internal chambers that give the window frames their rigidity and insulation value. Quality vinyl windows like the double-hung shown have heat-welded corners, and many windows have metal reinforcement in key areas.
SOURCE: Photo courtesy of Jeld-Wen, Inc.

the plastic, the stronger the window component. However, extrusions can gain strength from having a well-engineered profile, as well (many European windows use much heavier .125-inch plastic, but U.S. manufacturers feel this would make them less competitive).

The extrusions are then cut and joined to create frames and sash. Nowadays, most corners are heat-welded, producing the strongest joints, although some corners are still mechanically fastened. Some manufacturers add steel or aluminum reinforcing, particularly to larger windows, to stiffen them and help them meet structural load requirements. Typical places for reinforcing are sills, which have a tendency to sag in hot weather, lock rails, and mullions between mulled units.

Options. All vinyl windows have an integral nailing flange, simplifying watertight installations (see "Window Flashing," page 119). All standard window types, including tilt-turn, are available in solid vinyl today. All glazing types, except for true divided lites, are also available. For the look of divided lites, the options are either snap-in grilles or between-the-glass grills.

Typical exterior trims include brickmold or nominal 4-inch flat casings with either an integral or snap-on J-channel to receive the ends of wood, vinyl, or fibercement siding. Trims designed for stucco are widely available in the western states and Florida. Colors are generally limited to white and beige, since dark colors absorb heat and raise vinyl temperatures to near 165°F, where it begins to soften and sag. On the interior, most vinyl windows will accept either wood extension jambs or drywall returns.

Quality Issues. As with any building products, not all vinyl windows are created equal. Many early models were poorly made and subject to excessive thermal movement, often opening at corners, fading prematurely, and losing structural integrity from UV exposure, particularly in hot climates. Most window manufacturers today have overcome these problems with better vinyl formulations, improved extrusion design, and heat welding at corners. Many also use metal reinforcing at strategic points, such as meeting rails, sills of large units, and between mulled units.

Since the chemistry, extrusion design, metal reinforcing, and other determinants of quality are hidden from view, however, the best approach is to stick with an established manufacturer and to look for a certification label from the American Architectural Manufacturers Association (AAMA). AAMA established a separate standard for vinyl windows in 1997, which includes tests for strength, dimensional stability, strength of corner welds, heat and impact resistance, and weathering. The main features to look for include:

- Heat-welded corners
- Metal reinforcing, particularly on larger units
- Solid, heavy extrusions with little obvious flex
- Good quality hardware

Installation. In general, vinyl windows install the same as other flange-type windows, although the installer must take into account vinyl's high rate of thermal expansion. A 6-foot-wide window can expand as much as $\frac{5}{16}$ inch from 0°F to 100°F. To prevent problems, It is best to leave a $\frac{1}{8}$- to $\frac{1}{4}$-inch gap between the window and siding or wood trim (more in cold weather, less in hot). Good detailing of this joint is especially important with stucco, which can crack if set too tightly against the window or leak if the caulk joint fails.

Thermal expansion can also cause window sash to bind in hot weather if the rough opening is too small to accommodate the movement. And in very cold weather, some vinyl windows can bow inward due to temperature differences on either side of the window. Sturdy extrusion profiles with metal reinforcing can help prevent this. A related concern is sagging of the sill during hot weather—vinyl starts to soften and distort at about 165°F, a temperature easily reached on the surface of a dark building in direct sun. To prevent sagging, some manufacturers reinforce the sill, and all recommend specific shim spacing under the sill. Some require continuous support along the length of the sill, which is easiest to achieve by using a double 2x sill with leveling shims in between.

Nailing recommendations also vary among manufacturers. Some recommend driving nails tight; some suggest leaving the nail heads proud. Some recommend against nailing the head flange or corners; others require it. To avoid warranty problems, it is always best to closely follow the manufacturer's instructions regarding the rough opening, shimming, nailing, and other installation details. Other general recommendations that apply to all windows include:

- In hot weather, do not store vinyl windows in a container, such as a trailer, or lean them against a wall, as they can permanently deform.
- Seal around openings with flexible flashing before installing the window (see "Window Flashing," page 119).
- Set the nailing fin in a bead of high-quality caulking.
- Before nailing, make sure the side jambs are plumb and do not bow in or out in the center. Some installers use a wood spacer to hold the two side jambs parallel.
- Do not puncture the nailing fin when installing trim or siding.
- Do not use expanding foam to seal around the window frame on the interior.

Fiberglass Windows

Introduced in the early 1990s, a few manufacturers now offer windows built entirely of a tough composite called *pultruded* fiberglass. Unlike the layers of fiberglass cloth and resin used in boat and car construction, fiberglass window components are made from continuous glass fibers saturated with a thermoset resin and pulled through a heated die in a process called *pultrusion*. The result is a thin, strong composite that can be formed into detailed shapes and is used in a variety of high-tech applications.

Pultruded fiberglass is noted for its high strength, durability, and corrosion resistance. It is unaffected by temperatures up to 350°F and has an extremely low rate of thermal expansion—about the same as window glass. Because the frame and glass move at the same rate, temperature changes place less stress on the window frame and glass edge seals. Manufactures claim that pultruded fiberglass will not crack, peel, or warp and is impervious to moisture, insects, salt-air, and UV exposure (Figure 3-6).

FIGURE 3-6 | **Fiberglass Windows.**

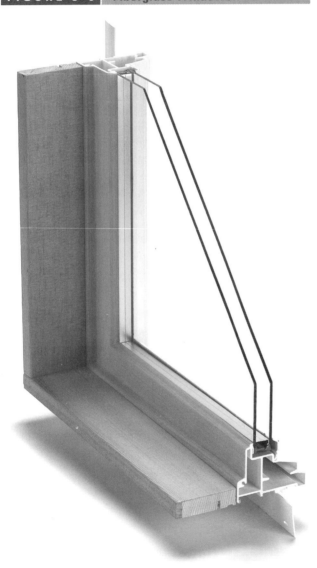

One of the newest window materials, pultruded fiberglass is noted for its high strength, durability, and dimensionally stability—it expands and contracts at the same rate as window glass. This model from Milgard has a durable baked-on paint finish on the exterior with a vertical-grain-fir veneer on the interior.

SOURCE: Photo courtesy of Milgard Windows.

Depending on the manufacturer, fiberglass frame components are either hollow or filled with foam or fiberglass insulation. The insulated frames are the most energy-efficient on the market. Most fiberglass frames are shipped with a high-performance baked-on factory finish, and they can be repainted on-site if desired. Installation is the same as for vinyl and other flange-style windows but without concerns related to sagging or thermal movement.

Manufacturers currently offering fiberglass windows include Milgard Windows and several Canadian manufacturers, including Fibertec and Thermotech Windows.

Aluminum Windows

While aluminum is strong, light, and durable, with an anodized or baked-on finish, it has been steadily losing market share since the early 1990s due to its poor energy performance. Fewer than 15% of windows sold today are aluminum, and these are mainly in lower-end housing in cooling-dominated climates. The poor insulating qualities of aluminum have less of an impact on cooling than they do on heating since the indoor and outdoor temperature difference is generally much smaller in cooling climates. Adding a thermal break can improve the energy performance of an aluminum window, but it still lags considerably behind wood and vinyl components.

Window Certification

It is always best to choose windows with third-party certification from American Association for Window Manufacturers (AAMA) or the Window and Door Manufacturers Association (WDMA, formerly the National Wood Window and Door Association, NWWDA). Both the AAMA and WDMA are trade associations representing manufacturers of windows, doors, and skylights. The WDMA is limited to wood windows and doors, while the AAMA encompasses a wide range of materials, from wood, aluminum, vinyl, and fiberglass to the newest composites.

Voluntary Standards. In 1997, the two groups joined forces to establish a unified standard for evaluating fenestration products, known as AAMA/NWWDA 101/I.S. 2-97 "Voluntary Specification for Aluminum, Vinyl and Wood Windows and Glass Doors." It establishes four performance requirements for a completed window or door.

- Structural ability to resist wind loads
- Resistance to air leakage
- Resistance to air infiltration
- Resistance to forced entry

Products that are certified under 101/I.S. 2-97 are designated by a four-part code that denotes the type of window, the performance class, and performance grade. For example, the code C-R15 indicates a casement window (C)

TABLE 3-2 Minimum Window and Door Performance Grades

Class	Performance Grade/Design Pressure (psf)	Structural Test Pressure (psf)	Equivalent Wind Speed (mph)*	Water Resistance Test Pressure (psf)	Equivalent Wind Speed (mph)*
Residential	15	22.5	95	2.86	34
Light Commercial	25	37.5	123	3.75	39
Commercial	30	45	164	4.50	42
Heavy Commercial	40	60	155	6.00	49
Architectural	40	60	155	8.00	56

*Equivalent wind speed is approximate and based on assumptions of building shape, exposure, and other conditions.

recommended for residential applications (R), with a performance grade of 15.

Performance Grade and Design Pressure. How well a window performs when subjected to heavy rains and high winds is indicated by its *performance grade* and *design pressure*. The design pressure is a structural rating only, while the performance grade also indicates that a window has met the water resistance and air infiltration standards for that grade (see Table 3-2).

The minimum recommended design pressure for residential doors and windows is 15 psf. A design pressure of 15 means a window has been tested to withstand sustained wind pressures of 22.5 psf, roughly equivalent to a 95-mph wind, applied to either side of the window, simulating both positive and negative wind pressures. The test pressure is always 150% of the rated design pressure to provide a safety factor. To earn a performance grade of 15, a window must also pass a water pressure test of 2.86 psf, which simulates rainfall of 8 inches per hour with a wind speed of 34 mph. In coastal areas or other areas prone to heavy winds or hurricanes, higher grade windows are recommended and may be required by code.

Storm-Resistant Windows

In response to the devastating impact of Hurricane Andrew in 1992, Florida enacted stringent codes to protect homes from severe storms. Other coastal states have followed suit in recent years, and now similar provisions in the International Residential Code (IRC) apply to coastal areas from Texas to Maine.

Protect the Openings. Researchers attributed much of Andrew's destruction to wind penetration into homes through broken doors and windows, leading to extensive water damage and, in many cases, roofs blown off and houses destroyed. The keys to preventing these problems were strengthening roofs and protecting windows and doors from wind and wind-borne debris. To protect windows, the new code allows three options: code-approved storm shutters, $\frac{9}{16}$-inch plywood panels screwed over windows at 8 inches on-center, or impact-resistant windows. The trend in new home construction is toward impact-resistant windows, sometimes marketed as "storm-resistant" or "hurricane-rated."

Miami-Dade County Standards. Miami-Dade County enacted the most stringent standard and test protocols, subjecting windows (and storm shutters) to a test in which a 9-pound 2x4 is hurled into the glass at 50 feet per second, followed by 4,500 cycles of positive and negative wind loads equivalent to a 146-mph wind. Miami-Dade also conducts AAMA/NWWDA testing for design pressure and water intrusion, but it conducts the water intrusion test after the structural test is completed rather than on a new window. Windows and doors that pass the Miami-Dade Product Control Standards are required throughout Miami-Dade County and most other coastal areas in Florida.

International Residential Code. The International Residential Code (IRC) requires impact-resistant windows in all hurricane-prone regions along the Gulf and Atlantic coasts from Texas to Maine. Depending on the wind-speed zones established in the IRC, windows need to meet design pressures ranging from 30 to 80 psf, and they must meet impact-resistance standards under ASTM E1886 or E1996. The design pressure required depends on both an area's design wind speed, found on IRC maps, and the building's exposure rating from A to D. Most buildings are rated Exposure B for "urban and suburban areas or wooded areas" or Exposure C for flat, open terrain with scattered obstructions of less than 30 feet. Waterfront buildings exposed to winds flowing over open water for at least a mile are rated Exposure D, the most severe.

Window Construction. Under pressure from both the building codes and insurance industry, most major window manufacturers have developed impact-resistant windows for residential applications that feature laminated glass along with heavier frames and hardware. The glass is similar to auto windshields with a plastic interlayer, but it is significantly heavier. Double-glazed units get a second layer of tempered glass either on the interior or exterior. Vinyl-framed windows are heavily reinforced with aluminum, and all windows use metal mullion bars anchored to the framing between mulled units. Window-to-framing attachment methods are also beefed up to comply with the new codes, and in some cases, metal clips are used to anchor the window to the frame. Storm-resistant windows cost from two to four times as much as standard windows; but under pressure from code agencies and insurance companies, these windows will soon become standard fare in coastal construction and other storm-prone areas.

WINDOW ENERGY EFFICIENCY

Windows have a large impact on a home's energy consumption, accounting for up to 25% of a typical home's heating bills in cold climates and up to 50% of cooling bills in hot climates, according to the Environmental Protection Agency's (EPA) Energy Star program. Even the best windows, with an R-value of 3 to 4, are thermal holes compared to today's average R-19 wall. In addition to conductive heat losses, older windows add substantially to a home's air leakage.

Beyond fuel bills, windows can also have a dramatic effect on occupant comfort. Sitting next to a leaky single-glazed window in winter will make an occupant feel cold regardless of the thermostat setting, due to both cold drafts and to radiant heat losses from his or her body to the window surface. Cold window surfaces also cause condensation, potentially leading to mold, peeling paint, and wood decay of window components.

Energy Certification

Evaluating a window's energy performance is a complex task that has been made a lot simpler by two programs developed in a collaborative effort between government and industry. The groups have developed standardized testing procedures and ratings, and provide simple recommendations based on climate zone.

The NFRC Label. The National Fenestration Rating Council (NFRC), with support from the U.S. Department of Energy, created test procedures and rating systems for the energy performance of windows, glazed doors, and skylights. Any window making energy claims without an NFRC label should be avoided. For every window, the NFRC label rates the U-factor, Solar-Heat-Gain Coefficient (SHGC), and Visible Transmittance (VT). Air Leakage (AL) and Condensation Resistance (CR) are optional ratings. The ratings are explained briefly in Figure 3-7 and in more detail in the following sections. Ratings should appear on the window label when delivered but can also be found on the NFRC website at www.nfrc.org.

Whole Window Ratings. It is important to note that NFRC ratings apply to the entire window, including the sash and the frame. NFRC uses a single standard size to simplify testing and to make it easier for consumers to make apples-to-apples comparisons between windows. The actual energy performance of windows significantly larger or smaller than the standard test size will vary somewhat from the label since the relative effect of the glass edge and frame is greater on smaller windows.

Glass-Only Ratings. Where glass-only ratings are needed, for example, for passive solar design, these can usually be obtained from the window manufacturer or the manufacturer of the insulated glass unit (IGU) installed in the window or door.

Energy Star Label. Launched by the U.S. EPA to promote the use of energy-efficient appliances and equipment, the Energy Star label was added to doors, windows, and skylights in 1998. An Energy Star Label certifies that the window or skylight meets the U.S. Department of Energy's (DOE) energy guidelines for the climate zones listed on the label (see Table 3-3). For windows without the label, you can still use the Energy Star guidelines in Table 3-3 as a selection guide. Remember that these guidelines are based on NFRC whole-window ratings, not just the glass. For cold-climate buildings designed to use passive-solar heating, look for a whole-window SHGC of .55 or above.

U-Factor

A window's ability to conduct heat (not including solar effects) is usually given as a U-factor or U-value. The lower the U-value, the more insulation value a window provides. Low U-values have the biggest impact in heating dominated climates but help reduce cooling loads as well. For climates with substantial heating or cooling loads, choose a total window U-value of .35 or less.

The U-value is the inverse of the more familiar R-value. For example, standard double glazing has a center-of-glass U-value of about .5, which equals an R-value of 2 (1/0.5). Typical glazing U-values are shown in Table 3-4. U-values for the entire window, however, must take into account the edge spacers, sash, and frame, as discussed below.

Gas Fill. Filling low-E coated glass with the inert gas argon or krypton will reduce heat loss through the glass by 10 to 15%. Krypton outperforms argon somewhat, but it is usually not enough to justify the higher cost. Since argon fill is now available on most low-E windows for little or no cost, getting the boost in R-value is always a good idea. In addition to reducing heat loss, it increases the temperature of the inside surface of the window, improving comfort and reducing condensation. Studies indicate that about 10% of the gas will leak out of a well-built sealed glass unit in about 20 years.

FIGURE 3-7 | Anatomy of an NFRC Label.

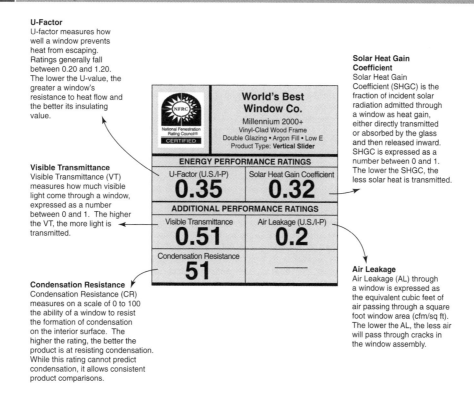

U-Factor
U-factor measures how well a window prevents heat from escaping. Ratings generally fall between 0.20 and 1.20. The lower the U-value, the greater a window's resistance to heat flow and the better its insulating value.

Visible Transmittance
Visible Transmittance (VT) measures how much visible light come through a window, expressed as a number between 0 and 1. The higher the VT, the more light is transmitted.

Condensation Resistance
Condensation Resistance (CR) measures on a scale of 0 to 100 the ability of a window to resist the formation of condensation on the interior surface. The higher the rating, the better the product is at resisting condensation. While this rating cannot predict condensation, it allows consistent product comparisons.

Solar Heat Gain Coefficient
Solar Heat Gain Coefficient (SHGC) is the fraction of incident solar radiation admitted through a window as heat gain, either directly transmitted or absorbed by the glass and then released inward. SHGC is expressed as a number between 0 and 1. The lower the SHGC, the less solar heat is transmitted.

Air Leakage
Air Leakage (AL) through a window is expressed as the equivalent cubic feet of air passing through a square foot window area (cfm/sq ft). The lower the AL, the less air will pass through cracks in the window assembly.

World's Best Window Co.
Millennium 2000+
Vinyl-Clad Wood Frame
Double Glazing • Argon Fill • Low E
Product Type: **Vertical Slider**

ENERGY PERFORMANCE RATINGS

U-Factor (U.S./I-P)	Solar Heat Gain Coefficient
0.35	**0.32**

ADDITIONAL PERFORMANCE RATINGS

Visible Transmittance	Air Leakage (U.S./I-P)
0.51	**0.2**

Condensation Resistance	
51	—

The NFRC label provides a reliable way to gauge a window's energy performance and to compare one window to another. The ratings are for the whole window, including the sash and frame, and are based on one standard size for each window type. Ratings for air leakage and condensation reduction are optional.

TABLE 3-3 | Recommended Glazing Types—(DOE Energy Star Program)

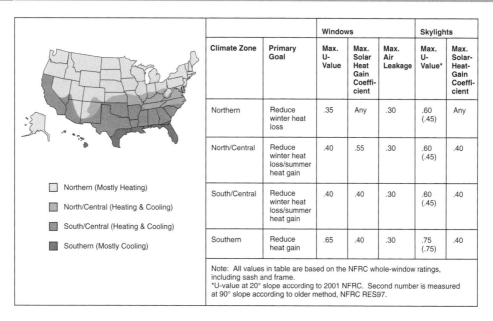

Northern (Mostly Heating)
North/Central (Heating & Cooling)
South/Central (Heating & Cooling)
Southern (Mostly Cooling)

Climate Zone	Primary Goal	Windows			Skylights	
		Max. U-Value	Max. Solar Heat Gain Coefficient	Max. Air Leakage	Max. U-Value*	Max. Solar-Heat-Gain Coefficient
Northern	Reduce winter heat loss	.35	Any	.30	.60 (.45)	Any
North/Central	Reduce winter heat loss/summer heat gain	.40	.55	.30	.60 (.45)	.40
South/Central	Reduce winter heat loss/summer heat gain	.40	.40	.30	.60 (.45)	.40
Southern	Reduce heat gain	.65	.40	.30	.75 (.75)	.40

Note: All values in table are based on the NFRC whole-window ratings, including sash and frame.
*U-value at 20° slope according to 2001 NFRC. Second number is measured at 90° slope according to older method, NFRC RES97.

Sash and Frame Effects. The U-value of the entire window, as reported on NFRC labels, includes the effects of the glass edge, sash, and frame. With high-R glass, standard edge and frame materials often lower than the overall R-value compared to the center-of-glass measure. Aluminum sash and frames without thermal breaks are the worst, contributing to both thermal losses and condensation in cold climates. Thermally broken metal frames are better but

TABLE 3-4 | **Typical Glazing Characteristics (center of glass)**

Type of Glazing	U-Value (R-Value)	Visible Light Transmittance	UV Light Transmittance*	Solar-Heat-Gain Coefficient	Recommended Applications
Single glazing, clear	1.0 (1.0)	90%	71% (85%)	.86	None
Double glazing, clear	.50 (2.0)	81%	56% (59%)	.76	None
Double glazing, low-E, high-solar gain	.35 (2.9)	75%	47% (51%)	.71	Cold climates; passive solar
Double glazing, high-solar gain, low-E, argon**	.29 (3.4)	75%	47% (51%)	.71	Cold climates; passive solar
Double glazing, moderate-solar gain, low-E, argon	.27 (3.7)	78%	23% (40%)	.58	Cold or mixed climates
Double glazing, spectrally selective low-E, argon***	.25 (4.0)	71%	16% (33%)	.39	Hot or mixed climates; west-facing glass
Double glazing (1 inch) with clear Heat film	.21 to .26 (3.8 to 4.8)	20 to 81% (varies with coating type)	<1% (28% to 53%)	.14–.57	Match coating to climate and design needs.

*Number in ( ) is "damage-weighted transmittance (T-dw)," which includes the portion of visible light that contributes to fading. Lower numbers indicate less fading.
**High-solar-gain glass uses "hard-coat" or pyrolitic coatings.
***Spectrally selective glass is also called "low-solar-heat-gain low-E."

TABLE 3-5 | **Frame Effect on Whole-Window U-Values**

	Whole-Window U-Value						
Center of Glass U-Value (low-E argon)	Aluminum Frame	Aluminum Frame with Thermal Break	Wood or Clad Wood Frame	Solid Vinyl Frame	Composite Frame	Insulated Vinyl Frame	Fiberglass Frame
U-Value .27	.60	.48	.35	.35	.35	.27	.27

NOTE: Numbers represent whole-window U-value with each frame type, based on an average of many windows.
SOURCE: Courtesy of Efficient Windows Collaborative.

should still be avoided in cold climates. Wood and hollow vinyl or fiberglass components all have moderately good thermal properties. Insulated vinyl and fiberglass frames offer the best thermal performance (Table 3-5).

Warm-Edge Spacers. Because of their high thermal conductivity, standard aluminum edge spacers lower the insulating value of insulated glass units (IGUs) and often cause condensation along the bottom of the window. The loss of insulation value is more pronounced in very high R-value windows and small windows where the window edge accounts for a larger proportion of the window area. Starting in the mid-1980s, manufacturers have responded with a variety of innovative solutions that are now used in about half of all new IGUs.

Solutions include using less conductive metals with an improved shape (PPG's Intercept Spacer) or switching to a plastic or synthetic rubber spacer with little or no metal content (TruSeal's Swiggle Seal or EdgeTech's Super Spacer). Warm-edge windows can raise the glass temperature at the perimeter of the window by 6°F to 8°F, significantly reducing the condensation potential. The biggest risk in using a new edge technology is that the seal

will fail prematurely, resulting in a fogged unit. To guard against this, it is best to stick with a technology that has proven itself in the marketplace and is backed by a good warranty and a reliable window manufacturer.

Low-E Coatings

Low-emissivity, or "low-E," coatings are microscopically thin metallic coatings applied to a glass surface, which reflect back radiant heat. Different low-E coatings transmit different amounts of visible light, short-wave and long-wave infrared, and ultraviolet radiation (Figure 3-8).

The most common type, called "soft coat" low-E, is applied to one of the inner surfaces of sealed insulated glass units (Figure 3-9).

Hard-coat, or "pyrolitic," low-E, which has a slightly lower R-value, is used in high-solar-gain glass and can also be used on storm windows and other removable glass panels exposed to the air. Low-E coatings can also be applied to a clear polyester film, called Heat Mirror, which is suspended between two panes of sealed glass, yielding insulation values as high as R-5 with one layer of film or R-8 with two layers.

FIGURE 3-8 Low-E Coatings.

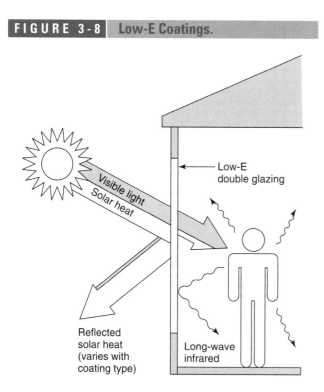

Low-emissivity coatings reduce heat loss in winter by reflecting long-wave heat energy back into the building. Their effect on solar heat gain (near-infrared radiation) depends on the coating type: High-solar-gain glass lets most of the available heat radiation into the building, while low-solar-gain glass rejects most of the sun's heat energy.

FIGURE 3-9 Anatomy of a Low-E Window.

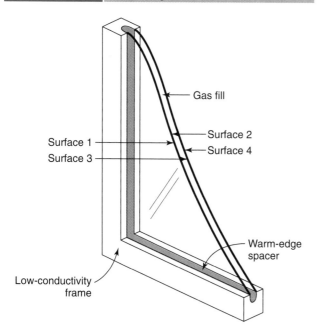

Soft-coat low-E coatings are placed on the inside of sealed-glass units. Placing the coating on surface 3 maximizes solar heat gains, while placing the coating on surface 2 minimizes solar gains.

Spectrally Selective Low-E. The newest generation of low-E glazing, often referred to as "spectrally selective," provides an ideal combination of high R-values, low heat gain, and high-visible-light transmittance. Spectrally selective windows generally outperform all other window types in mixed and hot climates, but they reap the greatest benefit in homes with significant cooling loads. Because of their high insulation value, spectrally selective windows even perform well in cold climates, particularly in homes with significant air-conditioning loads or large amounts of west-facing glass (see "Window Orientation," page 117). One exception is a house designed to use passive solar gain in winter, which would perform better with high-solar-gain glass.

Glazing and Climate. Due to its significantly lower U-value, low-E glass outperforms standard double glazing in all climates. However, which type of low-E glazing is optimal for a building depends on several factors, including the heating load, cooling load, and orientation of the glass. In general, the windows with the lowest U-values will yield the greatest savings in cold climates, while windows with the lowest solar heat gain will yield the greatest savings in hot climates. Performance comparisons of different window types in different climates, based on computer modeling, are shown in Figure 3-10.

Some window manufacturers market different glazing types in different parts of the country and may be able to provide different glazing types by special order. General recommendations from the EPA's Energy Star program are shown in Table 3-3, above.

- **Cold climates.** In climates dominated by heating loads, select a window with a low U-value (high R-value). Windows with high solar gain will slightly favor winter performance and windows with low solar gain will slightly favor summer performance, but annual energy costs are determined primarily by the U-value. Two exceptions to this are passive solar homes, which should use high-solar-gain glass, and homes with a lot of west-facing glass, which should use low-solar-gain (spectrally selective) glass.

- **Hot climates.** In climates dominated by cooling loads, choose a glazing type with low solar-heat gain. Spectrally selective coatings provide an ideal combination of high R-values, high visible-light transmission, and low solar-heat gain. UV radiation, which causes fading, is also cut significantly. Spectrally selective glass is a big improvement over tinted glass, which blocks solar gains, but also obscures views and creates glare and reflections when viewed from outside.

- **Mixed climates.** Spectrally selective glass or low-E glass with moderate solar gains are good choices in mixed climates. The greater the cooling load, the more important a low solar-heat-gain factor will be. However, the differences in annual fuel bills between using high, low, or moderate solar-heat gain glass in these climates will usually be small. Other issues like overall U-value, UV light transmission, cost, and durability might be the more important factors in choosing a window.

FIGURE 3-10 **Annual Energy Performance of Glazing Types by Climate.**

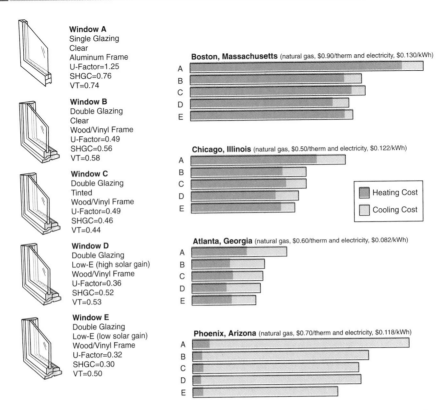

Window A
Single Glazing
Clear
Aluminum Frame
U-Factor=1.25
SHGC=0.76
VT=0.74

Window B
Double Glazing
Clear
Wood/Vinyl Frame
U-Factor=0.49
SHGC=0.56
VT=0.58

Window C
Double Glazing
Tinted
Wood/Vinyl Frame
U-Factor=0.49
SHGC=0.46
VT=0.44

Window D
Double Glazing
Low-E (high solar gain)
Wood/Vinyl Frame
U-Factor=0.36
SHGC=0.52
VT=0.53

Window E
Double Glazing
Low-E (low solar gain)
Wood/Vinyl Frame
U-Factor=0.32
SHGC=0.30
VT=0.50

Boston, Massachusetts (natural gas, $0.90/therm and electricity, $0.130/kWh)

Chicago, Illinois (natural gas, $0.50/therm and electricity, $0.122/kWh)

☐ Heating Cost
☐ Cooling Cost

Atlanta, Georgia (natural gas, $0.60/therm and electricity, $0.082/kWh)

Phoenix, Arizona (natural gas, $0.70/therm and electricity, $0.118/kWh)

NOTE: Annual energy-performance figures were generated using RESFEN software for a typical 2,000 square feet house with 300 square feet of window area, equally distributed on all four sides, with typical shading. Costs for heating with a gas furnace and cooling by air conditioning are based on typical energy costs for each location. U-factor, SHGC, and VT are for the total window, including frame.

SOURCE: Reprinted with permission from *Residential Windows, 2nd Edition* © 2000, John Carmody, Stephen Selkowitz, Dariush Arasteh, and Lisa Heschong. Published by W. W. Norton & Co.

Solar-Heat-Gain Coefficient

A window with an SHGC of .70 captures about 70% of the available solar energy falling on the window. Clear double glazing has an SHGC of about .75 versus .60 to .70 for standard low-E and about .40 for spectrally selective low-E. Which type of glazing is optimal for a given project depends on the climate, summer and winter fuels costs, and how glass is used in the house design.

- *Low SHGC.* Low-solar-gain glass blocks unwanted solar gain and provides significant savings in both peak and annual cooling loads in hot climates. For example, switching from clear double glass to low-SHGC glazing can reduce air-conditioning bills by 15 to 20% in a typical home in Phoenix or Miami (see Figure 3-10). Spectrally selective glass, introduced in the 1990s, combines very low solar gains with high visible light transmittance and high R-values (up to R-4 with gas fill). While this yields the greatest savings in hot climates, it is also a good choice in any climate with significant cooling loads or large amounts of unshaded west-facing glass (see "Spectrally Selective Glazing," page 118).

- *Moderate SHGC.* In northern cities like Boston or Chicago or mixed climates with more-or-less equal heating and cooling costs, moderate-gain glass is a reasonable choice, balancing moderate solar gains in winter with moderate blocking of solar gains in summer.

- *High SHGC.* High solar-heat-gain glass is a good choice in cold climate homes with enough south glass to take advantage of passive solar gain, called "sun tempering." Savings on winter heating bills will be partially offset by the increased cooling load in summer, however. To avoid overheating, south-facing glass should range from 4 to 7% of the total floor area (avoid sloped glass, which tends to overheat in summer and fall). With south glass in excess of 7% of floor area, thermal mass may be needed to prevent daytime overheating and to store heat for nighttime use. A designer with expertise in passive solar can help determine the right amounts of glass and thermal mass.

Visible Transmittance

People install windows primarily for daylighting and views, so the higher the percentage of visible light transmitted (VT), the better. Clear double-glazing has a VT of about 80% (see Table 3-4, above). With hard-coat low-E, that figure drops to 75%, and down to about 70% with the new spectrally selective coatings. All low-E coatings reduce visible light transmittance to some extent and some may appear slightly tinted or more reflective under certain light conditions. The new spectrally selective glazings are fairly color-neutral, but they may appear slightly darker compared to clear glass.

In general, most people do not notice tinting until the VT of the glazing falls below about 60%. The visible light transmittance ratings listed on NFRC window labels can be confusing since they include the sash and frame, not just the glass. The VT for the glass only should be available from the window or glazing manufacturer upon request. Beyond the numbers, it is always a good idea to examine a sample of the glass before purchasing. View the glass from both outdoors and indoors under different light conditions to check for tint and glare.

Air Leakage

In older homes, leaky windows contributed significantly to heating loads (less to cooling loads), and the drafts made occupants feel cold despite the thermostat setting. While windows built today are, in general, much tighter, the effect of air leakage can still be significant on cold, windy days, particularly on windows with direct wind exposure. Most windows today are built with a leakage rate of .30 cfm/sq ft of glass area or less, the minimum allowed under the AAMA/NWWDA standard. The best windows have leakage rates near .10 cfm/sq ft.

Windows with compression seals, such as casements and awnings, tend to be tighter than windows with sliding seals, such as double-hungs and sliders. Also slide-by weather-stripping is more prone to wear out over time and more likely to be breached by high winds that cause the window to flex. With any weather-stripping system, look for long-lasting materials such as EPDM and silicone and heavy-duty construction that can withstand years of use and exposure to water, freezing and burning temperatures, and UV radiation.

Fading

Most interior materials, including fabrics, carpeting, paint, and artworks, fade from exposure to sunlight. Although the most potent effect is from ultraviolet (UV) radiation, research has shown that the shorter wavelengths of the visible light spectrum also cause fading. To account for the relative effects of both UV and visible light on typical materials, researchers have developed an approach called "damage-weighted transmittance" (T-dw), which was recently standardized by the International Standards Association (ISO/CIE 89/3).

Typical T-dw numbers range from about 60% for clear double glazing to about 30% for spectrally selective glazing (see "UV Light Transmittance," Table 3-4, p. 114).

Lower ratings are available with triple glazing or tinted glass, primarily used in commercial construction. Low numbers for UV transmittance and T-dw indicate less fading potential, but some fading will still occur. The best approach with valuable rugs, artworks, and other light-sensitive furnishings is to place them in areas with minimal exposure to windows or to use shades or draperies that substantially cut light transmission.

Condensation Resistance

To rate a window's resistance to condensation, NFRC recently developed a method that evaluates the window's frame, glass, and glass edge at a standardized set of temperature and humidity conditions. Based on the coldest part of the window assembly, it is assigned a rating from 1 to 100. The higher the rating, the better the window is at resisting condensation, but the rating doesn't predict condensation under specific conditions. The voluntary minimum for a "thermally improved window" under the AAMA/NWWDA standard is 35.

The best protection against condensation is low-E glass with gas fill, combined with warm-edge spacers and a nonmetallic window frame, such as wood, vinyl, fiberglass, or one of the newer composites. Table 3-6 provides a general guide to when condensation is likely to form on different types of glazing. Without warm-edge spacers, condensation will occur at window edges first.

Window Orientation

Which way a window faces has a big impact on its contribution to comfort, heating and cooling loads, and daylighting.

- *West-facing glass* is the most problematic, because in summer the afternoon sun shines directly on the glass, causing glare and overheating the house already warmed by increasing afternoon temperatures. Provide shade with plantings or light-colored shades (exterior shades are most effective). Overhangs do not help much due to the low angle of the sun.

- *South-facing glass* (within 30 degrees of true south) gets direct sun exposure in the winter when it is desirable, at least in climates with significant winter

TABLE 3-6	Outdoor Temperature at which Window Condensation Occurs
Glass Type	Outdoor Temp. (Fahrenheit)
Single glazing	50° or less
Double glazing	20° or less
Double glazing low-E	−10° or less
Double glazing low-E, argon	−30° or less

NOTE: Assumes 50% relative humidity indoors. Condensation may form along glass edge at higher temperature unless window has warm-edge spacers.

heating loads. In summer, the sun is high on the south side, moderating the solar gain. Also, it is easier to block the high summer sun with appropriately sized overhangs or awnings. South-facing glass should not exceed 7% of the building floor area, unless thermal mass is used.

- **East-facing glass** provides desirable morning light and modest solar gains on cold winter mornings. Too much unshaded east glass, however, can cause over-heating in summer.
- **North-facing glass** provides diffused light that is free of glare and solar gains, which is ideal for daylighting and is sought after by artists for its consistent color and intensity.

In most cases, one type of glazing can work on all sides of the house. In houses with large amounts of west glass, however, it makes sense to use tinted or spectrally selective glass at least on the west face to reject the summer sun. This will dramatically improve comfort and reduce both peak and annual cooling loads. If the house is also designed to take advantage of passive solar heating, high-heat-gain windows are preferable on the south face. Mixing glazing types can get tricky, however, and should be handled by an experienced solar designer. One caution, also, is that the slightly different tints of the two glazing types might be objectionable to some clients.

Shading to Reduce Solar Gain

Shading of glass with overhangs, plantings, or shades will reduce cooling loads and increase comfort in any climate with significant cooling loads. It will also reduce glare (Figure 3-19, page 124), fading of furnishings, and localized overheating in rooms with south- or west-facing glass. Overhangs, plantings, or exterior shades that block the sun before it strikes the glass are the most effective approach since the heat never gets into the building. But light-colored interior shades can also substantially reduce heat gains.

Spectrally Selective Glazing. From a shading standpoint, using spectrally selective glass (SHGC below .40) is like having shades or blinds on standard low-E glass. However, adding good shading to spectrally selective glass can reduce cooling costs by another 10 to 15%. This would make sense in very hot climates or on houses with large expanses of glass on the south or west side. In many cases, the shading adds no cost or serves other design needs. For example, a porch on the east or west side of a house provides effective shading as well as outdoor living space.

Plantings. Deciduous trees can provide very effective summer shade on the south side but, depending on the type of tree, may block 20% or more of the solar radiation in winter. Because trees follow the local seasons rather than the calendar, the shading tends to occur when needed most. For example, leaves appear earlier in the spring and last longer in the fall in warmer climates, which need spring and fall shading. Trees also cool the area around them by their natural evaporative cooling—as water evaporates from the leaves.

Other options for shading south-facing windows include trellises with dense foliage or evergreen trees. Evergreens should be tall enough to block the summer sun but trimmed so their canopies allow the low winter sun to reach the windows.

On the east and west sides of the house, trees or large shrubs can provide very effective shading, since the problem times are morning and afternoon when the sun is low in the sky and easily blocked by a well-placed planting, either deciduous or coniferous.

Fixed Overhangs. These are commonly used on the south side of homes with clear glass or high-solar-gain glass. To be effective, the overhangs must be sized correctly to reject the high summer sun but allow in the low winter sun. In most temperate climates, a $1\frac{1}{2}$- to 2-foot-wide overhang is adequate for average size windows. However, to provide full shading from March to September in hot climates may require a 3-foot or wider overhang.

One limitation of fixed overhangs is that the shading will be the same on March 21 and September 21, although the heating and cooling needs at these times may be very different. The following guidelines for shading south-facing glass strike a balance between summer and winter performance:

- **Cold climates:** Above 6,000 heating-degree days (HDD), locate the shadow line at midwindow, based on the June 21 noon–sun angle (see Figure 3-11). This will shade the window 50% in mid-summer and provide full sun penetration from late September to late March. If more shading is required in summer, locate the shadow line closer to the window sill.
- **Moderate climates:** In climates with less than 6,000 HDD and less than 2,600 cooling-degree days (CDD), locate the shadow line at the window sill based on the June 21 sun angle at noon. This will allow full sun exposure from late October to mid-February.
- **Hot climates:** Above 2,600 CDD, locate the shadow line at the window sill using the March 21 sun angle at noon. This will provide full shading from late March to late September and about one-third shading in mid-winter.

Awnings and Shutters. Old-fashioned awnings are very effective at blocking solar gain—up to 65% on south-facing windows and up to 80% on east and west windows. Light-colored awnings are more effective, since they will reflect more solar radiation. To be most effective, the awning's "drop" should cover 65 to 75% of an east or west window and 45 to 60% of a south-facing window (see Figure 3-12).

Other low-tech, but very effective exterior options for windows that are difficult to shade include wooden shutters, bamboo shades, and rolling shutters.

FIGURE 3-11 | **Shading with Overhangs.**

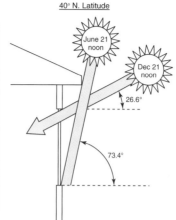

Sun Angles for Continental U.S.				
Latitude	Representative Cities	Sun Angle at Noon*		
		June 21	March 21/ Sept. 21	Dec. 21
28°	Tampa, San Antonio	85.4°	62.0°	38.6°
32°	Charleston, Dallas, San Diego	81.4°	58.0°	34.6°
36°	Albuquerque, Nashville, Raleigh	77.4°	54.0°	30.6°
40°	Columbus, Pittsburgh, Salt Lake City	73.4°	50.0°	26.6°
44°	Bangor, Sioux Falls	69.4°	46.0°	22.6°

* Based on solar noon when the sun passes directly overhead, usually within two hours of noon on clock time.
Sun angles provided courtesy of www.susdesign.com/sunangle

Use the sun angle on June 21 to determine the length of overhang needed to shade south-facing glass. In the example illustrated above, at 40° north latitude, a $1\frac{1}{2}$-foot overhang will shade the window 100% on June 21 but allow full sun exposure in midwinter.

FIGURE 3-12 | **Shading with Awnings.**

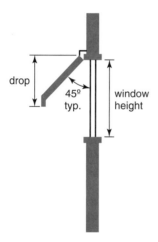

Awnings are very effective at blocking solar gain, even on hard-to-shade west windows. The awning's "drop" should cover about 70% of the height of an east or west window and 45% to 60% of a south-facing window.

- Light-colored interior shades can reduce heat gain by 15 to 20%, depending on the density and reflectivity of the material.

- Interior blinds typically reduce heat gain by 10 to 15%, depending on reflectivity.

Solar Screening. This dark plastic or fiberglass screening is mounted on the exterior of the window in a frame or retractable roller or, in some cases, applied directly to the glass. Depending on the weave, it can reduce a window's solar heat gain by 40 to 70%. Solar screening reduces glare and fading, but it also reduces daylight and obscures views. These are best used on difficult to shade areas, such as west-facing glass or skylights. Some are designed to also serve as insect screens.

Interior Shades. Though not generally as effective as exterior shading, light-colored drapes, shades, or blinds can reduce solar gain by 10 to 50% (for typical curtains or blinds) to as high as 70% for an insulated, reflective shade that seals tightly around the window perimeter. The performance of an interior shade depends on the reflectivity of its outer face, the density and R-value of the material, and whether it seals around the window. Between-the-glass pleated shades or miniblinds, available from Pella and some smaller manufacturers, are more effective in blocking solar gain than interior shades, because they block the heat before it enters the building interior.

WINDOW FLASHING

While modern flange-type window frames appear to simplify window installation, how best to integrate the nailing flanges with the sheathing wrap and siding has been a subject of debate, and recommendations vary among different window manufacturers, sheathing wrap manufacturers, and contractors. The approaches shown below represent a broad consensus of leading manufacturers and experienced contractors, but other approaches that follow the same basic principles can also work. These are:

- Always lap higher flashings over lower to shed water.

- Do not rely on caulking as a waterproof barrier, as it is likely to fail over time.

- With heavy exposure to wind-driven rain and snow, use pan flashing to protect the sill and use flashing tape to seal the head flashing to the sheathing.

- Leave horizontal joints unsealed at the top and bottom of the window frame to allow trapped water to escape.

FIGURE 3-13   Installing Flange-Type Windows Over Sheathing Wrap.

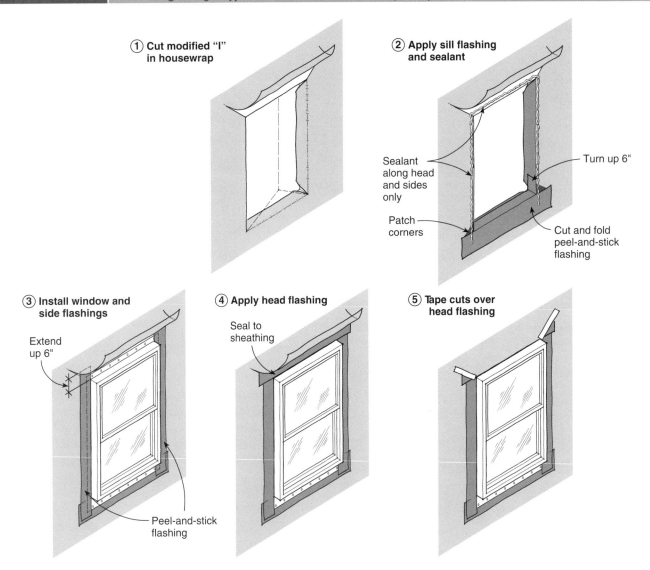

① **Cut modified "I" in housewrap**

② **Apply sill flashing and sealant**

Sealant along head and sides only

Patch corners

Turn up 6"

Cut and fold peel-and-stick flashing

③ **Install window and side flashings**

Extend up 6"

Peel-and-stick flashing

④ **Apply head flashing**

Seal to sheathing

⑤ **Tape cuts over head flashing**

The critical components for a weatherproof installation are the head flashing and sill pan. When the siding is installed, do not caulk the horizontal joints above and below the window, which must be left open so trapped water can drain freely.

Flange-Type Windows

All clad and solid-vinyl windows come with an integral nailing flange or one that is friction-fit into a slot in the frame. In addition to simplifying the nailing, the flange helps to create a weather-tight seal at the head and sides of the window. The most critical details are the head flashing, which should seal to the sheathing to pick up any dripping water from above, and the sill or pan flashing, which should freely drain to the exterior. The following approaches take advantage of the newer peel-and-stick flashing tapes, which have simplified the flashing of flange-type windows.

Installing Housewrap First. Figure 3-13 shows how to flash a window installed after the plastic housewrap is in place. After cutting a modified "I" in the plastic

housewrap, fold in the flaps, staple every 6 inches, and trim the excess.

- **Pan flashing.** The next step, the pan flashing, is often omitted but is critical for leak-free performance in harsh weather exposures. The pan flashing can either be a custom metal pan, an adjustable plastic pan, or one formed in place from flexible flashing tape, as shown (Figure 3-14). If using flashing tape, first add a piece of beveled wood siding to the rough sill, sloping to the exterior. Remember to increase the rough opening height by $\frac{1}{2}$ inch. Add patches of flashing tape to bridge the gap at the cut corners.

- **Install window.** Next install the window, slipping the top flange under the housewrap. Most manufacturers recommend bedding the side and top flanges in a bead of high-quality sealant to create a backup seal, a good

FIGURE 3-14 Formed-in-Place Pan Flashing.

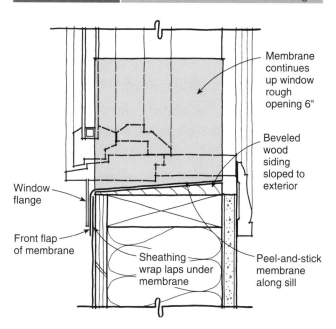

- Membrane continues up window rough opening 6"
- Beveled wood siding sloped to exterior
- Window flange
- Front flap of membrane
- Sheathing wrap laps under membrane
- Peel-and-stick membrane along sill

Slope the rough sill toward the exterior using a piece of beveled wood siding. Next, form a pan flashing from a 9-inch-wide strip of peel-and-stick flashing about one foot longer than the rough opening. A cut at each corner allows the front flap to fold down and side flaps to fold up the sides of the opening. Membrane patches are used to bridge any gaps at the bottom corners.

idea in harsh weather exposures. Do not caulk the bottom flange, however, which must be left unsealed so any trapped water can freely drain. To facilitate drainage at the sill, set the window on shims.

- *Side and head splines.* Side splines of peel-and-stick flashing tape go on next, sealing from the window flanges to the sheathing wrap, extending 6 inches above the top of the window opening and lapping over the pan flashing. Finally, the head spline is slipped under the housewrap, sealing the window's top flange to the sheathing. Patches of tape seal the diagonal slits in the housewrap.

- *Drip cap.* If the window is trimmed on the exterior with wood casings, use a metal drip cap on top of the head casing and seal the drip cap to the sheathing tape with peel-and-stick flashing tape. When the siding is installed, *do not caulk the horizontal joints* above or below the window, which would create a water dam. Leave all horizontal joints unsealed to drain away any trapped water.

Installing Windows Before Housewrap. When windows are installed before the building is wrapped, the key is to detail the pan flashing so it can properly lap over the housewrap. One approach is to leave the paper facing on the outer flap of the pan flashing so it can layer over the housewrap when installed. Another approach, shown in Figure 3-15, is to seal the pan flashing to a flap of sheathing wrap that will be layered over the housewrap when installed.

FIGURE 3-15 Installing Flange-Type Windows Before Sheathing Wrap.

① Lap pan flashing over flap of housewrap. Then set window into sealant on top and two sides.

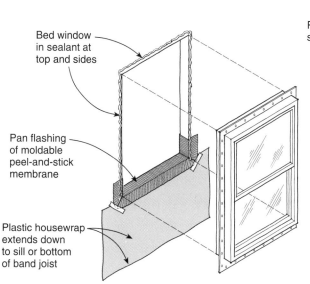

- Bed window in sealant at top and sides
- Pan flashing of moldable peel-and-stick membrane
- Plastic housewrap extends down to sill or bottom of band joist

② Install side splines, then head flashing. Later weave housewrap under flap and tape along side jambs.

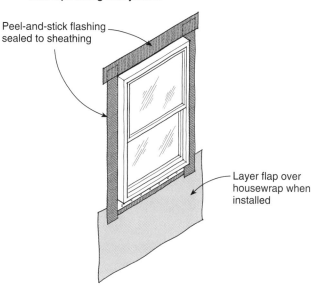

- Peel-and-stick flashing sealed to sheathing
- Layer flap over housewrap when installed

Make sure the pan flashing laps over the housewrap. This can be accomplished by leaving the release tape on the front flap of the pan flashing until the housewrap is installed, or by adding a flap of housewrap, as shown, to be woven in later. The moldable type of flashing tape, shown here, can be formed into a pan with no corner cuts.

Whether to seal the side window flanges directly to the sheathing or the housewrap is a matter of personal preference. Since few leaks originate at the sides of windows, either detail can work well as the flashings are detailed so each sheds water to the layer below.

Felt Paper. When using felt paper, paper up to the bottom of the window. Then install the pan flashing, window, and side and head flashings as shown in Figure 3-15. When installing the building paper later, layer successive courses over the side and head flanges, setting them into a bed of sealant at the sides of the window (see Figure 3-16).

Leave the building paper unsealed at the head flashing, however, to allow any trapped water from above to escape.

Round-Top Windows. To create a head flashing on round-top windows, one approach is to patch together several short pieces of flashing tape, making sure to start at the bottom and lap each upper piece over the preceding lower piece. Another option is to use a flexible membrane such as DuPont's Flexwrap® tape, which can be bent to conform to the curve without cutting.

Mulled Windows. Where multiple windows are mulled together in the field, treat the space between the windows like a small section of sidewall. Make sure horizontal mullions are lapped properly so the sill flashing above laps over the cap flashing below. Use peel-and-stick flashing tape on vertical mullions, overlapping the vertical tape onto any horizontal mullions that intersect.

Windows with Brickmold

If installing windows with integral brickmold, proceed the same way as shown for flange-type windows above. However, since there are no side flanges on the window, the brickmold goes over the side splines, which can be either felt paper or flashing tape. Flashing tape is of less value with brickmold since the windows have no flange to seal to.

Apply a continuous bead of sealant on the backside of the brickmold before installing the window (Figure 3-17).

As an extra precaution against wind-blown rain entering behind the side casings, you can fold back the inside

| FIGURE 3-16 | Installing Flange-Type Windows with Felt Paper. |

Set felt paper in sealant on sides only

#15 felt

Paper up to the bottom of the window; then follow the flashing-tape procedure shown in Figure 3-15. Finally, layer successive courses of building paper over the side and head flanges. The felt paper can be bedded into sealant up the two sides, but the top and bottom should remain unsealed for drainage.

| FIGURE 3-17 | Installing Windows with Brickmold. |

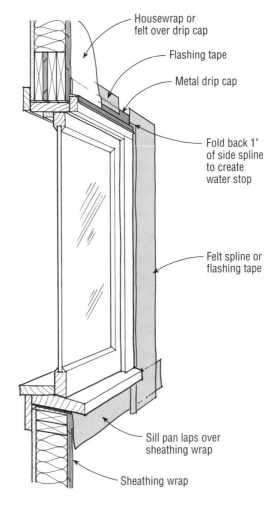

Housewrap or felt over drip cap

Flashing tape

Metal drip cap

Fold back 1" of side spline to create water stop

Felt spline or flashing tape

Sill pan laps over sheathing wrap

Sheathing wrap

Windows with brickmold should be installed over side splines of felt paper (flashing tape is less useful with no flange to seal to). Apply a continuous bead of sealant on the back side of the brickmold before installing the window. As an extra precaution against windblown rain entering behind the side casings, fold back the inside edge of the side splines to form a one-inch hem.

edge of the spline to form a one-inch hem, a detail developed by Pennsylvania contractor Carl Hagstrom.

At the head casing, a metal drip cap is required to protect the brickmold and provide a watertight seal. Slip the metal drip cap under the sheathing wrap and seal it to the sheathing with flashing tape.

SKYLIGHTS

A well-positioned skylight can help transform a dreary interior room into a pleasing sunlit space with a feeling of connection to the outdoors. In addition, venting skylights can play a significant role in exhausting hot, moist air from kitchens and bathrooms, and can enhance ventilation in any room with limited cross ventilation. However, skylights can also be a source of problems, such as roof leaks in winter and overheating in summer, if the installer does not pay attention to glazing type, installation, and flashing details.

Options

Like windows, skylights come in a wide variety of shapes, sizes, frame materials, and glazing options. In addition, they have an ever increasing variety of screens, shades, and motorized, automated, and computerized controls, providing convenience and good energy performance in almost any situation.

Glazing Types. Many lower-cost skylights use acrylic or polycarbonate glazing in single or double layers instead of glass. In general, plastic glazings resist breakage, but they can become scratched or brittle over time and are prone to yellowing. Unless specially coated to block UV transmission, plastic glazings allow high levels of UV radiation, which causes fading with many interior furnishings and finishes.

Most higher-end units use sealed insulated glass. Codes require the glass to be either tempered or laminated safety glass. When laminated glass is broken, the plastic interlayer holds the pane together. Tempered glass is harder to break, but it breaks into small, rounded fragments rather than dangerous shards. Double-glazed skylights with safety glass typically have the safety glass on the interior and tempered glass on the outside, combining high-impact resistance on the exterior with protection from falling glass below.

Sizes. Most manufacturers offer a wide variety of sizes, including narrow models designed to fit 24-inch rafter spaces. At least two manufacturers, Roto and Pella, make a 14-inch-wide model designed to fit between 16-inch on-center framing. Many standard widths are designed to fit in a double 16- or 24-inch bay. For narrow spaces, such as a short section of attic roof above a knee wall, Andersen offers several units wider than they are tall, measuring either 16 or 24 inches high by 38 to 72 inches wide. Some manufacturers are tooled up to offer custom sizes for a moderate up-charge.

Controls. Operable skylights use either metal arms that swing out or a concealed chain that unrolls and stiffens as the sash is cranked open. How many turns it takes to open the sash and how wide it opens vary considerably among units. For out-of-reach skylights, all manufacturers offer either extension poles or motorized controls. Some extension poles can be challenging to engage in the crank mechanism, making it a chore to open and close the skylight.

Motorized controls simplify the task, and manufacturers have been refining their offerings in this area. Some need hard wiring from the controller to skylight, while others need only a power connection and are controlled by a convenient hand-held remote. Other options include a battery backup, which could be useful during a power outage, and a rain sensor that automatically closes the skylights at the first drops of rain.

Shades and Screens. The tendency of south- and west-facing skylights to cause summer overheating can be greatly moderated with the new spectrally selective glazing (SHGC below .40). Still, shades and screens can be used to further reduce heat gains and UV radiation and to provide more diffused light with less glare. Numbers vary from one manufacturer to another, but typical shading effectiveness is as follows:

- Exterior solar screens can reduce heat gain by 40% or more.
- Light-colored interior shades can reduce heat gain by 15 to 20%, depending on the density and reflectivity of the material.
- Interior blinds reduce heat gain typically by 10 to 15%, depending on reflectivity.

In a bedroom, a client might also want shades for room darkening. Manufacturers offer a wide variety of shades, miniblinds, and solar screens, most of which can be controlled by the same motors that control the skylights. Pella's between-the-glass shades and blinds offer better energy performance than interior shades and never need cleaning, a big advantage with out-of-reach units.

Light Tubes. Introduced in the early 1990s, light tubes consist of a small plastic rooftop dome that conducts light to the interior through a rigid or flexible tube with a reflective interior. On a sunny day, the diffuser at ceiling level provides about as much light as a bright electrical ceiling fixture (see Figure 3-18).

Tubes range from 10 to 22 inches in diameter, and one manufacturer, Sun-Tek, offers a multitube model that supplies up to four tubes from a more conventional looking skylight panel. Used primarily in remodeling where it is too difficult or expensive to install a skylight, light tubes offer an economical way to bring daylight into bathrooms, walk-in closets, and other small interior spaces.

FIGURE 3-18   Light Tubes.

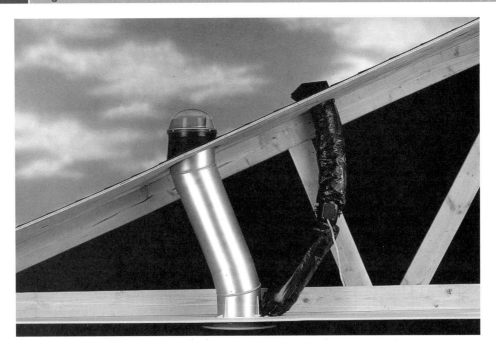

Light tubes, which bring daylight to the interior through a reflective duct, offer an economical way to add natural light to a dark bathroom, closet, or other hard-to-access space. The model shown includes an integral bathroom exhaust fan.
SOURCE: Photo courtesy of Solatube International, Inc.

Design Issues

Skylights can add a dramatic element to any room in addition to providing high levels of pleasing natural daylight. In the case of operable skylights, they can provide effective ventilation as well. On the other hand, too many skylights can produce uncomfortable glare and significant summertime overheating. A few simple guidelines can ensure a successful installation.

Sizing. A small skylight can go a long way toward brightening a space. However, too much direct sunlight can produce very uneven room lighting, excessive glare, and localized overheating. A rule of thumb developed by the Florida Solar Energy Center recommends that a skylight should be 4 to 6% of the illuminated floor area. This will provide a reasonable level of illumination (about 600 lux) during the morning and afternoon, and on days with overcast skies. So, for example, 5 to 7 square feet of rooftop glazing (measured horizontally) will provide a reasonable level of illumination to a 10x12 kitchen. For skylights with frosted lenses or high up in cathedral ceilings, use a larger skylight; in regions with predominantly sunny skies, a smaller skylight should suffice.

Glare. Glare can be caused by a bright light source bouncing off a work surface into your eyes, or from a bright source directly striking your eyes from straight ahead or from an angle (Figure 3-19).

FIGURE 3-19   Glare from Windows and Skylights.

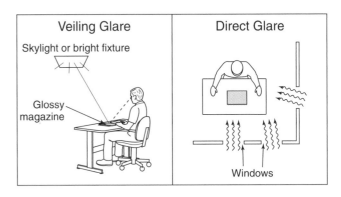

A skylight located ahead of a work area will cause "veiling" glare (left) when bright light reflects off a magazine page or shiny work surface. A light source from the front or side that is much brighter than the task lighting can cause direct glare (right) as the eye tries to adjust to conflicting light levels.

The best way to reduce glare from skylights is to provide diffused, even lighting over a larger area. This can be accomplished with frosted glazing or interior shades. However, both of these block views of the sky, and interior shades require maintenance. Another approach is to use a light-colored skylight well to reflect and diffuse the

FIGURE 3-20 Splayed Skylight Wells.

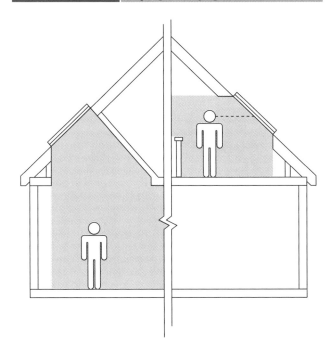

Splayed light wells bring light deeper into the interior space and provide better light distribution. Where skylights are near eye level (right), a horizontal angle at the top of the well improves both daylighting and views.

light over a larger area. Deep skylight wells can also reduce unwanted solar gains by as much as 25% by absorbing heat into the walls of the well area.

Splayed Openings. To maximize the daylighting potential of a skylight, it is best to paint the light well a light color and angle at least the top side of the well. Splaying the skylight well enlarges the opening at the bottom and brings the light deeper into the interior space, particularly in the winter when sun angles are lower.

Typically, the top side of the well runs at 90 degrees to the roof slope, and the bottom side, if splayed, is framed plumb (see Figure 3-20).

On very steep roofs where the top of the skylight is near eye level, a horizontal angle works well for the top side of the well. The sides are usually left plumb to simplify the framing (see "The Light Well," next page).

Ventilation. Ventilating through a skylight takes advantage of natural convection currents in a room that bring warm air toward the ceiling. As the air is exhausted out the skylight, cooler air is drawn in through windows or other rooms. In rooms that generate a lot of moisture, such as kitchens and bathrooms, a venting skylight helps to control moisture levels as well. Some skylights are equipped with flap-style vents that can remain open during rain without leaking.

Energy Efficiency

Overglazing with skylights can increase heat loss in cold, cloudy climates, but the biggest impact of excessive roof glass in most climates is overheating, increasing both peak and annual cooling loads and making people uncomfortably hot.

Solar Heat Gain. Because the sun is high in summer and low in winter, skylights tend to capture solar gains at just the wrong time. For example, an unvented skylight set at a 30- to 40-degree slope captures four times as much solar-heat gain in midsummer as a similarly sized vertical window. To put this in perspective, at noontime on June 21, two single-glazed, 6-square-foot skylights facing south produce nearly as much heat as a 1,000-watt electric space heater. West-facing skylights produce almost as much heat in summer.

Low-E Glazing. Fortunately, the new spectrally-selective coatings (see "Spectrally Selective Low-E," page 114), cut solar gains by nearly 50% compared with clear double glazing, greatly reducing the cooling penalty of skylights. Still, there is no sense in using more skylights than is needed to produce comfortable light levels, typically 5 to 7% of the floor area being lit. Where overheating is a possibility, use a venting skylight to change the glass angle and exhaust any heat buildup.

Orientation. While high-tech glass can significantly reduce solar gains, it is still possible to overglaze and overheat a room. For that reason, south- and west-facing skylights should be kept to a minimum. Also, south-facing skylights produce a lot of glare from direct summer sun. North-facing skylights consistently provide even, diffused lighting with little heat gain. East-facing skylights generate moderate heat gain on summer mornings, which could pose a problem in hot climates if overused.

U-Values. The same double-glazing has a higher U-value (lower R-value) when used in a skylight than in a vertical window. Increased convection between the two panes of glass in a horizontal or low-slope orientation causes the greater heat loss in skylights. Starting in mid 2004, all skylight manufacturers using NFRC labels started reporting U-values measured at a slope of 20 degrees. For older skylights rated at a 90-degree angle, multiply the U-value by 1.2 to get the 20-degree equivalent. This results in a 20% higher U-value.

Condensation. Because warm moist air is carried upward by convection currents, skylights are often one of the first places to develop condensation. This can lead to dripping and staining of the frame, well, or furnishings below. The best defense against condensation is to choose high

R-value glazing with warm-edge spacers. Good insulation of the skylight well also helps by keeping the surrounding area warmer. Several manufacturers offer skylights with integral condensation gutters, a helpful feature in cold climates.

Skylight Installation

Framing. Take care when framing skylight openings to make the opening as accurate and square as possible. This will simplify both the installation and trim, since most skylight designs leave little margin for error where the interior finish meets the frame. Unless using a skylight designed to fit between 16- or 24-inch on-center framing, you will need to head off the opening and use double trimmers on either side. For truss roofs, consult the truss designer or an engineer before modifying the truss plan or cutting an existing truss. In new construction, trusses are typically doubled on either side of the opening with 2x cross pieces framed in "ladder" fashion.

The Light Well. There are many ways to frame a light well. In a cathedral ceiling, the rafters and headers create the well. Framing the upper header square and lower header plumb splays the opening. With an attic, the top side of the well is typically framed perpendicular to the roof plane and the bottom and sides are framed plumb. For small wells, the opening can be created with $\frac{3}{4}$-inch plywood while larger openings are typically built like stud walls. Make sure the well is sealed tightly against air infiltration and well-insulated to guard against condensation on the skylight or sides of the well.

It is also possible to splay the two sides of a skylight well in addition to the top and bottom sides, although the framing and drywall are a lot more complicated. To keep the four sides of the well flat, the bottom of the opening must be shaped like a trapezoid, wider on the high side of the opening. To keep the opening rectangular at the ceiling plane, the two sides become curved like twisted strips of plastic. This requires framing with angled struts and bending the drywall. Either approach makes for a complicated framing job.

Underlayment at Skylights. To ensure a leak-free installation with any type of roofing, use peel-and-stick membrane to seal the skylight to the sheathing and underlayment, creating a primary barrier to water entry below the finished roofing. Any water that manages to penetrate the flashing will be stopped by the membrane. Hold the roofing felt back about 6 inches from the opening so the peel-and-stick membrane can bond directly to the roof sheathing.

After the skylight frame is in place, install 9- to 12-inch-wide strips of membrane around all four sides of the curb, starting with the bottom, and working uphill so the upper pieces always overlap the lower pieces (Figure 3-21).

FIGURE 3-21   **Skylight Underlayment.**

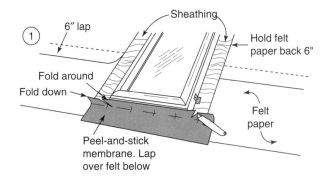

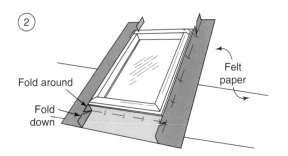

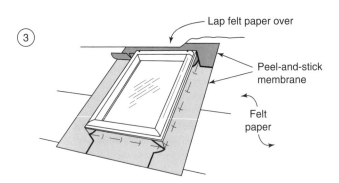

After the skylight frame is in place, install 9- to 12-inch-wide strips of membrane around all four sides of the curb, starting at the bottom and working upslope. Detail corners carefully, using patches of flexible membrane if necessary.

Detail corners carefully using patches of moldable flashing (see "Moldable Flashing," page 5) if necessary. Tuck the top piece of membrane under the roofing felt above, and lap the lower piece over the roofing felt below. With ganged units, use a single strip of membrane between each unit, running down one skylight frame, across the mullion, and up the next skylight frame.

Ice Dam Protection. Snow can build up above skylights and ice can build up below them as snowmelt from the glass area refreezes. As extra protection against leaks in areas subject to heavy snow and ice damming, some

FIGURE 3-22 | **Skylights on Low Slopes.**

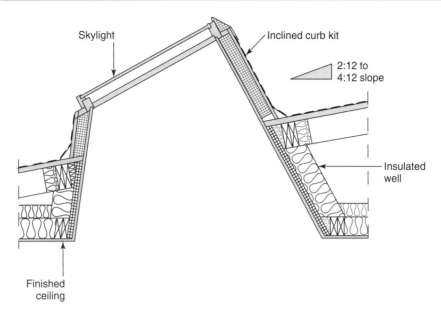

For roof slopes between 2:12 and 4:12, most skylight manufacturers will not guarantee a waterproof installation without raising the skylight angle. Some provide special inclined curbs or flashing kits, like the one shown, to complete the installation.

contractors run a wide band of the peel-and-stick membrane around the sides and top of the skylight and all the way down to the roof eaves (see Figure 2-5, page 57). On low-slope roofs, the membrane often covers the entire roof surface and continues up and around any skylight curbs for complete protection.

Skylight Flashing. Older-style skylights were notorious for leaking, but today's preformed flashing kits combined with bituminous peel-and-stick membranes have made leaky skylights mostly a thing of the past (see "Flashing Tapes," page 5). Most major skylight manufacturers provide flashing kits for asphalt shingles as well as specialty flashings for tile and metal roofing, typically with a wraparound head flashing and moldable apron flashing that conforms to the roofing profile.

With asphalt shingles, avoid skylights that provide only a continuous side flashing, which relies on roofing cement for a waterproof seal. Step flashing is much more reliable. With metal roofing, use the manufacturer-supplied kit or have the roofing contractor fabricate a matching flashing from the metal roofing material. While a custom flashing can create a more attractive installation, make sure the corners of the head flashing are properly sealed with solder or a high-performance caulk.

Low Slopes. For roof slopes between 2:12 and 4:12, most skylight manufacturers will not guarantee a waterproof installation unless the skylight angle is raised with a built-up curb or a special inclined flashing kit (Figure 3-22).

The raised curb should be sealed on all sides with peel-and-stick membrane carefully sealed at all corners, as in Figure 3-21.

Steep Slopes. For very steep slopes, such as mansards, skylight manufactures typically recommend a water deflector or Z-flashing to safely divert water around the unit. Check with the manufacturer for steep-slope requirements.

EXTERIOR DOORS

While some entry doors are well-protected from the elements by porches or recessed entries, many face harsh weather exposures in addition to the usual bumps and bruises from children, furniture movers, and others. In addition, doors must resist warping, shrinking, and swelling across a wide range of temperatures and moisture conditions in order to close tightly and to operate smoothly—all in all, a tall order met with an increasing degree of sophistication by manufacturers. For new construction, most entry doors are purchased prehung as an "entry system," which, in addition to the frame, hardware, weather-seals, and any sidelites, may also include integrated security systems, lighting, and keyless entry systems.

Note: Interior doors are covered under "Interior Finish," starting on page 186.

Materials

While solid-wood entry doors can last for decades and grace the fronts of many older homes, they are rapidly

TABLE 3-7	Exterior Door Materials				
Face Material	Core	Pros	Cons	Options	Recommendations
Wood	Solid, laminated, or engineered wood.	Beauty and heft of solid wood. Can be planed, if necessary.	Most expensive. Requires most maintenance. Vulnerable to warping or decay if not maintained.	Engineered wood core, insulated wood panels. Custom sizes, styles, and wood species. Decorative trim.	Select a durable factory finish and maintain as required. Should be sheltered by an entryway or overhang.
Steel	Wood or steel frame with polyurethane or polystyrene foam insulation.	Energy-efficient. Very durable and secure. Least expensive.	Relatively easy to dent (repairable with auto repair kit). Bottom edges may decay or rust. Requires painting or staining.	Vinyl coating for better weather resistance; stainable wood fiber finish.	Good choice for budget-minded project. Provides good security.
Fiberglass	Wood or steel frame with polyurethane or polystyrene foam insulation.	Energy Efficient. Won't rust, dent, warp. Wood-like appearance. Long warranties, moderate cost.	Not as strong as wood or steel. Requires painting or staining.	Smooth or wood-grain finish.	Good choice for humid or harsh climates.
Composite	Engineered wood frame with foam insulation or solid core of exterior grade particleboard or MDF.	Solid feel, strong, and dimensionally stable, moderate cost.	Long-term durability not well-established. Requires periodic painting or staining	Smooth or wood-grain finish	Use materials with proven track record on exterior. Avoid hardboard materials. Should be sheltered by an entryway or overhang.

giving way to a host of hybrid and composite products, some of which are difficult to categorize. While most budget-oriented projects use steel-faced doors, fiberglass and composite doors are the fastest growing market segment, promising greater durability at a price still well below solid wood. Exterior doors are typically classified by their facing material, but their performance and durability is more a function of their internal construction (Table 3-7).

Wood. Traditional frame-and-panel solid wood doors, once standard fare for residential entries, have become primarily a high-end specialty item. A few larger manufacturers, such as Jeld-Wen and Kolbe and Kolbe Millworks, still manufacture stock designs in solid wood, but many who have stayed in the business are niche suppliers of high-end custom doors in an endless variety of wood species, shapes, and styles from Shaker simplicity to 14-foot castle doors (see "Resources," page 132, at end of chapter).

To improve strength and stability in wood doors, Kolbe and Kolbe uses laminated-veneer lumber (LVL) for locking rails, and custom door makers Lamson-Taylor and Simpson build up their rails, stiles, and panels from two or more layers of wood. Lamson-Taylor laminates foam insulation between two solid wood faces to create a unique insulated wood door that the company says is immune to problems with temperature and humidity differences across the door.

To keep costs down and improve stability, many manufacturers offer simulated panel doors with a real wood veneer over an engineered wood core. These are sturdy and offer good value, but the veneered face is more vulnerable to damage than a solid wood model.

If well maintained and periodically repainted or stained, a high-quality wood door can last indefinitely. However, they are best suited to temperate climates and should be protected from direct weather exposure with a suitable overhang or inset.

Steel. A premium residential steel door has a 24-gauge or heavier galvanized steel skin over a wood or steel frame filled with foam insulation. This creates an extremely strong and durable product with an effective insulation rating of about R-8 for polyurethane insulation versus about R-5 for polystyrene.

Most doors contain a wood or composite lock-block for mounting the lockset, and may use wood, engineered lumber, or steel for the stiles and rails. Bottom rails, which get the greatest exposure to water, are sometimes made of waterproof composites. Steel is the strongest edge material but requires a thermal break to prevent condensation around the door's interior perimeter. Wood forms a natural thermal break, but it is prone to decay along the door bottom if not kept well painted (Figure 3-23).

Most doors come embossed with a wood-grain pattern and preprimed with a baked-on polyester finish, although smooth metal finishes are also available. For

FIGURE 3-23 | **Steel Door Edges.**

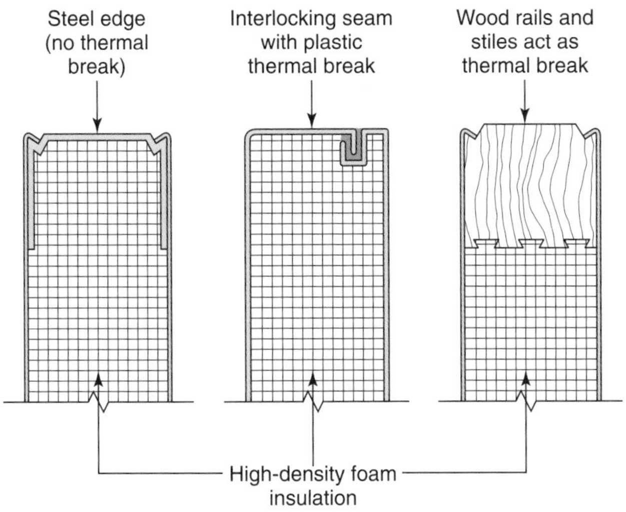

Steel edge (no thermal break)

Interlocking seam with plastic thermal break

Wood rails and stiles act as thermal break

High-density foam insulation

Steel edges (left) provide the greatest strength and security, but can lead to condensation in cold weather if not thermally broken (center). Wood edges (right) provide a good thermal break in steel doors, but are vulnerable to decay along door bottoms.

better protection, some manufacturers apply a vinyl coating with wood grain, which can be stained with a high-solids stain. The vinyl is fairly durable, but deep scratches are difficult to conceal. For a more realistic wood look, some manufacturers apply a stainable wood-fiber finish. Follow manufacturers' recommendations regarding prep work and finishing.

In general, premium steel doors are very strong and durable, although they require regular painting and over time will likely acquire a dent or two. Rust is generally not a problem with galvanized or stainless-steel facings, but condensation can damage finishes on doors without thermal breaks.

Fiberglass. Fiberglass doors, first introduced by Therma-Tru in the early 1980s, are built internally the same as a steel door with internal wood rails and stiles and a rigid foam core. The fiberglass facing is typically embossed with a stainable wood grain, but is also available with a smooth finish that when painted is hard to distinguish from a painted wood door. Fiberglass doors generally price midway between a steel and wood door and carry long-term warranties.

Fiberglass, while not as strong as steel, is very durable, stable, and energy-efficient. It will not warp, crack, or swell like wood and will not dent like steel, making it a good choice in harsh weather exposures. While fiberglass can be gouged or cracked if hit hard enough, repairs are no more difficult than for steel doors. Scratches, however, are difficult to sand out without destroying the wood-grain pattern on embossed panels.

Composites. A new breed of engineered wood doors are built of a variety of engineered wood materials, including laminated-veneer lumber (LVL), exterior-grade medium-density-fiberboard (MDF), and exterior-grade particleboard. Some are built with an engineered-wood skin over a foam or particleboard core, while others are milled from a single slab of MDF. Most come with either smooth or embossed wood-grain finishes ready to paint or stain.

Composites tend to price between steel and fiberglass and carry warranties up to 10 years. While many wood composites have established a good track record in exterior use, others, such as hardboard, have had problems with swelling and delamination if exposed to the weather and not protected by a good coat of paint. As with real wood, it would be prudent to use these products in a sheltered entryway and keep them well painted. Until long-term durability has been established, their use remains an open question.

Frames

In new construction, most exterior doors are purchased prehung in a frame complete with adjustable thresholds, sidelites or transoms, and, in some cases, high-tech electronics, such as motion-sensor lighting and keyless ignition systems. The frames come in a variety of materials from basic finger-jointed pine to low-maintenance frames clad in vinyl or aluminum.

The most critical piece of the frame is the sill and threshold. Most today are extruded aluminum, often with a treated wood or composite subsill (Figure 3-24).

Some have built-in channels with weeps to safely drain away water and many have an adjustable sill step, a helpful option since few of today's doors can be planed or easily adjusted.

When purchasing a complete entry system, make sure that the components all come from the same manufacturer, since many distributors mix and match door slabs from one company with more economical frames, hardware, or glazing systems from another, potentially voiding the warranty should some components fail.

Energy Efficiency

Insulation values for entry doors range from about R-2 for solid wood to about R-5 for a fiberglass or steel door filled with polystyrene foam. Doors with polyurethane foam average about R-8. The values are lower than for a solid slab of foam insulation due to internal blocking and frame materials. In doors with glazing, the numbers drop considerably.

However, because of a door's relatively small area, conductive heat loss has little effect on annual fuel bills. Thermal breaks are important with steel doors since they help eliminate condensation around the door's perimeter. Air leakage has the biggest energy impact since it can contribute to condensation, increased fuel bills, and discomfort due to drafts.

FIGURE 3-24 | **Aluminum Thresholds.**

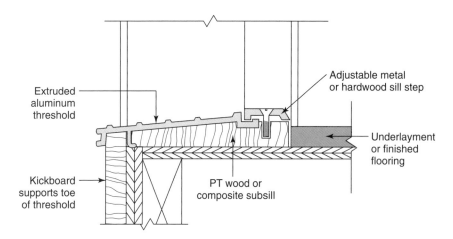

Most thresholds today are extruded aluminum, often supported by a treated wood or composite subsill. An adjustable aluminum or hardwood sill step guarantees a good fit at the door bottom.

Weather-Stripping. Look for airtightness ratings similar to windows, preferably below .10 cfm/sq ft. Equally important as the rating, however, is how the weather-stripping holds up over time. Magnetic weather-stripping generally performs well but is only available on steel doors. Compression bulbs form a tight seal, but some materials lose flexibility in the cold or take on a permanent "compression set." Silicone and EPDM both resist compression set and stay flexible in the cold. Neoprene and vinyl are less durable and less flexible in the cold. Another widely used weather-stripping material with a proven track record is Schlegel's proprietary Q-Lon, a thermoset plastic that outperforms thermoplastics, such as vinyl, TPE, and urethane foam.

Door Sweeps. Also look for a durable seal at the door bottom. There are many approaches to sealing at the threshold, including bulbs, sweeps, and interlocks that work in conjunction with the threshold to seal out water and air leakage (Figure 3-25).

One of the most effective approaches is an automatic sweep that retracts into a dado cut in the bottom of the door and drops down only when the door is closed. These are available as retrofits for wood doors and will even work without a threshold.

Flashing

Doors are flashed the same as windows on the sides and top, and similarly at the sill. Clad door frames are flashed like clad windows (see Figures 3-13, 3-15, 3-16, pages 120–122) and solid wood frames are flashed like traditional windows with brickmold (Figures 3-17). Unless a door is well-protected by a porch or large overhang, good pan flashing at the sill is critical to prevent water from seeping into the floor framing. Doors leading to patios

FIGURE 3-25 | **Sealing at Threshold.**

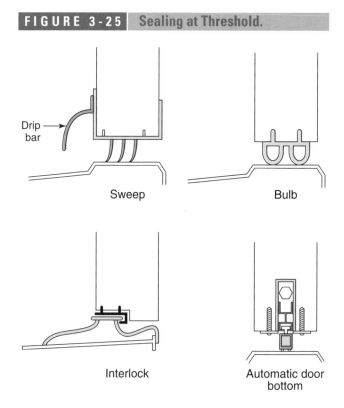

A raised threshold keeps out water and seals to the door's sweep, bulb, or interlock when the door is closed. An automatic door bottom (lower right) retracts into a dado in the door bottom and drops down only when the door is closed.

and decks are particularly vulnerable to wetting around the sill from splashback and, in cold climates, from snow buildup.

Pan Flashing. Prefab plastic door pans typically come in three sections that are fused together at the required length with solvent-based cement. Metal pans require a

FIGURE 3-26 **Flashing at Doors.**

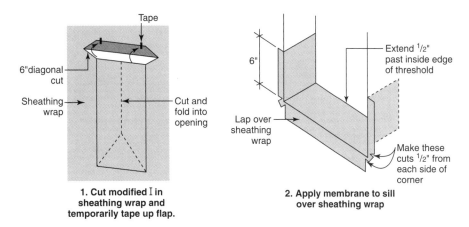

1. Cut modified I in sheathing wrap and temporarily tape up flap.

2. Apply membrane to sill over sheathing wrap

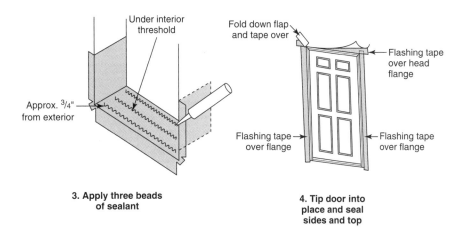

3. Apply three beads of sealant

4. Tip door into place and seal sides and top

Whether to use a custom metal pan, an adjustable plastic pan, or a peel-and-stick membrane, as shown, is a matter of personal preference. All pans should have a dam on the ends and along the interior edge. If using a flashing membrane, carry it up the side jambs at least six inches. For wide thresholds, two strips of membrane can be used with at least a one-inch overlap.

brake to form and should be caulked or, preferably, soldered at corners. Peel-and-stick membranes have become increasingly popular due to their ease of use and flexibility. Whether to use a metal pan, plastic pan, or peel-and-stick membranes is a matter of personal preference as all work well (see Figure 3-26).

Whatever material is used, all pans should have a dam on the ends and along the inside edge. On the exterior, the pan flashing should lap over the deck or masonry flashing below. If forming a pan with peel-and-stick membrane, carry it up the sides at least 6 inches, and turn up the inside edge so it is held in place by the underlayment or finish flooring (Figure 3-27).

FIGURE 3-27 **Pan Flashing Detail.**

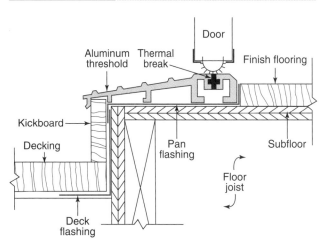

On the exterior, the door's pan flashing should lap over the deck or masonry flashing below. On the interior, turn up the inside edge to form a lip held in place by the underlayment or finish flooring.

RESOURCES

Manufacturers

Windows and Patio Doors

Andersen Windows and Doors
www.andersenwindows.com
Vinyl-clad windows and patio doors, including storm-resistant models

Atrium Companies Inc.
www.atriumcompanies.com
Vinyl and aluminum windows and patio doors

Certainteed Corp.
www.certainteed.com
Vinyl windows and patio doors

Crestline Windows and Doors
www.crestlinewindows.com
Wood, vinyl, and aluminum-clad windows and patio doors

Eagle Windows and Doors
www.eaglewindow.com
Extruded-aluminum-clad windows and sliders with LVL frames and steel entry doors

Fibertec Windows and Door Manufacturing
www.fibertec.com
Pultruded fiberglass windows and doors

Hurd Windows and Doors
www.hurd.com
Wood, vinyl, and aluminum clad windows and patio doors

Jeld-Wen Windows and Doors
www.jeld-wen.com
Wood, vinyl, aluminum-clad, and aluminum windows and patio doors

Kolbe Windows and Doors
www.kolbe-kolbe.com
Wood, vinyl, and aluminum-clad windows and patio doors

Marvin Window and Doors
www.marvin.com
Wood and extruded-aluminum-clad windows and patio doors, including true divided lites and storm-resistant models

Milgard Windows and Doors
www.milgard.com
Wood, aluminum, vinyl, and fiberglass-clad windows and patio doors

MW Windows
www.mwwindows.com
Wood, vinyl, and vinyl-clad windows and patio doors

Peachtree Doors and Windows
www.peach99.com
Vinyl-clad and aluminum-clad windows with optional hardwood interior; aluminum-clad, steel, and fiberglass patio doors with optional hardwood interior

Pella Windows and Doors
www.pella.com
Wood and aluminum-clad windows and patio doors with optional between-the-glass shades and blinds, including storm-resistant models

Thermotech Windows Ltd.
www.thermotechwindows.com
Complete line of fiberglass pultruded windows

Weather Shield Windows and Doors
www.weathershield.com
Wood, vinyl, vinyl-clad, and aluminum-clad windows and patio doors, including historic replacement windows and storm-resistant models

Windsor Windows and Doors
www.windsorwindows.com
Wood and vinyl windows and patio doors, including a line of wood windows with a cellular-PVC exterior

Doors

Benchmark Entry Systems (division of Therma-Tru Doors)
www.benchmarkdoors.com
Steel and fiberglass entry doors

Jeld-Wen Windows and Doors
www.jeld-wen.com
Wood, wood composite wood, fiberglass, and steel entry doors

Kolbe Windows and Doors
www.kolbe-kolbe.com
Wood, steel, and fiberglass entry doors with LVL core and optional extruded-aluminum cladding on frame

Lamson-Taylor Custom Doors and Millwork
www.lamsontaylor.com
Custom pine and hardwood entry doors with foam insulation core

Masonite Corp.
www.masonite.com
Steel, wood-edged steel, and fiberglass entry doors

Peachtree Doors and Windows
www.peach99.com
Steel and smooth and textured fiberglass entry doors

Pella Windows and Doors
www.pella.com
Fiberglass and steel entry doors

Phoenix Door Manufacturing Company
www.phoenixdoor.com
Softwood and hardwood entry doors up to 8 ft. high and custom designs

Simpson Door Company
www.simpsondoor.com
Douglas-fir, hemlock, oak, and mahogany entrance doors, including custom doors; also primed MDF, particleboard, and composite wood doors

Stanley Door Systems (division of Masonite)
www.stanleyworks.com
Steel and fiberglass entry doors

Weathershield Windows and Doors
www.weathershield.com
Wood and steel entry doors, with wood, vinyl, aluminum-clad, and vinyl-clad frames

Taylor Building Products
www.taylordoor.com
Steel (stainable finish) and fiberglass entry doors

Therma-Tru Doors
www.thermatru.com
Steel and fiberglass entry doors with optional vinyl-clad jambs

Skylights

Andersen Windows and Doors
www.andersenwindows.com
Skylights and roof windows with exterior sash clad with glass-fiber-reinforced material

Milgard Windows and Doors
www.milgard.com
Skylights with aluminum frames (thermal break optional) with vinyl subframes on operable models; optional motorized controls with rain sensor

Pella Windows and Doors
www.pella.com
Wood interior, aluminum exterior, optional motorized controls, and manual or motorized fabric-pleated shades

Roto Frank of America
www.roofwindows.com
Wood interior, aluminum exterior, optional motorized controls, and manual or motorized fabric-pleated shades; Sweet16 model fits 16 in. o.c. framing

Velux America Inc.
www.velux.com
Skylights and roof windows with wood interior and aluminum-clad exterior. Options include insect screens, blinds, motorized controls and shades with rain sensor, electrochromatic glass, and flashing kits for metal and tile roofs and mulled units

Light Tubes

SolaTube
www.solatube.com
Light tubes from 10 to 21 in. in diameter; options include electrical lighting, daylight dimmer, and integral bath fan

Sun-Tek Skylights
www.sun-tek.com
Light tubes from 10 to 21 in. in diameter; options include electrical lighting and multitube Spyder skylight

Velux America Inc.
www.velux.com
Sun Tunnel light tubes from 14 to 22 in. in diameter with flexible or rigid tubes

For More Information

American Architectural Manufacturers Association (AAMA)
www.aamanet.org

Efficient Windows Collaborative
www.efficientwindows.org

National Fenestration Rating Council (NFRC)
www.nfrc.org

Sustainable by Design
www.susdesign.com
Shareware calculators for sun angles, solar heat gain, and shading

Window and Door Manufacturers Association (WDMA)
www.wdma.com

| # Decks and Porches

Decks and porches are challenging applications for wood building products and finishes. The structural framework is exposed to the elements and may be in partial contact with the ground. Horizontal surfaces are exposed to rain, snow, and direct sun and, in the case of decking and stairs, to the abrasive effects of heavy foot traffic as well. As part of the growing trend toward low-maintenance exteriors, plastics and composites are starting to make significant inroads as decking and railing materials. Some of these products, such as Trex®, developed by Mobil Chemical in the late 1980s, have seen well over 10 years of use in the field and are proving themselves as durable, attractive alternatives to wood decking.

To simplify construction and reduce costs, many porches today are built essentially like exterior decks with added roofs and an open balustrade or partial-height walls. Since porch floors get less exposure to rain and sun and are therefore less prone to crack and warp, they can be built with less durable materials than open decks. A solid floor consisting of tongue-and-groove or square-edged decking, pitched for drainage, is sometimes used in porches, although spaced decking is more forgiving and requires less maintenance.

FRAMING MATERIALS

Most builders choose pressure-treated lumber for the structural framework because of its low cost and high durability. At this time there are few viable alternatives. However, composite and synthetic products are beginning to enter the market and are worth considering, especially for environmentally sensitive sites, such as wetlands or other applications where clients object to the use of treated lumber (see Table 4-1).

Pressure-Treated (PT) Lumber

The vast majority of residential decks were framed with lumber treated with chromated-copper arsenate (CCA) until 2004, when CCA was phased out and replaced primarily by alkaline copper quat (ACQ) and copper azole. The treated lumber is typically Southern yellow pine in the eastern United States and hem-fir in the West. The more expensive and stronger Douglas fir is also used in the West, but it is more likely treated with the waterborne treatment ammoniacal copper zinc arsenate (ACZA).

Incising. Both Douglas fir and hem-fir are typically "incised" with surface cuts for better penetration of the treatment chemical. Even with incising, however, full chemical penetration is rarely achieved with these species, so the center of that wood remains vulnerable to rot, particularly in 4x and larger material. With these species, effective field treatment of holes and cuts with a liquid preservative is essential.

Health Concerns. Despite CCA's track record as an effective, economical wood preservative, its safety has long been questioned by health and environmental advocates. Their primary focus has been CCA's heavy concentration of arsenic, a known carcinogen. Although most experts agree

TABLE 4-1   Deck Framing Materials

Material Type	Pros	Cons	Costs (1—lowest; 5—highest)	Recommendations
Pressure-treated (PT) yellow pine, hem-fir	Inexpensive, strong, and readily available.	Surface residue on new lumber toxic. Smoke from burned scraps toxic. Yellow pine tends to cup, twist, and check. New wood treatments very corrosive to metal fasteners and connectors.	1	Economical solution for most projects. Workers should wear gloves when handling PT materials, wear dust masks while cutting, and avoid burning scraps. Embedded posts must be rated for ground contact (retention of 40 lb/cu ft). Do not put cut, untreated ends in the ground. With hem-fir, treat all field cuts and holes with copper naphthenate solution. Use stainless-steel or heavily coated hot-dipped galvanized fasteners and hardware.
Pressure-treated Douglas fir	Very strong, stable	Does not take treatment well. Surface must be incised for good penetration. ACZA is very corrosive to metal fasteners and connectors.	1	Good choice for premium projects where high strength is required. Same worker precautions as above. Treat all field cuts and holes with copper naphthenate solution. Use stainless-steel or heavily coated hot-dipped galvanized fasteners and hardware.
Redwood, cedar	Natural decay resistance, attractive appearance	Decay resistance variable. Greatest in heartwood of old-growth timber, which is less available.	3–5	Good choice for premium projects where natural materials are preferred. Use structural grades in accordance with design loads.
Plastic structural lumber	Resists rotting, checking, and splitting. Works like wood.	Heavy material. Too flexible for long spans. May seem bouncy due to high elasticity.	2–3	Recommended near ponds, wetlands, salt water, other wet locations. Use with caution in structural applications as few industry standards exist for these products.

that leaching of arsenic from CCA lumber is minimal and poses negligible health risks to end users, the industry acknowledges that CCA does pose risks to workers who handle the wet wood or burn scraps, and significant pollution around treating plants has been well documented.

Phase Out. In response to these concerns, manufacturers began a voluntary phase out in 2003 of all CCA-treated lumber for noncommercial applications. Starting January 1, 2004, the Environmental Protection Agency (EPA) banned the manufacture of CCA-treated lumber intended to be used in residential settings, including retaining walls, decks, fencing, and playground equipment. Pressure-treated shakes and shingles were exempted. CCA treatment will also still be available for plywood and heavy timbers used in commercial, industrial, and marine applications. Existing stocks of CCA-treated lumber were mostly depleted by the end of 2004.

Existing Structures. The EPA has issued no warnings regarding existing installations of CCA-treated lumber. However, for homeowners who are concerned about potential exposure to chemicals leaching out of the wood, researchers at the USDA Forest Products Laboratory (FPL) recommend periodically treating the pressure-treated

lumber with a water-repellant or a semitransparent penetrating stain. Film-forming finishes, such as paints, are not recommended.

New Treament Chemicals. The two main chemicals replacing CCA are the waterborne compounds alkaline copper quat (ACQ) and copper azole. Copper azole Type B (CA-B) has largely replaced Type A (CBC-A) in the United States and Canada. Both ACQ and copper azole perform as well as CCA and are free of any EPA-listed hazardous compounds. As with CCA-treated wood, premium treated lumber is available with a factory-applied water repellant. With significantly higher copper content than CCA lumber, the new materials are 10 to 15% more expensive. Of greater concern is the fact that the higher concentration of copper makes the lumber more corrosive to certain metals and metal coatings (see "Increased Corrosion Potential," next page).

Health Precautions. Despite the lack of chromium, arsenic, or other hazardous chemicals, wood treated with ACQ and copper azole carry essentially the same handling instructions as CCA-treated materials. Workers handling ACQ and copper azole are still advised by the EPA to wear gloves or wash hands after contact, wear a dust mask when

TABLE 4-2	Pressure-Treated Wood Retention Levels (lb/cu ft)			
Application	CCA (chromated copper arsenate) and ACZA (ammoniacal copper zinc arsenate)	ACQ-C (alkaline copper quat) and ACQ-D (amine copper quat)	CBA-A (copper azole type A)	CA-B (copper azole type B)
Above ground	.25	.25	.20	10
Ground contact	.40	.40	.41	.21
Wood foundations, structural poles*	.60	.60	.61	.31
Pilings and columns (foundation/freshwater)	.80	.80	NR	NR
Pilings and columns (salt water)	2.50	NR	NR	NR

NR = not recommended

*Also recommended for contact with very wet ground or for applications where replacement of damaged wood would be difficult.

cutting, and not to burn the scraps. Like CCA-treated wood, it is not recommended for direct contact with food or drinking water.

Retention Ratings. While most CCA lumber was rated for ground contact, manufacturers are holding down costs with ACQ and copper azole by limiting treatment levels to the expected application of the lumber. For example, deck boards, 2x6s, and 4x4s at the lumberyard will typically have three different treatment levels (Table 4-2). In most cases, lumber will be stamped or tagged with a designation such as "decking," "above ground," "ground contact," or "PWF" (permanent wood foundation). Make sure the material purchased is rated for the intended application or one level higher.

Borate Treatment. Wood preservatives based on borate compounds have been used for decades abroad and are slowly becoming available in the United States. Borates are noncorrosive to metals and harmless to pets and humans, but they are very effective against insects and decay. Borate's main limitation is its tendency to leach out of wood that is buried in soil or exposed to regular wetting, making it unsuitable for decks or other exterior applications. New techniques to better fix the compounds into wood are under development, however, and may soon offer a viable alternative to copper-based treatments.

Increased Corrosion Potential. Because of their higher concentrations of copper, ACQ and copper azole are significantly more corrosive to aluminum, steel, and galvanized coatings than CCA (see "Galvanic Corrosion," page 83). Preliminary tests have also shown that formulations with ammonia-based carriers (used for better penetration in heartwood species such as Douglas fir) are more corrosive than those with an amine or hybrid bases. Many factors affect corrosion rates, but some studies have found ACQ-treated wood

to corrode untreated steel up to four times faster than CCA and to attack galvanized coatings at twice the rate of CCA (also see "Fasteners," page 141).

Decay-Resistant Species

For a price, redwood and cedar are available in structural grades. How rot-resistant the untreated wood is depends on the amount of extractives in the wood, which is greatest in the heartwood cut from dense, old-growth trees. To purchase all-heart, structural-grade redwood or cedar, expect to spend two to three times more than for pressure-treated lumber. It is also difficult to find away from the West Coast. Left untreated, even the heartwood of these species is not recommended for ground contact.

Plastic Alternatives

Structural lumber products made of recycled plastics are starting to make their way into the marketplace and are turning up as railroad ties, dock components, and park walkways. Since there are few standards available for these products, designers and installers will need to rely on manufacturer data for structural characteristics. One product, TriMax (TriMax of Long Island), is made of recycled plastic and fiberglass and has compression and horizontal-shear strength similar to treated yellow pine. Like most plastic products, however, the material is more flexible than wood due to a low modulus of elasticity. This means spans must be small and the structure may have a bouncy feel.

DECKING MATERIALS

Horizontal deck surfaces take a beating. Rain and snow, exacerbated by wet-dry and freeze-thaw cycles, open up widening cracks, and ultraviolet radiation breaks down wood surfaces. The more sun exposure and water a deck sees, the quicker it will deteriorate. In addition, foot traffic makes it hard to maintain protective finishes.

In choosing a decking material, look beyond its short-lived original condition to its appearance and maintenance needs down the road (Table 4-3). In all cases, use the best grade of material you can afford. With wood decking, select tight, straight grain, few if any knots, and low moisture content. Since most wood decking materials are graded only for appearance, the grading is unregulated and grade names may be confusing. It is best to see a sample before committing to a purchase.

TABLE 4-3	**Decking Materials**			
Decking materials	Pros	Cons	Costs (1—lowest; 5—highest)	Recommendations
Pressure-treated yellow pine	Inexpensive, strong, durable. Readily available in eastern states.	Pressure-treatment toxicity concerns. Prone to check, crack, and splinter. Corrosive to metal fasteners.	1	Economical solution for typical projects. Choose premium-grade, radius-edge decking (RED) for best performance. Use factory-sealed materials or seal soon after installation to prevent checking and cracking.
Pressure-treated hem-fir	Strong, durable. Readily available in western states.	Pressure-treatment toxicity concerns. Cuts and drilled holes need treatment. Corrosive to metal fasteners.	1	Economical solution for typical projects. Choose a visual grade intended for decking use. Treat all field cuts and drilled holes with copper-naphthenate solution. Treat wear surface with a water-repellant preservative.
Pressure-treated Douglas fir	Very strong, durable. Available in western states.	Pressure-treatment toxicity concerns. Cuts and drilled holes need treatment. Corrosive to metal fasteners.	2	For applications where high strength is required. Choose a visual grade intended for decking use. Treat all field cuts and drilled holes with copper-naphthenate solution. Treat wear surface with a water-repellant preservative.
Redwood	Heartwood is naturally decay-resistant. Attractive and easy to work with.	Splinters easily. Soft wear surface. Sapwood vulnerable to decay.	3–5	For high-end projects requiring natural materials. Use an all-heart grade, if available, which is increasingly difficult to locate away from the West Coast. Treat after installation with water-repellant preservative.
Cedar	Heartwood is naturally decay-resistant. Attractive, readily available.	Sapwood vulnerable to decay. Soft wear surface (western cedar).	3	For high-end projects requiring natural materials. Use an all-heart grade of western red, Port Orford, northern white, or Atlantic white cedar. Treat after installation with a water-repellant preservative.
Tropical hardwoods	Naturally decay-resistant. Hard, durable, and attractive. Tight grain resists water intrusion.	Availability varies. Environmental concerns. Must be predrilled.	3–4	High-end projects requiring natural materials. Ipe must be predrilled and should be treated upon installation with a UV-blocking sealer. For best performance, other tropical hardwoods should be treated periodically with a water-repellant preservative or penetrating oil finish. Contact supplier for specific recommendations.
Wood-plastic composites	Very durable, maintenance-free. Contains 50–100% post-consumer wood products and plastics. Looks and feels similar to wood.	Spans limited to 16 inches. Somewhat bouncy feel. Standard nails and screws leave pucker when set. Oil and grease can stain.	2	For any moderate budget project requiring low maintenance. Use as an alternative to treated lumber. Do not exceed recommended span for material.
Vinyl and other plastics	Durable and maintenance-free. Hidden fasteners. Systems designed for easy installation and may include integral railing.	Vinyl may fade or crack over time.	4–5	Good for DIY projects where look of wood not required. Look for materials with skid-resistant surfaces and high-quality, UV-resistant plastics and finishes. Follow manufacturer's instructions.

Pressure-Treated Decking

Pressure-treated decking is still the most common choice due to its low cost and ready availability. As with other pressure-treated lumber products, CCA has been phased out in favor of wood treated with copper quat (ACQ) and copper azole (see "Pressure-Treated Lumber," page 135).

Most builders in the eastern United States choose nominal $\frac{5}{4}$x6 radius-edge decking (RED), which is dressed to 1 inch thick and can span 16 inches. Western softwoods are typically sold as 2x4 or 2x6 stock and can span up to 24 inches. For fully exposed decks, particularly those with south-facing exposure, use the best grade available, generally called Premium in southern pine and Patio 1 or Dex in Western species. The radiused edges on RED stock make an attractive deck surface and help prevent splintering.

Many homeowners become disappointed with pressure-treated decking when they discover that it must be treated regularly to prevent cupping, checking, and warping. Except where the decking is under a roof and well protected from sun and rain, it must be treated with a water-repellant sealer or a semitransparent penetrating stain to prevent problems. It is important to finish the wood as soon as the surface is dry but before it has a chance to start cracking, allowing water to penetrate and do additional damage (see "Finishes for Decking," page 154). A few premium products come factory pretreated with a sealer that has penetrated the wood surface and should outlast on-site sealing.

Decay-Resistant Species

Until the recent introduction of synthetic decking materials, the only viable alternatives to pressure-treated lumber were naturally decay-resistant wood species, such as redwood and cedar. A number of tropical hardwoods are now available to deck builders as well.

Redwood and Cedar. The most decay-resistant materials are cut from the dense heartwood of old-growth trees, which is expensive and increasingly rare. In redwood, the lighter colored sapwood offers moderate resistance to decay. Cedar sapwood is even less resistant. If possible, choose all-heart grades for exposed areas (see Figure 4-1).

The two most commonly used cedar types are western red cedar and Alaska cedar, sometimes sold as Alaska cypress. Western red cedar is relatively soft and easily dented, but dimensionally stable. Yellow cedar is stronger and has a harder surface; but it is prone to shrinkage movement, so it is best purchased kiln-dried. Other less common cedar species suitable for decking include Port Orford cedar, a strong, dense wood from southern Oregon, and northern white and Atlantic white cedar.

Redwood and cedar should, at a minimum, be treated regularly with water repellants to protect against cupping, checking, and cracking. Occasional treatment with a water-repellant preservative will increase the service life. Even then, the wood will weather to a silver gray like nearly all wood decking products. For clients who want

FIGURE 4-1 **All-Heart Redwood.**

For those who can afford it, all-heart redwood, shown above, or all-heart clear cedar provide natural beauty and decay-resistance. The sapwood of these species, however, is vulnerable to decay.
SOURCE: Photo courtesy of California Redwood Association.

to preserve the original wood color, the best option is to use an oil-based semitransparent stain with pigments that resemble the desired color.

Tropical Hardwoods. A number of tropical hardwoods are becoming increasingly available as decking materials, typically sold as nominal 1x4 or $\frac{5}{4}$x6 stock, which can span 16 inches and 24 inches, respectively.

The most widely available hardwood decking now is Ipe, a group of dense teaklike woods sometimes marketed under the brand name Pau Lope and reported to have a life expectancy in outdoor use of up to 40 years. Another option, Cambara, is somewhat less dense and lighter in color than Ipe, but it also offers excellent resistance to decay and insects.

Both of these are typically knot-free with a tight grain pattern that helps keep out water. They are strong, dense, and highly resistant to decay and insects, making them ideal for deck surfaces. In general, these woods will outlast the redwood and cedar available today and should require only periodic treatment with a UV-blocking water-repellant or a penetrating oil finish such as Penofin (Performance Coatings Inc.). Like most hardwoods, Ipe and cambara cut slowly and must be predrilled for screws.

A third tropical hardwood group, sometimes marketed as *maranti,* is more commonly known as lauan or Philippine mahogany. This wood offers moderate resistance to decay and insects and requires periodic treatment with a water-repellant preservative.

With all tropical hardwoods, there are valid concerns about the wood's origin and the impact of its harvesting on the world's rainforests. Fortunately, there are established third-party certification organizations that track the "chain of custody" of hardwood products and certify that they were harvested using sustainable logging practices. Contact the Forest Stewardship Council or the Smartwood Program for more information (see Resources, p. 158).

Synthetic Decking

Manufacturers have introduced a wide range of synthetic decking products, most of which promise woodlike appearance and low or no maintenance. Most fall into a few categories discussed below, but each has unique characteristics and installation requirements. In all cases, review the product specifications and, if possible, look at an installation before purchasing.

Wood-Plastic Composites. Decking materials made from wood fiber with polymer resin have been in use for over a decade and have generally established a strong track record. Many, like Trex®, use a high percentage of recycled materials. Most have a solid profile and are sold in sizes that match and install like standard wood decking. The oldest solid composite decking on the market is Trex®, but competitors now include Boardwalk® (Certainteed), ChoiceDec® (Weyerhauser), and products from several smaller manufacturers (see Figure 4-2).

Other synthetic decking products are extruded into a hollow 2x6 profile, such as WeatherBest® (Louisiana Pacific) and TimberTech® (TimberTech Ltd.). These are generally lighter and stiffer than the solid materials, allowing spans up to 24 inches. These typically fit together in

FIGURE 4-2 **Composite Decking.**

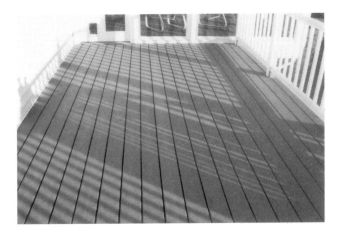

Decking materials made of wood fiber and polymer resin have established a strong track record in over a decade of use. Trex, made largely from recycled grocery bags and wood waste, was used here to replace three-year-old pressure-treated boards that were badly checked and warped.
SOURCE: Photo by author.

FIGURE 4-3 **Modular Decking Systems.**

While most synthetic decking materials have a solid profile and install like wood planks, others are extruded into hollow profiles that fit together in modular fashion, like Timber Tech's tongue-and-groove profile, shown above. Board ends are trimmed with skirting or proprietary end caps, and many systems include modular railings.
SOURCE: Photo courtesy of Timber Tech Ltd.

a tongue-and-groove fashion, have built-in drainage systems, and use proprietary fasteners. End caps and other accessories are used to trim out these products, and some include a modular railing system (Figure 4-3).

In general, composite decking materials are dimensionally stable, impervious to the elements, and can be worked more or less like wood. The solid products are installed like wood decking. Special screws designed for composite decking leave a clean hole without dimpling around the screw head. Most solid products cannot exceed 16-inch on-center framing, and they feel a little bouncy underfoot due to the material's greater flexibility (low modulus of elasticity) compared to wood. Some have an embossed wood-grain finish that may wear away over time.

With composite decking, no surface treatment is needed. Once the surface weathers, it bears a strong resemblance to weathered wood, but without the customary warping and checking. Although these surfaces are largely impervious to the elements and contain no food source for mold or mildew, manufacturers do point out that a dirty deck can support mold and mildew growth and recommend periodic cleaning with a deck cleaner to prevent this. Also synthetic decking is vulnerable to oil and grease stains, which can be difficult to remove if not cleaned right away with a degreasing agent.

Vinyl and Other Plastics. A variety of manufactured decking systems are aimed at the commercial and do-it-yourself (DIY) markets. Materials include FRP (fiber-reinforced plastic), recycled HDPE (high-density polyethylene), and polyvinyl chloride (PVC). Most of these products are available in a nominal 6-inch width and have an etched surface to provide traction and a wood-grain appearance. Most products are sold as a complete system with integral fasteners, end caps, and other accessories, making them well suited to the DIY market.

FASTENERS

Like other deck components, metal fasteners are subject to numerous stresses. The sealers and stains typically used on decks provide little protection to fasteners, and the copper-based chemicals used in most waterborne preservatives accelerate corrosion in many metals. When the fasteners corrode, they contribute to decay in the surrounding wood, further weakening the connection.

New Preservatives and Corrosion. Because the new wood preservatives ACQ (alkaline copper quat) and copper azole contain significantly more copper than the older CCA-treated lumber, they are estimated to be two to four times more corrosive to metals and galvanized coatings than the CCA they are replacing. Most treatment manufacturers recommend that fasteners and hardware in contact with the new treated wood be stainless-steel, heavily coated hot-dipped galvanized, or proprietary fasteners tested and approved by the manufacturer.

Hot-Dipped Galvanized. When using lumber treated with ACQ or copper azole, use hot-dipped galvanized fasteners that meet ASTM A153 with a minimum of 2 ounces of zinc coating per square foot. Some are marketed as "double-hot-dipped." Connectors, flashings, and other hardware in contact with treated wood should meet ASTM A653, Class G185 (1.85 ounces of zinc per square foot of metal surface). These have three times as much zinc coating as standard G60 connectors. Examples of G185 coatings include Simpson's Z-Max or USP Connector's Triple-Zinc. Galvanized fasteners may stain redwood and cedar, however.

Stainless Steel. In very wet or humid climates, or in areas subject to salt-water spray or deicing salts, stainless steel is the best choice. Stainless steel is also recommended for tropical hardwoods, which tend to cause staining with coated nails. Both types 304 and 316 stainless steel have been tested for use with the new wood preservatives ACQ and copper azole. Type 304 is suitable for above-ground applications. Type 316 is recommended in areas subject to salt or salt water.

Caution: *Never use stainless steel in contact with galvanized steel,* as the galvanized coating will quickly corrode. Where fasteners such as nails, bolts, or lags are in contact with metal connectors, use the same metal for both.

Proprietary Coatings. Many decking screws sold in home centers have newer proprietary ceramic and epoxy

coatings over an electrogalvanized core. Originally developed for use with CCA-treated lumber, many have now been tested and approved for use with ACQ and copper azole. If using one of these fastener types, make sure that it is recommended by the manufacturer for the specific type of decking being installed.

Aluminum. *Do not use aluminum fasteners, connectors, or flashings* in contact with pressure-treated wood. The copper-based waterborne preservatives will cause corrosion and premature failure.

Decking Fasteners

Whether using nails or screws, make sure to choose a product that is up to the task both structurally and aesthetically. In general, screws are more expensive to buy and install, but often they make for a neater job with fewer callbacks due to boards popping up. Construction adhesives specially formulated for treated wood may be used in conjunction with nails or screws.

Nails. If the decking is to be nailed, use either spiral-, twist-, or ring-shanked nails to resist pullout. When using pressure-treated wood, the fasteners should be either hot-dipped galvanized, meeting ASTM A153, or stainless steel Type 304 or higher (see "New Preservatives and Corrosion," above).

Galvanized nails sometimes cause discoloration with redwood, cedar, and tropical hardwoods, so stainless steel is a safer choice with these materials. Aluminum nails are also an option for redwood or cedar, but they should not be used with pressure-treated wood.

The bigger the head, the better the hold-down power. At a minimum, use a casing nail, preferably a common. Some manufacturers sell special decking nails with a head size in between the two. For $\frac{5}{4}$-inch decking use a 10d (3-inch) nail. For thicker decking, use a 16d ($3\frac{1}{2}$-inch) nail. Deck spans are shown in Table 4-4.

Screws. Power-driven screws are more expensive and a little slower to install, but they are the best choice since they do a better job of holding down the decking and are more easily removed for repairs. For treated wood, use either hot-dipped galvanized screws that meet ASTM A153, stainless-steel multipurpose screws, or specialty decking screws with proprietary coatings tested and approved for use with the new types of wood treatments.

Multipurpose screws have a thicker shank and courser thread than drywall screws, giving them greater strength and better holding power. Some specialized decking screws also have cutting nubs under the head for self-countersinking in wood. Others have special concave "pancake" heads for composite decking. The composite-decking screws leave a crisp hole in the decking surface without the typical pucker of excess material (mushrooming) around the screw head (see Figure 4-4).

TABLE 4-4	Decking Spans and Fastenings		
Wood Species	Nominal Dimension (in.)	Maximum Span (in.)	Fasteners (two per joist)
Southern yellow pine	$\frac{5}{4}$x4 $\frac{5}{4}$x6	24	3 in. nails or $2\frac{1}{2}$ in. screws
	2x4 2x6	24	$3\frac{1}{2}$ in. nails or 3 in. screws
Douglas-fir, hem-fir	$\frac{5}{4}$x4 $\frac{5}{4}$x6	16	3 in. nails or $2\frac{1}{2}$ in. screws
	2x4 2x6	24	$3\frac{1}{2}$ in. nails or 3 in. screws
Redwood, western cedar	$\frac{5}{4}$x4 $\frac{5}{4}$x6	16	3 in. nails or $2\frac{1}{2}$ in. screws
	2x4 2x6	16 24	$3\frac{1}{2}$ in. nails or 3 in. screws
Tropical hardwoods (Ipe, Cambara, maranti)	1x4	16	$2\frac{1}{4}$ in. screws (predrilled)
	$\frac{5}{4}$x6	24	$2\frac{1}{4}$ to $2\frac{1}{2}$ in. screws (predrilled)

NOTE: All lumber No. 2 or better

FIGURE 4-4	Composite Decking Screw.

The TrapEase composite-decking screw (left) has a concave pancake-type head that leaves a crisp hole in the decking surface without a pucker of excess material around the head (right).
SOURCE: Photos by author.

In addition to Phillips-head screws, decking screws come with square-drive and star-drive heads that allow higher torque driving without stripping the head.

Spacing the Decking. When securing the decking, it is important to leave adequate spacing between the boards for water to drain. The goal is to have about an $\frac{1}{8}$-inch gap (the diameter of an 8d nail) between boards after the decking has dried to its equilibrium moisture content. If the decking is installed wet, as is often the case for pressure-treated material, it is best to install the boards tight, letting gaps form as the wood dries.

When installing kiln-dried stock, use a 16d nail as a spacer to leave enough space for the boards to swell slightly and still leave an adequate drainage space. For wood that has partially air-dried, it is a judgment call. If in doubt, it is best to err on the side of leaving a little extra space for the wood to swell when wet.

Protecting the Joists. In decking applications subject to continuous wet conditions or the buildup of leaves and tree debris, it is best to protect the tops of the joists from moisture. This can be achieved by laying strips of felt paper or self-adhesive membrane (one approved for UV exposure) over the tops of each joist prior to laying the decking boards. A proprietary product for this application, Vycor Deck Protector™, is available from Grace Construction Products. Membranes such as Vycor Deck Protector can also be used as a barrier between pressure-treated joists and joist hangers, flashings, or other metal hardware to reduce corrosion.

Hidden Deck Fasteners

Over time, face-nailed deck fasteners may loosen, stain the wood decking, or lead to splitting and water penetration. Particularly with higher-end decking materials, such as tropical hardwoods, more customers are opting for hidden fastening systems (see Resources, page 157).

FIGURE 4-5 | **Hidden Deck Fasteners**

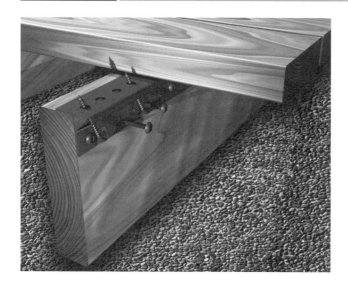

Hidden fasteners are often used with hardwoods and other high-end decking materials. Most systems either use a heavy steel track, like the Deckmaster (left), or proprietary clips, like the Eb-Ty, that fit between adjacent planks and nail or screw down into the joist (right).
SOURCE: Photo (left) courtesy of Grabber Construction Products. Photo (right) courtesy of Blue Heron Enterprises.

Each system is proprietary, and some require proprietary tools supplied by the fastener manufacturer. Some suppliers of hardwood decking recommend a specific fastener for their product and may sell the fasteners along with the decking. There are two types of systems. One, such as Deckmaster® (Grabber Construction Products), uses a right-angle bracket that fastens to the tops of the joists and screws into the underside of the decking. These are easiest to install if the installer has access from below the deck. The other type uses individual fasteners that fit between adjacent decking planks and screw down into the top of the joist. These typically attach to the edges of the decking planks with clips or prongs. A variation on this type called Eb-Ty (Blue Heron Enterprises) fits into slots cut into the edges of adjacent deck boards with a plate jointer (see Figure 4-5).

The biggest concern with hidden deck fasteners is whether they are strong enough to resist the tendency of deck boards to warp or twist. For that reason, they are best used with premium decking products, which are dimensionally stable. Tropical hardwoods and composite decking are good candidates for hidden fasteners.

Structural Fasteners and Connectors

At minimum, all structural hardware for decks should be hot-dipped galvanized steel. For the best protection, use stainless steel (see "Joist Hangers and Connectors," below).

At connections carrying structural loads, such as deck ledgers or railing posts, use through-bolts or lag screws. Through-bolts are stronger and should be used where possible. For the heaviest connections on a deck, such as where ledgers attach to the house or to posts, use $\frac{1}{2}$-inch

bolts or lags. Place large washers under the heads of lags and at both ends of through-bolts. Retighten bolts and lags after the first year and check periodically for tightness.

Bolts. Where both sides of the joint are accessible, bolts offer the strongest connections. Drill pilot holes $\frac{1}{32}$ to $\frac{1}{16}$ inch larger than the diameter of the bolt so it will slide through easily. After drilling, saturate the hole with preservative. Use large washers under both the head and nut. Retighten after the first year, since the wood may have shrunk.

Lags. For lags, drill a full-diameter pilot hole for the unthreaded portion and a smaller hole (65 to 75% of the lag's diameter) for the threaded portion. So, for example, a $\frac{1}{2}$-inch lag would get a $\frac{5}{16}$-inch pilot hole for the threaded portion; a $\frac{3}{8}$ lag would get a $\frac{1}{4}$-inch pilot hole. After drilling, saturate the pilot hole with wood preservative. It is also important that at least half the length of the lag is threaded into solid wood. For example, driving a 5-inch lag through a 4x4 post into a 2x joist will produce a weak connection with only $1\frac{1}{2}$ inches of anchoring. Instead, the lag screw should go through the 2x and be threaded into the thicker 4x4. Use a large washer under the head, and retighten after the first year in case materials have shrunk.

Joist Hangers and Steel Connectors. At a minimum, use hot-dipped galvanized hardware. With pressure-treated wood, hot-dipped galvanized steel should conform to ASTM A153 (for fasteners) or ASTM A653, G185 (for connectors). Stainless steel offers the best protection. Type 304 or higher stainless steel is recommended for very wet environments such as poolside decks; or Type 326 for exposure to salt or saltwater. *Also, do not mix metals: Use stainless-steel fasteners with stainless-steel connectors and galvanized fasteners with galvanized connectors* (see "New Preservatives and Corrosion," page 141).

CONSTRUCTION DETAILS

Ledgers

Most residential decks are supported on one side by a ledger that is bolted or lagged to the home's band joist. This connection is critical, since a failure here can cause a deck to collapse. Failure of the ledger can be caused by too few or undersized fasteners, or by decay in the ledger or band joist. Lags or bolts provide little support when fastened to rotted wood. So proper flashing of the ledger and band joist area is critical. It is also important that the band joist be nailed adequately to the surrounding structure, since the ledger is only as strong as the structural members it is attached to.

Band Joist-to-House Connection.
In new construction, if a deck is planned, make sure the band joist is pressure treated and adequately nailed to the sole plate above and the sill or top plate below, using stainless-steel or double-hot-dipped galvanized nails. Fastening with 16d common nails at 8 inches on-center is recommended. If the nailing cannot be confirmed in a retrofit, extra toenails driven through the exterior can help to reinforce this connection (see Figure 4-6)

Ledger-to-Band Joist Connection.
Through-bolts are the most reliable connection, but lag bolts are adequate as long as they are long enough to fully penetrate the band joist. For through-bolts, drill holes $^1/_{16}$ inch larger than the

bolt. For lags, drill a full-diameter hole for the unthreaded portion and a smaller hole (65 to 75% of the lag's diameter) for the threaded portion. Use washers under the head of the lag bolt or at both ends of through-bolts to keep the head from crushing the wood. Soak the holes with a preservative before inserting the bolts. Spacing for bolts and lags are shown in Table 4-5.

| FIGURE 4-6 | Strengthening the Band Joist. |

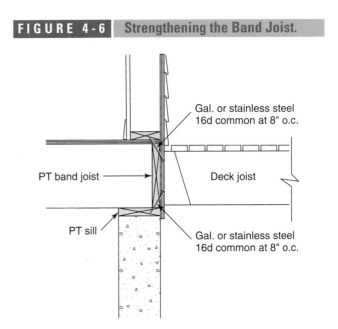

Where a ledger is bolted to the band joist, it is critical that the band joist be securely tied to the structure. In new construction, use a pressure-treated band joist and nail it to the sole plate above and sill or top plate below with 16d stainless-steel or double-hot-dipped galvanized nails. In retrofits, toenail 16d nails from the exterior, as shown.

| TABLE 4-5 | Fastener Spacing for Deck Ledger to Band Joist |

Joist Span	Spacing for $^1/_2$-inch lag screws	Spacing for $^1/_2$-inch through-bolts
6 ft.	30 in.	36 in.*
8 ft.	23 in.	36 in.*
10 ft.	18 in.	34 in.
12 ft.	15 in.	29 in.
14 ft.	13 in.	24 in.
16 ft.	11 in.	21 in.
18 ft.	10 in.	19 in.

Note: Table assumes 40 lb. live load, 10 lb. dead load, 2x8 PT southern pine ledger, $^{15}/_{32}$ in. plywood sheathing, and #2 SPF band joist. Lag screws must be predrilled, as shown at right, and the threaded portion must fully penetrate band joists. Holes for through-bolts should be $^1/_{16}$ in. larger than bolt. Washers are required at both ends of bolts, and are recommended for lag screws.

*These spacings have been reduced from the test values due to consideration of the bending strength of the ledger between bolts.

Data based on tests conducted by Joe Loferski, Frank Woeste, P.E., Mary Billings, of Virginia Tech University. Reprinted with permission from Professional Deck Builder Magazine © 2005 Dempsey Management Services, Inc.

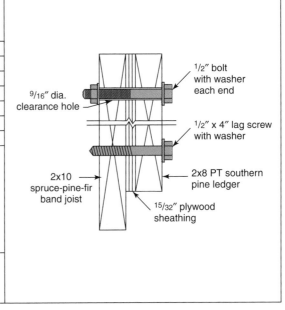

TABLE 4-6 | Fastener Spacing of Deck Ledger to Band Joist (with Spacers)

Joist Span	Spacing for 1/2-inch through-bolts
6 ft.	36 in.*
8 ft.	36 in.*
10 ft.	30 in.
12 ft.	25 in.
14 ft.	21 in.
16 ft.	19 in.
18 ft.	17 in.

Note: Table assumes 40 lb. live load, 10 lb. dead load, 2x8 PT southern pine ledger, 15/32 in. plywood sheathing, and #2 SPF band joist. Holes for through-bolts should be 1/16 in. larger than bolt. Washers are required at both ends of bolts.

*These spacings have been reduced from the test values due to consideration of the bending strength of the ledger between bolts.

Data based on tests conducted by Joe Loferski, Frank Woeste, P.E., and Mary Billings, of Virginia Tech University. Reprinted with permission from Professional Deck Builder Magazine © 2005 Dempsey Management Services, Inc.

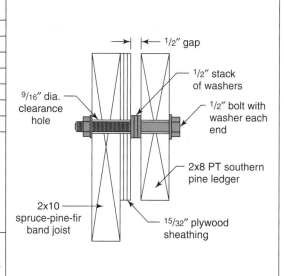

FIGURE 4-7 | Fastener Placement in Ledgers.

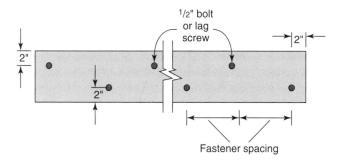

To prevent splitting, fasteners in the deck ledger should be held back 2 inches from edges and staggered, as shown.

In humid climates, some deck builders prefer to leave a $\frac{1}{2}$-inch air space between the ledger and house to assist with drying in the event that the ledger area gets wet. Because this weakens the connection, through-bolts should be used, as shown in Table 4-6. To prevent splitting, lag screws or bolts should be held back two inches from edges and staggered as shown in Figure 4-7.

Flashing. Proper flashing is critical since leakage at the band joists could lead to decay and failure of the deck connections. The siding should be removed over the band-joist area, and a wide band of peel-and-stick membrane or metal flashing should run over the band joist and up under the building felt or housewrap. A second cap flashing should direct water over the ledger and away from the house (Figure 4-8).

FIGURE 4-8 | Deck-Ledger Detail.

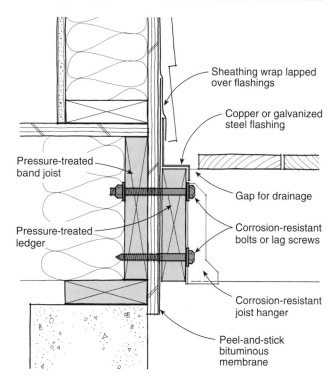

Though deck failures are rare, ledger connections are typically at fault when they occur. Through-bolts make the strongest connection, but adequately sized lag screws can also work. With either, it is critical to flash the ledger area and to only use metal components that are compatible with pressure-treated wood.

Some deck builders prefer to leave an air space at the ledger to assist with drying if the area gets wet. Flashing details are shown in Figure 4-9.

FIGURE 4-9 | **Spaced Deck-Ledger Detail.**

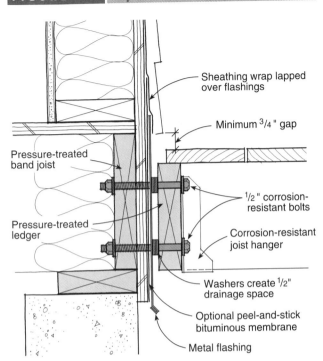

- Sheathing wrap lapped over flashings
- Minimum ³/₄" gap
- Pressure-treated band joist
- ¹/₂" corrosion-resistant bolts
- Pressure-treated ledger
- Corrosion-resistant joist hanger
- Washers create ¹/₂" drainage space
- Optional peel-and-stick bituminous membrane
- Metal flashing

Spacing the ledger away from the house helps prevent decay in the sheathing and band joist if the area gets wet, but this requires more bolts. A bituminous membrane across the band-joist area offers an extra layer of protection, useful in very wet or snowy regions.

Caution: *Do not use aluminum flashings with pressure-treated lumber* unless a durable barrier material, such as a bituminous membrane, separates the aluminum from contact with the wood. Preferably use membrane-type flashings, G185 galvanized steel, or copper. However, copper flashings should not contact galvanized hardware or fasteners.

Alternative to Ledger. Given all the problems inherent in supporting a deck with a ledger, one approach is to forgo the ledger altogether and support the deck on posts on all sides, keeping it structurally independent of the house. In this approach, the deck may be braced to the house to help it resist racking, but all vertical loads are carried to the ground by posts (Figure 4-10).

A conservative rule-of-thumb states that joists can cantilever one-fourth of their total length, assuming that the cantilevered end is not carrying any loads other than the normal uniform floor loading.

Posts

In frost-susceptible soils, all posts should sit on concrete piers that extend below the frost depth. The tops of the concrete piers should extend slightly above grade to keep the post ends out of standing water (see Figure 4-11).

It is also a good idea to use steel post bases to keep the wood out of direct contact with concrete. Where uplift

FIGURE 4-10 | **Free-Standing Deck.**

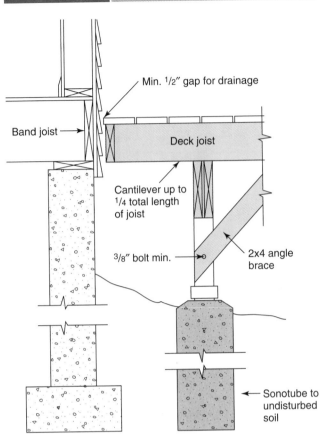

- Min. ¹/₂" gap for drainage
- Band joist
- Deck joist
- Cantilever up to ¹/₄ total length of joist
- ³/₈" bolt min.
- 2x4 angle brace
- Sonotube to undisturbed soil

One solution to ledger problems is to eliminate them altogether and support the deck independently on posts. The deck may be braced to the house to help it resist racking, but all vertical loads are carried to the ground by posts.

FIGURE 4-11 | **Post Bases.**

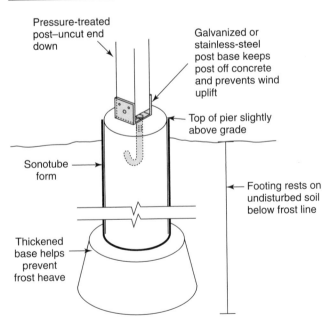

- Pressure-treated post—uncut end down
- Galvanized or stainless-steel post base keeps post off concrete and prevents wind uplift
- Top of pier slightly above grade
- Sonotube form
- Footing rests on undisturbed soil below frost line
- Thickened base helps prevent frost heave

Steel post bases keep the wood out of direct contact with soil or concrete. Where wind uplift is an issue, use anchor bolts rated for the necessary uplift load.

FIGURE 4-12 Diagonal Sway Bracing.

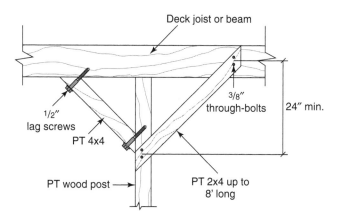

While some codes allow 4x4 deck posts up to 10 feet high, tall pressure-treated 4x4s are prone to warping and twisting. Diagonal sway bracing can help stiffen tall posts and provide resistance against racking.

FIGURE 4-13 Post-to-Beam Connections.

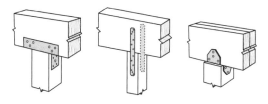

Steel Post-Cap Connectors

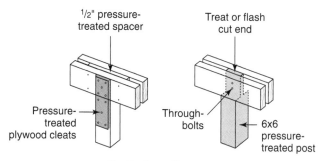

Post-to-Beam Connectors

For the strongest connection, place support beams directly atop the posts and reinforce the connection with steel strapping, a steel connector, or a treated-plywood cleat. Avoid notching 4x4 posts, but a 6x6 can be safely notched to accept a double 2x beam, as shown.

from wind is an issue, use structural post bases that are rated for the necessary uplift loads and connected to the concrete with anchor bolts. Many of these are adjustable laterally so the footing and anchor bolt do not need to be precisely placed. Use pressure-treated posts rated for ground contact with a minimum preservative level of .40 pounds/cubic feet. Treat any cut post ends and do not place cut ends in contact with soil or concrete.

While 4x4 posts can be used up to about 10 feet in height, depending on loads, tall pressure-treated 4x4s are prone to warping and twisting when they dry. Some local codes limit the use of 4x4s to 8 feet. Diagonal sway bracing can help stiffen tall posts and provide resistance against racking. The easiest approach is to run short 2x4 or 4x4 sway braces from posts to beams (Figure 4-12).

Support Beams

For the strongest connection of beam to post, place the support beam directly on top of the posts, rather than bolting them to the side, so the full load is transferred to the posts. To keep the post in place and to prevent any twisting or shifting, the connection should be reinforced with steel strapping, a steel connector, or a treated plywood cleat (Figure 4-13).

In general, notching a 4x4 post will leave too little wood for an adequate connection. A double 2x beam can rest on a notched 6x6 post, as shown.

For the strongest connection of joists to the support beam, the joists should sit on top of the beam. For a more streamlined appearance, however, joist hangers are acceptable. Make sure the hangers and nails are approved by the manufacturer for use with the new types of pressure-treated wood (ACQ or ACZA) and fill all the holes in the hangers with properly sized nails. In general, nail into the joists with $1\frac{1}{2}$-inch joist-hanger nails and nail into the beam with 10d to 16d common nails, as specified by the

hanger manufacturer. Sizes for joists and beams are shown in Tables 4-7 and 4-8.

Railings

Code Requirements. The International Residential Code (IRC) requires a minimum 36-inch-high guardrail for all decks, balconies, or screened enclosures more than 30 inches off the ground. For child safety, the balusters or other decorative infill must be spaced less than 4 inches apart (a 4-inch-diameter ball should not pass between the balusters).

The railing must be strong enough to resist horizontal loads from people leaning on it. The IRC requires that the railing be able to resist a 200-pound concentrated load applied along the top in any direction, while some local codes still in effect specify a smaller load of 20 pounds per linear foot. Under the IRC, the infill or balusters must resist a concentrated horizontal load of 50 pounds applied to a square foot area. The baluster requirement is easily met with standard fastening techniques, but meeting the IRC guardrail requirement is difficult without adding steel hardware. The majority of residential decks, which rely on notched posts lag-screwed into the band joist, do not meet the 200-pound requirement.

Post Connections. Posts that run continuously from footings to railings are the strongest, but these are often not practical. More commonly, the posts are attached to

TABLE 4-7   Maximum Beam Spans (ft) for Exterior Decks (40 lb live load; 10 lb dead load)

Wood Species*	Beam Sizes	Width of Deck Supported by Beam (feet)** (half of span for perimeter beams)								
		4	5	6	7	8	9	10	11	12
Southern yellow pine, Douglas fir-larch	(2) 2x6	7	6							
	(2) 2x8	9	8	7	7	6	6			
	(2) 2x10	11	10	9	8	8	7	7	6	6
	(3) 2x8	12	11	10	9	8	8	7	7	7
	(2) 2x12	13	12	10	10	9	8	8	7	7
	(3) 2x10	15	13	12	11	10	10	9	9	8
	(3) 2x12	16	15	14	13	12	11	11	10	10
Hem-fir, Spruce-pine-fir	(2) 2x6	6	6							
	(2) 2x8	8	7	6	6					
	(2) 2x10	10	9	8	7	7	6	6		
	(3) 2x8	11	10	9	8	7	7	7	6	6
	(2) 2x12	11	10	9	8	7	7	7	6	6
	(3) 2x10	13	12	11	10	8	8	8	8	7
	(3) 2x12	15	14	12	11	10	10	9	9	8
Ponderosa pine, redwood, western cedar	(2) 2x6	6								
	(2) 2x8	8	7	6	6					
	(2) 2x10	9	8	8	7	6	6	6		
	(3) 2x8	10	9	8	8	7	7	6	6	
	(2) 2x12	11	10	9	8	7	7	7	6	6
	(3) 2x10	13	11	10	9	8	8	7	7	7
	(3) 2x12	15	13	12	11	10	9	9	8	8

*All lumber No. 2 and Better grade; 10 lb dead load.
**Perimeter beams carry half the distance from beam to beam; interior beams carry full distance from beam to beam.

TABLE 4-8   Maximum Spans for Deck Joists (ft.-in.)*

Wood Species**	2x6			2x8			2x10			2x12		
	12 in. o.c.	16 in. o.c.	24 in. o.c.	12 in. o.c.	16 in. o.c.	24 in. o.c.	12 in. o.c.	16 in. o.c.	24 in. o.c.	12 in. o.c.	16 in. o.c.	24 in. o.c.
Douglas fir-larch	10-9	9-9	8-3	14-2	12-9	10-5	18-0	15-7	12-9	20-11	18-1	14-9
Hem-fir	10-0	9-1	7-11	13-2	12-0	10-2	16-10	15-2	12-5	20-4	17-7	14-4
Southern pine	10-9	9-9	8-6	14-2	12-10	11-0	18-0	16-1	13-2	21-9	18-10	15-4
Western red cedar**	9-6	8-3	6-9	12-1	10-5	8-6	14-9	12-9	10-5	17-1	14-9	12-1

*40 lb live load; 10 lb dead load. Deflection = span/360
**Western red cedar, No.1/No.2. All other lumber, No. 2 and Better.

the rim joist or beam, preferably with through-bolts (see Figure 4-14).

While 4x4 railing posts are often notched where they connect to the beam, this creates a weak point in the post that will not meet the load requirements. Another problem is that the rim joist needs to be reinforced to keep it from rotating when a strong force is applied to the railing. This can be achieved with lag bolts, steel strapping, or steel connectors tying the rim joist to the abutting joists. On sides where the rim joist runs parallel to the joists, solid blocking should be lagged in place to keep the rim joist from rotating. Additional steel connectors may also be required. Posts should be no more than 6 to 8 feet apart, depending on local codes.

Wood Railings. The top rail can be a 2x6 either flat or on edge. Use the longest pieces you can find—a continuous railing is best. Balusters can be nailed or screwed directly to the rim joist or attached to a bottom rail (Figure 4-15).

Use either one screw or two spiral-shank nails top and bottom on each baluster. If you use a flat rail on top, it is best to slope or chamfer the top surface to shed water.

Manufactured Railings. Many types of manufactured railing systems are also available, often from the same companies that provide composite decking products. Examples include SmartDeck's post and rail system made from an extruded wood-poly composite and a similar

F I G U R E 4 - 1 4   **Fastening Railing Posts.**

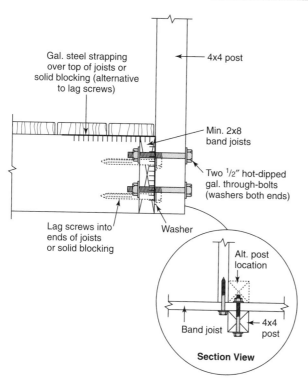

Gal. steel strapping over top of joists or solid blocking (alternative to lag screws)

4x4 post

Min. 2x8 band joists

Two ¹/₂" hot-dipped gal. through-bolts (washers both ends)

Lag screws into ends of joists or solid blocking

Washer

Alt. post location

Band joist

4x4 post

Section View

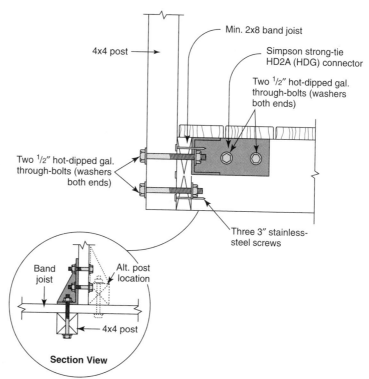

Min. 2x8 band joist

Simpson strong-tie HD2A (HDG) connector

Two ¹/₂" hot-dipped gal. through-bolts (washers both ends)

4x4 post

Two ¹/₂" hot-dipped gal. through-bolts (washers both ends)

Three 3" stainless-steel screws

Band joist

Alt. post location

4x4 post

Section View

While posts that run continuously from footings to railings are the strongest, more commonly the posts are bolted to the rim joists, preferably with through-bolts. To keep the rim joist from rotating when a load is applied to the railing, the rim joist should be reinforced with lag bolts or steel strapping (left). A more solid connection, tested to withstand the code-required load, can be achieved by using hold-down connectors, as shown (right).

SOURCE: Detail at right reprinted with permission from "Load Tested Guardrail Post Connections," by Joseph Loferski, Frank Woeste, P.E. Dustin Albright, and Ricky Caudill, published by *Professional Deck Builder Magazine* © 2005 Dempsey Management Services.

FIGURE 4-15 **Wood Railings.**

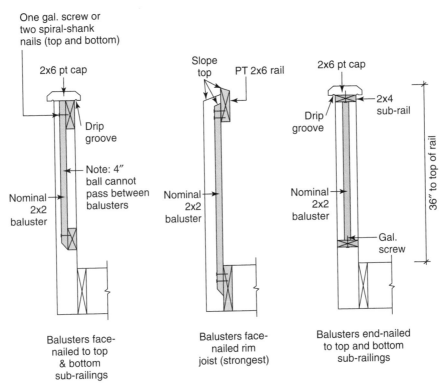

For a sturdy top rail, use long pieces of 2x6, either flat or on edge. If possible, use a single continuous board for each railing. Secure the balusters with either one screw or two spiral-shank nails at top and bottom, and slope or chamfer all top surfaces to shed water.

railing made of fiber-reinforced plastic (FRP) from Shakespeare Composites, best known for its FRP fishing rods (see Figure 4-16).

An advantage of the prefab systems, in addition to their easy assembly, is that most are engineered to meet the strength requirements of the model codes in the areas where they are marketed (see Resources, page 157).

ROOFTOP DECKS

Decks over living spaces can be detailed various ways, but all should have the following elements:

- A long-lasting membrane roof sloped at least $\frac{1}{4}$ inch per foot, designed to drain at the lower end

- Decking screwed to sleepers so that it can be removed for inspection or repair

- A sturdy railing system that, preferably, does not penetrate the roofing membrane

Single-Ply Membranes. A variety of roofing membranes have been used successfully under rooftop decks, but EPDM remains the most widely used on residential roofs. EPDM owes its popularity to its longevity and durability. It is dimensionally stable and strong over a wide

temperature range (it will not expand and contract or tug at flashings), is highly UV-resistant, and can stretch without tearing. In most residential jobs, EPDM is fully adhered to the roofing deck with a contact-type adhesive rolled onto both surfaces. Seams can be avoided on small jobs, since rolls come as wide as 50 feet. If necessary, however, seams are relatively easy to seal, using a special double-sided seam tape and lap caulk at the exposed joint. To seal around corners and penetrations, use flexible uncured EPDM, which will conform to irregular shapes and seals well to the main roofing membrane (see EPDM, page 96).

EPDM roofing membranes typically come in .045- and .060-inch thicknesses and carry at least a 10-year warranty in residential work. If properly installed, however, the .060 material should last for decades. EPDM's chief weakness is its vulnerability to petroleum products, such as oils, grease, and solvents. If used under a deck, warn the homeowners not to splash oil or grease from cooking or solvent-based wood finishes on the decking.

EPDM is often installed over a base layer of fiberboard or rigid insulation, but most EPDM membranes can bond directly to a sanded wood underlayment such as a $\frac{1}{4}$-inch AC plywood. If bonding directly to plywood or oriented-strand board (OSB), check with the roofing manufacturer regarding treatment of joints in the sheathing and

FIGURE 4-16   Modular Railings.

Most prefab railing systems assemble easily and are engineered to meet code requirements in the areas where they are marketed. The Armor-Rail system shown uses sturdy FRP tubing that can span up to 12 feet between posts.
SOURCE: Photo courtesy of Armor-Rail/Shakespeare Composite Structures.

the need for a primer. To allow for some movement at sheathing joints, some manufacturers recommend holding the adhesive back one inch from the joints.

Terminations and Flashings. Where the roofing membrane meets the house, run it 10 to 12 inches up the wall under the sheathing wrap. At outside edges, leave a 3- to 6-inch flap, depending on the edge treatment. A large metal drip-edge usually caps the roof along the fascia boards.

Decking. The decking sits on 2x sleepers, usually laid flat. If you want the decking surface to be level, the sleepers will need to be tapered to compensate for the slope in the roof deck. For a large deck where flat sleepers are too thin to taper, you can use 2x4s or 2x6s set on edge. To protect the roofing membrane, set the sleepers on strips of

EPDM or rooftop walkway matting (available from commercial roofing suppliers). Attach the decking with screws so it can be removed later for inspection or repairs.

Posts and Railings. As with other decks, guardrails must be a minimum of 36 inches high, infill balusters less than 4 inches apart, and the whole assembly strong enough to resist a 200-pound horizontal load (or 20 pounds per linear foot, depending on the local code). Use the longest pieces available for railings. If the deck is small enough, use a single continuous top railing on each side. Then tie the side railings to the house with steel angle brackets to create a rigid railing assembly.

From a waterproofing standpoint, it is desirable to keep the posts outside of the roofing membrane. This can be achieved by bolting the posts to the rim joists or subfascia in the roof framing (see Figure 4-17.).

If this is not practical or if it is unacceptable from a design standpoint, then the EPDM will need to be sealed around each post. This is best left to a professional roofer. Typically, the roofer will fashion a metal or membrane flashing collar around the base of each post and seal this to the roofing membrane, as with the boots used for plumbing vents.

COVERED AND SCREENED PORCHES

The detailing for decks and porches is very similar. Code requirements for railings are the same, whether a deck is open or closed in with screening. If the screening runs to the floor, railings will be required to protect the screening, even if the railing is not required by code. Porch decking can be spaced (like an open deck) or installed tight with a slope for drainage. Structurally, posts are easier to anchor on a porch, since they are tied in to the roof framing, which keeps them rigid.

Solid Decking. Where the decking will be protected by a roof, it can be either spaced like a typical open deck or sealed and painted with a decking enamel for a more formal appearance. Solid decking is typically tongue-and-groove 1x4, $\frac{5}{4}$x6, or 2x6 decking. For solid decking, choose kiln-dried stock and install it with tight seams. Solid decking should be sloped $\frac{1}{4}$ inch per foot to drain. Make provisions for drainage on the three exterior sides and, if exposed to significant wetting, on the house side as well. If the porch has a solid knee wall, leave minimum 1-inch scuppers at floor level to allow water to drain from the porch interior to outside.

Screened Enclosures. With a screened enclosure, use either solid decking or spaced decking with insect screening stapled to the underside of the joists. Furring strips tacked over the screening will help keep it from sagging and tearing.

FIGURE 4-17 Decks over Living Space.

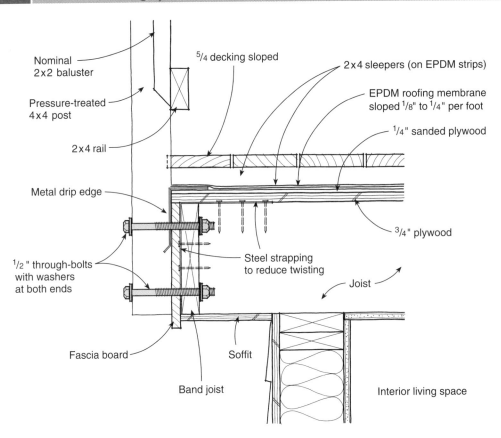

To simplify waterproofing, it is best to keep the posts outside of the roofing membrane. This can be achieved by bolting the posts to the rim joist or subfascia in the roof framing below. Do not notch the 4x4 posts by more than about an inch.

If the screening goes from floor to eaves, a system of rails and balusters will need to be installed to protect the lower half of the screen (Figure 4-18).

If the porch is 30 inches or more above grade, the railing will need to comply with code for guardrails (see "Railings," page 147). Another option is to build an enclosed knee wall and start the screening at the top of the knee wall (see Figure 4-19).

There are several options for screening. Whether you build your own or buy custom screens, aluminum screening is much stronger and more tear-resistant than fiberglass. Removable screens have the advantage of being easy to remove and repair. However, building screens from wooden screen mold is very time-consuming. Simpler options for screening are:

- Have a window supplier build custom wood or metal screens with vinyl splines to hold the screen in place.

- Use a manufactured screening system such as Screen Tight™ (One Better Way), a vinyl snap-in-place system for fiberglass screening that screws onto the exterior face of the porch framing.

- Staple aluminum screening to the exterior of the 4x4 posts and trim with 1x wood trim pieces screwed in place.

FIGURE 4-18 Screened-Porch Railings.

Where porch screening goes from floor to eaves, a system of rails and balusters is needed to protect the lower section of screen. If the porch is 30 inches or more above grade, the railing will need to meet code requirements for guardrails.
SOURCE: Photo by author.

FIGURE 4-19 Simple Screened-Porch Details.

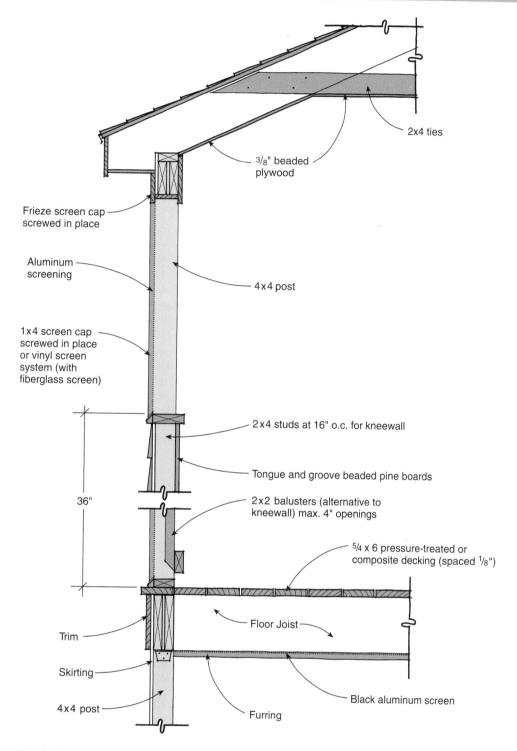

2x4 ties

³/₈" beaded plywood

Frieze screen cap screwed in place

Aluminum screening

4x4 post

1x4 screen cap screwed in place or vinyl screen system (with fiberglass screen)

2x4 studs at 16" o.c. for kneewall

Tongue and groove beaded pine boards

2x2 balusters (alternative to kneewall) max. 4" openings

36"

⁵/₄ x 6 pressure-treated or composite decking (spaced ¹/₈")

Floor Joist

Trim

Skirting

4x4 post

Furring

Black aluminum screen

The simplest approach to porch screening is to wrap a roll of screening around the exterior of the porch posts and then cover the heads, posts, and bottoms of each screen with 1x trim screwed in place for easy removal. Modular screening systems, such as Screen Tight, make installation even easier. If a porch is more than 30 inches above grade, a 36-inch knee wall or railing is required.

FINISHES FOR DECKING

All wood decking, whether pressure-treated or not, needs to be sealed at installation and periodically thereafter to prevent checking, warping, and deterioration of the surface due to exposure to water and sunlight. The chemicals in treated wood and the natural tannins in redwood and cedar resist decay and insect attack, but they will not stop checking and warping. There are a wide array of deck finishes on the market, but they all consist of one or more of the following: water repellants, preservatives, UV blockers, pigments, and a drying oil or varnish base (see Table 4-9.)

Water Repellants

At a minimum, all wood decks should be treated with a water-repellant coating, preferably a water-repellant preservative (WRP), which has an added mildewcide. Also called "sealers," these formulations typically contain a water-repelling wax and a varnish or drying oil, such as linseed or tung oil. The finish penetrates and seals the wood surface, reducing the amount of water absorption and thereby protecting against checking, splitting, and warping. After application, water should bead up as on a newly waxed car.

Some water repellants contain a small amount of wax (about 1% by volume) and are intended as a pretreatment for other finishes. Others contain up to 3% wax and are intended as a final coating. Some of these cannot be stained or painted over, so it is important to read the label.

WRPs, which have an added mildewcide, help prevent dark stains on natural woods like redwood and cedar, and on pressure-treated lumber as well. In addition, WRPs provide some protection against decay in the sapwood of redwood and cedar and in the cut ends of pressure-treated decking. Some sealers and WRPs also have UV-inhibitors, an important addition if the sealer is the final coating, since this will help protect against deterioration of the wood surface from sunlight.

If a sealer or WRP is the only treatment used, the homeowner should plan to recoat every one to two years or whenever water on the surface no longer beads up and is quickly absorbed. WRPs also make an excellent undercoat for semitransparent stains. The combination of a semitransparent stain over a WRP base coat provides the best long-term protection for decks. However, not all WRPs are suitable for use as an undercoat—so check the label or ask the manufacturer before proceeding.

Semitransparent Stains

Oil-based semitransparent stains contain many of the same ingredients as a WRP and penetrate the wood in the same manner. The main difference is the addition of pigments, which provide some color and help protect against UV radiation. Some, such as Penofin (Performance Coatings Inc.), are very lightly pigmented but add UV inhibitors to achieve a similar level of protection. Since oil-based stains penetrate the wood surface, they will not peel, blister, or chip like paint. Stains formulated specifically for decks may have improved resistance against abrasion as well.

Stains are a good finish for either treated wood or naturally decay-resistant species. The pigment provides good protection against UV radiation and extends the life of the finish beyond that of a simple water-repellant or WRP. Light-colored finishes will reflect more light and, therefore, tend to outlast darker colors on exposed surfaces.

For the stain to penetrate properly, the wood surface must be fairly dry when the stain is applied. If the decking material was factory-treated with a sealer or was recently sealed on-site, it may be necessary to wait two weeks or longer before staining. For best results, apply two coats of stain, with the second applied before the first coat completely dries. Once dried, the first coat will block the proper penetration of the second coat.

Paints and Solid-Color Stains. Paints and solid-color stains (also called "opaque" stains), whether latex or oil-based, all form films on the wood surface. While these provide excellent protection against water penetration and UV degradation, they are not recommended for decking for two main reasons: They do not protect against decay and they are prone to bubble, peel, and crack if moisture does get through. They can also peel or crack during the initial drying of the wood. While not recommended for the decking, paint may be applied successfully to other deck components, such as railings.

If the job calls for paint, take the following precautions: First, seal the wood with a water-repellant preservative formulated to serve as an undercoat. Make sure all end grain is sealed and primed prior to assembly, when it may become inaccessible. After two to three weeks, when the surface is dry enough to paint, prime and paint the rest of the wood. A better alternative, if the budget allows, is to buy kiln-dried pressure-treated lumber, which can be sealed, primed, and painted immediately. Kiln-dried pressure-treated lumber is marked KDAT (kiln-dried after treatment).

Woods such as redwood and cedar, which have a high level of extractives, require special stain-blocking primers, or the dark-colored extractives will bleed through and stain the painted surface. While painters have traditionally preferred oil-based primers on wood species prone to extractive bleeding, new latex primers specially formulated for stain blocking may also do the job.

TABLE 4-9   **Finishes for Exterior Decks**

Finish Type	Pros	Cons	Cost (1—least 3—most)	Best Uses	Recommendations	Service Life
Water repellants	Prevents water intrusion and resulting checking and warping. With UV blockers, also inhibits deterioration of wood surface. Easy application. No peeling of finish.	Needs frequent recoating. Limited UV protection.	1	Undercoating for paint or stain on pressure-treated (PT) wood or decay-resistant species. Final coating on decay-resistant species.	Surface of wood should be dry before application. Can be brushed, rolled, or applied with low-pressure sprayer. Dipping provides best coating.	1–2 years
Water-repellant preservatives	Same as above, but also inhibits mildew growth and provides some above-grade protection against decay in cut ends of PT wood and sapwood of decay-resistant species.	Needs frequent recoating. Limited UV protection.	1	Undercoating for paint or stain on all species. Final coat on naturally decay-resistant species.	Same as above. Also make sure cut ends are treated before installation makes them inaccessible.	1–2 years
Semitransparent stains (oil-based)	Same as above, but offers increased protection against UV degradation of wood surface. Outlasts clear finishes. Provides redwood and cedar with some pigmentation while allowing wood grain to show through.	Limited abrasion resistance for decking. Note: For best results, choose products specially formulated for decking.	2	Finishing pressure-treated wood. Also can help preserve the original look of redwood and cedar decking.	Wood surface must be dry enough to absorb finish before application. Can be brushed, sprayed, or rolled and back-brushed. If applying over sealer or factory-sealed lumber, allow to weather until surface will absorb finish. Apply a second coat before first dries completely.	2–3 years (single coat on new decking); 3–6 years (recoating with two coats).
Paints and solid stains	Offers best protection against UV degradation and moderate resistance to abrasion. Longest lasing finish with most color choices.	Prone to bubble, crack, and peel when applied to horizontal surfaces. Time-consuming to apply and difficult to recoat. Must be scraped and sanded.	3	For trim, railings, and other decorative surfaces not subjected to foot traffic or ponding of water. Wood must be dry at application. Primer required.	Wood should be fully dried or pressure treated before painting. Use over sealer for best finish. Seal and prime all end grain prior to installation. Generally not recommended for use on decking.	Decking: 2–3 years for paint; 1–2 years for solid stain. Trim: 3–6 years for paint; 3–4 for solid stain. (Assumes one primer coat and one top coat. Second top coat can double service life.)

RESOURCES

Manufacturers

Wood Treaters

Arch Wood Protection
www.wolmanizedwood.com
Copper-azole and borate-treated wood products with optional water repellent

Osmose
www.osmose.com
ACQ- and borate-treated wood products with water repellent

Chemical Specialties
www.treatedwood.com
ACQ-treated wood products with optional water repellent

Composite Structural Lumber

U.S. Plastic Lumber, Boca Raton, FL
www.usplasticlumber.com
Trimax and Durawood structural plastic lumber

Composite Decking Systems

Certainteed Corp.
www.certainteed.com
Boardwalk solid composite decking with hidden fasteners and optional railing system

Composite Building Products International
www.xtendex.com
Xtendex hollow composite decking system with optional railing

Correct Building Products
www.correctdeck.com
Solid composite decking with hidden fasteners and optional railing system

Fiber Composites
www.fibercomposites.com
Fiberon solid composite decking and optional railing system

Kadant Composites
www.geodeck.com
Geodeck hollow composite decking and railing system

Kroy Building Products
www.kroybp.com
Timberlast solid composite decking with optional hidden fastening system

Louisiana-Pacific Corp
www.weatherbest.lpcorp.com.
WeatherBest solid composite decking, railings, and accessories

Nexwood Industries Limited
www.nexwood.com
Hollow composite decking and railing systems

Tendura
www.tendura.com
TenduraPlank solid tongue-and-groove composite flooring for porches; natural finish or primed for painting

Thermal Industries
www.thermalindustries.com
Dream Composite solid tongue-and-groove composite decking system with optional vinyl railings

TimberTech Limited
www.timbertech.com
Floorizon hollow composite decking system, solid composite decking planks, and optional railing system

Trex Company
www.trex.com
Solid composite decking

Weyerhaeuser Building Products
www.choicedek.com
ChoiceDeck solid composite lumber and optional railings

Plastic Decking Systems

Kroy Building Products
www.kroybp.com
Classic Manor embossed vinyl decking with clip system

L.B. Plastics
www.lbplastics.com
Sheerline interlocking vinyl decking system and deck cladding systems

Renew Plastics
www.renewplastics.com
Solid recycled high-density polyethylene (HDPE) plastic decking with embossed wood-grain texture

Royal Crown Limited
www.royalcrownltd.com
Brock Deck and Deck Lok interlocking vinyl decking systems

Thermal Industries
www.thermalindustries.com
Dream Deck interlocking vinyl decking and railing system

U.S. Plastic Lumber Corp
www.carefreeexteriors.com
Carefree Xteriors recycled plastic HDPE decking with embossed wood grain and optional railings

Metal Decking Systems

FSI Home Products Division
www.lockdry.com.
LockDry aluminum decking and railing system

Prefabricated Railing Systems

Avcon Structural Railing Systems
www.avconrail.com
Thermoplastic and aluminum railings

CertainTeed
www.certainteed.com
EverNew PVC railing system

Global Dec-K-Ing Systems
www.globaldecking.com
DEC-K-ING aluminum railing system

DecKorators Inc.
www.deckrail.com
Decorative aluminum balusters and connectors for wooden railing systems. Also, tempered-glass balusters

Duradek
www.duradek.com
Durarail powder-coated aluminum railing system and walk-on vinyl decking membrane.

FSI Home Products Division
www.railingworks.com
Aluminum railing systems

Fypon
www.fypon.com
Polyurethane railing systems in classic architectural styles

HB&G
www.hbgcolumns.net
PermaPorch cellular-polyvinyl chloride (PVC) railings and posts reinforced with wood or aluminum; cellular-PVC or aluminum balusters

Kroy Building Products
www.kroybp.com
Classic Manor PVC-railing system

L.B. Plastics
www.lbplastics.com
Sheerline PVC-railing systems and PVC-post cladding

Royal Crown Limited
www.royalcrownltd.com
PVC railing system with steel reinforcing

Shakespeare Composites Structures
www.armor-rail.com
Armor-Rail structural fiberglass (FRP) railing system with turned balusters

Thermal Industries
www.thermalindustries.com
Dream Rail PVC-railing system with optional tempered glass balusters

U.S. Plastic Lumber Corp.
www.carefreeexteriors.com
Recycled HDPE railing system

Hidden Deck Fasteners

BEN Manufacturing
www.premier1.net/~ben69
Dec-Klips fit between deck planks with prongs into edges; nailed into top of joists; electrogalvanized steel

Blue Heron Enterprises
www.ebty.com
Eb-Ty UV-resistant polypropylene biscuit fits into slots in edge of decking, screws into top of joist

Grabber Construction Products
www.deckmaster.com
Deckmaster angle bracket screws to side of joist and up into decking; available in galvanized or stainless steel

Simpson Strong-Tie
www.strongtie.com
DBTC deck ties screw down to top of joist with prongs into edges of decking planks; triple-zinc-coated or stainless steel by special order; disposable plastic driving tool

Spotnails
www.spotnails.com
Tebo stainless-steel fasteners fit between decking boards with prongs into edges of decking; installed with mallet and proprietary tool

TY-LAN Enterprises Inc.
www.shadoetrack.com
Shadoe Track angle bracket nailed along top of joist and screws up into bottom of deck boards; available in galvanized, powder-coated, and stainless steel

USP Lumber Connectors
www.uspconnectors.com
Deck Clip screws into edge of one decking board and locks to next board; requires toenailing one edge of each board

Screen Systems

One Better Way
www.screentight.com
Screen-Tight vinyl porch screening system screws to exterior of porch framing; snap-on facings conceal screws and screen splines; available in white, beige, gray, and brown

DECK FINISHES

Amteco
www.mfgsealants.com/amteco.htm
Sealers, preservatives, and deck stains

Cabot
www.cabotstain.com
Clear sealers and deck stains

Cuprinol, a division of Sherwin Williams
www.cuprinol.com
Sealers, preservatives, and deck stains

The Flood Company
www.floodco.com
Clear sealers and deck stains

Penofin
www.penofin.com
Lightly tinted oil-based sealers and stains

Olympic, PPG Architectural Finishes
www.olympic.com
Clear sealers and deck stains

**Wolman Wood Care Products, division
of Zinsser Co.**
www.wolman.com
*Wolman sealers and deck stains and DAP Woodlife clear
sealers*

FOR MORE INFORMATION

American Wood Preservers Association (AWPI)
www.awpa.com

California Redwood Association
www.calredwood.org

Deck Industry Association
www.deckindustry.org

Forest Stewardship Council
www.fscus.org
Information on certified tropical hardwoods

Rainforest Alliance, Smartwood Program
www.rainforest-alliance.org
Information on certified tropical hardwoods

Southern Forest Products Association
www.sfpa.org

Southern Pine Council
www.southernpine.com

Western Wood Products Association
www.wwpa.org

Western Wood Preservers Institute
www.wwpinstitute.org

CHAPTER FIVE | Interior Finish

Interior finishes are the most visible and, on a square-foot basis, often the most expensive components in a house. However, since many of these products and materials are marketed directly to consumers, they are often not well understood by builders and designers. Making good decisions on such finish materials as flooring, carpeting, and lighting fixtures can make a critical difference to homeowner satisfaction. The builder or designer can play a key role in helping the homeowner choose finishes that are well-suited to the intended use, as well as providing the structural support and prep work the materials require for good performance.

DRYWALL

Single-layer, $\frac{1}{2}$-inch drywall is the default wall and ceiling treatment in most residential construction. Done well, it goes largely unnoticed. Nail pops and cracks, however, are very conspicuous and remain the leading cause of builder callbacks. With wet or poor-quality framing, there are bound to be problems in the drywall finish. With dry lumber and proper detailing, however, drywall problems can be kept to a minimum.

Types of Drywall

Drywall consists of a gypsum core covered by two layers of treated paper. The long sides are tapered for easy finishing with joint compound. The short or "butt" ends are not tapered.

Standard Drywall. This material comes in four thicknesses: $\frac{1}{4}$ inch, $\frac{3}{8}$ inch, $\frac{1}{2}$ inch, and $\frac{5}{8}$ inch. A single layer of $\frac{1}{2}$-inch drywall covers most residential walls and ceilings. For a stiffer wall and better sound deadening, use $\frac{5}{8}$-inch drywall or a double layer of $\frac{1}{2}$-inch drywall, with all joints staggered between layers and the second layer glued to the first for best performance. The $\frac{3}{8}$-inch panels are useful for covering existing walls and ceilings in remodeling. One-quarter-inch board, installed in layers, is useful for curves. Special $\frac{1}{4}$-inch bending-type drywall has the smallest bending radius.

Fire-Resistant. Fire-code drywall has special additives, including glass fibers, to increase its fire resistance. Residential building codes typically require Type X $\frac{5}{8}$-inch fire-code drywall with a one-hour rating for party walls, ceilings over furnaces, and common walls between living space and garages.

Moisture-Resistant. Moisture-resistant (MR) board, sometimes called "green board" because of its green paper facing, has limited water resistance from asphalt additives, and is recommended for high-humidity areas such as bathrooms and laundries. The material is denser and less rigid than regular drywall, so it is prone to sag on ceilings unless the framing is 12 inches on-center or less. Also it will fall apart, like regular drywall, if it gets soaked. For that reason it should not be used as a tile substrate in any application where it might get wet.

Mold-Resistant Drywall. This is a relatively new product that uses an inorganic fiberglass matt instead of paper facing, since the paper facing readily supports mold growth. Panels are available with the fiberglass matt on one side or two. Glass fibers in the gypsum core add strength as well.

Installation

To prevent problems, use good quality framing lumber and follow these recommendations:

- For walls 8 feet high or less, run drywall sheets perpendicular to the studs. This is stronger, bridges framing irregularities, and results in fewer joints.
- Make sure the drywall is tight against the framing before fastening.
- Install with screws, which have better holding power than nails and are less likely to tear or weaken the drywall facing.
- Use the correct length fastener. Either longer or shorter than recommended can lead to nail pops (see Table 5-1).
- Maintain the temperature at 55°F to 70°F during and after applying the joint compound.
- If using fiberglass mesh rather than paper tape, use setting-type joint compound to embed the mesh, since it is not as strong as paper tape.
- Install drywall on ceilings before walls and use floating corners to allow for some movement.
- Keep butted joints (short ends) to a minimum, and butt the sheets together loosely. Because they are untapered they are more visible.

Framing. Drywall should be installed over straight and level framing. If the framing is excessively wet, it will crack the drywall and cause nail pops as it shrinks. If the framing is twisted, bowed, or out of alignment, it will cause weak points in the surface and possible cracking. Moderately bowed studs can be fixed by cutting a kerf at midheight, straightening the stud, and scabbing a section of 1x4 or plywood on either side. Repair or replace problem studs before installing the drywall.

On ceilings, it is common practice in some parts of the country to install 1x3 furring strips at 16 inches on-center perpendicular to the ceiling joists before installing the drywall. The furring is shimmed to even out irregularities in the ceiling joists and creates a more stable substrate for the drywall with less chance of cracking. Also, the furring provides a wider nailing surface for hanging drywall.

Spans. On walls and ceilings, it is best to install drywall perpendicular to the framing. This ties together more framing members and provides greater racking strength. On walls, $\frac{1}{2}$-inch or $\frac{5}{8}$-inch drywall can span up to 24 inches whether it is installed parallel or perpendicular to the framing.

On ceilings, $\frac{1}{2}$-inch or $\frac{5}{8}$-inch drywall can span 24 inches only if it is installed perpendicular to the joists and supports less than 1.3 pounds per square foot (psf) of insulation. Otherwise, 16-inch on-center spacing is recommended. With latex spray textures or airless spraying of latex paints, perpendicular installation over 16-inch on-center framing is recommended to prevent sagging.

Adhesive Installation. Another way to minimize nail and screw pops is to minimize the number of fasteners. Gluing the drywall to the studs with construction adhesive allows the installer to eliminate 75% of the fasteners (Table 5-2). Using adhesives also helps to even out minor irregularities in the framing and results in a much stronger and stiffer wall. Use a construction or drywall adhesive that meets ASTM C557.

Apply a $\frac{3}{8}$-inch bead down the center of each stud or joist, stopping about 6 inches from each end. Where two drywall panels meet, apply two $\frac{3}{8}$-inch beads so each panel gets full contact with adhesive. No adhesive is needed at inside corners, top and bottom plates, or at bridging, diagonal bracing, or other miscellaneous framing. Also do not use adhesive over polyethylene sheeting or insulation batts with paper flanges stapled over the stud faces.

To ensure a good bond, drywall manufacturers recommend prebowing the drywall by stacking several sheets face up with a 2x4 under each end. Left overnight, this will leave a permanent bow, forcing the center of the sheet tight against the adhesive (except in very humid weather, when the boards may remain flexible).

Push drywall panels into the adhesive with hand pressure along joists or studs. Do not apply more adhesive than can be covered in 15 minutes, or it may skin over.

TABLE 5-1 Drywall Fastener Types				
	$\frac{1}{2}$ in. Drywall Over Wood Framing	$\frac{5}{8}$ in. Drywall Over Wood Framing	$\frac{1}{2}$ in. Drywall Over Steel Studs	$\frac{5}{8}$ in. Drywall Over Steel Studs
Nails	$1\frac{1}{4}$ in. annular ring	$1\frac{3}{8}$ in. annular ring	—	—
Screws	$1\frac{1}{4}$ in. Type W bugle head	$1\frac{1}{4}$ Type W bugle head	1 in. Type S bugle head	1 to $1\frac{1}{8}$ in. Type S bugle head

*Do not install water-resistant (WR) board on ceilings.

TABLE 5-2 Drywall Fastener Spacing

	Max. Fastener Spacing on Walls (no adhesive)	Max. Fastener Spacing on Ceilings (no adhesive)	Max. Fastener Spacing Over Resilient Channel (wall/ceiling)	Max. Fastener Spacing on Walls (with adhesive)	Max. Fastener Spacing on Ceilings (with adhesive)
Nails	8 in.	7 in.	—	16 in. on perimeter. 1 fastener per stud in field (perpendicular installation)*	16 in. on perimeter. 24 in. in field (parallel installation); 1 fastener per joist (perpendicular installation)
Screws	16 in. (12 in. for WR board)*	12 in.	12 in.	"	"

*If boards are prebowed and installed parallel to framing with adhesive, fastening at top and bottom only is allowed.
SOURCE: Table courtesy of USG Corporation.

Allow the panels to dry at least 48 hours before adding joint compound or skim coating.

Floating Corners. Inside corners at walls and between walls and ceilings are stress points for drywall and common places for cracks or nail pops. Leaving one side of the joint free to move without fasteners will eliminate most of these problems.

On ceilings, place the first screws 7 to 12 inches in from the corner and support the ceiling drywall with the wall panels. Also, do not fasten the top 8 inches of the wall panels. No screws should go into the top plate, where shrinkage may occur. Similarly, leave one side unfastened at wall-to-wall corners, but make sure it rests against solid wood backing or drywall clips (see Figure 5-1).

Control Joints. While control joints in drywall are not commonly used in residential construction, they are a good idea in surfaces over 30 feet long or at changes in floor level, such as stairway walls. On a stairway wall, locate the control joint at the top of the first-floor wall where the top plate meets the ceiling joists. The $\frac{1}{4}$-inch joint can be painted with the wall and left as a reveal. Another option is to omit the metal control joint and simply leave a small gap between the upper and lower drywall, and cover the joint with wood trim.

Corner Bead. Outside corners fashioned with metal corner bead are also prone to cracking and nail pops. To avoid problems, do not nail into the top plate, and leave a gap at the bottom of the wall to accommodate any settling. Nail with drywall nails at 9 inches on-center on both sides of the corner, with nails opposing each other.

Newer "mud-on" or "tape-on" corner beads are less prone to edge cracking than traditional metal corner beads and, with no nails, eliminate nail popping. The corners are metal or plastic and are held in place by paper or vinyl flanges embedded in joint compound. Some of the vinyl corner beads can also be installed with spray-on contact cement. In general, tape-on corner beads require fewer coats and less joint compound than traditional metal corner beads, speeding up the finishing process.

FIGURE 5-1 Floating Corners.

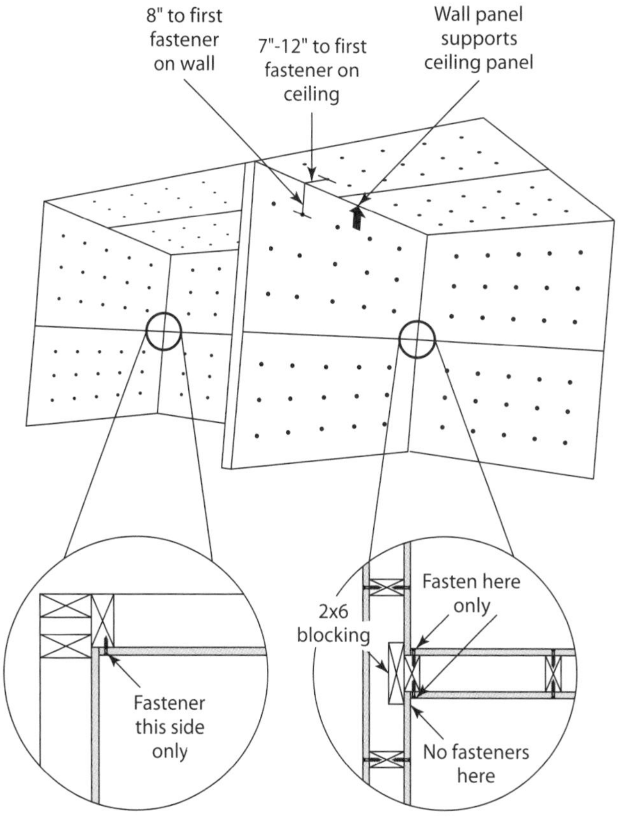

Inside corners at walls and ceilings are common places for cracks and nail pops. Leaving one side of the joint free to move without fasteners will eliminate most of these problems. Make sure the floating side is supported by solid wood backing or drywall clips.

Curves. For radiused walls, the easiest approach is to use two layers of $\frac{1}{4}$-inch drywall, preferably the "high-flex" type, if available. If not available, it is possible to wet the side of the drywall that will be compressed (the inside of the curve) with a garden sprayer or short-nap roller. Then stack the boards with wet face to wet face and cover with plastic sheeting. After an hour, install the panels with their

TABLE 5-3 **Minimum Bending Radii of Drywall**

Thickness of drywall (in.)	Dry gypsum board		Wet gypsum board	
	Min. radius	Max. stud spacing (in.)*	Min. radius	Max. stud spacing (in.)*
1/4	5'-0"	6	2'-0"	6
1/4 (flexible-type)	1'-8"	9	0'-10"	6
3/8	7'-6"	6	3'-0"	8
1/2	20'-0"	6	4'-0"	12

* Measured on-center along outside (long side) of plate.
Table courtesy of USG Corporation.

long dimension across the studs. Minimum bending radii are shown in Table 5-3.

Taping and Finishing

The building should be heated before finishing begins and maintained at 55°F to 70°F throughout taping and finishing. The cooler and more humid it is, the longer it will take ready-mixed joint compounds to dry. If necessary, use supplemental heaters and provide adequate ventilation to remove excess moisture. Too much moisture can soften and weaken the bond between the drywall and the paper facing. Conversely, if the weather is too hot and dry, paper drywall tape may not bond well and joints may experience excess shrinkage and cracking.

Mesh vs. Paper Tape. Mesh tape is easier to apply but not as strong as paper tape. It should never be used in inside corners, where it can tear or be cut by the trowel. However, if combined with setting-type compound, mesh tape is nearly as strong as paper tape and can produce a quality job. Mesh tape is also very useful for repairing cracks in older plaster walls or ceilings.

Joint Compound. Many residential jobs are taped with premixed all-purpose joint compound for all three coats. While this is acceptable, according to U.S. Gypsum guidelines, installers can produce stronger joints less prone to cracking by using special *setting-type* compounds for the first coat to embed the tape and corner beads and patch any big holes. Do not use setting-type compound on nail or screw indents.

Setting compounds are mixed on-site and set up by a chemical reaction, rather than evaporation of the water. They dry rock-hard and do not shrink. Setting times range from 20 minutes to several hours, and the compound can be recoated as soon as it sets, rather than the next day as is typical for ready-mix compound. Durabond® 90 is the most commonly used. A new type of setting-type

compound from USG, called Easysand®, overcomes the chief liability of setting-type compounds—that they are nearly impossible to sand.

For nail and screw indents, and the *fill* and *topping* coats on seams, most contractors use premixed all-purpose compound; although special topping compounds are available. Premixed compound should be stored, applied, and allowed to dry at between 55°F and 70°F, preferably over 60°F. If allowed to freeze, manufacturers claim that ready-mix compound can be reused if thawed and remixed thoroughly with an electric mixer, but it is probably wiser to just throw it away.

Specialty Trims

Drywall window and door trim, acute and obtuse angles, bullnose corners, and arches have been greatly simplified by the introduction of specialty drywall trims and accessories (see Resources, page 206). Most profiles are available in metal and plastic in either nail-on or tape-on styles.

Off–90 Degree Angles. Standard nail-on or tape-on corner bead provides the strongest 90-degree corner. However, for acute or obtuse outside corners—for example, around skylight wells—you are better off with flexible tape-on corner trims reinforced with metal or plastic that can be set at any angle. These help at inside corners as well, such as between intersecting roof planes, where standard paper tape tends to leave a wavy line.

Rounded Corners. A variety of trim products simplify creating rounded inside (cove) or outside (bullnose) corners. These come in both tape-on and nail-on styles, but the tape-on type are less prone to cracking. Some of the plastic profiles can also be applied with spray-on contact adhesive. In general, one finish coat of joint compound is applied to the flanges only (after the embedding coat dries), so these profiles are quicker to finish and dry than standard metal corner bead (Figure 5-2).

FIGURE 5-2 Tape-On Corner Trim.

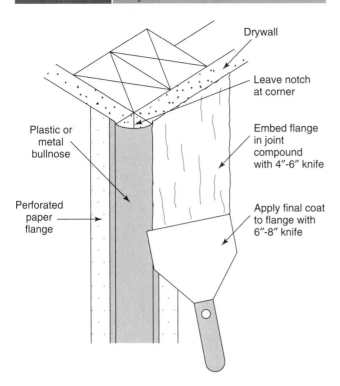

New tape-on trims, like the bullnose shown, are bedded in joint compound and then feathered on the paper flange only, taking less mud and drying time than standard metal corner bead and resulting in fewer corner cracks.

FIGURE 5-3 Drywall Returns at Doors and Windows.

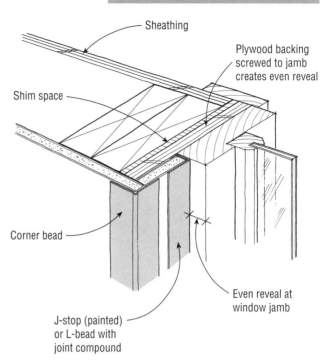

Trim the inside edge of drywall returns with a J- or L-bead to create a clean joint and to protect the edge from door or window movement. Attaching a plywood backing to the jamb before installation creates an even reveal along the jamb.

For miters and three-way corners, some suppliers provide special trim pieces. If these are not available, the trim will need to be miter-cut with a carbide or abrasive blade, depending on the trim material.

Windows and Doors. For a simple, contemporary detail, you can return the drywall directly to the window or door jamb and trim the edge of the drywall with a J-bead or L-bead, available in both plastic and galvanized steel. J-bead must be slipped on the end of the drywall before it is installed; it creates a separation from the wood frame, which is useful where movement in the door or window might otherwise crack the finished joint. The reveal type of J-bead, called J-stop, is not mudded, as it serves as finish trim.

Shimming around the rough opening to get an even reveal around door or window jambs can be tricky. Enlarging the rough opening and attaching a plywood or pine backing to the jamb simplifies the task (Figure 5-3).

Arches. Before the introduction of specialty beads designed for curves, building drywall arches meant snipping metal corner bead every inch and bending it as well as possible to conform to the contour of the arch. A variety of bendable corner beads have simplified the task. Two of the more popular are Archway L Bead (Trim-Tex Inc.), a vinyl installed with a spray-on adhesive, and Arch-Flex (Con-Form International), a vinyl tape-on bead (see "Resources," page 206).

Use 2x stock, or plywood with 2x blocking in between, to frame the curve of the arch. To form a smooth curve, use two layers of flexible $\frac{1}{4}$-inch drywall around the curve. If this is not available, score the back face of a strip of $\frac{1}{2}$-inch drywall every inch and form it to the curve (Figure 5-4).

Veneer Plaster

Veneer or skim-coat plaster has, for the most part, replaced traditional three-coat plaster in residential work. It consists of a single coat of plaster $\frac{1}{16}$ to $\frac{1}{8}$ inch thick over a special type of gypsum board, commonly called *blueboard*, which is treated to bond well to plaster. Although the finished job costs 30 to 50% more than standard drywall, veneer plaster has a number of advantages over drywall:

- A pleasing, smooth texture that is very similar to traditional three-coat plaster.

- A harder, brighter surface that resists dents and nail pops.

- No raised seams, tape bubbles, or other imperfections associated with drywall.

- Greater mass reduces sound transmission.

- Quicker installation: about two days for a typical house, and it can be painted 24 hours later.

- Requires no sanding, making it particularly good for remodeling.

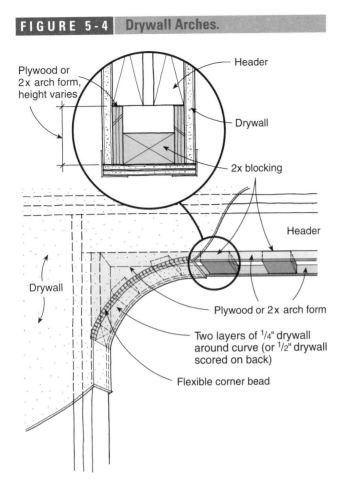

FIGURE 5-4 **Drywall Arches.**

Plywood or
2x arch form,
height varies

Header

Drywall

2x blocking

Drywall

Header

Plywood or 2x arch form

Two layers of $\frac{1}{4}$" drywall
around curve (or $\frac{1}{2}$" drywall
scored on back)

Flexible corner bead

Bendable corner beads have simplified the task of creating drywall arches. Use two layers of $\frac{1}{4}$-inch drywall to conform to the curve or use $\frac{1}{2}$-inch drywall, scored every inch along the back face before bending it.

Preparation. Skim coat prep work is similar to drywall with a few exceptions. Because the finish coating is less than $\frac{1}{8}$ inch thick, the boards must lie flat in a plane. Other than that, the board can be hung pretty quickly with few concerns. The screws can be left flush with the surface, and butt joints can fall anywhere. Expanded metal corner bead goes on all outside corners and self-sticking mesh tape goes over all seams.

Some plasterers prefer to apply the finish with baseboards and door and window casings already in place, protected with masking tape, so the plaster can fill in any waviness in the board behind the trim. Otherwise, install $\frac{1}{8}$-inch plaster grounds at the baseboard and around all door and window openings to guide the trowel and produce an even finish.

Application. Using a 12- to 16-inch plaster trowel, a first *scratch* coat goes over all flat seams and then the finish coat is applied right away. If the seams are allowed to dry overnight, they will need to be wetted first or the dried plaster will suck too much moisture out of the finish coat, leaving a weak joint. The same is true of cold joints along a wall. If a wet edge is allowed to dry out, it should be

rewetted. Otherwise it will be difficult to blend the new plaster into the cold joint.

Different brands and types of veneer plaster get slightly different treatments, but in general, the finish coat is troweled on in one or two passes and troweled smooth. Once dried, small imperfections or voids can be misted with water and fixed with standard joint compound.

WOOD FLOORS

Wood floors provide a natural warmth and beauty like no other flooring material. And new developments in finishes and engineered products have expanded their durability, versatility, and ease of installation. Still, control of moisture levels in the flooring and structure around it remains the biggest issue influencing the success of a wood flooring installation, particularly with unfinished strip flooring, but with many of the engineered products as well.

Solid Wood Strip and Plank

Traditional unfinished $\frac{3}{4}$-inch hardwood strip flooring in oak or maple remains the most common wood flooring type and the best choice where heavy use and frequent refinishing are likely. While the most common species are still oak and maple, an amazing variety of domestic and imported woods have become available in recent years.

Species. Flooring species are rated for hardness and dimensional stability (see Table 5-4). Hardness is rated on the Janka scale, which measures the force required to push a small steel ball into the wood surface. The results are often compared to red oak, which is used as a benchmark. Unless a floor with lots of "character" caused by dents and wear marks is desired, avoid woods significantly softer than oak.

Also consider the wood's dimensional stability. Less stable woods are likely to lead to gaps, cupping, or other problems with wider plank flooring and in regions like the Northeast, which has big seasonal swings in relative humidity. Other locations where moisture movement might be a problem include below-grade spaces, slabs-on-grade without vapor barriers, or rooms over crawlspaces. In these environments, choose a stable wood and a narrow profile to avoid problems. Laminated floors, discussed below, are often the best choice for these applications.

A number of exotic hardwood imports are also now available. Many of these are plantation-raised or logged with sustainable forestry practices, but some are not. To be sure, work with a reputable importer and look for third-party certification of sustainable logging practices. For more information, contact the Forest Stewardship Council or the Smartwood program (see Resources on page 208).

Hardwood vs. Softwood. While hardwoods are harder and more durable than softwoods in general, this is not always the case. For example, heart-pine flooring, whether antique or new (cut from the centers of longleaf southern

TABLE 5-4 Domestic and Imported Woods

			Domestic Woods		
Species	Appearance	Grain	Hardness (relative to red oak)*	Dimensional Stability (relative to red oak)**	Durability
Ash	Similar to white oak with yellow tinge. Heartwood tan to dark brown; sapwood creamy white.	Moderately open. Occasional wavy figuring. Strong contrast in plainsawn boards.	102%	26% less movement	Hard, elastic, shock-resistant.
Beech	Heartwood light to dark reddish brown. Sapwood pale white.	Mostly closed, straight grain, uniform texture.	101%	17% more movement	Hard, elastic, shock-resistant. Wears well.
Birch (yellow)	Light reddish brown heartwood, white sapwood.	Mostly closed, straight grain with fine, uniform texture. Some wavy figuring.	102%	8% less movement	Hard, strong, stiff, shock-resistant.
Black cherry	Heartwood reddish brown (darkens with exposure). Sapwood light yellow.	Fine with a uniform texture. Frequently wavy.	74%	33% less movement	Strong, moderately hard, shock-resistant.
Douglas fir	Young rapid-growth trees have reddish heartwood and are called red-fir. Tight-grained old trees may be yellowish brown and are sold as yellow-fir. Sapwood light tan.	Mostly straight with some wavy figuring. Nearly all flooring is vertical grain.	51%	28% less movement	Durable but dents easily. Prone to splinter.
Heart pine (antique longleaf)	Heartwood rich colors from pinkish tan to reddish brown. Sapwood yellowish white.	Dense with high figuring. Plainsawn-swirled; quartersawn primarily pinstriped.	95%	29% less movement	Strong, hard, very stiff with moderately high shock resistance.
Hickory and pecan (typically mixed in flooring)	Heartwood tan to reddish brown. Sapwood white tinged with brown.	Hickory closed with somewhat rough texture. Pecan open, occasionally wavy.	141%	Hickory 11% more movement Pecan 15% less movement	Strong, hard, tough, and very shock-resistant.
Maple (hard, sugar)	Heartwood is light reddish brown, but sometimes much darker. Sapwood white with reddish-brown tinge.	Fine, uniform texture. Generally straight-grained but also figure known as birds-eye, curly, and fiddleback.	112%	4% less movement	Very dense, strong, stiff. Resists shock and abrasive wear. Used in bowling alleys and sports floors.
Mesquite	Heartwood from yellow-brown to reddish brown, with dark stripes, darkens with exposure. Sapwood light yellow.	Coarse to medium texture, with wavy to interlocked grain	182%	65% less movement	Dense and very strong. End-grain highly resistant to shock and abrasive wear.
Oak (red)	Heartwood reddish brown. Sapwood white to very light brown. Slightly redder tone than white oak.	Open, course texture with heavy, straight grain. Plainsawn has flared pattern; quartersawn has flake pattern.	Benchmark	Benchmark	Dense, hard, strong, and very stiff. Highly shock and wear-resistant.
Oak (white)	Heartwood light to dark brown. Sapwood white to cream. Less variation than red oak.	Open, course texture with longer rays than red oak. Some burled or curly grain. Plainsawn has flared pattern; quartersawn has flake pattern.	105%	Same as benchmark	Dense, hard, strong, and very stiff. Highly shock and wear-resistant. More durable than red oak.
Walnut (American Black)	Heartwood light to dark, chocolate brown. Sapwood nearly white. Great variety in lower grades.	Mostly open and straight with some burled or curly grain.	78%	26% less movement	Moderately dense, hard, and stiff with good shock resistance.

(continued)

| TABLE 5-4 | Domestic and Imported Woods (*Continued*) | | | | |

Imported Woods

Species	Appearance	Grain	Hardness (relative to red oak)	Dimensional Stability (relative to red oak)**	Durability
Brazilian walnut (Ipe)	Heartwood yellowish tan blackish brown. Sapwood whitish or yellowish. Darkens over time to medium to dark brown.	Very dense with fine to medium graining. Grain straight to very irregular. Oily looking.	185%	7% less movement	Very dense and hard.
Jarrah	Heartwood uniformly pinkish to dark red ages to deep brownish red. Sapwood pale.	Texture even to moderately course. Grain interlocked or wavy.	148%	7% more movement	Very dense, strong, resistant to wear.
Santos mahogany (Balsamo)	Heartwood reddish brown aging to deep red or somewhat purplish. Fairly uniform to striped. White sapwood.	Medium texture. Grain typically interlocked.	171%	36% less movement	Very hard and durable.
Merbau (Ipil)	Heartwood yellowish to orange brown. Ages to dark orangey brown.	Texture rather coarse. Grain straight to interlocked or wavy. Yellow dust in pores appears as gold flecking when finished.	149%	57% less movement (more movement observed in actual use)	Strong, hard, moderate density.
Padauk (African)	Heartwood medium to dark orange with darker striping, aging to very dark reddish or purple brown. Sapwood cream colored. Very uniform.	Texture coarse; grain straight to interlocked.	134%	52% less movement	Medium-hard with average to high durability.
Purpleheart	Heartwood brown aging to deep purple. Sapwood cream color.	Very dense with medium to fine texture. Grain usually straight, sometimes wavy or irregular.	144%	43% less movement	Very dense and strong.
Teak (Thai-Burmese, true teak)	Heartwood pale yellow to orange browns and very variable, aging to more uniform golden brown. Sapwood pale yellowish.	Grain straight, sometimes wavy. Texture coarse, uneven. Dull with an oily feel.	78%	50% less movement	Very durable and stable. Similar to oak.

*Based on Janka hardness test, which measures the force required to dent the wood surface. The benchmark species, red oak, is rated at 1290.

**Comparisons are based on plainsawn flooring and are compared to red oak (dimensional coefficient = .00369). Movement in quartersawn flooring is 30%–50% less than indicated in chart. To find the dimensional change of red oak in inches, use the formula: .00369 × width of board (in.) × percent change in moisture content. For example, from 10% to 6%, the percent change in wood moisture content = 4.

DATA SOURCES: USDA, Forest Service; National Wood Flooring Association; and various wood-flooring manufacturers.

yellow pines) is nearly as hard as oak, while black cherry, a popular hardwood flooring, is 26% softer than oak.

Other traditional softwood choices are white pine, popular for Colonial reproductions in the Northeast, and fir flooring in the Northwest. While fir flooring is dimensionally stable, wide white pine boards can be expected to swell and shrink significantly, leaving gaps in the winter months. Both are relatively soft and easy to dent, creating a rustic appearance.

Strip vs. Plank. Narrow flooring boards up to $3\frac{1}{4}$ inches wide are called *strips* and boards 4 inches and wider are called *planks*. The wider the board, the greater the seasonal movement will be and the fewer the number of fasteners to resist movement. Plank flooring over 4 or 5 inches wide has a greater tendency to shrink and leave gaps, or to swell, causing the flooring to cup or curl over time. For that reason, some manufacturers recommend additional fasteners in the face of plank flooring, either nails or countersunk screws. Also the wider the board, the more critical it is to monitor and control moisture conditions of the flooring and structure (see "Moisture and Wood Flooring," p. 167).

Grades. The best grades of wood flooring have longer pieces, fewer variations in color (more heartwood), and

FIGURE 5-5 **Flooring with Character.**

To save money on wood flooring, consider using lower grades of cherry, maple, and other hardwoods, like the No. 3 cherry flooring show above. The greater variation of grain, color, and figure in lower grades can create an interesting and attractive floor. SOURCE: *Photo by author.*

FIGURE 5-6 **Equilibrium Moisture Content for Wood at 70°F (dry bulb).**

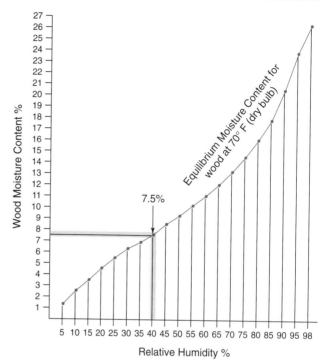

The moisture content of wood slowly changes with the relative humidity of the surrounding air. Kiln-dried hardwood flooring is typically delivered with a 7.5% moisture content. This is the equilibrium moisture content for wood at 70°F and 40% relative humidity, typical indoor conditions throughout most of the United States.

fewer knots and other defects. Quartersawn grades will have much better dimensional stability, with 30 to 50% less movement than plainsawn boards.

In better grades, pieces are also likely to be straighter, making life easier for the installer. With many species, however, the visual variations in lower grades can be attractive as well as economical. A No. 2 or No. 3 maple or cherry floor can be distinctive and striking (Figure 5-5).

Moisture and Wood Flooring

Understanding and controlling moisture levels is the key to success with wood flooring. The conventional wisdom of acclimating wood flooring to job-site conditions can cause more harm than good if the job site is not sufficiently dry when the flooring arrives.

Wood flooring installed in very dry conditions and later exposed to high moisture levels can cause problems such as cupping, particularly with wider planks. In extreme cases, the swelling planks crush the wood fibers along their edges, leaving a permanent "compression set." Gaps appear when the flooring returns to its normal moisture content.

Moisture Content. Wood is a hygroscopic material, meaning that it picks up or gives off moisture to the air until it reaches equilibrium with the relative humidity. As it absorbs or releases moisture, the wood swells or shrinks (see Figure 5-6). Finishes and sealers on the wood slow this process, but do not stop it.

Most hardwood flooring is kiln-dried and delivered with a moisture content (MC) of about 7.5%, which is approximately the equilibrium moisture content for wood

at 70°F and 40% relative humidity—typical indoor conditions for most of the U.S.

Acclimatization. While much has been written about acclimating wood flooring to the job site before installation, in most cases it is the job site that should be dried out before the wood is delivered. If dry wood flooring is brought onto a wet job site, the flooring will swell as it adjusts, creating unsightly gaps when it shrinks back to normal levels.

Before the flooring is delivered, the building should be closed in, and all concrete, masonry, drywall, paint, and other wet work should be thoroughly dry. The basement should be dry and the ground sealed in any crawlspaces. The goal is to have the indoor relative humidity and the moisture content of the subflooring close to the levels they will be after the home is occupied. To sufficiently dry out the site, it may be necessary to run the heating or air-conditioning for a week or more prior to delivery of the flooring.

As a rule of thumb, the subflooring moisture content should be no more than 2% over the maximum normal level for that region based on the map in Figure 5-7, and the flooring and subflooring should be within 2 percentage points of each other. A moisture meter is necessary to

FIGURE 5-7 | **Average Wood Moisture Content in Homes.**

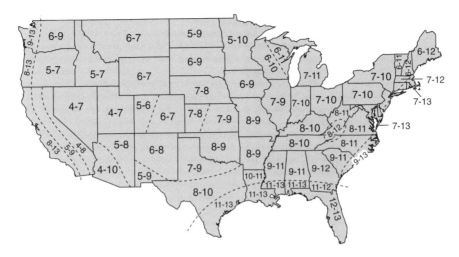

Each pair of numbers shows the average moisture content of wood inside a home in January (lower number) and July (higher number).
SOURCE: USDA Forest Products Laboratory.

determine these levels. Checking the relative humidity with a hygrometer is also a good idea.

With the exception of extremely humid regions such as the Gulf Coast, or extremely dry regions like the arid Southwest, wood delivered at 7.5% moisture content will be suitable for installation in a dry home. If the flooring needs to be acclimated, unbundle the boards and spread them out in the rooms where they will be installed until they reach a moisture content within the range shown in Figure 5-7.

Installation

Standard $\frac{3}{4}$-inch strip or plank flooring is nailed through the tongue into a sound, dry wood subfloor—either plywood, oriented-strand board (OSB), or solid planks. If installed over a slab, the subfloor can either be floated or nailed to the slab.

Wood Subfloors. In new construction, the best subfloor for wood flooring is nailed and glued $\frac{3}{4}$-inch T&G plywood, with the finish flooring installed perpendicular to the joists if possible. Research conducted by the National Oak Flooring Manufacturers Association (NOFMA) has shown that $\frac{5}{8}$-inch plywood or $\frac{3}{4}$-inch OSB also have adequate nail-holding ability for hardwood flooring, although OSB can swell if it gets wet.

Before installing the flooring, nail or screw any loose spots, and shim or sand down any uneven spots to prevent squeaks. Then lay down a layer of 15-pound asphalt felt, which reduces the flow of water vapor into the flooring. The added friction also helps restrict movement in the flooring. Leave a $\frac{3}{4}$-inch expansion space along the edges on the long side of the flooring to accommodate any movement. The expansion space can be concealed with baseboard and shoe molding or by cutting back the drywall (Figure 5-8).

FIGURE 5-8 | **Hardwood-Floor Expansion Space.**

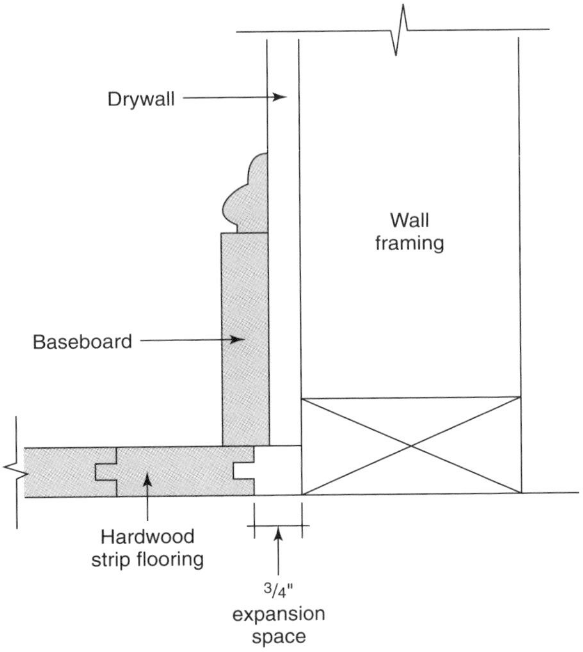

Leave a $\frac{3}{4}$-inch expansion space along the long side of the flooring to accommodate any movement. Conceal the space with baseboard and shoe molding or by cutting back the drywall, as shown.

Installing Over Concrete. For below-grade installations, use a laminated flooring product. For slabs-on-grade, a plywood subfloor is required—either nailed to the slab or floated on top. The slab should be poured over granular backfill with a vapor barrier and must be dry before installation. To test for dryness, duct-tape a one-square-foot piece of polyethylene film to the floor for 24 hours. If the film is clouded or beaded up with moisture, the slab is too wet.

FIGURE 5-9 | **Nailed Plywood Subfloor Over Concrete.**

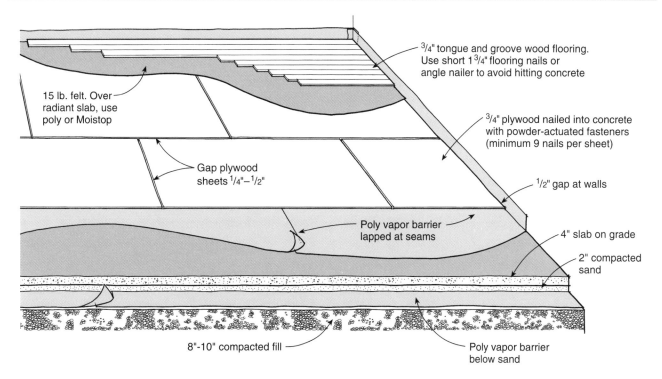

On a dry slab without moisture problems, the simplest approach is the nail-on method, as shown. Install the plywood over a poly vapor barrier with at least nine power-actuated nails per panel. Shorter flooring nails or an angled adapter is required for the finished flooring.

Slabs less than 60 days old are usually too wet. Use heat and ventilation, if necessary, to speed up the drying time.

The slab should be level to $\frac{1}{4}$ inch in 10 feet. Level any uneven spots with clean mason's sand or a floor leveling compound. Next, lay down a 6-mil polyethylene vapor barrier.

On a dry slab where moisture problems are not anticipated, the nail-on method is preferred. Nail $\frac{3}{4}$-inch plywood to the concrete with powder-actuated fasteners using at least nine nails per panel (Figure 5-9).

Leave $\frac{1}{4}$ to $\frac{1}{2}$ inch between sheets and $\frac{1}{2}$ inch around the room perimeter for expansion. Start alternating courses with half sheets so the joints are staggered. Lay 15-pound felt over the plywood and install the flooring. To avoid puncturing the vapor barrier and hitting the concrete, use shorter $1\frac{3}{4}$-inch flooring nails or an angled adapter on the floor nailer.

If there is any question about moisture coming up through the slab, use the floating method (Figure 5-10).

As an extra precaution, tape the laps in the poly vapor barrier and run it a few inches up the wall. Next lay down, but do not nail, 4x8 sheets of $\frac{1}{2}$-inch plywood with the long edge oriented along the length of the room. Leave a $\frac{1}{4}$- to $\frac{1}{2}$-inch gap between sheets and $\frac{3}{4}$ inch around the room perimeter. Next, lay another layer of $\frac{1}{2}$-inch plywood oriented at 45 degrees to the first layer with the same spacing, and staple, screw, or nail ($\frac{7}{8}$-inch ring-shank nails) the top layer to the bottom, being careful not to puncture the vapor barrier. Finally, cover the plywood with 15-pound

felt and install the flooring. To insulate the floor, a layer of compression-rated foam insulation can go between the poly vapor barrier and the plywood.

Nailing. In general, the more nails in wood flooring, the less likely there is to be movement or squeaks. The recommended nailing schedule for $\frac{3}{4}$-inch-thick strip flooring is every 8 to 10 inches with a 7d or 8d flooring nail (see Table 5-5). If the subfloor is less than $\frac{3}{4}$ inch thick, nail into the joists with one nail between each joist. Stagger the ends of strip flooring at least 6 inches.

For plank flooring 4 inches and wider, the minimum nail spacing is 8 inches; closer is better. With boards over 5 inches wide, if the ends are not end-matched (with T&G), the ends will tend to cup or curl unless face-nailed or screwed and plugged with two to three fasteners. It is also a good idea to secure the flooring along its length with face-nails or screws and plugs. If nailing, use wedge or screw-shank flooring nails set below the surface, or decorative nails left exposed for a traditional appearance. Drive the face nails about 30 degrees away from the center to help reduce cupping. Use two to three nails across for planks up to 5 inches, three to four nails for planks up to 8 inches.

Wood Floor Finishing

If possible, allow the home's heating, ventilating, and air-conditioning (HVAC) system to run for two weeks after

FIGURE 5-10 Floating Plywood Subfloor Over Concrete.

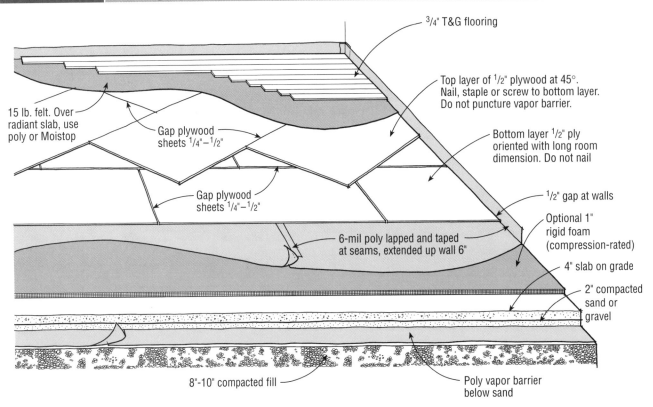

³/₄" T&G flooring

Top layer of ¹/₂" plywood at 45°.
Nail, staple or screw to bottom layer.
Do not puncture vapor barrier.

Bottom layer ¹/₂" ply
oriented with long room
dimension. Do not nail

15 lb. felt. Over
radiant slab, use
poly or Moistop

Gap plywood
sheets ¹/₄"–¹/₂"

Gap plywood
sheets ¹/₄"–¹/₂"

¹/₂" gap at walls

Optional 1"
rigid foam
(compression-rated)

4" slab on grade

6-mil poly lapped and taped
at seams, extended up wall 6"

2" compacted
sand or
gravel

8"-10" compacted fill

Poly vapor barrier
below sand

If there are concerns about moisture coming up through the slab, use the floating method with the vapor barrier taped and run up the walls a few inches. Be careful not to puncture the vapor barrier when nailing together the plywood layers.

TABLE 5-5 Wood Flooring Nailing Schedule

Flooring Size	Fastener Sizes	Spacing
T&G strip ³/₄x1¹/₂ to 3¹/₄ in.	7d or 8d flooring or casing nails; 2-in. 15-gauge staples with ¹/₂-in. crowns; or 2-in. barbed flooring cleats.*	8 to 10 in. o.c., 1 to 3 in. from end (min. 2 nails per strip)
T&G plank ³/₄x4 to 8 in.	7d or 8d flooring nails; 2-in. 15-gauge staples with ¹/₂" crowns; or 2-in. barbed flooring cleats.*	8 in. o.c.; 1 to 3 in. from each end (min. 2 nails per plank). Also face-nail or screw/plug each end with 1 to 4 fasteners, if not end-matched. Face-nail or screw/plug 1 to 4 fasteners 16 to 24 in. o.c. along plank, or per manufacturer's recommendations.
T&G strip ¹/₂x1¹/₂ to 2 in.	1¹/₂-in. barbed flooring cleat; or 5d screw-type, cut-steel, or wire casing nail	10 in. o.c.
T&G strip ³/₈x1¹/₂ to 2 in.	1¹/₄-in. barbed flooring cleat; or 4d bright-wire casing nail	8 in. o.c.

*Use 1¹/₂-in. fasteners with ³/₄-in. plywood subfloor on a concrete slab.
SOURCE: Courtesy of NOFMA, The Wood Flooring Manufacturers Association. © NOFMA 1997.

the flooring is installed before sanding and finishing. The most common site-applied floor finish today is oil-based polyurethane, although waterborne urethanes are rapidly gaining market share due to their fast drying, low level of volatile organic compounds (VOCs), and nonyellowing appearance (although "ambering" formulas are now available). For those seeking a more rustic, lower-gloss appearance and willing to wax and buff periodically, traditional oil sealers and wax remain an option. Finishing options are summarized in Table 5-6.

TABLE 5-6	Finishes for Wood Flooring			
Finish Type	Pros	Cons	Application	Recommendations
Penetrating oil sealers and wax	Easy to apply and to recondition worn areas without sanding and refinishing entire floor.	Least durable finish. Needs regular touch-up and periodic removal of wax. Water spots. Cannot be wet-mopped. Flammable (sealers) during application.	Wet surface, allow to soak in, and wipe excess. Use two or more coats of oil sealer; then wax. Buff to restore sheen, and rewax as needed.	Use in light-traffic areas not subject to wetting. Provides rustic look. Low sheen tends to hide wear and scratches*.
Oil-modified urethanes	Easiest urethane to apply. Durable finish. Easy to recoat and blend with light sanding.	Darkens wood over time with amber tone. Noxious and flammable fumes. Slow to dry and cure.	Dry overnight between coats and lightly buff or screen. Use mineral-spirits on tack cloth to remove any dust. Air blowers okay to speed drying. Apply 2 to 3 coats.	The higher the solids content, the more durable the finish. Allow one week, or per manufacturer's instructions, before heavy traffic or furniture placement.
Water-borne urethanes and urethane/acrylic blends	Does not darken wood. Durable finish. Minimal fumes and odors. Dries and cures quickly. Can apply two or more coats in one day. Nonflammable.	Raises grain, particularly on open-grain woods, such as oak. Requires more coats than oil-modified type. Most are two-part. Takes practice to avoid bubbles, thick spots, and other defects.	May require special sealer and applicator. Make sure wood surface and sealer are dry before applying. Follow manufacturer's instructions about buffing between coats. Never use steel wool. Keep wet edge to prevent visible lines. Apply 3 to 4 coats.	With blends, higher urethane content more durable. Avoid high temperatures and direct sun while drying. Allow one to two weeks, or per manufacturer's instructions, before heavy traffic or furniture placement. Use experienced applicators.
Moisture-cure urethanes	Generally harder and more moisture-resistant than other site-applied finishes.	Extremely difficult to apply properly. Noxious, highly flammable fumes. Skin irritant. High-gloss only.	Requires specific level of relative humidity for proper cure. Follow manufacturer's instructions and safety procedures.	Should only be applied by highly skilled applicators. Suitable for kitchens and high-traffic areas.
Acid-cure urethanes (Swedish finishes)	Clear, fast drying. Resists yellowing. Generally harder than oil-based urethanes. Recoatable.	Difficult to apply properly and unforgiving of mistakes. High VOC content with formaldehyde. Noxious, highly flammable fumes.	Closely follow manufacturer's instructions. Requires special application techniques and safety procedures.	Should only be applied by highly skilled applicators
UV-cured urethanes (factory-applied)	Highly durable. Fully cured during manufacture. With aluminum oxide or other additives, creates hardest surface finish.	On porous woods such as oak, may not fully cure beneath surface, leaving odor. Aluminum oxide may result in slightly rough surface. Difficult to blend in refinished section.	Factory-applied only. Cured with UV radiation during application.	Look for 1.5- to 2-mil-thick finish regardless of number of coats. With aluminum oxide, ceramic, or other additives, warranties up to 25 years.
Acrylic impregnation (factory-applied)	Highly durable, impact resistant. Injected into cell structure of wood, providing uniform color and increased hardness and dimensional stability. Tends to conceal scratches.	Only available on porous woods, such as red oak or ash. Subject to green oxidation stains from metal table legs or other metals left on damp floor.	Factory-applied only. Fully penetrates top $\frac{1}{8}$-inch ply in $\frac{3}{8}$-inch laminated flooring.	Some manufacturers cure finish with gamma radiation to provide greater abrasion resistance. Scratches can be buffed out or spot repaired. Warranties up to 25 years.

*In general, higher gloss finishes show more scratches and scuff marks and are not recommended for high-traffic areas.

Penetrating Sealer and Wax Systems. These finishes are generally made from mixtures of linseed or tung oil, sometimes with synthetic polymers and additives to improve hardness and drying time. Usually two coats of a penetrating sealer are applied, followed by a coat of wax, providing a low-sheen, rustic appearance. While easy to apply, the finish is fairly high-maintenance, requiring periodic buffing and rewaxing to keep it looking good. Over time, the wax will become discolored from dirt and grime and will need to be stripped. In its favor, high use areas of the floor are easy to touch up without sanding and refinishing the entire floor.

Surface Finishes. Most site-applied finishes today are oil-based or water-borne urethanes applied to the surface of the wood in three or more coats. In general, any high-quality urethane applied properly will provide a durable, moisture-resistant surface. While water-borne finishes had some quality problems when first introduced for residential use in the late 1980s, they have continually improved and now offer durability equal or superior to traditional oil-based urethanes. While most floor finishes claim in their marketing to be the toughest, hardest, and longest lasting, moisture-cured urethane is generally considered the toughest site-applied finish. The most durable finishes, some with warranties of up to 25 years, are available only on factory-finished flooring.

Whatever finish is applied, follow the manufacturer's instructions closely. The following recommendations apply to most site-applied finishes:

- After sanding, wipe the floor with a lint-free tack cloth and allow to fully dry. Use mineral sprits for oil-based finishes, water for water-borne finishes.

- If you use a stain or penetrating sealer, make sure it is completely dry before applying the top coats.

- Make sure the temperature is right—65°F to 85°F for most finishes. Colder will delay drying; hotter will dry too fast and the finish will not level.

- Keep indoor humidity from 45% to 75%. Higher humidity will delay drying and can harm the finish.

- Provide exhaust ventilation during drying. Poor ventilation can cause a soft or rough finish.

- Allow each coat to fully dry before applying the next coat. Multiple thin coats make the best finish.

- Lightly abrade finish between coats as recommended by manufacturer. (Not required with some water-borne finishes if recoated within 3 to 6 hours—follow manufacturer's recommendations.)

Prefinished Strip and Plank Flooring. Many solid wood floor products are now available prefinished in thicknesses from $\frac{1}{2}$ to $\frac{3}{4}$ inch for roughly twice the price of comparable unfinished wood. In addition to speeding up the installation and eliminating the dust and fumes associated with job-site finishing, these finishes have several other advantages:

- The most durable finishes, such as UV-cured urethane and acrylic impregnation, can only be applied in a factory (see Table 5-6).

- Finishes are applied in a dust-free, controlled environment.

- Factory finishes typically carry long-term warranties.

- There are no surprises in the color of the finished floor.

The main disadvantage of this approach is that the installer must take extra care not to mar or scratch the finishes during installation, adding to the installation cost. Also, most prefinished flooring is chamfered to some degree to hide the inevitable differences in thickness from one board to the next, called *overwood,* which results from small variations in water absorption, swelling, and shrinking among flooring boards. Debris or irregularities in the subfloor can also cause an uneven surface.

The grooves in deeply beveled flooring tend to trap dirt and can cause problems when it is time to sand and refinish, particularly with stained flooring. Unless the bevels are sanded away, they are hard to strip and difficult to match, leaving dark lines in the refinished floor. One approach is to refinish before the old finish is completely worn through by just lightly abrading the surface prior to recoating or using one of the no-sand refinishing products.

To address these concerns, most flooring manufacturers now offer prefinished flooring products with no bevel (square edged) or nearly invisible "microbevels" that minimize the effects of a beveled edge.

Maintenance and Reconditioning. A sealed and waxed floor typically needs rewaxing no more than once or twice a year to keep it looking good.

With surface coatings, most manufacturers do not recommend waxing. If a surface-coated floor gets lightly scratched over time but has not worn through to bare wood, it can be recoated in most cases without complete resanding. After thoroughly cleaning the floor with a non-residue cleaner, rough up the old finish with steel wool, light sandpaper, or a sanding screen, then apply a new coat of finish. Many coating manufacturers now offer no-sand refinishing products as well, formulated to bond to the old finish without sanding.

No wood floor should be flooded with water during cleaning. Either use a dry mop or a wet mop that has been squeezed dry. Water can find its way between floor boards and through scratches, swell the wood, and undermine the finish.

Engineered Wood Floors

The main advantage of engineered wood floors is dimensional stability. Because most engineered floors consist of cross-laminated plies of wood, they are less likely to swell, shrink, cup, or warp. This makes them the best choice for

FIGURE 5-11 Wear Layer of Wood Floors.

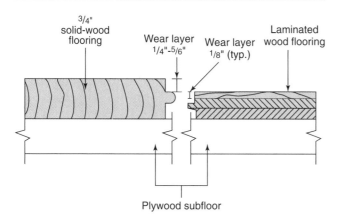

The wear layer of the best laminated wood flooring products is nearly as thick as in traditional strip flooring, and it can be sanded up to three or four times over its service life.

FIGURE 5-12 Longstrip Flooring.

Longstrip flooring, primarily used in floating floors, uses several short pieces of strip flooring in the top ply to give the appearance of random-length strip flooring. However, butt joints line up at panel ends.
SOURCE: Photo courtesy of Tarkett Wood Floors.

applications subject to wide changes in moisture levels—below grade, over radiant heating, or over concrete slabs with poor drainage or no vapor barrier. Laminated floors are also the best choice for glue-down applications because of their inherent stability. Another advantage is that engineered wood floors do not require a beveled edge. Most have tight-fitting, square-edged joints.

Materials. Laminated floors come in a wide variety of sizes, from $\frac{1}{4}$ to $\frac{3}{4}$ inch thick. In general, the more plies a flooring has, the stronger and more stable it will be. Three plies is typical. Look for a top *wear layer* of at least $\frac{1}{8}$ inch if the floor is to be sanded and refinished in the future. Typical laminated floors can be sanded and refinished once or twice; the best, up to three or four times. The sanding thickness in some laminated flooring is nearly as thick as in traditional hardwood strip flooring, which although $\frac{3}{4}$-inch thick, can be sanded down only about $\frac{1}{4}$ inch (see Figure 5-11).

Longstrip flooring has several short pieces of strip flooring in the top ply to give the appearance of random-length strip flooring. Single panels are as large as 8x96 inches (see Figure 5-12). It is used primarily in floating floors, but some products may also be nailed or glued. Since each panel is two or three strips wide, the end joints of these strips line up at panel ends, unlike in a true strip floor. Single-strip products, on the other hand, are visually indistinguishable from a traditional strip floor.

Veneers may be either sliced or rotary cut. Rotary-cut veneers make better use of the tree with less waste, but sliced veneers are harder and less prone to denting. Also, look for interior veneers that are the same wood as the face veneer or at least as hard. Soft interior veneers make a weaker flooring that is more prone to denting.

Installation. Laminated wood floors can be nailed, stapled, or glued with mastic to any dry wood subfloor.

Many can also be glued directly to dry concrete (see "Installing over Concrete," page 168, for how to test dryness). If the concrete is below grade, check with the manufacturer to see if the product is guaranteed for that application. A floating floor may be a better choice for below-grade applications.

In general, glue-down products are $\frac{1}{4}$- to $\frac{3}{8}$-inch thick parquet tiles, strips, or planks. Strips or planks are generally less than 2 feet long, since longer pieces are too difficult to straighten with glue. The only solid wood flooring that is glued is parquet, which gains stability from the short pieces and different orientations of the grain.

As with unfinished flooring, the building should be closed in, with all wet work completed and dried, before installing engineered wood flooring. Make sure the concrete or subfloor is sufficiently dry and the indoor humidity level is close to the level it will be when the building is occupied. Keep the flooring materials packaged until installation.

Floating Floors. Another option is to float the finish flooring. You can float a floor over virtually any stable substrate, including concrete, wood, smooth tile, or even short-nap carpet. With most floating floors, the T&G pieces are edge-glued to one another with PVA wood glue and installed over a thin layer of closed-cell, high-density foam and a vapor barrier.

A floating floor is more resilient underfoot than one glued to concrete, but feels less solid than a nailed or glued floor. Some manufacturers offer a harder premium foam underlayment, which is recommended for those seeking a more solid feel underfoot. Still, customers should be aware that a floating floor will feel different from a nailed or glued floor.

Installers must leave a $\frac{1}{2}$-inch expansion gap around the edges of floating floors, typically hidden by baseboard, or use special T- or L-shaped moldings at door openings and other transitions to accommodate movement (Figure 5-13).

Existing door casings can be undercut to allow for movement. Restraining the flooring ends at doorways or the room perimeter can lead to open joints or buckling.

Because there are no mechanical fasteners to the substrate, floating floors rely on good quality flooring and a very flat slab or subfloor to produce a smooth, trouble-free floor. In shopping for the flooring, choose materials that are straight and uniform in thickness, fit together snugly, and lay flat with few visible gaps. The subfloor should be level to within $\frac{1}{8}$ inch over 10 feet. If necessary, shim low points with clean mason's sand or felt or rosin paper layered in the low spots to create a tapered shim (do not use asphalt felt over radiant floors, however, to avoid fumes).

While floating floors cannot tolerate the heavy vibration caused by standard floor sanding equipment, most floating wood floors can be lightly sanded and refinished or coated with sandless finishes. Follow the manufacturer's recommendations regarding refinishing.

No-Glue Floors. No-glue longstrip flooring is available from Alloc, Inc. and BHK of America (see "Resources" on page 206). Each company uses its own interlocking edge design to snap the 8x48-inch or 8x96-inch panels together in place of adhesive. These products were developed in Europe where people often take their floors with them when they move.

The Denmark-based company Junckers Hardwood Inc. manufactures the only solid-hardwood floating floor. The 6-foot-long strips are held together with special metal clips that snap in place on the underside of the floor (Figure 5-14).

The clips, along with adhesive at butt joints only, work together to create a strong monolithic floor with the appearance of traditional strip flooring but the ability to move with moisture and temperature changes, making it ideal for use over radiant slabs. The 5-inch-wide boards

FIGURE 5-13 | **Floating Floor Installation.**

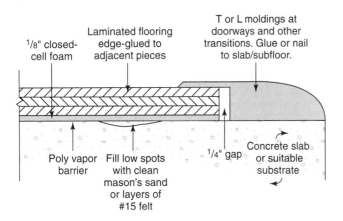

Floating floors must be detailed to accommodate expansion and contraction at the room perimeter, and at doorways and transitions. Restraining the flooring ends can lead to open joints or buckling.

FIGURE 5-14 | **No-Glue Wood Flooring.**

Junckers Clip System

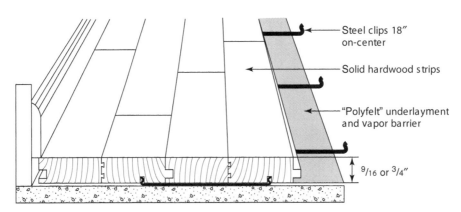

No-glue laminated wood floorings are popular in Europe, where people often take their floors with them when they move. The Junckers system uses solid hardwood strips, end-glued and locked together with steel clips, to produce a strong, monolithic floor well-suited to radiant slabs.

SOURCE: Courtesy of Junkers Hardwood.

are either a single plank in width or two strips dovetailed together. The $\frac{9}{16}$-inch product is guaranteed for two sandings, the $\frac{3}{4}$-inch for seven.

Bamboo. A recent introduction to the flooring market, bamboo is not really a wood, but a type of grass that matures in three to five years on plantations, making it an environmentally friendly alternative to premium hardwoods. To make bamboo into flooring, thin strips are laminated to form planks from $\frac{3}{8}$ to $\frac{3}{4}$ inch thick. The familiar nodes that separate bamboo stalks into short sections create darker cross markings, giving the product an attractive and unusual appearance. A more homogenous color is also available from some manufacturers by using laminated strands. Engineered bamboo flooring is as hard as maple and more stable than oak, and comes either unfinished or prefinished with the same types of finishes as used on hardwood flooring.

Wood Floors over Radiant Heating

Radiant heating is a challenging application for wood floors due to the high temperatures, excessive drying, and greater temperature cycling to which the wood and finish are subjected. Careful monitoring and control of the moisture levels of the flooring and structure at the time of installation are critical for success. Also, because a $\frac{3}{4}$-inch wood floor over $\frac{3}{4}$-inch plywood has an R-value of almost 2, similar to plush carpeting, wood systems generally must run at higher water temperatures than tile or vinyl floors. Large area rugs make it more difficult for the heating engineer to design a system that will heat the room without overheating the flooring.

Wood flooring can be installed over radiant slabs, or over dry systems where the hydronic tubes are stapled directly to the subflooring (staple-up) or laid on top in grooved panels. Dry systems are more common in retrofits and generally require water 10°F to 20°F higher than thin slabs, leading to reduced efficiency, and often ruling out low-temperature heat sources like heat pumps or solar. Also, with less thermal mass than slab-based systems, dry systems are more prone to temperature fluctuations. All systems are designed to heat floors to about 80°F. Floors heated above 85°F are uncomfortable for occupants and may be damaged from the heat.

With any approach, the radiant slab or subfloor must be dry prior to installation. With slab systems, run the heating system for at least a week, up to three weeks if necessary, to dry the slab to a moisture content of 8 to 12% before installing the subfloor. The subfloor and hardwood floor should be acclimated to the average annual moisture levels for the region and be within 2 percentage points of each other (see "Acclimatization," page 167). Flooring that is installed too wet can lead to shrinkage cracks; flooring installed too dry can lead to expansion problems or cupping in humid summer weather.

To steer clear of problems, also follow these recommendations:

- If possible, choose an engineered wood product rather than solid wood. Floating floors are best, since they are designed to accommodate movement.
- If solid wood is used, avoid flooring wider than 3 inches—the narrower the better. If possible, use quartersawn wood, which swells and shrinks 30%–50% less than flat-sawn.
- Choose a prefinished flooring coated on all faces. Prefinished flooring with chamfered edges (microbevels) will help conceal shrinkage cracks.
- Increase nailing of strip flooring to 4 to 5 inches on-center.
- If finishing on site, use a recommended sealer to reduce the chance of edge-bonding when the flooring shrinks, leaving large gaps every several courses.
- Always use a vapor barrier under the flooring, but **do not** use asphalt-impregnated felt, which will give off fumes.
- Avoid glue-down flooring. If used, make sure both the flooring and adhesive manufacturers approve the application.

Slabs-on Grade. Over traditional concrete radiant slabs at least 4 inches deep, use either a floating hardwood floor or install subflooring and nail on $\frac{3}{4}$-inch strip flooring. The subflooring can consist of two layers of $\frac{1}{2}$-inch plywood floating over the slab (see "Installing Over Concrete," page 168), or a single-layer $\frac{3}{4}$-inch subfloor nailed to the slab with powder-actuated fasteners. Because it is thicker, the floating subfloor (Figure 5-15) will take slightly longer to heat up, but it does not risk puncturing the hydronic tubing.

Thin Slabs. In wood-frame construction, use a minimum $1\frac{1}{2}$-inch-thick slab of Gyp-Crete® or lightweight concrete, which provides thermal mass for the radiant floor. Above the slab use a floating hardwood floor, or nail strip flooring to $\frac{3}{4}$-inch sheathing installed over the lightweight concrete. Fasten the sheathing to 2x4 sleepers placed 12 inches on-center, with the lightweight concrete and tubing in between (see Figure 5-15). A two-layer floating subfloor, as described above, is also an option for larger rooms where the subfloor will be heavy enough to stay solidly in place without nails.

Staple-up and Panel Systems. There are a variety of dry radiant systems that install just under or over the subflooring, making them ideal for retrofits. The tubing is either stapled to the underside of the subflooring, laid over the joists (with spacers to fur up the sheathing), or placed over the subflooring in grooved plywood panels. Engineered wood floating floors are best with these systems, but nail-on hardwood flooring can work if installed with care.

FIGURE 5-15   Wood Floors Over Radiant Slabs.

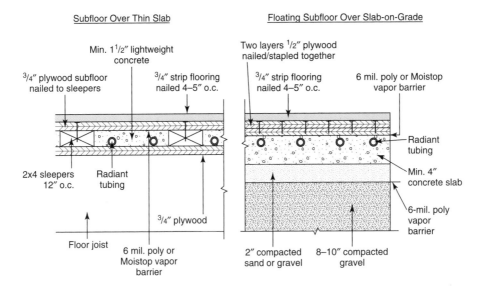

While laminated floating wood floors are preferable with radiant slabs, nail-on strip floor-ing can also work over thin slabs (left) or slabs-on-grade (right) if installers pay close attention to details: use narrow-width flooring, preferably quartersawn, install with extra nails, and provide a continuous vapor barrier. Also avoid heating the wood above 85°F.

RESILIENT FLOORING

Multilayer sheet vinyl is by far the most common material used in resilient floors. It comes in a variety of grades and a vast array of colors and patterns and, if installed well and maintained properly, should last 10 to 20 years. Solid vinyl tiles are another popular option; but, with multiple seams, they are more vulnerable to intrusions from water and dirt. Natural alternatives to vinyl that are growing in popularity include cork, in sheet or tile form, and old-fashioned linoleum, which is making a comeback in residential installations with new colors and marbleized patterns.

Sheet Vinyl

Sheet vinyl is manufactured to be either fully adhered to the substrate with mastic or bonded only at the edges, called a *perimeter-bond* system. Flex-type vinyl flooring, made for perimeter-bond installation, tolerates minor un-evenness and movement in the substrate better than fully adhered systems, but fully adhered systems are more durable overall and less likely to be damaged from stresses like a heavy piece of furniture being dragged across.

All sheet vinyl has three layers:

- *Backing layer.* Typically, this is felt or vinyl. Better grades generally have vinyl backing that resists denting better than felt. To improve the tear resistance and toughness of layered flooring, some manufactur-ers laminate an additional layer between the core layer and the backing.
- *Core layer.* This is a foam interlayer that gives the ma-terial its resilience and resistance to denting. The color

and pattern are either printed on this layer or "inlaid." With inlaid construction, the color and pattern run through the material from the wear layer to backing, making it less prone to show a small nick in the surface.

- *Wear layer.* The top layer protects the flooring from wear and is generally either clear vinyl or a more durable urethane (PUR) finish, sometimes enhanced with aluminum oxide or other additives to increase wear resistance. Many finishes are marketed as "no-wax," but these are not maintenance-free and require periodic application of an acrylic dressing. Lower gloss finishes are recommended for high-traffic areas.

Better quality vinyl floors tend to be thicker overall and have a thicker and higher-quality wear layer. As the wear layer gets abraded from dirt and grime, it becomes duller and harder to clean. The thickness of the wear layer can range from 5 to 25 mils, and the flooring thickness from about $\frac{1}{16}$ inch to $\frac{1}{8}$ inch. Better quality products offer better resistance to stains and scratches than lower-end floors, and some of the top quality floors are guaranteed not to rip or permanently dent.

Solid Vinyl Tile

Similar to inlaid sheet vinyl, the color and pattern in solid vinyl tiles run through the full thickness of the tile, mak-ing them very durable. Because the color and pattern extend through the tile, they do not wear away with heavy use, but choices are limited. Solid vinyl tiles are cut from a solid block of material and come with a low-gloss finish. One type, vinyl composition tile or VCT, is essentially the same product as solid vinyl, but with other binders and fillers. Both types require waxing and buffing, both to seal any gaps between tiles and to create an easy-to-clean surface.

Vinyl Installation

Vinyl flooring can be installed over approved wood-based underlayments, dry concrete, or existing vinyl or linoleum if it is in good condition, clean, and free of wax or grease. However, any imperfection in the underlayment will telegraph through the finished floor, so if there are any questions, it is best to install new underlayment. Most problems with vinyl are caused by problems with the underlayment, such as nail pops and swelling or delamination due to moisture. Adhesive failures at edges or seams can also be a problem.

To avoid these types of problems, use only underlayments and adhesives that are recommended by the flooring manufacturer. Also, if possible, avoid seams—most sheet vinyl comes in 6- and 12-foot rolls, so many rooms can be done without a seam. If seams are required, darker colors and textured pattern are preferable and help hide dirt and scuff marks as well. All seams should be sealed with an approved sealer to keep dirt out and to keep water from penetrating and undermining the adhesive bond.

Concrete. If installing over a concrete slab, make sure it has a proper vapor barrier and has cured for at least 60 days. A concrete sealer is recommended. Existing slabs should be wire brushed, swept clean, and primed with an approved primer before gluing down resilient flooring.

Acclimatization. Because vinyl shrinks and expands with room temperature, it should be allowed to adjust to the room temperature before installation. In general, the room should be heated or cooled to its normal temperature and the vinyl allowed to acclimate for 24 hours.

Underlayments. For a problem-free floor, sheet vinyl must be installed over a smooth, hard, and dry surface approved for use with vinyl.

- *Plywood.* The most reliable underlayment, accepted by all vinyl flooring manufacturers, is sanded plywood with solid inner plies (no voids) that resist denting or puncturing. This is usually either designated "Underlayment with sanded face" or "C-C Plugged with sanded face." (Other possible grades include "Plugged crossbands under face" or "Plugged inner plies"). Avoid plywood with plastic or resin fillers on the surface, as these may stain the vinyl.

- *Lauan.* Type 1 exterior-grade lauan plywood is sometimes used as an underlayment and is approved by some vinyl flooring manufacturers. If using lauan, use the best grade available, which is often labeled B-B.

- *Particleboard.* This is discouraged by most manufacturers but is sometimes used in areas with limited

exposure to moisture, since particleboard has the potential to swell at edges if wet. Also, the particleboard surface can tear when installers pull back the vinyl to spread adhesive at seams.

Make sure the subflooring is dry before installing the underlayment. Use minimum $\frac{1}{4}$-inch-thick panels so that the underlayment plus subfloor is at least 1 inch thick. Stagger joints in the underlayment so they are offset from joints in the floor sheathing by at least 2 inches (see Figure 5-16).

Most flooring manufacturers specify a $\frac{1}{32}$-inch gap between sheets, filled with a quick-setting latex-based cementitious filler. The filler restrains the edges of the underlayment and helps prevent ridging from movement or the absorption of flooring adhesive at panel edges.

The nailing schedule is shown in Table 5-7. Fasteners should approximately equal the thickness of the underlayment and subfloor and should not be driven into the framing. Many contractors prefer staples to nails, because they do not leave dimples in the underlayment. Before using staples, however, make sure that they are approved by the resilient-flooring manufacturer. Nails should be ring-shank or spiral-shank and driven flush or just below the surface, but the heads should not be filled. Other holes, gaps, and voids should be filled with a latex-based cementitious filling compound before laying the floor.

Natural Alternatives to Vinyl

Homeowners who want a resilient floor covering but are looking for an alternative to vinyl should consider the new cork products as well as traditional linoleum, which is enjoying a comeback in residential applications.

Cork. Cork is a renewable resource that is harvested every 9 or 10 years from the outer bark layer of cork oak trees in Portugal and other Mediterranean countries. Cork has a number of desirable attributes for a flooring material: its air-filled, watertight cells are strong, soft to walk on, and insulating, making it a good choice over a concrete slab. To make it into flooring, manufacturers grind up the cork, mix it with a chemical binder, bake the material, and slice it into sheets. Cork flooring products range in thickness from $\frac{3}{16}$ to $\frac{7}{16}$ inch for some laminated products.

Most cork flooring is sold as tiles and installed with adhesive, similarly to other resilient tiles. Tiles are available either unfinished or prefinished with carnauba wax or a more durable polyurethane or acrylic coating. Tiles tend to have natural color variation and can be purchased in light, medium, or dark tones. As with wood floors, wax finishes need regular buffing and periodic rewaxing, depending on use. Polyurethane-finished cork typically needs recoating in four to eight years. One advantage of purchasing unfinished tiles and finishing in place is better protection against moisture penetration between tiles. The cork itself is moderately water-resistant.

FIGURE 5-16 Underlayment for Resilient Flooring.

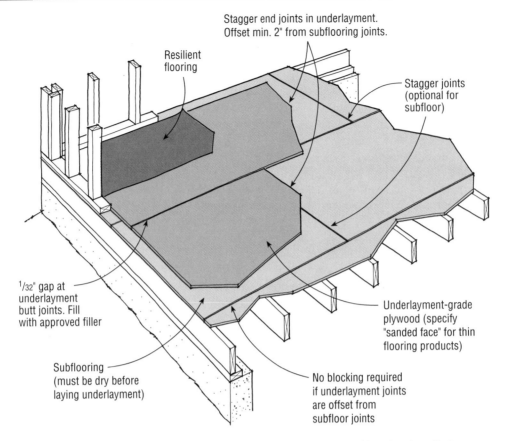

Stagger end joints in underlayment.
Offset min. 2" from subflooring joints.

Resilient flooring

Stagger joints (optional for subfloor)

¹/₃₂" gap at underlayment butt joints. Fill with approved filler

Underlayment-grade plywood (specify "sanded face" for thin flooring products)

Subflooring (must be dry before laying underlayment)

No blocking required if underlayment joints are offset from subfloor joints

Most resilient flooring problems start with the underlayment. For a problem-free installation, use void-free plywood approved for resilient flooring. Underlayment nails should be spiral- or ring-shank and driven flush with the surface. Nails should fully penetrate the subflooring but not the floor framing.

SOURCE: Adapted from *Builder Tips: Proper Selection and Installation of APA Plywood Underlayment*, © 2002 APA. Courtesy of APA, the Engineered Wood Association.

TABLE 5-7 Nailing for Resilient Flooring Underlayment

Application	Minimum Plywood Thickness	Fastener Type and Size	Maximum Fastener Spacing[1]	
			Panel Edges	In Field
Over smooth subfloor	$\frac{1}{4}$ in.	3d ($1\frac{1}{4}$ in.x12 ga.) ring or screw-shank nails; 4d ($1\frac{1}{2}$ in.x$12\frac{1}{2}$ ga.) for panels $\frac{19}{32}$ to $\frac{3}{4}$ in. thick. For staples, see note below.[3]	3 in.	6 in.
Over lumber subfloor or other uneven surface	$\frac{11}{32}$ in.	Same as above	6 in.	8 in.[2]

[1]Space fasteners $\frac{3}{8}$ in. from panel edges, and place so they do not penetrate framing.
[2]12 in. for 5-ply plywood underlayment and panels thicker than $\frac{1}{2}$ in.
[3]Although not recommended by the APA, staples are accepted by most resilient flooring manufacturers for underlayment up to $\frac{3}{8}$ in. thick. Use $\frac{7}{8}$ to $\frac{11}{8}$ in. galvanized staples and align crown of staple with face grain. Polymer-coated staples provide greater withdrawal resistance.
SOURCE: Courtesy of APA, the Engineered Wood Association.

A variety of other cork and cork composite products are now on the market, including tongue-and-groove (T&G) floating floors and cork and vinyl laminates. A number of manufacturers now offer 12x36-inch floating T&G planks with an MDF core sandwiched between a cork underlayment and aggregate cork wear layer. Manufacturers include Korq, Inc., American Cork Products Company, and Nova Distinctive Floors, which offers a unique no-glue option. Some manufacturers offer a composite product with an inner layer of cork sandwiched between a thick vinyl wear layer and vinyl backing (see "Resources," on page 207).

While cork products appeal to healthy-house advocates, the binders and adhesives used with tiles, and the fiberboard or vinyl layers used in laminated products, may not provide the completely nontoxic, nonoffgassing material desired. Using solid-cork (aggregate) tiles with a low-VOC adhesive is probably the best choice for those seeking natural, nontoxic materials.

Linoleum. For the last 50 years or so, linoleum has been used almost exclusively in commercial settings, but it is making a comeback in residential settings, due largely to its use of all-natural ingredients and reputation for durability. Linoleum is made by boiling oil to form a thick cement paste that is mixed with pine rosin, wood flour, and other fillers such as clay or limestone to make a durable, resilient sheet flooring that wears well and resists indentation. The backing is typically jute fabric, a natural fiber. Other than relatively minor initial off-gassing from the linseed oil base, linoleum is considered nontoxic by most healthy-house advocates. It is also naturally antimicrobial and antistatic, making it well suited for hospitals, schools, and rooms with electronic equipment. If well maintained, a linoleum floor can provide a 20- to 30-year service life.

In response to new demand for the product in recent years, manufacturers have responded with a wide variety of solid and marbleized colors and attractive checkered patterns, available in sheet form as well as 19x19-inch tiles that can be mixed to create borders and other designs. Unlike vinyl, linoleum colors go all the way through the product, making scratches and wear spots less noticeable than on vinyl. Also, scratches, cigarette burns, and other surface wear can be removed with steel wool or a nylon abrasive pad and buffed out.

However, since linoleum does not have a separate wear layer like vinyl flooring and is slightly porous, it requires somewhat more maintenance than vinyl. Applying a sealer or polish to the new floor will help it resist stains and make it easier to clean. Also, portions of a linoleum floor not exposed to light will tend to darken or yellow due to the natural oxidation of the linseed oil base. This coloration will disappear upon exposure to light, and the original linoleum color will be restored, or "bloom."

All linoleum flooring is now manufactured in Europe. The largest supplier in the United States is European-based Forbo Linoleum, Inc., but U.S.-based flooring companies such as Armstrong are beginning to offer linoleum products as well. A unique floating linoleum plank floor that can be installed with or without glue is available from Nova Distinctive Floors.

LAMINATE FLOORING

Plastic laminate flooring was introduced to the U.S. market in the mid-1990s and now competes with vinyl as a low-maintenance floor covering. Similar to the high-density plastic laminate used on countertops, the flooring is protected by a clear melamine layer, in some cases reinforced with aluminum oxide. A high-density fiberboard core provides stability and resilience, and a melamine layer on the bottom provides some protection against moisture.

The product comes as either tiles that resemble stone or ceramic tiles or planks that simulate wood flooring. Both are floating products, edge-glued with PVA adhesive, but not attached to the subfloor. As with floating wood floors, a layer of $\frac{1}{4}$-inch-thick high-density foam goes under the floor to provide a cushion and even out small irregularities in the subfloor. Most manufacturers offer a higher density premium foam, which is recommended for a more solid feel underfoot.

If the product is installed properly, with sufficient glue to squeeze out along all joints, laminate flooring is moderately waterproof. The main problems occur at edges. To prevent water infiltration, seal with silicone any edges that might be exposed to water in kitchens, bathrooms, or other wet areas. Even with these safeguards, many manufacturers will not warranty laminate flooring in bathrooms. Radiant heat applications are generally acceptable.

Durability. In general, laminate flooring has a very hard surface that resists dents, scratches, and other damage. However, over time, high-traffic areas will lose their sheen and show signs of wear. Other than to replace the planks in those areas, there is little that can be done to restore the original finish.

Installation. Laminate flooring installs over a vapor barrier and thin layer of foam like floating wood floors (see "Floating Floors," page 173). As with floating wood floors, the substrate must be very level. Fix any low spots with latex- or acrylic-based cementitious compound or building paper layered in progressively larger pieces.

Concrete must be dry enough that it will not fog a one-foot square of polyurethane taped overnight. Thoroughly clean the slab or subfloor, and lay down the vapor barrier and foam provided by the laminate flooring manufacturer. Lap the vapor barrier 8 inches or tape the foam seams, if that is also serving as the vapor barrier, as allowed in some systems.

The tiles or planks are then glued together with a PVA glue provided by the manufacturer. Pieces are tapped into place and clamped with special strap clamps. Even squeeze-out of glue along the entire glue line indicates sufficient glue, which is necessary for a solid, waterproof floor.

As with other floating floors, special T-shaped or L-shaped threshold, end, and transition moldings conceal the ends of flooring at transitions while allowing movement. A $\frac{1}{4}$-inch gap at the perimeter of the room is typically concealed by the baseboard or quarter-round molding nailed along the bottom of the baseboard (nailed only into the baseboard).

CARPETING

Materials

Over 90% of the carpet installed in the United States is tufted, meaning that loops of yarn are stitched through a fabric backing, usually polypropylene, and glued in place with styrene-butadiene (SB) latex adhesive. This carpet is backed by a thick layer of SB latex or, in higher-end products, a secondary layer of fabric. The loops of yarn are either left in place for loop-style carpets, such as Berbers, or cut with blades for cut-pile carpet (Figure 5-17).

FIGURE 5-17   **Tufted Carpet Construction.**

Cut Pile

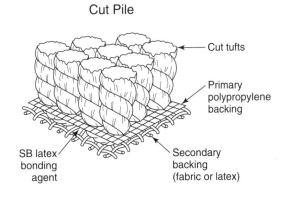

- Cut tufts
- Primary polypropylene backing
- SB latex bonding agent
- Secondary backing (fabric or latex)

Looped Pile

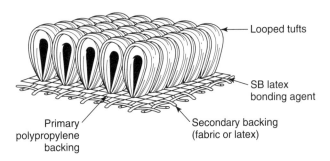

- Looped tufts
- SB latex bonding agent
- Primary polypropylene backing
- Secondary backing (fabric or latex)

Tufted carpet, which accounts for over 90% of the carpeting made in the United States today, consists of loops of yarn stitched though a backing fabric and glued in place with latex adhesives. The loops of yarn are left in place for loop-style carpets such as Berbers, or cut with blades for cut-pile carpets.

Traditional woven carpeting, representing only about 2% of U.S. production, is costly but creates a dimensionally stable and durable carpet including velvet, Axminster, and Wilton. With modern manufacturing techniques, however, nearly any style can be created using tufted construction. Common styles and their wear characteristics are shown in Table 5-8.

Nylon, considered the most durable synthetic carpet, accounts for about 60% of all pile carpeting. Most of the remaining are made of olefin and polyester, with wool accounting for less than 2% due to its high cost. Nylon is popular because of its good resilience (springs back rather than crushing) and overall durability (Table 5-9). Additives can give nylon good stain resistance.

Because olefin (polypropylene) is prone to crushing, it is generally used for low-pile designs, such as Berbers. Olefin is also widely used for indoor/outdoor carpeting used in high-moisture and recreational environments because of its resistance to moisture, mildew, and stains. Polyester carpeting is very soft to the touch but not as durable as the other synthetics.

Quality

Other than the material, the durability of a carpet depends on several factors: density of the tufts, twist of the yarn, and heat setting.

Density.   Density refers to how much yarn is used in the pile. The more tufts of yarn per square inch, the more yarn there is to wear and provide a resilient surface that resists crushing. The denser a carpet, the harder it is to push through the carpet to the backing with your fingers. Also, when bent back in a U-shape with the pile facing outward, a denser carpet will show less of the backing.

Density is measured in *stitches per inch* or *face weight*, which is the weight of the fiber in the pile per square yard of carpet. When divided by the pile height, this gives the *average density* per inch of pile. These numbers are useful for comparing similar products that use the same materials, but otherwise can be misleading.

Twisting.   Twisting the yarn enhances the durability, particularly in cut-pile carpets. In most nylon, olefin, and polyester cut piles, the twist is set by heat or steam to help the carpet retain the twist. The cut ends of the carpet pile should be neat and tight.

Pile Height.   Higher piles create a softer feel and more luxurious appearance but tend to crush more easily and are more difficult to clean.

Color and Pattern.   Most carpeting today is very colorfast. Solution-dyed carpet, in which the dye is added to the fibers when they are made, is extremely colorfast. Yarn-dyed carpet, which is dyed after the yarn is made, provides some color variation and is also very colorfast.

In general, light-colored carpets show dirt and stains, while dark colors show lint. Mottled colors such as tweeds

TABLE 5-8	Tufted Carpet Types and Styles

Carpet Types	Styles	Pros	Cons	Uses	
Loop Pile (tufts form loops)	Level loop	Wears well since tips of fabric not exposed.	May be difficult to conceal seams	High-traffic areas. Berbers and other casual styles.	
	Multi-level loop	Wears well. Texture helps hide wear, dirt.	May be difficult to conceal seams	High-traffic areas. Berbers and other casual styles.	
Cut Pile (tufts are cut)	Velvet plush	Feels soft. Smooth level surface.	Less durable than loop piles. Shows dirt, footprints.	Low-traffic areas. Formal settings.	
	Saxony	Smooth, dense surface. Twisted fiber adds resilience, hides footprints.	Less durable than loop piles.	Moderate traffic areas. Less formal.	
	Frieze	Textured with highly twisted tufts adds resilience, hides footprints.	Curly, textured surface less smooth, plush.	High-traffic areas. Informal settings.	
Cut-Loop (combines cut and uncut tufts)	Cut-loop	Islands of high cut loops and low uncut tufts create sculpted pattern that helps hide wear and dirt.	Uneven surface, less plush than cut piles.	Family rooms, stairways, other high-traffic areas. Informal settings.	

TABLE 5-9	Carpet Fibers

Fiber Type (tradenames)	Market Share	Pros	Cons	Uses
Nylon (Anso, Ultron, Zeftron)	56%	Very wear-resistant and resilient (springs back). Colorfast. Can be made stain-resistant.	Prone to static electricity buildup unless treated.	Good for all residential traffic areas.
Olefin or polypropylene (Genesis, Lana, Marquesa)	36%	Very wear-resistant. Excellent resistance to moisture, mildew, and stains. Nonstatic.	Relatively low resilience (prone to crushing)	Low, loop-pile commercial carpeting and Berbers. Indoor and outdoor, kitchens, baths, basements, playrooms.
Polyester (Pentron, Trevira)	7%	Soft to touch. Excellent color clarity and colorfastness. Resists water-soluble stains.	Less resilient and durable than nylon. Less durable than olefin.	Thick, luxurious cut-pile textures. Bedrooms and other low-traffic areas.
Wool	0.5%	Soft, attractive, very resilient.	Expensive. Only moderate resistance to stains and wear.	Moderate traffic areas.

and textured patterns tend to disguise dirt and wear, and are good choices for high-traffic areas and rooms where spills or stains are likely.

Ratings and Warranties.

Many manufacturers rate the durability of their carpeting on a numeric scale or with descriptions such as low, medium, and high durability. These are a useful gauge of performance, but the proof is in the warranty. Look for a 7- to 10-year wear-and-stain warranty. Find out if the warranty is prorated or covers the full replacement cost. Also, read the fine print, as certain kinds of stains, such as pet stains, are often excluded.

Carpet Pad

By absorbing much of the impact of foot traffic, carpet padding helps prevent the carpet fibers from getting crushed and wearing out prematurely. The cushioning effect also makes the carpet more comfortable underfoot. Good padding is sufficiently firm and resilient to absorb foot traffic, and durable enough that it will not break down or collapse over time. Good padding also increases insulation and soundproofing and makes carpeting easier to vacuum by allowing air to circulate through the carpet.

For residential applications, pads should generally be no more than $\frac{7}{16}$ inch thick for high piles and no more than $\frac{3}{8}$ inch thick for Berbers or low piles. In general, softer, thicker pads are used in bedrooms, dens, and other rooms with light traffic. Thinner, firmer pads are recommended for living rooms, family rooms, hallways, stairs, and other high traffic areas. Berber-style carpets also require thinner, firmer cushions for support.

If too thick, the pad can cause too much flexing in the carpet, weakening the backing and opening seams. A pad that collapses, or starts out too thin, can cause carpeting to wrinkle or wear out quickly. Seams in the pad should run perpendicular to the carpet seams or be offset by at least 6 inches.

Foam Padding. Prime urethane pads are the least expensive, but have a tendency to compress with use, particularly in high-traffic areas. As the pad compresses, the carpet backing can break down from too much flexing. For that reason, prime urethane pads are not recommended for carpeting subject to moderate or heavy traffic. One exception is a proprietary urethane called Omalon (E. R. Carpenter Co.), which has a special cell structure that resists crushing and is guaranteed for the life of the carpet.

Rebond. Bonded or rebonded pads, made of multicolored scraps of high-density polyurethane foam bonded together, are the most common in residential construction. The denser the foam, the better the feel underfoot and the durability. The Carpet and Rug Institute (CRI) recommends that rebond be a minimum of 5 pounds per cubic foot and $\frac{3}{8}$ inch thick for light-traffic areas, such as a bedroom, and 6.5 pounds and $\frac{3}{8}$ inch thick for heavy-traffic areas, such as hallways. For longer wear in high-traffic areas, use a 7- to 8-pound rebound. For a more plush feeling, choose a $\frac{7}{16}$ inch thickness.

Fiber. Natural and synthetic fiber pads are sometimes used under area rugs, commercial carpets, and some Berber carpets. They are made of jute or recycled synthetic carpet fiber and are among the densest and most resilient pads. Synthetic fiber pads are the best choice for potentially damp concrete floors. With synthetic fiber pads, look for a minimum density of 7.5 pounds per cubic foot or 12 pounds for jute. The thickness should range from $\frac{3}{8}$ to $\frac{7}{16}$ inch.

Special Pads. Some Berber carpets require special padding. In general, the bigger the loop in the Berber, the firmer the padding should be. Woven carpet may also require special padding, typically an extra-dense fiber pad or, in some cases, a heavy frothed foam.

Installation

Stretch-in installations using tack strips along the room perimeter are the most common approach in residential carpeting. Glue-down installations are primarily used in commercial work but are used residentially over slabs-on-grade and in basements. Glue-down installations can either use carpeting with an attached cushion backing or the "double-glue" method in which the pad is glued to both the concrete and the carpet. For installations over concrete, the concrete should be fully cured and surface free of dirt, dust, and any curing agents.

Subfloor. A good carpet installation starts with a properly prepared subfloor. The minimum recommended subfloor is $\frac{3}{4}$ inch T&G plywood, nailed and glued. For a higher quality job, an $\frac{1}{4}$- to $\frac{3}{8}$-inch underlayment should be installed over the plywood with the seams offset from the subfloor. Follow the underlayment specifications for resilient flooring, discussed above. Check for loose or squeaky spots and nail with spiral or ring-shank nails before installing the carpet.

For a level transition, the top of the underlayment should sit about $\frac{1}{2}$ inch below the finished height of adjacent solid flooring materials, such as wood, tile, or resilient flooring.

Carpet and pad can also go over hardwood floors or tightly glued resilient flooring. Repair any loose areas or damage in the existing flooring before installing the pad and carpet.

Seams. Most residential carpeting in the United States is available in either 12- or 15-foot-wide rolls, but the installer needs a few inches of waste on each end for stretch-in installations, limiting the size of a room that can be done with no seams.

Since all seams are visible to some extent, they should be placed where they are the least visible and get limited traffic, such as inside of closets. Seams should always run with the pile in the same direction. Where a room is lighted from windows, the seams should go perpendicular to the windows. In hallways, place any seams along the length of the hall. If a seam must be between rooms, make sure it is hidden when the door is closed. As the fibers are compressed from wear over time, seams become more conspicuous.

Seams are easiest to conceal in deep, dense, cut-pile carpeting. With short loop-pile carpets, such as Berbers and other loop-pile carpets with heavy textures and irregular rows of tufts, it can be difficult to hide seams. Also carpets with pads hide seams better than glue-down installations. Where seaming problems are anticipated, use wider 6-inch hot-melt tape at seams rather than the standard 3-inch tape. The wider tape helps avoid a high spot at the seam.

Stretching. To avoid problems with wrinkling, carpeting should be warmed up to the normal room temperature for about 24 hours before it is installed. This can take place in the home or in a heated warehouse. The building should also be heated to normal temperatures before and during the installation and be free of excess moisture. If the carpet is installed cold, it can expand and wrinkle when heated to normal conditions.

Wrinkling and ridging at seams can also result from carpeting that is not adequately stretched during installation. While manual stretching was adequate for older carpeting with natural jute backing, the polypropylene backing used today requires the greater force of power stretching. In fact, many manufacturers will not warrant their carpet on rooms larger than 12x12 feet unless it is power stretched.

The stretched carpet is held in place with tack strips nailed around the perimeter of the room about $\frac{1}{2}$ inch in from the baseboard. Standard 1-inch-wide tack strips are adequate for most carpeting, but some heavy woven and Berber-style carpets require 2-inch strips (or two 1-inch strips) to hold them securely in place.

Health Effects of Carpeting

In recent years, a number of homeowners and advocacy groups have attributed a variety of health problems to exposure to new carpeting. Although studies have been inconclusive, the carpeting industry has taken steps to reduce exposures of certain chemicals and has established a certification program for low-emitting carpets. For more information, see the section on "Carpeting," page 292).

INTERIOR TRIM

Once the domain of premium softwoods, such as clear pine, poplar, and other easily machined woods, interior trim is just as likely now to contain a mix of finger-jointed stock, medium density fiberboard (MDF) molded urethane for decorative trim, and flexible polyester moldings that must bend around curved surfaces.

Wood moldings and other finish lumber are graded for visual properties only. In general, the higher the grade, the more uniform the grain and color will be, and the fewer the defects, such as small knots, pitch pockets, and other natural markings. In some species, there is also a marked color difference between heartwood and sapwood. Some customers might like the natural variation found in lower grades; others find it objectionable.

Wide Range of Stock Profiles. Most lumberyards stock only a few molding profiles in pine and even fewer in hardwoods. Specialty molding suppliers, however, offer a far wider variety of stock profiles in both softwoods and common hardwoods. Molding suppliers also stock a variety of architectural ornaments, such as rosettes and plinth blocks, that can dress up a job or match a traditional style without the cost of custom millwork.

Most wide, flat moldings are recessed or "backed out" a little to reduce the tendency to cup. Cutting kerfs in the back of flat board stock will accomplish the same effect (see Figure 5-18).

Hardwoods

While some lumberyards stock small quantities of milled hardwood boards and a few molding profiles, most larger jobs require the purchase of rough stock from a hardwood supplier or millwork shop. Hardwood trim characteristics are shown in Table 5-10.

| FIGURE 5-18 | Common Molding Profiles. |

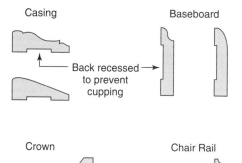

Specialty molding suppliers offer a much wider variety of stock profiles than local home centers. Most flat profiles are recessed in the back to prevent cupping.

Grading. If a job requires all clear stock that is "color-matched" with minimal color variation from board to board, you will probably need to purchase the highest grade available, often FAS (firsts and seconds), and may still need to cull some pieces. For jobs where more grain variation is acceptable, No. 1 Common or No. 2 and 3 Common may suffice. FAS is at least 80% clear stock with minimum boards 6 inches wide by 8 to 16 feet long. No. 1 is at least 65% clear with narrower boards, and No. 2 and No. 3 are 50% and 33% clear, respectively.

Ordering. Providing the shop with a specific cut list of finished pieces is the best way to guarantee that they deliver the pieces needed for the job. For a premium, you can obtain all-heartwood, all-sapwood, or color-matched boards for uniform color in glue-up work and throughout the job. Also, the millwork shop can plane the stock on one or both sides, joint one or both edges, and sand one or both faces as needed. Generally, the millwork shop can dress the boards far more economically than a contractor can in the field or in a small shop.

Custom Profiles. A job with hardwood trim may also require profiled moldings, such as baseboard, chair rail, or crown. Custom hardwood moldings require a substantial lead time and a setup fee to make the cutter knives. Many shops keep cutters on hand for standard profiles, as well as custom profiles from prior jobs. Using an existing cutter can significantly cut costs and lead time.

Finger-Jointed Moldings

Finger-jointed stock is widely used for paint-grade door and window jambs, as well as profiled moldings. Finger-jointed stock generally performs well, but in some cases, joints between the individual pieces will "telegraph" through the painted finish due to minute differences in the swelling and shrinking of the individual pieces of wood.

TABLE 5-10	Common Domestic Hardwood Characteristics			
Species	Appearance	Working Properties	Dimensional Stability	Recommendations
Ash	Sapwood nearly white; heartwood varies from grayish brown to light brown. Straight grain and coarse texture similar to oak.	Machines well. Good for nailing, screwing, and gluing. Stains and finishes well.	Little movement	All-sapwood sometimes selected for light appearance and may be marketed as "white ash." Heartwood sometimes called "brown ash."
Birch, American Yellow	White sapwood, light reddish brown heartwood. Generally straight-grained with a fine, uniform texture and wavy pattern.	Machines fairly easily. Nails and screws adequately with pre-boring, and glues well. Stains and finishes very well.	Moderately high shrinkage when drying. Susceptible to movement from moisture changes.	All-heartwood sometimes selected for woodwork.
Cherry, American Black	Heartwood rich red to reddish brown; darkens significantly with age. Sapwood creamy white. Uniform straight grain, smooth texture. Dark streaks common from pith flecks and gum pockets.	Machines easily, but prone to surface burns from cutters. Nails and glues well. Sands, stains, and finishes very well.	Moderately high shrinkage but is dimensionally stable after kiln-drying.	All-heartwood may be preferred due to sharp contrast between sapwood and heartwood.
Hard Maple	Sapwood creamy white; heartwood varies from light to dark, reddish brown. Fine, uniform texture, generally straight-grained, but it can also have curly figure.	Because of hardness, tends to machine slowly and dull blades. Preboring needed when nailing and screwing. Glues satisfactorily. Stains and finishes very well.	High shrinkage when drying. Susceptible to movement with moisture changes.	Light-colored sapwood often preferred for woodwork. Both sapwood and heartwood can contain dark staining from pith.
Oak, Red	Sapwood white to light brown; heartwood reddish brown. Similar to white oak, but with less pronounced figure due to the smaller rays. Mostly straight-grained with coarse texture.	Machines well. Nailing and screwing are good with preboring. Stains and finishes well.	High shrinkage when drying, with tendency to split and warp. Prone to movement from moisture changes.	Most common hardwood. Some pieces have dark streaks due to mineral staining.
Oak, White	Sapwood light-colored; heartwood light to dark brown. Mostly straight-grained with a medium to coarse texture. Longer rays and more figure than red oak.	Machines well. Nails and screws well with preboring. Gluing properties variable. Stains and finishes well.	High shrinkage with tendency to check. Prone to movement from moisture changes.	Galvanized nails recommended as the wood reacts with iron.

To avoid this problem, sand any uneven joints before applying any finish. Also, back-priming the material will reduce any moisture movement after installation, minimizing problems with telegraphing.

MDF

Medium-density fiberboard (MDF) is a fine-grained composite material made from wood particles and resin bonded under heat and pressure. The resin is generally urea-formaldehyde, a known lung irritant, but a few manufacturers offer alternative products made with the more stable phenol formaldehyde or other low-emission resins. SierraPine Composite Solutions makes Medex, a moisture-resistant MDF product, and Medite II, an interior panel, both using a formaldehyde-free resin called MDI (methylene diisocyanate).

In many markets, MDF has become the material of choice for trim and casework due to its low cost, ease of machining, and excellent appearance when painted. It is uniform in consistency and dimensionally stable. MDF trim is available preprimed in a number of standard molding profiles, and 4x8 MDF panels are easy to cut to size and can be routed or shaped to a clean, crisp profile. However, a $\frac{3}{4}$-inch 4x8 panel weighs 95 pounds versus 75 pounds for birch plywood, making MDF sheets a challenge to maneuver.

While MDF offers many benefits, it is not problem-free. Cutting and milling creates a super fine dust, which requires workers to wear tight-fitting respirators. Shops should have a good dust-extraction system as well. The urea-formaldehyde makes the dust more irritating to eyes and lungs and off-gasses to some extent after installation, making the product unacceptable to some (see Chapter 7, "Formaldehyde," page 287).

Because of hardness, MDF moldings must be installed with pneumatic nailers, which tend to pucker the material around the nail. These "mushrooms" must be chiseled off prior to filling the nail holes. And although it holds paint well,

FIGURE 5-19 | **Flexible Moldings.**

Newer formulations of flexible polyester moldings (inset) are easy to nail, resistant to cracking, and come in a wide variety of profiles, in both paint and stain grades. Most pieces such as crowns and arch-top casings need to be preformed by the manufacturer for the specific radius needed.
SOURCE: Photos courtesy of Flex Trim.

cut and routed edges of MDF will absorb water-based primer and swell. To avoid these problems, edges should be sealed with a shellac-based or oil-based primer or painted with special finishes formulated for use with MDF. Due to its potential for absorption at edges, MDF is not a good choice for wet areas. Edge nailing is also not recommended, so MDF is not well suited to applications such as jamb extensions.

Urethane Moldings

Although pricey, polyurethane foam moldings (also called polymer moldings) are popular for ornate decorative work. The leading manufacturer, Fypon, makes a wide range of large crown and cornice moldings, as well as architectural ornaments for mantles, decorative ceilings, and other decorative elements.

Urethane foam moldings are sold preprimed, and they can be cut, planed, and sanded like wood—only more easily because of their lighter weight. The moldings are installed with proprietary caulk or adhesive rather than nails, although a few finish nails are often used to hold them in place while the glue dries. Butt joints and miters are bonded with the same adhesive. Larger moldings are limited in length to 10 to 12 feet, requiring multiple joints on long runs.

Flexible Moldings

Flexible moldings made from dense polyester resin have been available since the late 1960s, but they have improved

a lot in recent years. Newer formulations are easier to nail, more resistant to cracking, and come in a wide variety of profiles, in both paint and stain grades (Figure 5-19).

Most manufacturers offer thinner profiles and softer formulations for tighter curves, as well as fire-retardant formulations. Less expensive rigid versions are also available for straight runs. While originally developed for interior use, many of these products are suitable for exterior applications as well.

The stain-grade material has an embossed grain, but must be stained after installation due to the stretching of the surface and requires a heavy pigmented oil-based or gel-type stain with a clear topcoat.

Most flexible moldings are made to order and can perfectly match typical finger-jointed or MDF profiles if specified correctly when ordered—manufacturers have thousands of molds matched to various manufacturers' stock moldings. Simple curves such as baseboard or chair rails generally do not need preforming, but crowns, arch-top casings, and most small-radius curves must be preformed by the manufacturer for the specific radius needed. Manufacturers can accommodate ovals, ellipses, and other irregular curves if provided with accurate design specs.

The material cuts easily with standard woodworking tools, but it needs to be held in a jig or sandwiched between wood blocks for difficult cuts. Most manufacturers recommend installation with construction adhesive, panel adhesive, or gel-type super glue, with a few finish nails to hold the molding in place while the glue dries. Pneumatic pin nailers work well. However, nailing too close to the edge may distort or crack the rubber material. Large moldings such as crown need wood backing or triangular blocks to prevent the molding from bowing in. (See "Resources," on page 208 for a list of suppliers.)

Casework

For shelving, built-ins, and other casework, contractors can choose from a wide array of panel products. The most widely used are veneer-core plywood, MDF, and particleboard. MDF and particleboard are available either unfinished or with a wood veneer or melamine facing. Medium-density overlay (MDO) is a good option for cabinets exposed to very high humidity or exterior uses.

Veneer-Core Plywood.
Cabinet-grade plywood typically has five inner plies (more for better grades), plus the face veneers, and in most cases uses phenol-formaldehyde adhesive, which has negligible off-gassing. Plywood is strong and dimensionally stable. For paint-grade cabinets, birch plywood remains an excellent choice. Baltic birch plywood uses all birch for the inner plies, is free of voids, and can be edge sanded, making it ideal for drawer sides and similar applications. For stain-grade work, hardwood plywood can be special ordered with matched veneers. Where screwing into edges is required, 7-ply material is less likely to split.

MDF.
Medium-density fiberboard is a fine composite material made from fine wood fibers and resin, usually urea-formaldehyde (see description on page 184). MDF is available as both a paint-grade panel or faced with wood veneer or melamine. Because of its competitive pricing and good workability, MDF is now the dominant panel product in many markets. In paint-grade work, the edges need to be sealed or banded due to high absorption of paint. Concerns about off-gassing of formaldehyde could be a concern to customers with allergies or chemical sensitivities. However, if laminated on all faces with an impervious facing, such as melamine, or finished with two or more coats of varnish or an oil-based paint (or paint rated as a vapor barrier) on all faces, the off-gassing will be minimized.

Particleboard.
Particleboard is similar to MDF, but with larger fibers, so it doesn't machine to a crisp edge and leaves a noticeable texture when painted. Also, edges and corners are more prone to chipping than with MDF. Like MDF, it off-gasses urea-formaldehyde. Sealing all surfaces will minimize the problem.

MDO.
Medium-density overlay is an exterior grade plywood with a durable resin-treated paper facing that takes paint exceptionally well. It is widely used for sign making as well as concrete forms. Though not typically used in casework, it is an ideal material for cabinets that will be exposed to extreme moisture or exposed to weather on porches, patios, or other outdoor locations.

Shelving.
Typical shelf spans for simple shelves sitting on cleats at both ends are shown in Table 5-11. These

TABLE 5-11	**Maximum Bookshelf Spans**
Material	Span (inches)
$\frac{3}{4}$ inch particleboard	21
$\frac{3}{4}$ inch MDF	20
$\frac{3}{4}$ inch plywood	30
$\frac{3}{4}$ inch softwood	30
$\frac{3}{4}$ inch hardwood	34

NOTE: Assumptions: 10-in.-wide shelves supported only at ends. Load = 35 lb. per foot. Deflection = L/360.

assume a load of heavy books and minimal deflection, although long-term deflection under a constant load may be greater. To stiffen shelving, it can be supported along the back edge or reinforced in front with solid-wood facing, glued and nailed in place. For example, a $1\frac{1}{4}$ inch solid-wood apron along the front edge will increase the span for plywood shelving to about 36 inches.

INTERIOR DOORS

Over 90% of interior doors today are either flush or molded. In either case, a facing of wood veneer or hardboard is glued to a core, providing the door with its strength. Traditional rail-and-style construction is still used, primarily for stain-grade work, although composites and veneered construction are widely used with this type of door as well.

Frame-and-Panel Construction

These are the most expensive doors and are used mostly for stain-grade work. They gain their stability by allowing the flat or raised panels to float in the frame without increasing the door's overall width (see Figure 5-20). Rails and stiles are typically dowelled and glued to make a rigid connection at corners.

For both cost saving and increased dimensional stability, many doors now use composite materials. Rails and stiles may have a wood veneer over a core of finger-jointed wood, particleboard, or MDF. Other doors build each rail and stile from a solid strip of appearance-grade solid wood sliced in half, reversed upon itself, and reglued so the opposing grain patterns help resist warping.

On paint-grade doors, the raised or flat panels are often MDF, which does not move with humidity changes or leave an unpainted strip when the panels shrink. If painting doors with solid wood panels, order them preprimed to help reduce problems with the paint line.

On stain-grade doors, the panels are either solid wood or veneered MDF, which offers greater dimensional stability and the appearance of solid wood to all but the most discerning eye.

Flush and Molded Doors

The standard choice for modern homes in the 1950s and 1960s, flush doors have a $\frac{1}{8}$-inch wood or composite veneer

FIGURE 5-20 Six-Panel Door.

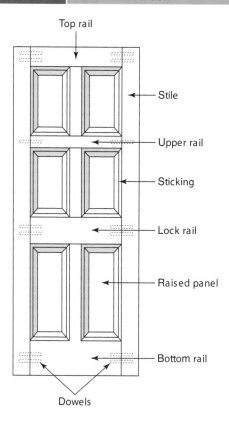

Top rail

Stile

Upper rail

Sticking

Lock rail

Raised panel

Bottom rail

Dowels

Traditional frame-and-panel doors allow the wood panels to swell and shrink in the frame without changing the door's overall width or height. Rails and styles are typically dowelled and glued to make a rigid connection at corners.

glued to either a solid or hollow-core frame. Molded doors are constructed the same way, but with a hardboard facing molded to simulate a frame-and-panel wood door.

All flush and molded doors have solid rails and stiles and a solid area (the lock block) where the lockset is installed. The rails and stiles are either solid wood, finger-jointed stock, or MDF in lower-end doors. Wood stiles may be combined with MDF or particleboard rails to save money. MDF stiles may not perform well in bathrooms or wet areas due to their tendency to absorb moisture.

Hollow Core. A corrugated cardboard grid fills in between the rails and stiles and keeps the facings rigid on a hollow-core door. The lock block where the lock set is drilled may be solid wood, particleboard, or MDF. The rails and stiles are often wider than on solid-core doors to provide structural stability. Despite their light weight, hollow-core doors are dimensionally stable and problem-free as long as the installer does not remove too much material during installation.

Solid Core. In residential doors, the rails and stiles are typically not fastened to one another or to the core material, but are held together by the wood or hardboard facing. The core is typically particleboard, MDF, or low-density fiberboard, which reduces the weight by about 25%. A standard 2'6"x6'8" hollow-core flush door weighs about 30 pounds versus 75 to 80 pounds for a solid-core version. The price difference is

modest, but most homeowners prefer the solid feel and better sound blocking of a solid-core door. However, the extra weight can put a strain on MDF jambs, which are now finding their way onto job sites. Driving one long hinge-screw into the framing at each hinge will help avoid problems.

Molded Doors. A molded door is built like a flush door, except the hardboard facing is molded to simulate the appearance of a traditional frame-and-panel door. Most are available with an embossed wood grain. As with a flush door, the core may be either hollow or solid. How well the molded surface simulates a wood panel door varies from one manufacturer to another. Look for a product with crisp, well-defined details at the panels and molded edges around them, called *sticking*. The solid-core version also feels like a solid wood door when operated and provides better sound blocking than a hollow model.

Cutting and Trimming Doors

Trimming too deep into a door's stiles or outer rails can destroy its structural integrity. How much material can be safely trimmed depends on the specific door, so pay attention to the manufacturers' recommendations. As a general rule, do not cut more than $\frac{3}{4}$ inch off the top or bottom rails of traditional frame and panel doors, although some doors can be trimmed by 2 or more inches on the bottom rail.

With flush or molded doors, how much can be trimmed depends upon the width of the rails and stiles and, with solid-core doors, whether they are glued to the core material. With doors that comply with WDMA (Window and Door Manufacturers Association) specs, follow the minimum widths in Table 5-12.

Door Standards and Warranties

With laminated doors, look for products in compliance with the WDMA Standard I.S.1-87. Under this standard, door samples must withstand multiple wetting and drying cycles without significant delamination. Products in compliance typically carry a one- to five-year warranty against delamination. Most warranties also cover any warping and twisting in excess of $\frac{1}{4}$ inch across the length or width of the door but require that the door be sealed on all six edges. Oversized doors may have more limited protection against warping.

TABLE 5-12 Minimum Door Rail and Stile Widths After Trimming*				
Door Type	Hinge Stile (inches)	Lock Stile (inches)	Top Rail (inches)	Bottom Rail (inches)
Solid Core (standard)	1	$\frac{13}{16}$	$\frac{13}{16}$	$\frac{13}{16}$
Solid Core (stiles and rails glued to core)	$\frac{3}{8}$	$\frac{3}{8}$	$\frac{3}{8}$	$\frac{3}{8}$
Hollow Core	1	$\frac{13}{16}$	$1\frac{1}{4}$	$1\frac{1}{4}$

*For flush doors certified by WDMA.

STAINS AND CLEAR FINISHES

Finishing stain-grade trim is equal parts art and science. There are a wide range of products and application techniques. With all finishes, careful prep work and control of dust on the job site are critical for a professional quality finish.

Sanding and Prep Work

Starting with coarse grits and working to finer grits, sand all cabinets, doors, and other woodwork to remove any milling marks or chatter, scratches, dirt, or other imperfections. Highly visible surfaces like cabinets and doors should be taken down to a 180 or 220 grit. Use a dusting brush to clean off any visible dust between sandings, and thoroughly clean up after the final sanding. With solvent-based finishes, use a tack cloth to remove any residual fine dust.

Open-Grain Woods. With open-grain woods, such as oak, ash, mahogany, and walnut, it may take many coats of clear finish to fill the wood pores and achieve a glassy, smooth surface. Where a premium finish is desired, one approach is to apply a paste filler to the sanded wood, which is a thick, pastelike varnish with finely ground quartz or talc to add bulk, and usually a pigment as well to match the wood tone. It is typically applied with a rag and sanded clean the following day. If using a filler that is darker than the wood, first seal the wood with a sanding sealer or thinned coat of the clear finish to keep the wood from being overly darkened. Generally, stains are applied after the filler has been applied and sanded.

Water-Based Finishes. Water-based stains and finishes tend to raise the wood grain when applied, creating a rough surface. The best way to avoid problems later is to intentionally raise the grain and sand it down before applying the finish. To accomplish this, after sanding the work, wet the wood surface with a sponge or cloth, and allow to dry overnight. Then knock down the raised grain with 180 to 220 grit sandpaper.

With some of the newer water-based formulations, this step may not be required. Instead, a light sanding after the first coat may be all that is needed. Whatever approach is taken to sanding, never use steel wool with water-based finishes, as leftover steel particles can rust and stain the work. Also, do not use a solvent-type tack cloth with water-based finishes, as the solvent residue can interfere with the finish. A clean cloth lightly misted with water can be used to remove any dust or sanding residue.

Stains and Dyes

Stains for interior trim are either pigmented stains or penetrating dyes. Many ready-made stains at the lumberyard combine both pigments and penetrating dyes. The penetrating dyes work for the small-pore areas and the pigments add contrast to the larger pores.

Stains. Oil-based pigmented stains tend to highlight distinctive grain patterns, particularly in wood with large pores, such as oak and ash, but they also highlight any scratches or defects in the wood. Wood with uneven absorption will look blotchy. Also, because the pigments are large, opaque particles, they tend to act like watered-down paints, obscuring the wood itself.

Dyes. Dyes, which must be mixed by the applicator, are very transparent and tend to get absorbed equally into the wood surface, resulting in a more uniform color. They tend to give the wood an even, transparent color while letting the grain pattern show through. Over time, they will fade from exposure to natural light. Dyes are either dissolved in a water or oil solution and must be precisely mixed to obtain controlled colors.

Use of Sealers. Softwoods, like pine, and light-colored hardwoods, such as maple or birch, tend to absorb stain unevenly, so they benefit from sealing prior to staining. Depending on the desired appearance, you can use a shellac-based sealer with a pigmented stain, obscuring the underlying wood, or a prestain sealer with a penetrating stain. Prestain sealers allow stain to penetrate the wood surface but with more even absorption. Prestain sealers can also be useful when staining birch veneer, which tends to absorb stain unevenly, creating a blotchy appearance.

Stains and dyes may be oil, alcohol, or water-based. They may be applied with a sprayer, brush, roller, or rag and are typically applied to the surface, allowed to sit, then wiped off. Whatever type of stain is used, it should be completely dry before application of the clear topcoat. If using a water-based topcoat, check for compatibility with oil-based stains. Using a stain and clear finish from the same manufacturer will help guard against compatibility problems.

Clear Finishes

The best clear finish depends on the look desired, hardness required, and whether it must resist water (Table 5-13). Some finishes are best sprayed on, but most may be brush-applied. Oil-based finishes are generally wiped on with a rag and create a low-luster, hand-rubbed appearance, but provide the least protection. With most surface finishes, it is best to lightly abrade the finish between coats with 220-grit paper or No. 00 steel wool to increase the bond between coats. After sanding, wipe with a tack cloth for oil- or solvent-based finishes and a water-dampened cloth for water-based finishes. Most professional painters apply three to four coats of clear finish.

SOUND CONTROL

Because of its stiffness, wood framing readily transmits low-frequency sounds and impact noises through wood-frame houses. This is particularly a problem in floors and walls separating two housing units, but it can also be an

TABLE 5-13	Clear Finishes for Woodwork				
Finish Type	Ingredients	Application	Pros	Cons	Best Uses
Wiping Oils	Tung and/or linseed oil. Danish type has varnish added for greater protection.	Brush or wipe on. Let sit, then wipe off excess. Recoat periodically as needed.	Easy application. Soft, rich, hand-rubbed look. Easy to touch up or add coats.	Minimal protection against scratches. Can water spot.	Furniture, cabinets.
Lacquer	Resin made from cotton fibers in a volatile solvent.	Generally spray-applied, but careful brushing OK for small jobs. Apply 3–4 coats. Brush carefully to avoid brush marks.	Fast drying—can apply multiple coats in one day. Moderately durable to heat and solvents.	Solvents flammable and toxic. Can be difficult to apply in humid conditions.	Furniture, cabinets, trim.
Alkyd Varnish	Polyester (alkyd) resin combined with alcohol and acid and cooked with vegetable oils.	Spray or brush. Can be applied over alkyd sanding sealer. Apply 3 or more coats.	Relatively easy to apply. Good resistance to abrasion, heat, solvents, and water vapor. Rich luster.	Dries slowly so prone to dust contamination. Yellows over time.	Doors, windows, interior trim.
Polyurethane	Alkyd varnish with added polyurethane resin.	Spray or brush 3 or more coats.	Improved scratch resistance and moisture resistance over alkyd varnish.	Dries slowly so prone to dust contamination. Yellows over time.	Doors, windows, interior trim, handrails, floors, countertops
Water-based Varnish	Acrylic and/or urethane resins suspended in water with surfactants, solvents, and other additives.	Spray or brush 4 or more coats. Sanding only required on first coat unless long delay between coats. Brush carefully to avoid brush marks.	Low odor. Fast drying. Does not darken wood. Similar to polyurethane in durability.	Needs more coats to achieve same thickness. May raise grain and require extra sanding. Difficult to apply in cold, hot, or humid weather. Learning curve for new users.	Doors, windows, interior trim, handrails, floors, countertops.

issue within a single-family home. For example, a person with a home office or music room might want to isolate it acoustically from the surrounding rooms so meetings or music proceed in private and so outside noises will not intrude. Bedrooms located under living spaces can also require special treatment to reduce impact noises from above.

Another kind of noise control is important where a house sits by a highway or under a flight path. The goal here is to keep outdoor noises from entering the house by reducing sound transmission through windows, doors, and exterior walls and ceilings. Special acoustical windows rated for low sound transmission are often required for substantial reductions in outside noise.

Principles of Sound Transmission

Sound can travel through both air (airborne sound) and solid materials (structure-borne sound). Structure-borne sound can be directly imparted to the building structure by a vibration, such as a humming compressor, or by direct impact, such as a boot stepping on a hardwood floor. As sound energy travels through a building, it changes from one type of transmission to the other and back, losing energy in each transition. Because of its rigidity, wood framing is a very good transmitter of low-frequency sound and hollow wall cavities and thin doors do little to reduce sound transmission.

TABLE 5-14	Typical Sound Levels (Decibels, dB)
Audible sound threshold	3 dB
Normal breathing	10 dB
Whisper	20 dB
Normal conversation	30–50 dB
Busy street	60–70 dB
Subway, person shouting	80 dB
Lawn mower, loud stereo	90 dB
Table saw, auto horn	100 dB
Elevated train, thunder	120 dB

NOTE: Continuous exposure to noise levels above 85 dB can impair hearing in most people.

Sound Levels. Sound levels are measured in decibels (dB), which are on a logarithmic scale. A sound increase of just 10 dB indicates an increase of ten times the intensity, although our subjective experience is that the sound is twice as loud. Decibel levels for common sounds are shown in Table 5-14. Continuous exposure to sounds above about 85 dB can cause hearing loss in most people.

Absorption vs. Isolation. Sounds in an acoustically "live" room with all hard surfaces will seem loud and harsh due to the sound reverberating off the hard surfaces. Adding sound-absorptive materials, such as carpeting and soft furniture, will make sound softer and more pleasant within the room, but will do little to reduce the transmission of sound to adjacent rooms. To reduce transmission requires sound isolation strategies, typically using high-mass materials, double-framed walls, or resilient connections between the drywall and framing.

Isolation Strategies. To keep airborne sound from passing through walls and floors, there are four main strategies:

• *Add Mass:* Increase the mass of the wall or floor by using heavy, dense materials.

• *Decoupling:* Break the path of vibration with a break in the framing or a resilient connection to the drywall.

• *Absorption:* Provide sound-absorptive material, such as fiberglass batts, in the cavity.

• *Sealing:* Block airborne sound from leaking through gaps and cracks.

A cavity with fiberglass is far more effective at blocking sound if the two wall surfaces (or ceiling and floor surfaces) are mechanically decoupled as in a double-stud or staggered-stud wall. Resilient channel works essentially the same way by breaking the vibration path from the stud or ceiling joist to the drywall.

The hardest sounds to block are low frequency, such as the thumping of a stereo bass. Using decoupled construction, such as double walls or resilient channels, is effective. Where that is impractical, adding mass can also be effective. Very massive, nonrigid materials such as lead or sand are ideal, but doubling or tripling the drywall is also helpful.

Flanking Paths. Sound takes the path of least resistance between rooms, through any air leaks or through rigid connections in the structure itself. These routes that by-pass efforts at sound insulation are called *flanking paths*. These can significantly reduce the effectiveness of sound-proofing efforts. Building walls with high STC ratings will do little good if sound can pass easily though electrical outlets or a thin, loosely fitting door. For example, an un-gasketed door or the equivalent of a one-inch-square hole in a wall can reduce an STC 50 wall to STC 30. Common flanking paths include:

• Air leakage around partitions; around doors; and through plumbing penetrations, back-to-back medicine cabinets, unsealed electrical outlets or recessed lights (Figure 5-21).

• Shared ductwork between two rooms.

• Hollow-core doors and single-pane glass, which are good sound transmitters.

• With resilient channels, a few drywall screws that penetrate into the ceiling joists, undermining the decoupling system

• With decoupled framing, a solid path through a band joist or drywall panel that provides a bypass for structure-borne sound.

FIGURE 5-21 **Common Flanking Paths.**

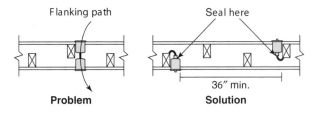

Electrical Outlets

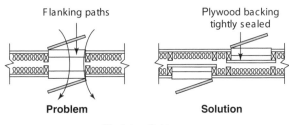

Medicine Cabinets

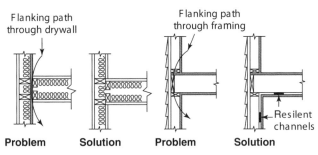

Partition Corner–Plan View   Floor-Ceiling–Section View

Soundproofing efforts are often undermined by sound leaks, called "flanking paths." Common airborne leaks include back-to-back electrical outlets and medicine cabinets (top and middle). Structure-borne sound can pass through drywall or framing that connects unbroken from room to room (bottom).

Minimizing flanking paths requires both good planning and workmanship. Common strategies include:

• Avoid back-to-back holes for electrical and mechanical equipment.

• Along partition bottoms, leave a $\frac{1}{4}$-inch gap between the drywall and subfloor, and fill with acoustical sealant.

• On bathroom partitions, install drywall all the way to the floor before installing the tub and seal all plumbing penetrations through walls with a flexible sealant.

• Upgrade to solid-core doors and add weather-stripping.

Addressing obvious flanking paths is often the most cost-effective step in soundproofing a home. Strategies such as sealing air leaks between rooms, upgrading doors, and adding weather-stripping may provide adequate sound isolation without the need for more exotic and expensive measures.

STC Ratings. How effectively a wall or floor reduces airborne sound is measured by STC ratings (sound transmission class). Roughly speaking, the STC rating equals the reduction in decibel levels across the partition. So, for example, a 50 dB noise on the other side of an STC 35 wall will sound like a 15 dB noise to the average listener (see Table 5-15). Walls and floors in the field often measure lower than in laboratory ratings due to variations in workmanship as well as leaks and bypasses. The higher the STC rating, the more likely it is to be compromised by site conditions. For that reason, it is best to select a building assembly rated at least 5 points above the design goal.

Walls

In single-stud walls, the most cost-effective upgrade is to double the drywall on one side and add insulation to the cavity, increasing the STC from 33 to 40 (see Table 5-16).

TABLE 5-15	STC Ratings and Noise Reduction	
STC Rating	Speech Heard Through Wall or Floor	Noise Control Level
25	Normal speech understandable	Poor
30	Loud speech understandable	Marginal
40	Loud speech audible as murmur but unintelligible	Good
50	Loud speech barely audible	Very Good
55 and up	Loud speech not heard	Excellent

NOTE: Assumes background noise of 30dB on the listening side.

TABLE 5-16	STC Ratings of Typical Wall Assemblies		
Wall type: 2x4 wood studs, 16″ on-center (except as noted)	STC Rating (with insulation)	STC Rating (no insulation)	
Single layer ¹/₂″ drywall each side	38	35	
Single layer ¹/₂″ drywall one side, double layer ¹/₂″ drywall other side	40	38	
Double layer ¹/₂″ Type X drywall each side	45	39	
Resilient channel one side, single layer ¹/₂″ Type X drywall each side	47	39	
Staggered studs with 2x6 top and bottom plates. Single layer ¹/₂″ drywall each side	49	39	
Resilient channel with single layer ¹/₂″ drywall one side, double layer ¹/₂″ drywall other side	52	44	
Staggered studs 24″ o.c. with 2x6 top and bottom plates. Double layer ¹/₂″ drywall one side. Single layer ¹/₂″ Type X drywall other side	53	47	
Resilient channel one side with double layer ¹/₂″ Type X drywall each side	55	52	
Staggered studs 24″ o.c. with 2x6 top and bottom plates. Double layer ¹/₂″ Type X drywall each side	55	52	

(continued)

TABLE 5-16	STC Ratings of Typical Wall Assemblies (*Continued*)

Double studs and plates, single layer 1/2" Type-X drywall each side.	57 (two layers insulation)	46	
Double studs and plates, double layer 1/2" Type-X drywall one side, single layer Type-X drywall other side.	60 (two layers insulation)	48	
Double studs, double layer 1/2" Type-X drywall each side.	63 (two layers insulation)	54	

Wall type: 3 5/8" metal studs (25 gauge) 24" on center	STC Rating (with insulation)	STC Rating (no insulation)	
Single layer 1/2" Type-X drywall each side.	47	39	
Double-layer 1/2" drywall one side, single layer 1/2" drywall other side.	52	42	
Double layer 1/2" Type-X drywall both sides.	56	50	

Note: Actual performance in field may be lower by 5 dB or more due to "flanking paths" and quality of workmanship.

The joints on the second layer of drywall should not line up with the first layer.

To achieve substantially higher STC ratings requires adding a resilient channel to one side of the wall or decoupling the two sides of a wall with double framing. With no rigid connection bridging the two sides of the wall, sound transmission is significantly reduced. Decoupling and also increasing mass, such as doubling the drywall layers, will help cut transmission of low-frequency sounds as well.

For higher STC values required for special situations, such as a music room or home office, additional upgrades include increasing the mass on either side of the cavity, enlarging the cavity, or adding fiberglass batts or other sound-absorbing materials. Filling the gap more than three-quarters of its width with insulation provides little additional benefit. In fact, stuffing the cavity too tightly could reduce the benefit of the fibrous insulation by creating a solid bridge. In general, polystyrene and other closed-cell insulations are poor sound absorbers and provide little benefit.

In general, doors should be within 10 STC points of the surrounding wall. Solid-core doors are recommended for bedrooms and bathrooms. Where higher-level sound isolation is required, you will need to add high-quality gasket-type weather-stripping and a sealed threshold. Also the gap between the door jamb and studs should be caulked or grouted to avoid sound leaks around the door.

TABLE 5-17	Sound Transmission of Interior Doors

Type of Door	STC Rating (unsealed)	STC Rating (well sealed)
Hollow-core door	17	20
Solid-core wood	20	28
Two hollow-core doors, 4 inch space	22	26
Two solid-core doors, 3 inches space	28	40

For even higher ratings, which might be needed for a music room, for example, double doors are required (see Table 5-17).

For party walls between adjacent living units, STC ratings should be a minimum of 50. Recommended STC levels between bedrooms and adjacent rooms in single-family homes and apartments are shown in Table 5-18. Where privacy and quiet are of concern to clients, a minimum STC rating of 45 is a reasonable target for bedroom and bathroom partitions. Closets along a wall can help buffer sounds as long as doors are not louvered.

TABLE 5-18	Recommended Sound Control for Bedroom Partitions in Single-Family Dwellings
Adjacent Area	Minimum STC Rating*
Bedrooms	44
Living Room	46
Bathroom	48
Kitchen	48

*Recommended levels for average noise environment.
SOURCE: Adapted from *A Guide to Airborne, Impact and Structureborne Noise Control in Multifamily Dwellings*, 1974, U.S. Dept. of Housing and Urban Development (HUD).

Floors

The STC rating of a floor measures only the reduction in airborne sound transmission. A floor, however, also transmits structure-borne sound, such as footsteps or a slammed door, directly through the materials. The ability to reduce impact sound is rated by the Impact Isolation Class (IIC) rating. The most cost-effective technique to reduce impact noise is to add a carpet and pad. For example, adding a carpet and pad to a conventional plywood subfloor over a gypsum ceiling increases the IIC rating from 37 to 65. By comparison, it increases the STC rating by only 4 points.

Where higher STC and IIC ratings are needed, a resilient channel can be added to the ceiling below. Where this is not possible, for example when the joists are exposed below, you can use a floating floor over a layer of soundboard or a high-mass floor over a layer of sand or lightweight concrete (see Table 5-19).

IIC levels are of greatest concern in stacked multifamily dwellings or in a single-family dwelling with bedrooms below other living spaces. Acoustical experts recommend a minimum IIC rating of 50 to 55 in ceiling/floor construction, separating living units in multifamily construction. HUD recommendations for bedrooms under living spaces are shown in Table 5-20. While these recommendations were developed for multifamily dwellings, they provide reasonable targets for single-family homes where sound privacy is desired.

Plumbing Noise

One of the most common noise complaints in single-family construction is the sound of water gushing through PVC waste pipes. The best solution, short of using cast iron, is to box in the pipes and fill the cavity with fiberglass insulation. Then enclose the cavity with one or two layers of drywall.

Water supply and heating pipes can also radiate noise through the framing if there is rigid contact between pipes and framing or finish materials. This can be a particular problem when heating pipes expand and contract. To avoid these problems, make sure pipe runs are not tight against framing.

Soundproofing Materials and Workmanship

Like weatherization work, effective soundproofing requires careful detailing and workmanship. Small holes and bypasses can lower field STC values to 15 to 20 points below laboratory values. Leaky edge joints, unsealed doorways, interconnecting ductwork, and unsealed electrical and plumbing penetrations all degrade acoustical performance.

Acoustical Sealant. While special nonhardening acoustical sealants are often specified in commercial work, any high-quality sealant that remains flexible can be effective in blocking sound transmission. Butyl, silicone, and urethane caulk can all be used. To prevent sound leaks, use sealant around electrical boxes, plumbing penetrations, and any other penetrations in the wall or ceiling surface. For walls with STC ratings in excess of 35, apply a flexible sealant at the joint where the drywall meets the floor. Acoustical sealant is also used to seal around the perimeter of walls or ceilings hung from resilient channel.

Resilient Channel. Resilient channel is installed perpendicular to the studs or joists and needs at least 3 inches of free space in the cavity behind it to be effective. It is not effective if attached to sheet materials, such as drywall. It is also important to use the right length screws, so they do not penetrate into the wood framing. Just a few screws into the wood can undermine the resilient connection and substantially lower the STC and IIC ratings. Leave a $\frac{1}{4}$- to $\frac{1}{2}$-inch gap around the perimeter of a ceiling or wall hung from resilient channel and fill with an acoustical or other nonhardening sealant.

Insulation. Ordinary fiberglass insulation is an effective sound absorber in cavities and increases the STC rating of walls by 3 to 5 decibels. The insulation needs to fill only about three-quarters of the thickness of the cavity to be effective. Adding more adds little additional sound protection, and stuffing insulation in too densely could actually increase sound transmission. Cellulose insulation has about the same sound deadening characteristics as fiberglass. Foam insulation is not particularly effective for sound control. Foam is too light to add mass to the wall and is not resilient enough to absorb sound.

Gasketing. Flexible, heavy rubber gasketing makes an effective seal against sound leaks as well as thermal leaks around doors and windows. Either bulb- or magnetic-type weather-stripping is effective as long as it makes an airtight seal between the frame and door or window.

TABLE 5-19 **STC and IIC Ratings for Typical Wall/Ceiling Assemblies**

Wood Floor

2x10 joists, 5/8" plywood subfloor, 3/8" particleboard underlayment. 1/2-inch Type-X drywall ceiling.

Variations	IIC	STC
Add vinyl floor covering	38	38
Add vinyl, R-11 fiberglass batts, and resilient channels	49	50
Add carpet and pad	50-60*	40
Add carpet and pad, R-11 batts, and resilient channels	63-73*	53

*Note: Low-pile carpet on fiber pad lowest. High-pile on thick foam pad highest.

Floating Floor

2x8 joists, 1/2" plywood subfloor, 1/2" sound-deadening board nailed to subfloor, 2x3 furring strips (unnailed), 5/8" plywood subfloor, Type-X drywall ceiling.

Variations	IIC	STC
Add vinyl flooring, R-11 fiberglass batts	49	52
Add carpet and pad, R-11 fiberglass batts	68-78*	51

*Note: Low-pile carpet on fiber pad lowest. High-pile on thick foam pad highest.

Lightweight Concrete Floor

2x10 joists, 5/8" plywood subfloor covered with 1-1/2" lightweight concrete. Type-X drywall ceiling.

Variations	IIC	STC
Add carpet and pad	50-59*	47
Add carpet and pad, R-11 fiberglass batts, resilient channel	64-74*	58
Add vinyl flooring, R-11 fiberglass batts, resilient channel	47	50

*Note: Low-pile carpet on fiber pad lowest. High-pile on thick foam pad highest.

Wood Floor

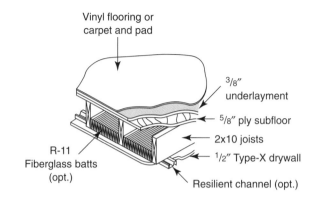

Vinyl flooring or carpet and pad

3/8" underlayment
5/8" ply subfloor
2x10 joists
1/2" Type-X drywall
Resilient channel (opt.)
R-11 Fiberglass batts (opt.)

Floating Floor

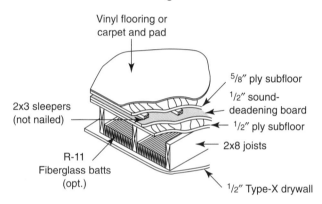

Vinyl flooring or carpet and pad

5/8" ply subfloor
1/2" sound-deadening board
1/2" ply subfloor
2x8 joists
1/2" Type-X drywall
2x3 sleepers (not nailed)
R-11 Fiberglass batts (opt.)

Lightweight Concrete Floor

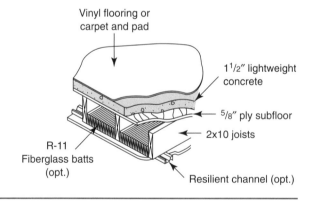

Vinyl flooring or carpet and pad

1 1/2" lightweight concrete
5/8" ply subfloor
2x10 joists
Resilient channel (opt.)
R-11 Fiberglass batts (opt.)

Duct Insulation. Use fiberglass ductboard or fiberglass duct liners to quiet the noises of fans and moving air. Avoid sharing a common duct between two rooms that need sound privacy.

LIGHTING

With the exception of purely decorative lighting fixtures, all lighting fits into one of three main categories: *ambient, task,* and *accent.* Most rooms use a mixture of lighting types to create visual interest and to meet the functional needs of the space. A space lit only by indirect light sources can seem visually flat, while a space lit only by directed light from spots and floods can seem harsh and cast dark shadows. Similarly, a space lit only by accent lighting can look like an art museum and leave people in the dark. A balanced combination of strategies works best.

Ambient Lighting is the general background illumination that is bright enough to allow people to move about safely and perform simple tasks. Ambient lighting can be

TABLE 5-20	Minimum Sound Insulation for Ceiling/Floor Assemblies Above Bedrooms	
Room Above Bedroom	STC	IIC
Bedroom	52	52
Living Room	54	57
Family Room	56	62
Kitchen	55	62
Hall	52	62

NOTE: Recommended levels for an "average" noise environment.
SOURCE: Adapted from *A Guide to Airborne, Impact and Structureborne Noise Control in Multifamily Dwellings*, 1974, U.S. Dept. of Housing and Urban Development.

FIGURE 5-22   Lighting a Desktop.

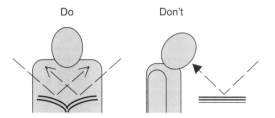

Whether coming from a desk lamp, an overhead fixture, or from under a wall cabinet, lighting for desk work should come from the side or from slightly behind a worker. Lighting in front will cause glare.
SOURCE: Courtesy of Lightolier, a division of Genlyte, Inc.

achieved by directly lighting the lower part of the room (direct lighting), or by reflecting light off the ceiling and upper half of the room (indirect lighting).

Reflecting light upward off the ceiling and upper walls tends to give a room a spacious feeling and soften shadows on objects and faces. It can be achieved with built-in coves, wall sconces, and pendants that direct light upward, or freestanding torchiere-style floor lamps.

Lighting just the lower part of the room can create a more intimate feeling. Options include recessed lights and wall and ceiling fixtures that direct light downward. Fixtures with diffusers will help prevent glare. Evenly illuminating a wall with downlights called "wall washers" is another way to provide soft ambient light and also makes a room feel larger. Many surface-mounted and hanging luminaires project light in more than one direction, providing both uplighting and direct lighting in a single fixture.

Task Lighting is bright light directed to a specific surface, like a countertop or desk, to illuminate activities such as reading, homework, meal preparation, or laundry.

For reading and desk work, task lighting should be bright and well diffused and come from the side or from over the shoulders. Overhead light often casts shadows from a person's head and body onto the work surface. Also, light from directly overhead or in front of a person is prone to cause glare (veiling reflection) on shiny work surfaces, such as a glossy magazine. Light coming from one or both sides of the work reduces glare. For example, a table or floor lamp on the side is effective for reading or deskwork. Undercabinet lighting can also be effective for desk work if placed toward the front of the cabinet on either side of the occupant (see Figure 5-22).

Where a computer screen is used, avoid bright sources of overhead light that reflect off the screen. Also keep the screen at right angles to windows if possible to avoid glare.

For kitchen work, laundry, or hobbies, concentrated light from above can be effective as long as the fixtures are placed so the occupants do not shade the work surface.

Undercabinet lighting is another effective strategy for placing bright task lighting on a kitchen counter or workbench.

Accent Lighting, sometimes called "object lighting," directs light to specific objects, such as artwork, furniture, plants, or architectural features. When lighting a single object or work of art, use a directional source, such as a PAR (parabolic aluminized reflector) or BR (bulged reflector) lamp in a track, or an adjustable recessed fixture, such as an "eyeball." Position the fixture so the light strikes the wall at a 30-degree angle from the vertical.

When lighting a large picture or grouping of pictures, it is often best to illuminate the entire wall section with a wash of light from multiple track lights or adjustable recessed fixtures. For consistent lighting across the wall without a "scalloping" pattern, use special "wall washer" fixtures or nondirectional lamps (A-bulbs or compact fluorescents) to diffuse the light beams (Figure 5-23).

Decorative Lighting includes candlestick chandeliers and sconces, decorative table lamps, and other fixtures whose main function is to provide luminous "sparkle" to a room.

Glossary of Common Lighting Terms

*   ***Ballast:*** A device that regulates the flow of electricity to start and operate fluorescent and other discharge-type lamps.
*   ***Beam angle:*** The angle of the central cone of light (the beam spread) cast by a reflector-type lamp, beyond which the beam intensity is less than 50% of the maximum. Generally, a lamp with a beam angle of less than 25 degrees is considered a *spot;* with an angle greater than 25 degrees, a *flood.*
*   ***Diffused light:*** Light that is dispersed in a wide pattern with no directional quality, similar to outside conditions on an overcast day. Glare is reduced, but such light may tend to wash out and flatten objects.
*   ***Diffuser:*** A glass or plastic lens over a lamp that scatters light in all directions. In fluorescent

FIGURE 5-23 **Accent Lighting.**

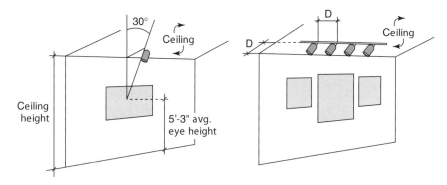

Ceiling Height	Distance (D) From Wall to Fixtures and From Fixture to Fixture
8 ft.	20 in.
9 ft.	27 in.
10 ft.	33 in.
11 ft.	40 in.
12 ft.	48 in.

To light artwork on a wall, use track lighting or adjustable recessed fixtures, such as "eyeballs." Locate the fixtures at a 30-degree angle to the artwork. For large works or multiple pictures, use multiple fixtures spaced apart by the same distance (D) as they lie from the wall.

downlights, a plastic or metal grid is often used to block glare and direct the light downward. Deep-cell parabolic louvers are the most efficient grid configuration.

- *Efficacy:* The light output of a lamp per electricity consumed, expressed as lumens per watt. Fluorescents are three to five times more efficient than standard "A" lamp incandescents.
- *Footcandles:* A measure of the total light falling on a surface. One footcandle (fc) is equal to one lumen per square foot, originally based on the illumination of one candle held one foot from a surface.
- *Lamp:* A light bulb or fluorescent tube.
- *Lumens:* A measure of the light output of a lamp (light bulb) or other light source. A candle provides about 12 lumens, a 60-watt soft light bulb about 850 lumens.
- *Luminaire:* Any lighting fixture or built-in lighting system, such as a cove or valence.
- *Wattage:* A measure of the energy consumption of a lamp or other electrical device. More efficient lamps produce more lumens for the same wattage consumed.

Recommended Lighting Levels

Many factors affect the illumination needed for a specific task. An often overlooked factor is the age of the occupants. At 60 years old, we need two to three times the light we needed at age 20, and also more shielding and diffusers since older eyes are more sensitive to glare. The other main factors in determining lighting requirements are how detailed the work is and the level of contrast and reflectance of the work surface.

Table 5-21 shows the recommended lighting along with common strategies for each type of room. For task lighting, the low numbers in each range represent the light needed for simple tasks with high contrast (reading large black type on white paper). The high number is for tasks with more detail or lower contrast (reading the newspaper). For very detailed, low-contrast work or for older persons, light levels of 100 footcandles are often needed.

Rules of Thumb. The illumination level on a surface depends on many factors, including the colors of the room and furnishings and the type of lamp and fixture. High ceilings, dark colors, and diffusers on fixtures all reduce light levels. The commonly used black baffles in recessed lights reduce output by up to 40%. Tightly focused spots produce much higher light levels than wide floods. The distance from the light source is also critical. Doubling the distance to a lamp reduces the lighting level by a factor of four. So moving the light closer to the task is often the simplest way to provide a big boost in lighting levels.

As a starting point for design in kitchens, baths, home offices, and other brightly lit spaces, provide at least 2 watts of incandescent light or $\frac{3}{4}$ watt of fluorescent light per square foot of floor area. In larger spaces, using multiple fixtures will provide more even lighting and reduce

TABLE 5-21	**Suggested Lighting Strategies and Levels by Room.**		
	Ambient	Task	Accent
Dining Room	Chandelier or sconces for traditional look. Downlights for contemporary. (5–10fc)	Chandelier over table with optional central downlight. (10–20fc)	Track, eyeball, or wall washers to light art or sideboard. (10–20fc)
Living Room	Reflect light off walls and furnishings with multiple fixtures around the room. (5–10fc)	Downlights for game table or piano. Table lamp and small pendant for reading. Undercabinet lighting at bar. (20–50fc)*	Downlights close to fireplace or stone/brick wall provide "grazing" light. Track, eyeball, or wall washers to light art, book cases, or display cabinets. (10–20fc)
Kitchen	Downlights or surface-mounted ceiling fixtures. (20–30fc)	Undercabinet lighting at counters; downlight over sink; pendant over table. Downlights or minipendants over island or bar. (30–50fc)*	Track, eyeball, or wall washers to light art, note board, or cookware. (20–50fc)
Utility/Laundry Room	Fluorescent ceiling fixture with acrylic diffuser or parabolic louver to cut glare. (10–20fc)	Track lighting for sewing, ironing, crafts. (20–50fc)*	NA
Bathroom	Sealed, recessed fixture over tub and shower. Separate ceiling fixture or downlights for larger bath. (20–30fc)	Vanity lights both sides of mirror. Recessed or wall-mounted fixtures near tub and toilet for reading. (20–50fc)	NA
Office/Study	Fluorescent ceiling fixture with parabolic louver to cut glare. Alternative: Torchiere to bounce light off ceiling. (10–20fc)	For desktop from one or both sides with undercabinet lights, ceiling fixtures, or tabletop lamp with flexible arm and opaque shade. Small track or recessed wall washers for bookshelves. (20–50fc)*	NA
Hobby Area	Fluorescent ceiling fixture with acrylic diffuser or parabolic louver to cut glare. (10–20fc)	Track lighting for work bench. Undercabinet lighting for counters. 50–200fc	NA

*100fc for older person working on small detail.

glare. Also, since lighting needs change throughout the day with changes in daylight and usage, it is good to provide flexibility by separately switching groups of lights and adding dimmers.

Increase these minimums by 50 to 100% for:

- Valences or other indirect lighting schemes
- Rooms with dark surfaces
- Lighting placed in cathedral ceilings or other high ceilings
- Recessed lighting with diffusers, black baffles, or other light-blocking trim

Lamp Types

There is a vast array of choices in light bulbs, known in the lighting industry as "lamps." For residential lighting, the main choices are incandescent, halogen, low-voltage, tubular fluorescent, and compact fluorescent. Which lamp to choose for a given application will depend upon the amount of light needed (lumens), color of light desired,

TABLE 5-22	**Fluorescent Lamp Equivalency**
Standard Incandescent	Compact Fluorescent
40W	10W
60W	15W
75W	20W
100W	25W
150W	40W
200W	55W

NOTE: Values are approximate and vary with individual luminaires.

type of fixture (luminaire), and whether the application calls for a directed beam or a diffused light source.

Also, some lamps are more energy-efficient, providing more lumens for the same amount of electricity consumed. Fluorescents are the most efficient, using up to 70% less energy than an equivalent incandescent bulb (see Table 5-22).

Incandescents include the familiar nondirectional "A" lamps, as well as a variety of directional flood and spot lamps designated by an "R" or "BR." Incandescents have a low color temperature of around 2700 K, which produces a warm light with lots of red and yellow tones that make skin, natural wood, and other warm colors look good. To some extent, things look good to us under incandescent light because it is what we are most accustomed to. Incandescent lamps are inexpensive and are easy to dim, but they are also the least efficient type of bulb and the shortest lived.

Halogen, also known as tungsten-halogen, is actually a kind of incandescent with more blue and less red light (3000 K), giving it a whiter appearance than standard incandescent lamps. Halogen lamps provide good color rendition and good light for reading and fine detail work. When dimmed, however, halogen light becomes more yellow, like standard incandescent lighting. Also, dimming can cause a halogen lamp to darken due to tungsten evaporation. Turning the lamp to full illumination for about 10 minutes will restore its full power.

Halogen lamps tend to be smaller, produce 10 to 15% more lumens per watt than standard incandescents, and last about twice as long. They come in a wide range of beam spreads and wattages. However, since halogen lights burn very hot, they must be shielded from contact with other materials or they can create a fire hazard. Also, the bulbs should not be touched without wearing a glove (since the oil from your skin can create a weak spot on the bulb), and should be cleaned with alcohol. Halogen PAR (parabolic aluminized reflector) lamps are enclosed in a protective glass casing, which allows them to be handled like ordinary bulbs.

Low-Voltage lamps are tungsten-halogen, incandescent, or the newer xenon lamps, operating at 12 volts DC. Their small size makes them ideal for undercabinet lighting, and their very precise beam control makes them well-suited to accent lighting of artwork. Many low-voltage fixtures allow the lamps to rotate within the housing to precisely aim the beam.

Low-voltage lights use a step-down transformer to convert 120V line voltage to 12 volts DC. Most newer fixtures use solid-state electronics, which are more energy-efficient and longer lasting than the older magnetic type. Transformers are either attached to the fixtures or installed remotely. Since the transformers, as well as the lamps and dimmers, emit a slight hum, remote location can be an advantage. However, locating the transformer too far from the fixtures can result in a loss of power and dimming of the lamps. When using dimmers with low-voltage lighting, make sure they are specifically designed for low-voltage systems and for the specific type of transformer.

Fluorescent lamps produce light by energizing the phosphor coating on the inside of a glass envelope. A device called the ballast regulates the power needed to start the lamp and keep it going. Older magnetic ballasts caused humming and flickering, but new electronic or solid-state ballasts have eliminated these problems.

Fluorescents produce three to five times the output as incandescent lamps, last about ten times as long, and stay very cool. Because they reduce lighting bills by as much as 75%, and reduce cooling loads as well, they are heavily promoted by model energy codes and mandated in some areas. For example, the California Energy Code requires that the main lighting in kitchens and baths be fluorescent.

The downside of fluorescents has always been their poor color rendering. Standard fluorescents emphasize the blue range of the spectrum, giving skin an unflattering, pale appearance. Manufacturers have worked hard over the years to improve the light quality. So-called "deluxe" fluorescents offer CRI (color rendering index) values in the 85 to 90 range but with a 25% loss of efficiency. To achieve CRIs in the high 90s without sacrificing energy efficiency, manufacturers use more expensive rare earth phosphors, creating *triphosphor* and *quad-phosphor* lamps.

Fluorescent lamps with high CRIs, and color temperatures within the range of 2700 to 3500K, create pleasing light for skin tones and natural wood and can blend in with incandescent lighting. In applications where color accuracy is important, such as laundry areas, lighting artwork, and certain hobbies, full-spectrum daylight lamps may be preferred. These lamps, which produce light similar to natural daylight, include General Electric's Chroma 50 and Chroma 75.

Dimming also used to be a challenge with fluorescents. However, using solid-state dimming ballasts and special dimmers designed for fluorescents can eliminate any humming sounds. These also allow a single dimmer switch to dim groups of fixtures with different length tubes.

Compact Fluorescents have created a lot more flexibility, allowing fluorescents to be used in recessed downlights, wall sconces, pendants, and just about any type of luminaire. Early compact fluorescents were noisy, slow to start, and had a limited selection of color temperatures. Newer products, however, are quiet and typically have rapid-start ballasts. Dimmable ballasts are also available for compact fluorescents, but are costly. As with tube fluorescents, look for high CRIs and lower (warmer) color temperatures from 2700 to 3500K to blend in with incandescent and halogen lighting. All compact fluorescents have a minimum 80 CRI.

While some compact fluorescents have been introduced that mimic R and PAR-type reflector bulbs, directional lighting is best achieved with incandescent or halogen lamps. Fluorescents are better used for ambient lighting, indirect lighting, and lighting of closets and storage areas. Although they cost $5 to $20 per bulb, depending on the wattage and configuration, they generally pay for themselves within two to three years in both energy savings and longevity of the bulbs.

Common Lamp Abbreviations

There are literally thousands of lamps to choose from, but the most common in residential lighting are standard incandescent A lamps, 120-volt **BR** and **PAR** directional lamps, and low-voltage PAR and MR lamps, along with a variety of tubular and compact fluorescents. While different lamp manufacturers use different codes and abbreviations to label their lamps, most list the wattage first, followed by the bulb shape, width of the bulb (in eighths of an inch), and additional information about the shape and beam angle. For example, a 50PAR36/H/NSP8° is a 50-watt PAR lamp, $\frac{36}{8}$ ($4\frac{1}{2}$) inches across, halogen with an 8-degree narrow spot beam. Common abbreviations include the following:

- *A*: General incandescent.
- *BR*: Incandescent, bulged reflector lamps, which replaced the older "R" lamps. These produce up to twice the light in footcandles on the subject as A lamps.
- *F*: Fluorescent.
- *G*: Globe, incandescent.
- *H*: Halogen.
- *IF*: Inside frost.
- *MR-16*: Multifaceted reflectors – low-voltage halogen lamp with faceted mirrors that provide superior beam control. Available in numerous beam widths.
- *NSP*: Narrow spot.
- *PAR20, 30, 38*: Parabolic aluminized reflector—A halogen lamp protected by a heavy glass lens. PAR lamps provide excellent beam control and produce up to four times the light on the subject as A lamps.
- *PAR36*: Low-voltage halogen lamp with superior beam control over longer distances.
- *R*: Reflector.
- *SB*: Silver bowl, which indicates that the bottom of the lamp is opaque and reflects the light upward.
- *T*: Tubular fluorescent.
- *VWFL*: Very wide flood.
- *WSP*: Wide spot.
- *WFL*: Wide flood.

Color Temperature and CRI

Color temperature and color rendering index (CRI) are two different ways to characterize how colors appear under a light source.

Color Temperature is expressed in degrees Kelvin, and for incandescent lights equals the temperature of the metal filament. For fluorescents and other bulbs without filaments, it is the theoretical equivalent temperature. Lower color temperatures indicate "warmer" light with more yellow and red tones, which complement skin and natural wood finishes. Higher color temperatures indicate "cooler" light with more blue and green tones, which renders faces harshly and tends to make skin look pale (Table 5-23).

Skin tones look best under lamps rated from 2700K (standard A bulb) to 3500K and with a CRI over 80. Residential lamps range as high as 7500K for continuous spectrum

TABLE 5-23	Color Temperatures of Common Lamps
2700°K	**Standard Incandescent** Reddish yellow light complements warm colors and skin tones. Best light for low light levels.
3000°K	**Tungsten-Halogen, Warm Fluorescent** More blue, less red light. Appears whiter than standard incandescent.
3500°K	**Color-Corrected Fluorescent** Compromise provides good skin tones and good daylight color matching. Compatible with incandescent or daylight.
4100°K	**Cool White Fluorescent** Good daylight color matching with adequate skin tones. Neat and clean look popular in offices, schools, meeting rooms.
5000°K	**Full-Spectrum Fluorescent** Excellent color matching that simulates daylight. Skin tones slightly greenish. Popular in art galleries, jewelry stores.
6500°K	**Daylight**

fluorescents, such as GE's Chroma 50 or 75. These simulate daylight and are good for detailed work where color accuracy is critical, but they give skin an unflattering greenish tone.

CRI is a rating on a scale of 1 to 100 of how accurately a lamp shows colored objects. The higher the CRI, the closer the colors look to a standard reference. For incandescent lamps and all others with a color temperature of 5000K or less, the reference is an incandescent or halogen bulb, which are both assigned CRIs of 100. For lamps with a color temperature of over 5000K, the reference is natural daylight, which also has a CRI of 100.

CRI numbers are best used to compare lamps with color temperatures within about 300K of each other. Colors will look very different under a 3000K lamp and a 6000K lamp with the same CRI.

Luminaires

While there are thousands of different luminaires on the market, they all fall into a few basic categories. Many mix more than one lighting strategy within a single fixture. All luminaires can be categorized as either direct lights, downlights, accent lights, or indirect lights. Many luminaires combine two or more of these strategies. For example, many dining room chandeliers include a downlight that provides accent or task lighting to the table top in addition to the fixture's ambient lighting. Common fixture types and placement are covered below.

Direct Lights. These include most surface-mounted fixtures on walls and ceilings, often with a diffusing globe or lens to reduce glare. In general, these are very efficient sources of light, but may also produce a lot of glare. Common types include surface-mounted ceiling fixtures, pendants, chandeliers, and sconces.

- *Surface-mounted.* Either incandescent or fluorescent fixtures mounted directly on the ceiling are a very efficient source for ambient lighting. Some, such as fluorescent "clouds," use rounded diffusers that cast light on the ceiling as well as downward.

- *Pendants.* Pendant fixtures are often used to provide task lighting above kitchen tables or eating counters and may also project light sideways and upward to the ceiling. Select a fixture at least 12 inches less in diameter than the table's smallest dimension and mount the fixture 27 to 36 inches above the table or counter (see Figure 5-24). Mini-pendants with halogen spots are often used to accent breakfast bars and kitchen islands.

- *Chandeliers.* When used over dining room tables, select a fixture at least 12 inches less than the width of the table and locate them 27 to 36 inches above the table, as in Figure 5-24. Some fixtures contain a central downlight to provide task lighting to the table.

FIGURE 5-24 Pendants and Chandeliers.

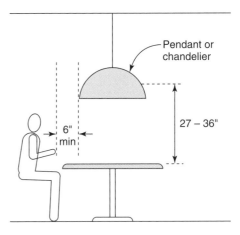

Mount pendants or chandeliers 27 to 36 inches above a table. Pendants with translucent shades provide ambient as well as task lighting.

FIGURE 5-25 Wall Sconces.

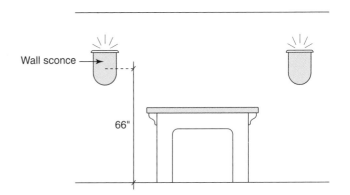

Sconces are often placed in pairs at about 66 inches high to provide soft direct and indirect lighting to living areas and hallways.

- *Sconces.* Wall sconces provide soft lighting in living and dining areas and hallways. Most provide some direct lighting as well as an indirect "wash" of light upward onto walls and ceilings. Often placed in pairs, they are typically located about 66 inches from the floor to the center of the sconce (Figure 5-25).

Downlights. These are predominantly recessed ceiling lights that create a dramatic effect by casting pools of bright light on floors and work surfaces while leaving the ceiling in shadow. Used with A lamps, floods, or compact fluorescents, and spaced properly, downlights can create even general lighting. With more focused spot bulbs and special trims, they can function as task lighting, accent lighting, or wall washers.

Accent Lights. When lighting a picture or single object, use a directional spot lamp in a shielded fixture. These are often track-mounted or adjustable recessed fixtures, such as "eyeballs." To create even lighting over a large picture or group of pictures, it is best to use special "wall-washer" fixtures, or nondirectional lamps such as A-bulbs or compact fluorescents (Figure 5-23, page 199).

Indirect Lights. Bouncing light off light-colored walls and ceilings creates a soft and diffused illumination with little glare and gives a room a feeling of spaciousness. Examples include upward directed floor lamps and wall sconces, as well as site-built coves and valences, which can make use of cost-effective fluorescent tubes. Coves reflect light off the upper walls and ceiling and dramatize a high or cathedral ceiling. Brackets provide downlighting as well to emphasize wall surfaces or artwork. Typical cove and bracket details are shown in Figures 5-26 and 5-27. The shield should be designed to protect the bulbs from view within the room.

Recessed and Track Lighting

Recessed lighting can provide either ambient, task, or accent lighting, depending on the lamp type, its beam spread, and the type of reflector and trim used. Where recessed fixtures are used for ambient lighting, they should be spaced to provide even lighting without dark spots.

Track lighting follows the same design principles as recessed, but is best used for accent or task lighting in certain situations. It is particularly well-suited to situations where flexibility is required since fixtures may be easily moved as lighting needs change.

Lamps. Beam spreads for directional lights vary depending on the lamp and fixture. For general lighting, choose a wide flood with a beam spread of at least 50 degrees. BR lamps are the most economical directional lamp and provide good enough beam control for general lighting. Standard A lamps with Alzak trim or compact fluorescents also provide good general lighting.

Halogen PAR lamps offer more precise beam control suitable for task or accent lighting. Low-voltage M-16 and PAR36 lamps offer very precise beam control, making them well-suited to accent lighting. Because of their narrow focus, spots produce higher illumination levels than floods but over a smaller area. Beam spreads and lighting levels for common directional lamps are shown in Table 5-24.

Recessed Housings. Typical residential recessed lights come in 4- to 7-inch diameters and can take a variety of different trims that significantly affect light output and glare.

For general lighting, a 5- to 7-inch diameter housing is commonly used. For accent lights, smaller 4-inch housings are available for both line-voltage and low-voltage figures. Special recessed housings are also available for compact fluorescents, sloped ceilings, and retrofit installations.

Standard recessed housings must be left uninsulated above. For insulated ceilings, use a can rated IC for "insulation contact." Also make sure the housing is rated "airtight,"

FIGURE 5-26 Cove Lighting.

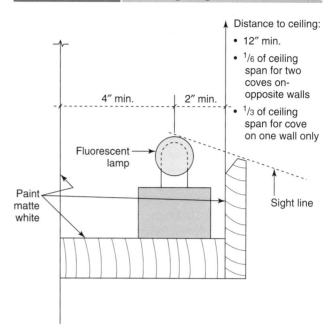

Built-in coves with fluorescent tubes can provide low-cost indirect lighting, making a room feel more spacious or highlighting a tall or cathedral ceiling. The wider the room, the greater the distance should be from the bulb to the ceiling.

FIGURE 5-27 Lighting Bracket.

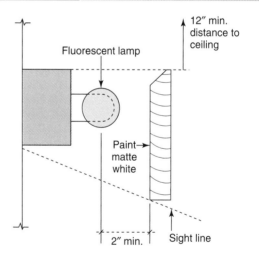

Built-in brackets provide indirect lighting reflected from upper walls and ceiling, as well as direct lighting of the lower wall or artwork.

which is not true of all IC units. Air leaks through recessed lights can be a significant source of heat loss and moisture problems in cathedral ceilings.

Recessed Lighting Trims. The common black or white step baffles are designed for use with a PAR or BR lamp, although homeowners often put in the less expensive A19 bulbs. Baffles reduce glare, but also cut light output by 50% or more for A lamps and up to 40% for directional lamps.

TABLE 5-24   Performance of Common Directional Lamps

		5 ft.		8 ft.		10 ft.	
		Foot-candles	Beam Spread/Spacing (ft.)	Foot-candles	Beam Spread/Spacing (ft.)	Foot-candles	Beam Spread/Spacing (ft.)
BR30							
Flood	65W	19	8.4	7	13.4	5	16.8
Spot	65W	50	2.3	20	3.7	13	4.6
BR40							
Flood	75W	18	7.8	7	12.5	5	15.6
Flood	100W	36	7.8	14	12.5	9	15.6
Spot	100W	200	1.9	78	3.1	50	3.9
PAR20 Halogen							
Flood	50W	56	2.7	22	4.3	14	5.4
Spot	50W	128	1.4	50	2.2	32	2.8
PAR30 Halogen							
Wide Flood	50W	20	5.5	8	8.9	5	11.1
Spot	50W	260	1.1	102	1.7	65	2.1
Flood	75W	88	3.6	34	5.8	22	7.3
Spot	75W	420	1.1	164	1.7	105	2.1
PAR38 Halogen							
Wide Flood	60W	50	5.0	20	8.0	13	10.0
Spot	75W	480	1.1	188	1.8	120	2.7
MR16 Halogen Low-Voltage							
Wide Flood	50W	46	5.2	18	8.3	12	10.4
Flood	50W	74	3.6	29	5.8	19	7.3
Spot	50W	408	1.2	159	2.0	102	2.5
PAR36 Halogen Low-Voltage							
Flood	50W	24	3.4	9	5.5	6	6.9
Narrow Spot	50W	368	0.9	144	1.4	92	1.7

SOURCE: Courtesy of Lithonia Lighting.

Black baffles cut light output significantly more than white (Figure 5-28).

For maximum light output, use a clear or gold *specular* reflector, also known as *Alzak* trim. To reduce glare, which can be a problem with these highly efficient reflectors, it is best to use a deep-profile Alzak trim, offered by most recessed lighting manufacturers. These work well with standard A19 bulbs as well as BR lamps (Figure 5-29). Gold Alzak is about 10% less efficient than the clear style.

For accent lighting, eyeballs and similar adjustable trims allow the homeowner to direct the light to the artwork or architectural feature being lit (Figure 5-30).

These are typically used with a narrow spot to provide bright focused light on a small area. Slotted wall wash trim is used to splash diffused light on broad areas of wall or bookcases. Nondirectional A lamps or compact fluorescents work well in this application. General recommendations are given in Table 5-25 (page 204).

Spacing. The general rule for ambient or task lighting is to space recessed ceiling fixtures approximately the same distance apart as the beam spread at the work height, typically assumed to be 30 inches above the floor (36 inches for kitchen counters). The beam spread is the central cone of light, where the beam is at least 50% of the brightness at the center of the beam. Most manufacturers publish beam spread data for their recessed lights with different trim options. Beam spreads and lighting levels for some common fixtures and lamps are shown in Table 5-26.

FIGURE 5-28 **Recessed Light With Baffle.**

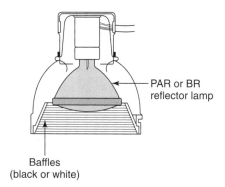

PAR or BR
reflector lamp

Baffles
(black or white)

The commonly used black baffles reduce glare but also cut light output by as much as 40% for the reflector lamps typically used, and even more with ordinary A-bulbs. White baffles increase light output by 15 to 30% compared to black.

FIGURE 5-29 **Deep Alzak Recessed Light.**

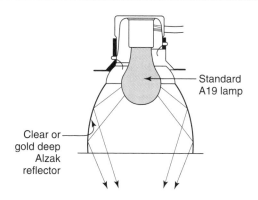

Standard
A19 lamp

Clear or
gold deep
Alzak
reflector

Recessed lights with highly reflective specular or "Alzak" trim have the greatest light output and can be used with standard A-bulbs. To reduce glare, use deep Alzak trim, which has the bulb deeply recessed.

FIGURE 5-30 **Recessed Eyeballs.**

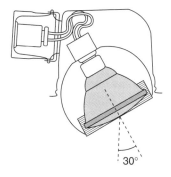

30°

Eyeballs and other adjustable recessed fixtures work well for accent lighting. Most are designed to rotate at least 30 degrees from the vertical, the recommended angle for lighting artwork on a wall.

For ambient lighting, choose a compact fluorescent, A lamp, or wide flood with a beam angle of at least 50 degrees. Typical spacing for ambient lighting with recessed lights is 6 to $7\frac{1}{2}$ feet for an 8-foot ceiling, or 7 to $8\frac{1}{2}$ feet

for a 9-foot ceiling. Spacing from the first row of lights to the wall is half this distance.

For accent lighting, space recessed or track fixtures so their light hits the wall at about 30 degrees. For lighting a large wall area, the distance between fixtures should be equal to or less than their distance from the wall (see Figure 5-23, page 196).

Closet Lighting

Due to risk of fire, the International Building Code and the National Electrical Code require that all fixtures installed in closets must be either surface-mounted or recessed and must completely enclose the bulb. Only incandescent or fluorescent lamps are allowed.

In addition, the fixture must be installed either in the wall above the door or on the ceiling and have the following clearances:

- For surface-mounted incandescent fixtures: 12 inches to the nearest point of clothing or storage space.
- For surface-mounted fluorescent fixtures: 6 inches to the nearest point of clothing or storage space.
- For recessed fluorescent or incandescent fixtures (with a completely enclosed lamp): 6 inches to the nearest point of clothing or storage space.

Kitchen Lighting

Kitchens require general ambient lighting as well as task lighting on sinks, ranges, counters, and eating areas. Given the high lighting needs of a kitchen, the energy savings from fluorescent lights can be substantial. Look for fluorescent bulbs with a CRI over 80 and a color temperature near 2800K to match standard incandescent lights, or 3500K to match halogen lights.

Ambient Kitchen Lighting. For efficient general lighting, use one or more enclosed ceiling fixtures with a white diffuser that illuminates the ceiling as well as the space below. In a very small kitchen, placing the ceiling fixture near the sink and counter can provide effective task lighting as well. For a softer glow in a kitchen, indirect lighting can also work nicely with lights placed in coves and above the cabinets to illuminate the ceiling.

Although not the most energy-efficient, recessed lighting has become a popular choice for kitchen lighting because of its sleek appearance and dramatic effect. For even lighting, use fixtures and lamps with wide beam spreads and spacing based on a 36-inch work plane (Figure 5-31). Also see the discussion on "Spacing," page 202.

As a rough guide, the American Lighting Association suggests the following minimum lighting levels:

- ***Small kitchens of under 75 square feet:*** 150 watts of direct incandescent lighting from up to three bulbs, or about 60 watts of fluorescent.
- ***Kitchens up to 125 square feet:*** 200 watts of incandescent from up to four bulbs, or about 80 watts of fluorescent.

TABLE 5-25   Recessed Lighting Recommendations

Lighting Application	Recommended Lamps	Fixture Spacing			
General ambient Level: 5-20 fc	• A19, BR30 (ceilings 10 ft. or less) • BR40, PAR30/38 (ceilings over 12 ft.)	**Ceiling Height** 8 ft. 9 ft. *Assumes 30-inch workplane	**Distance From Wall** 3 ft. $3^{1}/2$ ft.	**Fixture Spacing*** 6 to $7^{1}/2$ ft. 7 to $8^{1}/2$ ft.	
Task: Counters, desks, and other work surfaces. Center over work surface or position over counter edge. Level: 35-55 fc	• PAR30 (10 ft. ceiling or less) • PAR38 (ceilings over 12 ft.)	**Ceiling Height** 8 ft. 10 ft. 12 ft.	**Fixture Spacing** 3 ft. 4 ft. 5 ft.		
Accent: Artwork on wall. Shine light at 30° from vertical with eyeball or other adjustable trims. Level: 60-100 fc	• PAR30, MR16 (ceilings 10 ft. or less) • PAR38/36 (ceilings over 12 ft.)	**Ceiling Height** 8 ft. 10 ft. 12 ft.	**Distance From Wall** 18-24 in. 30-36 in. 40-48 in.		
Wall washing: Bookcases and walls of art. Level: 35-55 fc	• A-19, compact fluorescents	**Distance From Wall or Bookcase** 36 in.		**Fixture Spacing** 36 in.	
Wall grazing: Highlight rough textures such as brick or stone. Level: 20-25 fc	• BR30, PAR30	**Distance From Wall** 6-9 in.	**Fixture Spacing** 12-18 in.		

SOURCE: Courtesy of Lithonia Lighting

- *Kitchens over 120 square feet:* two watts of incandescent lighting or $\frac{3}{4}$ watt of fluorescent per square foot of floor area.

These numbers should be increased by 50 to 100% for indirect lighting, dark surfaces, lighting placed high in cathedral ceilings, or use of recessed lights with diffusers, baffles, or other light blocking trim.

Kitchen Task Lighting. Work counters, sinks, and cooktops all need high lighting levels. Where wall cabinets are present, undercabinet lighting provides excellent illumination for counters. Place lights as close as possible to the front of the cabinets to avoid glare reflecting off the work surface (Figure 5-31). Low-voltage xenon "festoon" lamps provide bright, even light similar to halogen but without the high temperatures and pressures, eliminating the safety concerns associated with halogen. Also, xenon lamps can be touched with bare skin and provide 10,000 hours of service.

An alternative for lighting at counters is to place a row of recessed fixtures directly over the outer edge of the counter. If used for task lighting, place fixtures about 36 inches apart for 8-foot ceilings or 48 inches apart for 10-foot ceilings (see Task Lighting in Table 5-25).

Sinks, cooktops, islands, and counters without cabinets above can be lit by small recessed downlights or track lighting. Mini-pendants with 12-volt halogen bulbs offer an attractive and functional way to illuminate islands, peninsulas, and eating counters (Figure 5-32).

TABLE 5-26 Performace of Common Recessed Fixtures

5" Deep Alzak - 100W A19		
Ceiling Height	Footcandles (fc)*	Beam Spread (in.)
	at 30" above floor	
8 ft.	15	8
10 ft.	8	12
12 ft.	5	16

*36 fc avg level at 30" workplane for multiple fixtures, based on 10x10 ft. room, 8-ft. ceiling, 4 fixtures placed 4 ft. on-center.

5" Step Baffle (Black) - 100W A19		
Ceiling Height	Footcandles (fc)*	Beam Spread (in.)
	at 30" above floor	
8 ft.	5	8
10 ft.	3	9
12 ft.	2	10

*12.5 fc avg level at 30" workplane for multiple fixtures, based on 10x10 ft. room, 8-ft. ceiling, 4 fixtures placed 4 ft. on-center.

5" Step Baffle (Black) - 75W BR30		
Ceiling Height	Footcandles (fc)*	Beam Spread (in.)
	at 30" above floor	
8 ft.	13	5
10 ft.	8	7
12 ft.	5	9

*17 fc avg level at 30" workplane for multiple fixtures, based on 10x10 ft. room, 8-ft. ceiling, 4 fixtures placed 4 ft. on-center.

5" Step Baffle (White) - 75W BR30		
Ceiling Height	Footcandles (fc)*	Beam Spread (in.)
	at 30" above floor	
8 ft.	15	5
10 ft.	9	7
12 ft.	6	9

*22 fc avg level at 30" workplane for multiple fixtures, based on 10x10 ft. room, 8-ft. ceiling, 4 fixtures placed 4 ft. on-center.

Tables courtesy of Randy Blanchette, applications engineer, Lightolier, a division of Genlyte Group, Inc.

Table Top. Choose a pendant at least 12 inches less in diameter than the table's smallest dimension and mount the fixture 27 to 36 inches above the table. A 120-watt incandescent or 40- to 50-watt fluorescent fixture will generally provide sufficient illumination (see Figure 5-24, page 200).

Bathroom Lighting

Bathroom Mirror. Good lighting is critical at the bathroom mirror for shaving, makeup, and other tasks of personal

FIGURE 5-31 Kitchen Lighting.

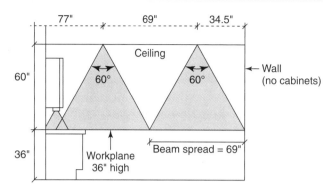

Mount undercabinet lights close to the front of the cabinets, as shown. If using recessed fixtures for ambient lighting, space them apart by a distance roughly equal to their beam spread at the work plane. The first row of recessed lights is pulled out from the wall so they illuminate the counters rather than the cabinet fronts.

FIGURE 5-32 Islands and Peninsulas.

Small downlights or mini-pendants provide a stylish way to light islands, peninsulas, and eating counters.

hygiene. For optimal lighting, place strip lights or globe-type light bars at least 16 inches long on each side of the mirror centered at 61 to 64 inches (about the average eye height). Wall sconces on either side are also an option for smaller mirrors. These provide even cross lighting without shadows or glare (see Figure 5-33).

For small mirrors under 30 inches wide, use about 75 watts of incandescent lighting or 20 watts of warm-white fluorescent on each side. For larger mirrors, use up to 150 watts of incandescent or 40 watts of fluorescent on each side. Additional lights across the top of larger mirrors are also helpful. If using fluorescents, select lamps with high CRIs and warm color temperatures in the 2700K to 3000K range.

Lighting from above the mirror only using globe-type light bars, a pair of recessed downlights, or a lighting soffit is acceptable as long as the vanity top is a light color.

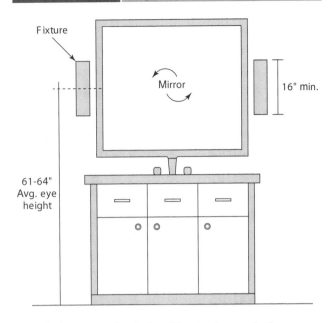

FIGURE 5-33   **Lighting at Bathroom Mirror.**

Cross lighting provides the best illumination at a bathroom mirror, with minimal glare and shadows. Between 75 and 150 watts of incandescent light per side will provide good illumination.

Otherwise, areas under the eyes, nose, and chin will be in shadow. If recessed fixtures are used, choose an A lamp, flood, or compact fluorescent for a diffused beam.

General Lighting. As a rule of thumb, provide one watt of incandescent or $\frac{1}{3}$ to $\frac{1}{2}$ watt of fluorescent light per square foot of floor space. Increase this by 50 to 100% for recessed lights, indirect lighting, or a room with dark surfaces.

In a small bathroom, the mirror lights can also provide the ambient light. For larger baths, a separate ceiling fixture mounted near the tub and toilet can be useful for ambient light and reading. Finally, in a room with a high ceiling, indirect lighting with coves or uplights can create a feeling of spaciousness in a bathroom, along with a pleasing, soft glow.

Lighting Over Tub and Shower. A recessed light with a white diffuser mounted over the tub or shower will be appreciated by bathers. Electrical codes require that these fixtures be totally enclosed and rated for use in a damp location (tub area) or wet location (shower). Most require GFCI protection for their UL rating. In addition, fixtures must be at least 6 feet above the water line and switches must be a minimum of 5 feet from the edge of the bathtub or shower. Check with local codes for specific requirements.

RESOURCES

Manufacturers

Drywall Trims and Accessories

Clinch-On Products, A Deitrich Metal Framing Company
www.dietrichindustries.com
Nail-on and clinch-on galvanized metal corner beads

Con-Form International/Strait Flex
www.straitflex.com
Strait-Flex fiber-composite mud-on corner bead for inside and outside off-90 degree angles

Drywall Systems International
www.no-coat.com
No-Coat prefinished drywall tapes for inside and outside corners, off angles and bullnose trims

Flex-Ability Concepts
www.flexc.com
Curved metal top and bottom plates for curved wood or metal stud walls

Grabber Construction Products
www.grabberman.com
Drywall screws, corner clips, and fiberglass mesh tapes

Insta Arch Corp.
www.instaarch.com
Galvanized steel preformed and custom arches for drywall

National Gypsum Co.
www.nationalgypsum.com
ProForm tapes and finishing compounds

Pla-Cor
www.pla-cor.com
ABS corner trims, bullnose, 3-way corner caps, and flexible arches

Phillips Manufacturing Co.
www.phillipsmfg.com
Metal and vinyl corner beads, bullnose trim, and flexible bullnose and angled arch trim

Trim-Tex
www.trim-tex.com
Vinyl drywall beads, flexible arch beads, and finishing accessories

U.S. Gypsum
Beadex and Sheetrock-brand tape-on metal corner beads and trims. Complete line of drywall finishing compounds

Vinyl Corp., A Deitrich Metal Framing Company
www.vinylcorp.com
Full line of vinyl beads and trim

Prefinished Wood Flooring

Alloc Inc.
www.alloc.com
Snap-together, no-glue long-strip and single-strip laminated flooring

Anderson Hardwood Floors
www.andersonfloors.com
Laminated strip, nail, glue, and floating

Armstrong World Industries
www.armstrong.com
Bruce, Hartco, and Robbins prefinished plank and engineered strip

BHK of America
www.bhkuniclic.com
Snap-together, no-glue laminated wood flooring

Columbia Wood Flooring
www.columbiaflooring.com
Prefinished solid strip

Duro-Design
www.duro-design.com
Floating click-lock oak flooring

Gammapar
www.gammapar.com
Engineered wood strip with oil, UV-cured urethane, or acrylic-impregnated finish

Junkers Hardwood
www.junkershardwood.com
Floating solid strip with metal clip installation

Kahrs
www.kahrs.com
Laminated strip with UV-cured acrylic urethane, nail, glue, or floating

Lauzon
www.lauzonltd.com
Prefinished strip, laminated strip, and click-lock laminated plank

Mannington Mills
www.mannington.com
Laminated strip and plank with polyurethane and aluminum-oxide finish

Medallion Hardwood Flooring
www.medallionhardwood.com
Prefinished solid hardwood strip and plank with aluminum-oxide finish

Tarkett Wood Floors
www.harris-tarkett.com
Prefinished solid hardwood, laminated, and long-strip flooring

Mercier Wood Flooring
www.mercierdurable.com
Prefinished solid strip and laminated strip with aluminum-oxide finish

PermaGrain Hardwood Flooring
www.permagrain.com
Acrylic-impregnated laminated strip and plank

Zickgraf Hardwood Flooring
www.zickgraf.com
Prefinished solid strip with UV-cured urethane with diamond and aluminum oxide

Bamboo Flooring

Bamtex (a division of Wood Flooring International)
www.bamtex.com
Laminated bamboo and palm flooring

Duro-Design
www.duro-design.com
Laminated bamboo flooring with durable water-based finish

Gammapar
www.gammapar.com
Acrylic-impregnated bamboo flooring

Hawa Bamboo Flooring
www.hawabamboo.com
Prefinished bamboo flooring with aluminum-oxide finish

Natural Cork
www.naturalcork.com
Prefinished glue-down or nail-down 3-ply bamboo planks with aluminum-oxide finish

Resilient Flooring

Congoleum
www.congoleum.com
Vinyl sheet flooring and tiles

Domco (division of Domco Tarkett Group)
www.domco.com
Vinyl sheet flooring and tiles

Forbo
www.forbo-flooring.com
Linoleum sheet and tiles

Mannington Mills
www.mannington.com
Vinyl sheet flooring and tiles

Nova Distinctive Floors
www.novafloorings.com
Laminated linoleum surface over fiberboard and cork planks, floating installation

Tarkett
www.tarkettna.com
Vinyl sheet flooring and tiles

Cork Flooring

American Cork Products Co.
www.amcork.com
Prefinished parquet tiles and floating floor planks

Amorim Revestimentos (formerly Ipocork)
www.wicanders.com
Floating or glue-down laminated cork tiles with UV-acrylic or oil finish

BHK of America
www.bhkuniclic.com
Snap-together, no-glue, laminated cork flooring with UV-acrylic finish

Expanko Cork Inc.
www.expanko.com
Cork tiles with wax or polyurethane finish

Korq Inc.
(212) 758-2593

Natural Cork
www.naturalcork.com
Glue-down cork tiles and floating laminated planks with UV-cured acrylic finish

Nova Distinctive Floors
www.novafloorings.com
Laminated cork planks with glue-down and floating click-lock installation

WECork
www.wecork.com
Cork tiles, sheets, and floating floors

Polymer (Urethane), MDF, and Vinyl Trim

Burton Mouldings
www.burton-mouldings.com
MDF(medium-density fiberboard), polymer, flex, and wood

Fypon
www.fypon.com
Polymer moldings and components

Nu-Wood
www.nu-wood.com
Polymer moldings and components

Outwater Plastics
www.outwater.com
Polymer moldings and components

RAS Industries
www.rasindustries.com
Polymer moldings and components

Royal Mouldings (formerly Marley Mouldings)
www.royalmouldings.com
Polymer, polystyrene, expanded-PVC, CPVC, and acrylic molding profiles and components

Flexible Trim

Flex Trim
www.flextrim.com
Flexible polymer moldings

Resin Art
www.resinart.com
Flexible polymer moldings

For More Information

American Lighting Association
www.americanlightingassoc.com

Association of the Wall and Ceiling Industries
www.awci.org

Carpet and Rug Institute (CRI)
www.carpet-rug.org

Drywall Finishing Council
www.dwfc.org

Forest Stewardship Program
www.fscus.org

FloorFacts
www.floorfacts.com

The Gypsum Association
www.gypsum.org

National Oak Flooring Manufacturers Association (NOFMA)
www.nofma.com

National Wood Flooring Association
www.woodfloors.org

Painting and Decorating Contractors of America
www.pdca.org

Smartwood/Rainforest Alliance
www.smartwood.org

CHAPTER SIX | Kitchen and Bath

KITCHEN DESIGN BASICS

Whether designing a small galley kitchen or an expansive space for multiple cooks and entertaining, the same rules apply regarding clearances and relationships between key work centers so that work in the kitchen flows smoothly and efficiently. While the traditional American kitchen developed around three main appliances—the sink, range, and refrigerator—today's kitchen may have many more centers of activity, including the following list adapted from the National Kitchen and Bath Association (NKBA):

- *Primary clean-up center:* Includes the main sink, dishwasher, recycling center, and waste disposer.

- *Secondary sink center:* May also serve cleanup functions. Often associated with the food preparation center.

- *Food preparation center:* A clear space at least 16x36 inches typically located between the sink and cooktop or sink and refrigerator. A two-cook kitchen requires two such spaces.

- *Cooking center:* Revolves around the cooktop and may also include a separate built-in oven or microwave.

- *Microwave center:* Because of its frequent use, this should be near the main activity areas.

- *Pantry center:* Tall storage cabinets work well to store food and cooking supplies near the preparation area. Tall cabinets may also store dishes in the serving or dining area.

- *Serving center:* This area stores dishes and other serving items and may be in the kitchen or closer to the dining area.

- *Dining center:* Many kitchens include either an eating counter or a separate dining area.

- *Socializing center:* A casual seating area adjacent to the kitchen work space allows other family members or friends to visit and socialize with the cook.

- *Home office center:* A space for the telephone, mail, household records, and cookbooks is often incorporated into the kitchen. Concealing the desktop visually from the kitchen is appreciated by many clients. (List adapted with permission from John Wiley & Sons from *Essential Kitchen Design Guide,* © NKBA, 1996.)

Kitchen Design Guidelines

In 1992, the National Kitchen and Bath Association (NKBA) introduced new design guidelines based on research conducted at the University of Minnesota. These have been expanded and revised over time to reflect the continuing evolution of kitchen design and usage. The key design rules are shown below. Accessibility recommendations are listed separately here, but they are now incorporated into all NKBA guidelines.

Walkways and Work Aisles. Work aisles with counters or appliances on both sides should be at least 42 inches wide for a one-cook kitchen and 48 inches wide for a two-cook

FIGURE 6-1   **Walkways and Work Aisles.**

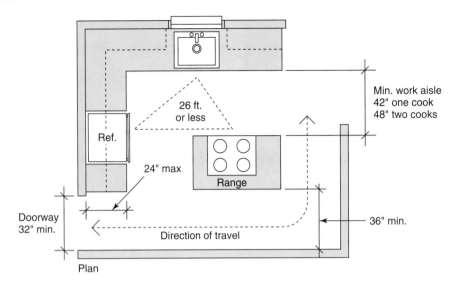

Plan

Make work aisles with counters or appliances on both sides at least 42 inches wide (48 inches wide for a two-cook kitchen). Passageways, which may have counters or appliances on one side, should be at least 36 inches wide and should not cross the work triangle.

SOURCE: Reprinted with permission from *Architectural Graphic Standards, Residential Construction,* © 2003 John Wiley & Sons.

kitchen. Walkways, which may have a work counter on one side, should be at least 36 inches wide and should not cross the work triangle (see Figure 6-1).

Work Triangle. The shortest walking distance between the refrigerator, primary sink, and primary cooktop should be 26 feet or less, as shown in Figure 6-1. Each leg of the triangle should range from 4 to 9 feet long. No major walkway should pass though the triangle, and no corner of an island or peninsula should intersect the triangle by more than 12 inches.

Two-Cook Work Triangles. In a two-cook kitchen, each person should have his or her own work triangle of less than 26 feet. The two triangles may share a leg, but they should not cross one another (Figure 6-2).

Conflicting Doors. All entry doors, appliance doors, and cabinet doors should swing freely without interfering with another door.

Cabinet Frontage. Provide the minimum cabinet frontage shown in Table 6-1. Do not count difficult-to-reach wall cabinets over hoods or refrigerators unless special access is provided. A pie-cut lazy Susan base counts as 30 inches. Tall cabinets 72 inches or higher can count as either base or wall cabinets as follows: for 12-inch-deep cabinets, multiply frontage by one to count as base cabinets and by 2 to count as wall cabinets. Double these amounts for 21- to 24-inch-deep tall cabinets.

Counter Heights and Edges. Provide at least two counter heights in the kitchen with one 28 to 36 inches high and the other 36 to 45 inches high. Varied heights

FIGURE 6-2   **Two-Cook Work Triangles.**

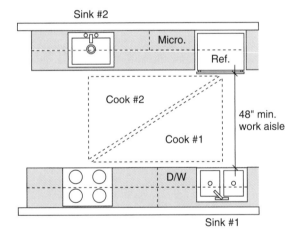

In a two-cook kitchen, each person should have his or her own work triangle. The two triangles may share a leg, but they should not cross one another.

SOURCE: Adapted from *Architectural Graphic Standards, Residential Construction,* with the permission of John Wiley & Sons © 2003.

create work spaces for various tasks and for cooks of different heights, including seated cooks. Also, clip or round over countertop corners and edges to eliminate sharp edges.

Dishwasher Work Center. Locate the dishwasher within 36 inches of the sink and allow at least 21 inches of clearance between the dishwasher and any counters, cabinets, or appliances placed at a right angle to the dishwasher. If possible, allow 30 inches of clear floor space on each side of the dishwasher so two people can work at the same time (Figure 6-3).

TABLE 6-1 Minimum Recommended Cabinet Frontage		
	Small Kitchens (<150 sq. ft.)	Large Kitchens (>150 sq. ft.)
Wall Cabinets	144 in.	186 in.
Base Cabinets	156 in.	192 in.
Drawers or roll-out shelves	120 in.	165 in.
Usable countertop*	132 in.	198 in.

*To be counted, countertops should be at least 16 inches deep with wall cabinets at least 15 inches above. Do not count space at inside corners.

FIGURE 6-3 Dishwasher Work Center.

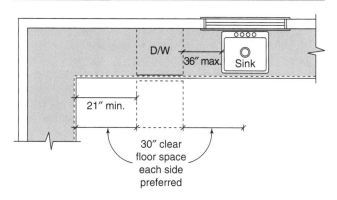

Locate the dishwasher within 36 inches of the sink and at least 21 inches away from a wall or right angle in the counter. If possible, allow 30 inches of clear floor space on each side of the dishwasher so two people can work together.

Sink Work Center.
Locate the primary sink between or across from the cooking surface, food preparation area, or refrigerator (Figure 6-4).

- *Counter space:* Allow 24 inches on one side of the sink and 18 inches on the other. If the sink is within 3 to 18 inches of a corner, provide at least 21 inches of additional space on the return counter.

- *Cabinet space:* Provide at least 60 inches of wall cabinet frontage within 72 inches of the primary sink centerline. Alternate: Use one tall cabinet within 72 inches of the sink.

- *Floor space:* A 30x48–inch floor space centered in front of the sink will make it wheelchair accessible.

Secondary Sinks.
Provide a minimum of 3 inches of countertop frontage on one side of a secondary sink and at least 18 inches on the other side.

Food Preparation Area.
Provide 36 inches of continuous countertop, at least 16 inches deep, immediately adjacent to a sink. For a two-cook kitchen, provide either two separate 36-inch spaces or one 72-inch space adjacent to a sink (see Figure 6-5).

FIGURE 6-4 Sink Work Center.

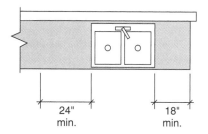

Min. counter frontage

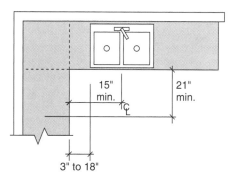

Sink adjacent to corner

Provide at least 24 inches of countertop on one side of the main sink and 18 inches on the other. If the sink is within 3 to 18 inches of a corner, provide at least 21 inches of additional space on the return counter.
SOURCE: Adapted from *Architectural Graphic Standards, Residential Construction,* with the permission of John Wiley & Sons © 2003.

Recycling Center. Unless provided elsewhere in the plan, provide at least two waste receptacles in the kitchen, one for garbage and one for recyclables.

Cooktop Work Center

- *Counter space:* Provide 15 inches on one side of the cooktop and 9 inches on the other. Or if placed against an end wall, leave at least 3 inches of clearance to the wall and cover it with a flame-retardant material. Where there is no backsplash, as in an island or peninsula, provide a minimum of 9 inches behind the cooktop for safety reasons (Figure 6-6).

- *Clearances:* Allow at least 24 inches of clearance between a cooking surface and a protected surface above, such as a range hood, or 30 inches to an unprotected surface.

- *Ventilation:* Ventilate all major appliance cooking surfaces with a minimum 150 cfm exhaust fan. Gas appliances must vent to the exterior.

Oven Landing Space. Provide at least 15 inches of landing space, a minimum of 16 inches deep, next to or above the oven. If the oven does not open into a traffic area, the landing space can be directly across from the oven by no more than 48 inches.

FIGURE 6-5   Food Preparation Area.

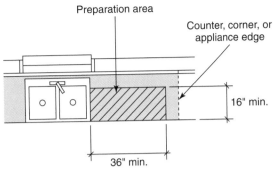

One person area

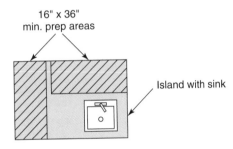

Two person area

For food preparation, provide 36 inches of continuous counter-top, at least 16 inches deep, immediately adjacent to a sink. For a two-cook kitchen, provide either two separate 36-inch counter spaces or one 72-inch space adjacent to a sink.
SOURCE: Adapted from *Architectural Graphic Standards, Residential Construction,* with the permission of John Wiley & Sons © 2003.

Microwave Work Center. Locate stand-alone microwave ovens so that the bottom of the appliance is 24 to 48 inches above the floor. Provide at least 15 inches of landing space, a minimum of 16 inches deep above, below, or to the side of the microwave oven.

Refrigerator Work Center. Provide at least 15 inches of counter space as a "landing area" adjacent to the handle side of the refrigerator or on both sides of a side-by-side refrigerator. Alternately, provide 15 inches of countertop directly across from the refrigerator and no more than 48 inches away. With a side-by-side unit, provide easy access to a counter from the fresh food side (Figure 6-7).

Overlapping Work Centers. Where countertop areas of two work centers (e.g., sink, refrigerator, food preparation) overlap, the minimum counter frontage between the centers should equal the longest of the required two lengths plus 12 inches.

Tall Cabinets Between Work Centers. Do not separate two primary work centers (primary sink, refrigerator,

FIGURE 6-6   Cooktop Work Center.

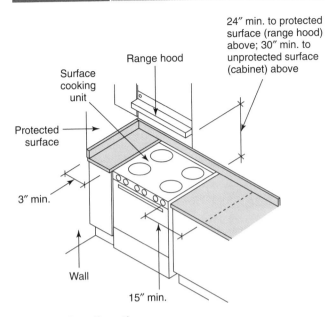

Enclosed configuration

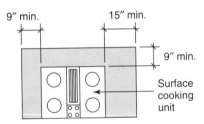

Open (island) configuration

The main cooking surface should have at least 15 inches of countertop on one side and 9 inches on the other. If placed against an end wall, leave at least 3 inches of clearance to the wall and cover it with a flame-retardant material. For island or peninsula cooktops, provide at least 9 inches of counter space behind the cooktop for safety reasons.
SOURCE: Adapted from *Architectural Graphic Standards, Residential Construction,* with the permission of John Wiley & Sons © 2003.

preparation area, or cooking center) by a full-height, full-depth tower such as an oven cabinet, pantry cabinet, or refrigerator. One exception is a corner-recessed tall tower if knee space is planned to one side.

Kitchen Eating Areas

- *Eating counter heights.* Heights and capacities for tables, eating counters, and bars are shown in Figure 6-8. Seating widths have been increased to 30 inches in the 30-inch-high seating area to accommodate wheelchairs.

- *Clearances to walls.* Allow a minimum clearance of 36 inches from the edge of a counter or table to a wall or obstruction. Increase this to 65 inches if the space also serves as a walkway (Figure 6-9).

FIGURE 6-7 Refrigerator Work Center.

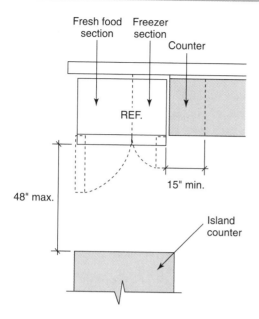

Provide at least 15 inches of countertop "landing area" adjacent to the handle side of the refrigerator or 15 inches on both sides of a side-by-side refrigerator. The landing area can also be across from the refrigerator, but no more than 48 inches away.

SOURCE: Adapted from *Architectural Graphic Standards, Residential Construction,* with the permission of John Wiley & Sons © 2003.

TABLE 6-2 Seating Capacity of Dining Tables

Rectangular Tables	
Size (in.)	Seating
30x48	4
30x60	4-6
36x72	6
36x84	6-8

Round Tables	
Diameter (in.)	Seating
30	2
36	2-4
42	4-5
48-54	5-6

- *Table sizes.* Many kitchens feature small or full-size dining tables (Table 6-2). When selecting a table, pay close attention to whether leg placement will interfere with the number of chairs planned.

Electrical Devices. Install ground-fault circuit interrupters (GFCIs) on all receptacles within the kitchen.

FIGURE 6-8 Eating Counter Typical Dimensions.

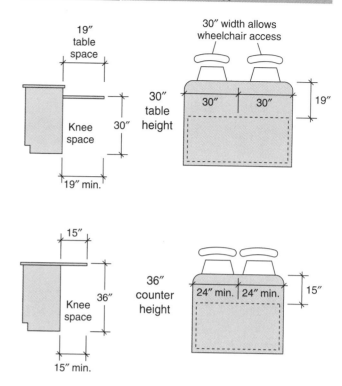

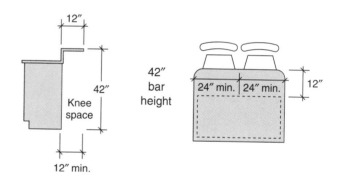

The minimum width for dining at a counter or table is 24 inches. Thirty inches is preferable for full-sized chairs at table height and is required for wheelchair access at a 30-inch-high table (top).

SOURCE: Adapted with permission of John Wiley & Sons, from *Essential Kitchen Design Guide,* by the National Kitchen & Bath Association, © 1996 NKBA.

Locate wall-mounted room controls, including electrical receptacles, switches, thermostats, telephones, and intercoms, between 15 to 48 inches above the finished floor.

Fire Protection. A fire extinguisher should be visibly located in the kitchen away from cooking equipment and 15 to 48 inches above the floor. Smoke alarms should be installed near the kitchen.

Natural Lighting. The combined area of windows and skylights should equal at least 10% of the square footage of the kitchen. Also, every work surface should be well illuminated by appropriate task or general lighting (see "Kitchen Lighting," page 203).

(see "Kitchen Lighting," page 203).

FIGURE 6-9 Passage at Seating Areas.

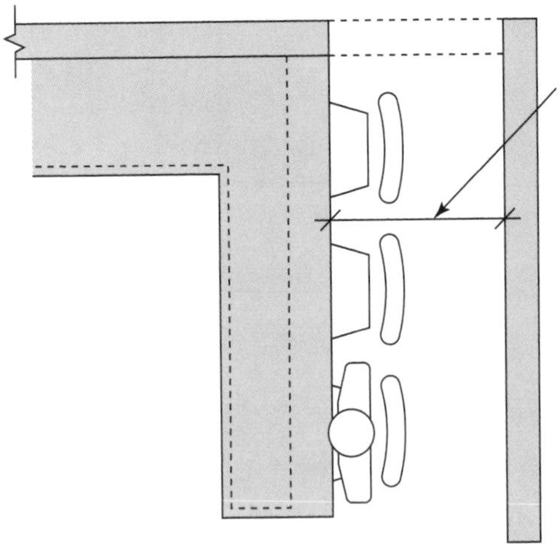

• 36" min.
• 54" for wheelchair access
• 65" if used as walkway

Allow a minimum clearance of 36 inches from the edge of a counter or table to a wall or obstruction. Provide 65 inches of clearance if the space is also used as a walkway.
SOURCE: Reprinted with permission of John Wiley & Sons, from *Essential Kitchen Design Guide,* by the National Kitchen & Bath Association, © 1996 NKBA.

Typical Kitchen Layouts

Kitchen plans should follow the work flow from the garage or side entrance where food enters the home to the storage area or refrigerator. From there, work typically flows to the sink and food prep area, then on to the cooktop or oven, and eating area. Finally, dishes move back to the sink and dishwasher for clean up. All kitchens from the simplest to most complex should be designed to smoothly accommodate this basic workflow (see Figure 6-10).

Single-Wall. The simplest kitchen plan is designed for one cook and primarily used in small apartments. If the kitchen has a second wall, use it for storage with pegboards or shelving narrow enough to leave adequate clearance.

Galley. A corridor or galley-style kitchen can provide an efficient layout for a single cook. If household traffic must regularly pass through, provide at least 48 inches of clearance between counters.

L-Shaped. This layout provides adequate counter space and creates an efficient work triangle away from the household traffic flow. The large open space can often accommodate a dining area.

U-Shaped Plan. Many designers consider this the most efficient plan, since the cook is surrounded on three sides

FIGURE 6-10 Typical Kitchen Layouts.

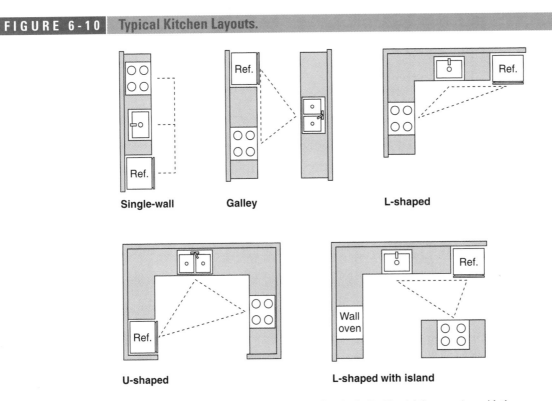

Single-wall   **Galley**   **L-shaped**

U-shaped   **L-shaped with island**

SOURCE: Adapted from *Architectural Graphic Standards, Residential Construction,* with the permission of John Wiley & Sons © 2003.

by countertop, work centers, and storage areas. Also, household traffic is naturally directed around the work area, keeping it clear for kitchen tasks.

L-Shaped with Island. This combines the compact efficiency of a U-shaped plan with the benefits of a more open plan. The island invites interaction between the cook and visitors or helpers because more than one person can work at the open counter.

Accessible Kitchen Design

To make a kitchen fully functional for wheelchair users and other seated occupants requires simple commonsense changes, like placing knobs within reach, as well as more significant changes, such as lowering counters and providing knee space below. The guidelines below, based on ANSI (American National Standards Institute) A117.1 standards, are a good starting point in design, but they should be tailored to the size, reach, and specific capabilities of the occupants.

Work Aisles and Passageways. Clear space at doorways and passageways must be at least 32 inches wide and no more than 24 inches long in the direction of travel. Eliminate any thresholds at doorways.

The minimum work aisle with counters or appliances on both sides should be 40 inches. Walkways with counters or appliances on only one side can be 36 inches wide, but if a walkway turns a corner, as in Figure 6-1 (page 210), one leg should be widened to 42 inches for a wheelchair to make the turn. From a table or eating counter to a wall, leave 54 inches for wheelchair access. In a U-shaped kitchen the minimum clearance between counters is 60 inches (Figure 6-11).

Knee Space. Wherever possible, provide knee space for a seated user below or adjacent to sinks, cooktops, ranges, dishwashers, refrigerators, and ovens. To accommodate

a seated user, below-counter knee space should be a minimum of 30 inches wide, 27 inches high in front, and 19 inches deep, with a minimum 9-inch-high toe space, which will accommodate most wheelchair footrests. Protect users from exposed pipes and mechanicals with a protective panel and insulation (Figure 6-12).

Clear Floor Space. To make work centers universally accessible, provide a clear floor space of 30x48 inches or 48x30 inches, centered in front of the sink, dishwasher, cooktop, oven, and refrigerator. These floor spaces may overlap, and the long dimension can include up to 19 inches deep of knee space below counters (Figure 6-13).

FIGURE 6-11 **Accessible U-Shaped Kitchen.**

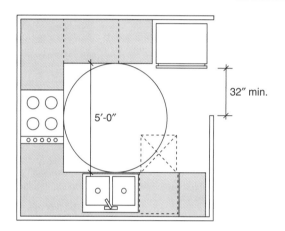

For full accessibility, a U-shaped kitchen should have 5 feet of clearance between opposing counters and a 5-foot-diameter circle for reversing direction in a wheelchair. Doorways and passageways should be at least 32 inches wide, and narrow passageways should not exceed 24 inches in length.
SOURCE: Adapted from *Architectural Graphic Standards, Residential Construction,* with the permission of John Wiley & Sons © 2003.

FIGURE 6-12 **Knee Space.**

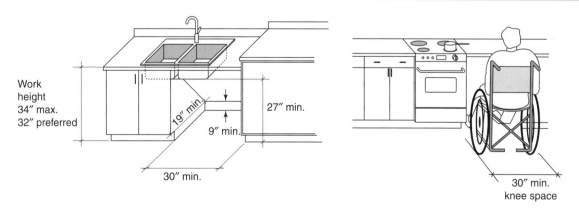

For wheelchair accessibility, provide knee space where possible adjacent to sinks (left), cooktops (right), ranges, dishwashers, refrigerators, and ovens. Protect users from exposed pipes and mechanicals with an insulated protective panel.
SOURCE: Reprinted with permission of John Wiley & Sons, from *Essential Kitchen Design Guide,* by the National Kitchen & Bath Association, © 1996 NKBA.

FIGURE 6-13 Clear Floor Space.

Provide a clear floor space of 30x48 inches either parallel or perpendicular to the front of the sink, dishwasher, cooktop, oven, and refrigerator. These spaces should be centered on the fixture or appliance and may overlap. The clear spaces can include toe-kick space (minimum 9 inches high to accommodate most wheelchair footrests) or up to 19 inches deep of knee space.
SOURCE: Adapted with permission of John Wiley & Sons, from *Essential Kitchen Design Guide,* by the National Kitchen & Bath Association, © 1996 NKBA.

Counter and Appliance Height. The optimal height for most seated occupants at counters, sinks, and cooktops is about 32 inches and should be no higher than 34 inches.

Storage Height. Most seated users can fully reach shelving located from 15 to 48 inches high. Storage located from about 20 to 44 inches is considered optimal. Use open shelving, shelf racks on pantry doors,

FIGURE 6-14 Accessible Storage Space.

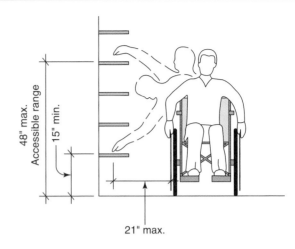

Storage from 15 to 48 inches high will be accessible to most seated users reaching sideways or forwards. Shelving from 20 to 44 inches high is considered optimal. Use shallow open shelving, drawers, or roll-outs for easy access.
SOURCE: Adapted from *Architectural Graphic Standards, Residential Construction,* with the permission of John Wiley & Sons © 2003.

and drawers or roll-out shelving for easy access (Figure 6-14).

Handles and Controls. Controls, handles, and door and drawer pulls should be operable with one hand, require minimal strength, and not require tight grasping, pinching, or twisting of the wrists. Lever-action handles work well for doors and faucets. A simple test is to try to operate the controls with a closed fist.

Mount wall cabinet doors at the bottom of the cabinets and base cabinet pulls at the top of the cabinets.

Sink and Dishwasher Work Center. Use a shallow sink mounted at 32 to 34 inches (32 preferred) with the drain in the rear so it does not interfere with knee space (Figure 6-15). The garbage disposal must also be offset so

FIGURE 6-15 Accessible Sink and Dishwasher Work Center.

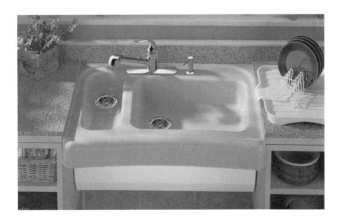

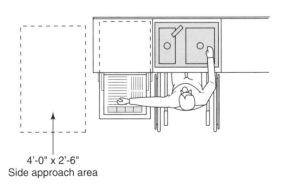

4'-0" x 2'-6"
Side approach area

For seated users, install a shallow sink mounted 32 to 34 inches high with the drain in the rear where it will not interfere with knee space (photo, left). Locating the dishwasher close to the sink is useful for users of all abilities (right).
SOURCE: Photo courtesy of Kohler. Illustration adapted from *Architectural Graphic Standards, Residential Construction,* with the permission of John Wiley & Sons © 2003.

it does not interfere with knee space. A tall faucet and pull-out spray attachment are recommended to simplify work at the sink. Locate the dishwasher adjacent to the sink or no more than 12 inches away.

Lighting. Lighting levels should be up to twice normal levels. Using light-colored floors, walls, ceilings, and counters will help keep all areas well illuminated. Light colors on the insides of cabinets and drawers will help make items more visible.

Cooking Work Center. If possible, place the cooktop and sink on the same wall so users do not have to carry heavy pots across the room. Electric cooktops with a smooth surface and controls on the front work best so the user does not have to reach over the top. Look for units with staggered burners for easier access to back burners. Use a separate wall-mounted oven, not an under-counter design. An oven with a side-hinged door rather than the usual pull-down style works well.

Refrigerator Work Center. Side-by-side units with doors that swing back a full 180 degrees are preferable to up-and-down models. Provide at least 18 inches of counter space adjacent to the refrigerator.

BATHROOM DESIGN BASICS

A well-designed bathroom is comfortable to use, safe, durable, and easy to clean. Space planning revolves around the main fixtures and their required clearances. Proper clearances are critical to avoid problems such as banged elbows at a sink placed too close to a wall or difficult access to the tub faucet.

Safety concerns should be paramount in design decisions and material choices. For example, choose only nonskid flooring types and select tub and shower controls

with foolproof antiscald protection. Avoid designs with sunken tubs or tub surrounds with steps, both of which are hazards. Also remember that following code is not a guarantee of safety. For example, while it is legal to place bathroom lighting circuits downstream from the GFCI outlet, it is unwise since anything that trips the GFCI will also plunge the bathroom into darkness.

Bathroom Design Guidelines

The following recommendations are based on guidelines first published by the National Kitchen and Bath Association in 1992. While accessible design principles are provided separately below, NKBA now incorporates these principles into their recommendations for all projects.

Lavatories

- **Clearances.** Locate each sink so its centerline is at least 15 inches from a wall and 30 inches from the centerline of a second sink. The minimum walkway shown in front of the sink may not allow full accessibility (see Figure 6-16).
- **Height.** While the standard vanity or sink height is 30 to 32 inches, 34 inches is a better compromise between shorter and taller users. If a bath has more than one vanity, set one at 30 to 34 inches and the other at 34 to 42 inches high.

Mirror Height. The bottom edge of a mirror over a vanity should be no more than 40 inches above the floor, or 48 inches if the mirror is tilted forward.

Showers

- **Clearances:** Allow a minimum 21-inch walkway (30 inches preferred) from the front of the shower

---

FIGURE 6-16 | **Lavatory Clearances.**

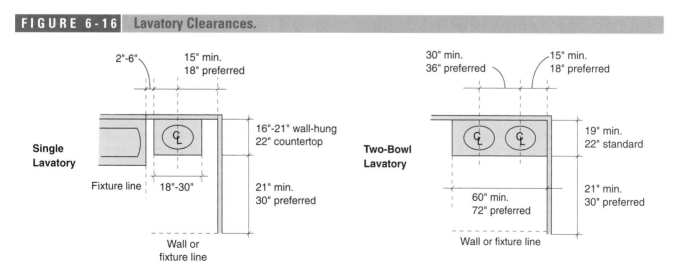

Locate bathroom sinks so their centerline is at least 15 inches from a wall (left) and 30 inches from a second sink's centerline (right). While the standard vanity or sink height is 30 to 32 inches, 34 inches is a better compromise between shorter and taller users.

FIGURE 6-17   Shower Dimensions.

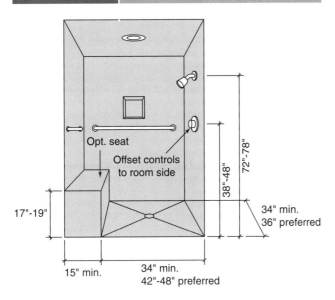

Showers smaller than 36x36 inches are acceptable by code, but not very comfortable. A shower 36 inches deep by 42 to 48 inches wide allows space for the user to step out of the stream of water, enhancing both comfort and safety. A seat also improves both comfort and safety, but should not encroach on the minimum 34x34–inch floor space.

SOURCE: Reprinted with permission of John Wiley & Sons, from the *Essential Bathroom Design Guide* by the National Kitchen & Bath Association, © 1997 NKBA.

FIGURE 6-18   Bathtubs and Tub/Showers.

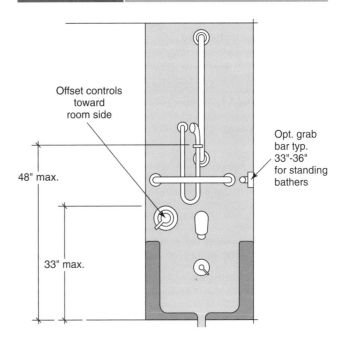

Offset the bathtub or tub/shower and shower controls toward the outside of the tub, so they are accessible from both inside and outside the fixture. Bathtub controls should be no more than 33 inches high. If a handheld showerhead is used, it should be no higher than 48 inches at its lowest position.

stall to a wall or fixture. These clearances may not allow full accessibility.

- *Size:* Provide a minimum clear floor space inside the shower stall of 34x34 inches, preferably 36x36 inches or larger. For optimal comfort and safety, increase the size to 36x42 inches to 48 inches, which allows space for the user to step out of the stream of water to adjust the temperature (Figure 6-17).

- *Neoangles:* Neoangle showers are popular space savers, but the showering area is reduced in size due to the cut-off corner. The size of the corner cut varies from one model to another, with some cutting significantly into the shower space. For comfort, neoangles should be at least 42x42 inches.

- *Showerhead:* Locate the showerhead supply pipe 72 to 78 inches above the finished shower floor. The installed showerhead will be 4 to 6 inches lower. If a handheld showerhead is used, it should be no higher than 48 inches at its lowest position.

- *Seating:* Shower stalls should include a bench or seat that is 17 to 19 inches high and a minimum of 15 inches deep. The seat should not encroach on the minimum 34x34-inch floor space.

- *Doors:* Shower doors must open into the bathroom, not into the shower stall.

- *Controls:* Locate controls 38 to 48 inches off the floor and offset toward the room so they are accessible from both inside and outside the fixture.

Bathtubs

- *Clearances.* Allow a minimum 21-inch walkway (30 inches preferred) from the open side of the tub to a wall or fixture. These clearances may not allow full accessibility.

- *Steps to a tub.* Do not build steps leading to a bathtub or raised tub platform. These create a serious hazard. It is much safer for users to sit on the lip of the tub or platform and swing their legs in. Sunken tubs are also a hazard. Safety rails should be installed to help users get in and out of any tub configuration.

- *Controls.* Offset controls toward the outside of the tub so they are accessible from both inside and outside the fixture. Bathtub controls should be no more than 33 inches high. If a handheld showerhead is used, it should be no higher than 48 inches at its lowest position (Figure 6-18).

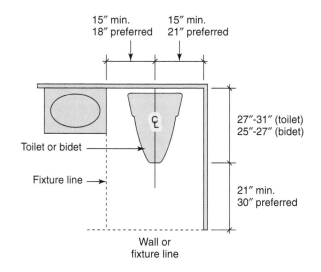

FIGURE 6-19 **Toilet and Bidet Clearances.**

15" min.
18" preferred

15" min.
21" preferred

27"-31" (toilet)
25"-27" (bidet)

Toilet or bidet

Fixture line

21" min.
30" preferred

Wall or
fixture line

Provide a minimum of 15 inches of clearance (17 to 18 inches preferred) from the centerline of a toilet or bidet to a wall or another fixture. Provide 21 inches of leg room in front of the fixture as a bare minimum.

Antiscald Protection. Protect all tubs and showers with a pressure-balancing valve or thermostatically controlled valve to limit water temperatures at a faucet or showerhead to 120°F or less. Recommend that homeowners set water heaters to no more than 120°F as an added precaution.

Toilets and Bidets

- *Clearances.* Fifteen inches is the allowable minimum from the centerline of toilets and bidets to a fixture or wall, while 17 to 18 inches will create a more comfortable space (Figure 6-19). Sixteen inches is the minimum for full accessibility as long as there is sufficient transfer space in front or on the side (see "Accessible Bathroom Design," below).

- *Separate compartment.* Compartmental toilet areas should be a minimum of 36x66 inches with a swing-out or pocket door.

- *Toilet paper holder.* Locate with the center 26 inches above the floor, about 8 inches forward from the front of the toilet.

Flooring. Make all bathroom flooring slip-resistant.

Ventilation. Provide mechanical ventilation to the exterior (see "Bathroom Ventilation," page 261, for sizing information).

Electrical. All bathroom receptacles must be GFCI protected. All light fixtures above a tub or shower must be rated for damp locations (tub) or wet locations (shower). Switches must not be reachable from within a tub or shower. Many bathrooms are wired so that all the lights go out if a GFCI is tripped. Although this is allowed by code, it is neither safe nor convenient for the homeowners.

Lighting. The vanity area should include both overhead and side lighting. Place side lighting centered at eye level (see "Bathroom Lighting," page 205). Where possible, provide natural lighting as well from a window or skylight area equal to at least 10% of the floor area.

Glass Safety. All glass used in a tub or shower enclosure or other glass applications within 18 inches of the floor should be safety glazing, such as laminated glass, tempered glass, or an approved plastic.

Typical Bathroom Layouts

Bathrooms are divided into three main centers of activity: lavatory/grooming, toilet/bidet, and bathing/showering. In smaller bathrooms, these all share one common space, while in more spacious rooms, the grooming area or toilet area may be separated to allow greater flexibility and privacy for multiple users. Larger spaces also allow for greater storage, such as a linen closet, within the bathroom space. Typical bathroom layouts with minimum dimensions for comfortable use are shown in Figure 6-20.

Accessible Bathroom Design

To make a bathroom fully functional for wheelchair users and other seated occupants requires commonsense changes, such as using universal controls and placing them within reach, as well as some significant changes, such as lowering sinks and providing knee space below. In some cases, the room will need to be enlarged to accommodate a roll-in shower or to allow room for wheelchair users to reverse direction. The minimum guidelines below, based on ANSI Standard A117.1, are a good starting point in design, but they should be tailored to the size, reach, and specific capabilities of the occupants.

Doors and Passageways. Clear space at doorways and passageways should be at least 32 inches wide and no more than 24 inches long in the direction of travel. Walkways between vertical objects (walls, cabinets, fixtures) greater than 24 inches long in the direction of travel should be at least 36 inches wide.

Pocket doors or doors that swing outward are preferred, since they do not encroach on bathroom space and will not get blocked in an emergency. Eliminate any thresholds at doorways.

FIGURE 6-20   Typical Bathroom Layouts.

Half Baths

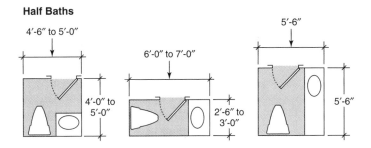

Three Quarter Baths

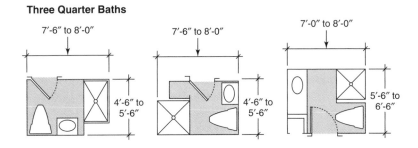

Full Baths

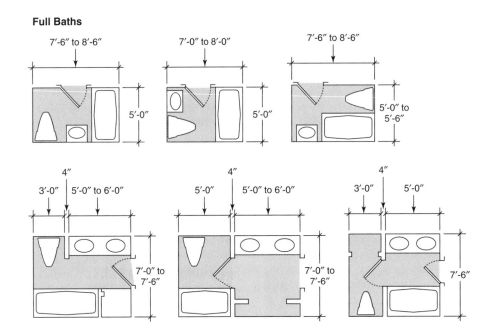

While the minimum dimensions shown will provide functional spaces, the more generous dimensions are preferred where space allows.

Lavatory Height and Knee Space. For most seated users, the recommended sink height is 32 inches (Figure 6-21). Provide knee space at the sink at least 27 inches high at the opening and 19 inches deep, with adequate toe space. Protect users from exposed pipes and mechanicals with insulation and a protective panel.

Floor Space at Lavatory. Provide a minimum clear floor space of 30x48 inches centered in front of the sink either parallel or perpendicular to the sink. Up to 19 inches of the 48-inch dimension can extend under the sink if knee space is provided (Figure 6-22).

FIGURE 6-21   Accessible Lavatory.

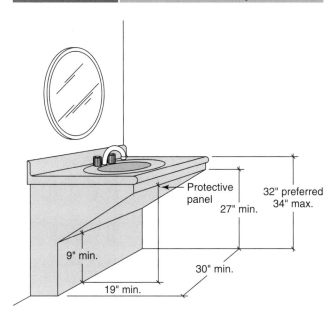

Provide knee space at the sink for seated users, as shown. The optimal sink height for most seated users is about 32 inches.
SOURCE: Reprinted with permission of John Wiley & Sons, from the *Essential Bathroom Design Guide* by the National Kitchen & Bath Association, © 1997 NKBA.

FIGURE 6-22   Floor Space at Accessible Lavatory.

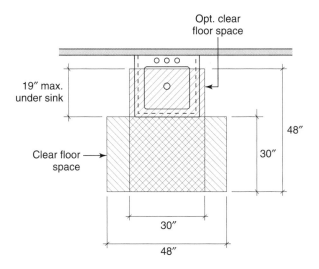

Provide a 30x48-inch floor space either parallel or perpendicular to the sink. Up to 19 inches of the 48-inch dimension can extend under the sink if adequate knee space is provided.

Floor Space at Toilet or Bidet.
Provide a minimum clearance of 18 inches on either side of the toilet or bidet centerline to a wall or fixture. Also provide a minimum 30x48-inch clear floor space (preferably 48x48 inches) in

FIGURE 6-23   Floor Space at Accessible Toilet or Bidet.

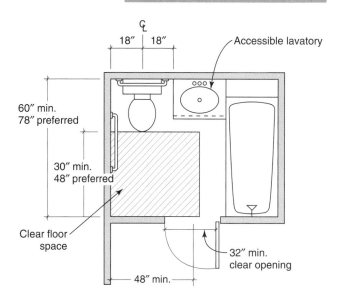

Provide a minimum 18-inch clearance on either side of a toilet or bidet and clear space in front, as shown. If possible, provide additional clear space on one side to ease transfers and provide space for a helper to stand.

front of the toilet or bidet. If necessary, the clear floor space may include up to 12 inches of knee space under an adjacent sink. Remember that these are minimum clearances. To simplify transfers, leave as much free space on one side of the toilet as possible (Figure 6-23).

Floor Space at Tub. Provide a minimum clear floor space of 60 inches along the length of the tub by 30 inches deep for a parallel approach or by 48 inches deep for a perpendicular approach. An additional 12 to 18 inches of clear space beyond each end of the tub is also desirable for access to controls and to ease transfers (see Figure 6-24).

Floor Space at Shower. For people who shower standing, provide a minimum 36x36–inch shower with appropriate grab bars, and provide a minimum clear floor space in front, 36 inches deep by the width of the shower plus 12 inches. People who cannot leave their wheelchair require wider roll-in showers of at least 30x60 inches. The minimum access space should be the full length of the shower by 36 inches. For either type of shower, an additional 12 to 18 inches beyond each end is desirable for better access to controls and to ease transfers (Figure 6-25).

Overlapping Floor Spaces. Clear floor spaces in front of fixtures may overlap and may include up to 12 inches deep of knee space below the sink.

FIGURE 6-24 Floor Space at Accessible Tub.

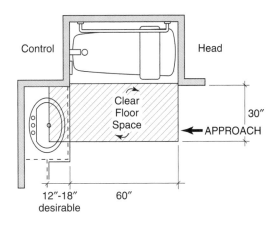

Parallel Approach

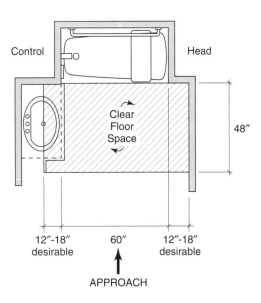

Perpendicular Approach

The 60-inch clear space shown above, based on ANSI standards, is a bare minimum. Additional floor space at one or both ends is helpful for transfers and access to controls.
SOURCE: Adapted with permission from *NKBA Essential Design Guide*, John Wiley & Sons, 1996.

Turning Space. A bathroom clear space for reversing direction in a wheelchair should be either a circle, 60 inches in diameter, or a T-shaped space of 36x36x60 inches (see Figure 6-26).

Grab Bars. These should be installed in the bathtub or shower and toilet areas for full accessibility.

- **Reinforcing.** Reinforce attachment points at the time of construction with $\frac{3}{4}$-inch plywood or solid

FIGURE 6-25 Floor Space at Accessible Shower.

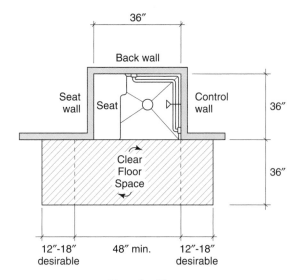

Transfer Shower

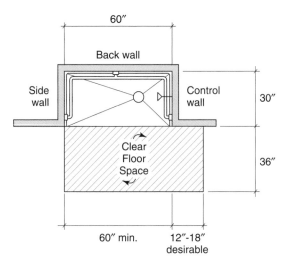

Roll-In Shower

Both standing and roll-in showers require adequate grab bars and clear floor space, as shown. For either type of shower, 12 to 18 inches of floor space beyond the minimum is desirable for better access to controls and to ease transfers.
SOURCE: Reprinted with permission of John Wiley & Sons, from the *Essential Bathroom Design Guide* by the National Kitchen & Bath Association, © 1997 NKBA.

2x6 blocking designed to bear a static load of 300 pounds.

- **Size.** Grab bars should be from $1\frac{1}{4}$ to $1\frac{1}{2}$ inches in diameter with a slip-resistant surface and sit $1\frac{1}{2}$ inches away from the wall.

FIGURE 6-26 **Wheelchair Turning Space.**

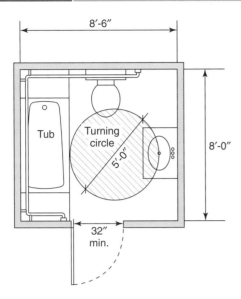

Turning Circle

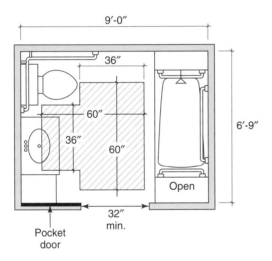

Space for T-Turn

Allow space in the bathroom for a wheelchair to reverse direction. Where a 60-inch-diameter circle is not possible, a T-shaped turning space is an option. Where space is tight, toe and knee clearance can be included in the turning circle or at the end of one arm of the turning T.

• *Location.* The optimal location of grab bars will depend on the users' specific needs, such as whether they will be sitting or standing and which types of movements they can and cannot perform. One vertical bar placed at the entry point to a shower or bath enclosure is generally useful to anyone getting in or

out. A horizontal bar on the control side is useful for people who stand in the shower.

ANSI guidelines specify grab bars at 33 to 36 inches above the floor. However, accessibility experts often place them higher or lower based on an individual's specific needs. In toilet areas, install one grab bar behind the toilet and one on the side wall closest to the toilet. Some people require grab bars on both sides. ANSI guidelines for toilet and tub areas are shown in Figure 6-27.

Storage. Locate storage for toiletries, linens, and bathroom supplies within 15 to 48 inches from the floor. Locate towel racks, soap dishes, and other personal hygiene items within the same height range.

Controls. Controls, dispensers, and outlets should be located from 15 to 48 inches high, and all devices should be operable with a closed fist. Offset controls in showers and tubs toward the room side, as shown in Figure 6-27. This makes them easier to reach for all users.

KITCHEN AND BATH FLOORING

The leading choices for kitchen and bath flooring installed in new homes are sheet vinyl and ceramic tile, chosen for their durability, ease of maintenance, and tolerance of water (Figures 6-28 and 6-29). Vinyl offers a resilient surface underfoot and is generally less expensive than tile, although the high-end vinyl products can cost nearly as much as lower-end tiles.

Tile is often chosen for its durability in both wet and dry environments. Although a ceramic tile installation is not completely waterproof without the addition of a waterproofing membrane, the tile itself, along with the cement backerboards and thinset mortars used in quality installations, are unaffected by water.

Hardwood, too, has become popular in kitchens as kitchens have evolved into primary centers for recreation and socializing. Although not the most practical choice for wet areas or high-traffic zones, new harder finishes make it more resistant to scratching and better able to tolerate the occasional wetting. The toughest finishes are available only on prefinished flooring, but seams between the boards could allow water to penetrate if exposed to standing water (see "Wood Floor Finishing," page 169). A satin finish is preferable in a high-traffic area like a kitchen, since it shows scratches less than a glossy finish. With any wood floor, the homeowners should wipe up spills quickly and use water sparingly when cleaning.

FIGURE 6-27 **Recommended Grab Bar Locations.**

Grab Bar Dimensions

Grab Bars at Toilets

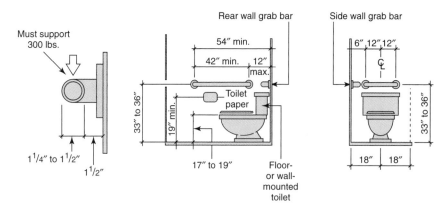

Grab Bars at Tubs (Seat in Tub)

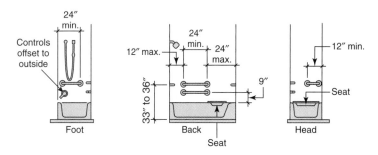

Grab Bars at Tubs (Seat at Head of Tub)

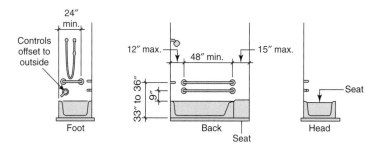

ANSI guidelines specify grab bars at 33 to 36 inches above the floor. However, accessibility experts may place them higher or lower based on an individual's specific needs. Note that tub and shower controls are offset toward the outside of the tub for easier access.

SOURCE: Details adapted from ADA guidelines, ANSI 117.1, and recommendations of accessibility experts.

For consumers seeking a resilient walking surface but reluctant to use vinyl for environmental or health reasons, newer options include a variety of cork products, newly introduced lines of traditional linoleum, and a variety of laminated bamboo products that perform essentially the same as solid hardwood flooring. With proper installation and care, any of these materials can provide a durable, attractive floor in a kitchen. In a bath, the best option is a waterproof surface with few joints or seams to allow water penetration (Table 6-3).

FIGURE 6-28	**Kitchen Flooring in New Homes.**

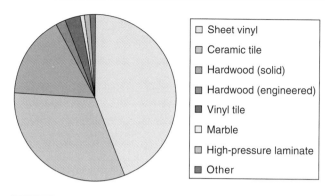

Legend:
☐ Sheet vinyl
☐ Ceramic tile
▨ Hardwood (solid)
▨ Hardwood (engineered)
▪ Vinyl tile
☐ Marble
▨ High-pressure laminate
▨ Other

Sheet vinyl	44%
Ceramic tile	32%
Hardwood (solid)	16%
Hardwood (engineered)	2%
Vinyl tile	3%
Marble	1%
High-pressure laminate	1%
Other	1%

SOURCE: Adapted from *Assessment of K&B Materials in Residential Construction, Assessment of K&B Materials in Residential Construction,* 2003, courtesy of NAHB Research Center.

FIGURE 6-29	**Bathroom Flooring in New Homes.**

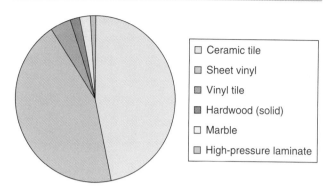

Legend:
☐ Ceramic tile
▨ Sheet vinyl
▨ Vinyl tile
▨ Hardwood (solid)
☐ Marble
▨ High-pressure laminate

Ceramic tile	46%
Sheet vinyl	43%
Vinyl tile	4%
Hardwood (solid)	2%
Marble	2%
High-pressure laminate	1%

SOURCE: Courtesy of NAHB Research Center, *Assessment of K&B Materials in Residential Construction.*

TABLE 6-3	**Kitchen and Bath Flooring Options**				
Type	Material Cost (1—Low; 5—High)	Suitability for Radiant Heat	Pros	Cons	Recommendations
Sheet vinyl (page 176)	1–3	Good. Consult mfg for installation instructions.	Many colors and patterns. Few seams. Resilient, stain-resistant, low-maintenance. Impervious to water. Premium types have "inlaid" color-through patterns and thick wear layers; some are toughened with aluminum oxide, etc.	Dents, gouges difficult to repair. Can stain if spills not wiped quickly. Can fade or yellow over time. Subfloor problems may telegraph through or seams may open.	Buy premium grade with thick wear-layer. Install over approved subfloor. Use fully adhered installation for best durability. Sweep and mop to remove grit.
Solid vinyl tiles (page 176)	2–3	Good. Consult mfg for installation instructions	Color and pattern run through thickness of tile. Very durable. Impervious to water. Damaged tiles can be replaced.	Requires periodic waxing and buffing. Joints may leak water or attract dirt. Can stain.	Maintain wax coating to protect tiles and keep joints well sealed. Sweep and mop to remove grit.
Ceramic tile (page 223)	3–4	Excellent. Avoid organic mastics.	Many color and texture choices. No-slip textures available. Very durable. Impervious to water, stains (except unglazed nonvitreous). Low maintenance.	Grout may stain, even if sealed. Unglazed porous tile needs sealing. Glossy types slippery when wet. Hard surface under feet. Requires stiff subfloor.	Choose at least Group II for bathrooms, Group III for kitchens (PEI scale). Choose no-slip type for bathrooms. Use gray or dark grout on floors and seal grout.

(continued)

TABLE 6-3	Kitchen and Bath Flooring Options (*Continued*)				
Type	Material Cost (1—Low; 5—High)	Suitability for Radiant Heat	Pros	Cons	Recommendations
Natural stone tile (page 231)	4–5	Excellent	Can take a high polish or honed finish. Granite is very hard and stain-resistant. Stains can usually be removed with some effort.	Marble is porous and stainable, and will etch from mild acids. Oils will stain granite. All stone must be periodically sealed to prevent stains. Requires very stiff sub-floor to prevent cracking.	Used polished or honed on bathroom floors. Honed on kitchen floors. Seal to prevent stains. Dry mop to remove abrasive grit. Clean with pH-neutral cleaner.
Hardwood (solid) (page 164)	3–5	Fair. Must use narrow, kiln-dried strips of stable species. Quartersawn preferred.	Warm, natural appearance. Somewhat resilient. Newer finishes require less maintenance. Can be refinished.	Not recommended in bathrooms or wet locations. Subject to water damage, scratches, wear. Prefinished flooring might allow water penetration at joints.	Finish in place with multiple coats of hardest finish available (not penetrating oil). Use satin finish in kitchens. Sweep and damp mop as per mfg recommendations.
Hardwood (engineered) (page 172)	4–5	Good. Floating installation best. Consult mfg. about max. temperature.	Warm, natural appearance. Floating type is resilient. Very hard reinforced finishes available. Most products can be refinished.	Not recommended in bathrooms or wet locations. Subject to water damage, scratches, wear. No finish coat to seal seams.	Select hardest finish available. Use satin finish in kitchens. Most OK to damp mop (consult mfg). Select edge-glued flooring or product with tight edge seals.
High-pressure laminate (page 179)	2–3	Generally acceptable. Consult mfg. for max. temperature.	Hard, durable surface that resists denting, stains, scratching, fading. Low-maintenance. Tight seams resist water. Damaged planks replaceable.	In baths, must be sealed at edges and penetrations to prevent water damage. Finish can chip, dent. High-traffic areas show wear. No refinishing.	Damp mop only. Do not flood surface.
Linoleum (sheets or tiles) (page 179)	2–3	Movement due to temperature change and sensitivity to concrete moisture may cause problems.	All natural materials. Resilient, durable. Also naturally nonallergenic and antistatic. Mottled colors hide stains. Little off-gassing. Presealed type does not require waxing.	Limited color choices. Requires occasional resealing and polishing to prevent stains and make cleaning easier. May darken or yellow.	Recommended for clients seeking natural materials with low off-gassing. Use pH-neutral cleaner.
Cork (page 177)	3–5	High R-value may cause problems, particularly with thick materials (over $\frac{3}{8}$ inch).	Natural renewable resource. Resilient, insulating, sound-dampening, and water-resistant. Available as tiles or planks. Variety of finishes available, some with low VOCs.	Limited colors. Wax finish needs regular buffing and rewaxing. Seams and edges need sealing to prevent water swelling. Can fade, dent. Has mild odor.	For a watertight floor, finish tiles in place with polyurethane. Damp mop only with mild cleaner.
Rubber tiles	2–3	Not suitable.	Many colors and surface textures available. Durable, resilient, insulating, sound-dampening, and impervious to water and most chemicals. Textured surface nonslippery. No waxing required.	Matte finish only. Oil or grease can stain some products.	Use grease-resistant product in kitchen. Sweep, mop with neutral pH cleaner. Buff at low speed.

TABLE 6-4	Tile Bisque Characteristics			
Tile Type	Characteristics	Common Examples	Best Uses	Installation
Nonvitreous	7% or more water absorption (by weight). Low strength.	Glazed wall tile, handmade Mexican pavers.	Interior applications not subject to frequent wetting or heavy loads. Avoid exterior use, particularly in cold climates.	Organic mastic or thinset adhesive. Use latex-modified thinset and grout in wet applications. Mist unglazed portions of tile before grouting.
Semivitreous	3 to 7% water absorption. Medium strength.	Floor tile, machine-made pavers and quarry tile.	Dry or wet interior applications. Exteriors not subjected to freeze-thaw cycles.	Organic mastic or thinset adhesive. Use latex-modified thinset and grout in wet applications. Mist unglazed portions of tile before grouting.
Vitreous	0.5 to 3% water absorption. High strength and impact resistance.	Floor tile, mosaic tile, machine-made pavers, quarry tile.	All interior or exterior residential applications. (Some vitreous tiles have soft glazes not suitable for floors; check with mfg.)	Organic mastic or thinset adhesive. Use latex-modified thinset and grout in wet applications. Sponge tiles before grouting.
Impervious	Less than 0.5% water absorption. High strength and impact resistance.	Porcelain tile, porcelain pavers.	All interior or exterior applications, including sanitary, commercial, and industrial.	Use latex-modified thinset or epoxy. Sponge tiles before grouting in hot weather.

CERAMIC AND STONE TILE

A quality ceramic tile job starts with proper framing to support the tile and the selection of tile materials that are right for the application and compatible with each other. The substructure must be stiff enough to support the tile without excess movement or deflection, and the tile, backerboard, adhesive, grout, and any waterproofing membrane must be compatible with one another. If all these products are installed following the manufacturer's instructions as well as the specifications of the Tile Council of America (TCA), the result should be an attractive and durable job. Finally, it is critical that the installer leave the required expansion joints at the room perimeter, tub lips, and other places the tile is restrained—the source of many tile callbacks.

Selecting Tile

A wide array of tiles are readily available. In addition to aesthetic concerns, tiles vary in strength, water absorption, scratch resistance, ease of cleaning, and slip resistance. In general, look for harder tiles for floor and counter applications, and tiles low in water absorption for wet applications. Beyond looking at the specifications, it is a good idea to test a sample of tile for scratch resistance, scuffing, and ease of cleaning, using real pots and pans, shoes, and household cleansers.

Strength and Water Absorption. The body of a tile, also called the *bisque,* is made by heating a mixture of clay and other additives in a kiln. In general, the longer the clay is fired and the higher the temperature, the denser and stronger the tile will be and the more impervious to water absorption. Nonporous tiles that absorb little water will perform better in wet applications than porous tiles. The tile bisques manufactured according to ANSI standards are rated from nonvitreous to impervious (see Table 6-4).

Glazed vs. Unglazed Tile. With the exception of quarry tile, terra-cotta, and some porcelains and mosaics, most tiles come glazed. The glaze consists of a mix of silica and pigments that is fused to the surface of the tile at high temperatures, creating a glasslike coating. Glazes provide decorative color and protect the surface of porous tiles from absorbing water and stains. How well a glaze resists abrasion and shows scratches depends on several factors:

- *Temperature:* Glazes fired hotter and longer tend to be harder and more scratch-resistant.
- *Color:* Light-colored glazes tend to be harder than dark colors and conceal scratches better.

TABLE 6-5	PEI Tile Wear-Rating System
Group	Suitable Applications
Group 1	Light duty: Residential interior walls and bathroom or bedroom floors where soft footwear is worn.
Group 2	Light traffic: Residential bedroom and bathroom floors. General residential traffic, except kitchens, foyers, and other high-traffic areas.
Group 3	Light to moderate traffic: All residential floor, countertop, and wall applications, and light-commercial areas.
Group 4	Moderate to heavy traffic: All residential and commercial floors where normal footwear is worn, including restaurants, hotels, hospital lobbies and corridors.
Group 5	Heavy to extra-heavy traffic: Heavy traffic and wet areas where safety is a concern, such as food service areas, exterior walkways, shopping malls, and swimming pools.

- *Gloss level:* Matte-finished glazes tend to be harder than high gloss and also conceal scratches better.

Unglazed tiles show the natural color of the clay, although some unglazed mosaics have pigment added to the clay. Unglazed tiles may need to be sealed to prevent staining during grouting or in use on floors, counters, and other applications prone to staining. Sealing is generally done before grouting. If used on a counter, make sure the sealer is suitable for use around food.

Many manufacturers now rate the abrasion resistance of their tile using the guidelines of the Porcelain Enamel Institute (PEI). The PEI system rates tiles from 1 to 5 as shown in Table 6-5. Select Grade 3 or higher where scratching of the tile surface is a concern.

Slip Protection. Many glazed floor tiles become dangerously slippery when wet. This is a concern wherever floors are subject to wetting, but particularly on shower floors and bathroom floors near tubs and showers. In general, unglazed tiles or textured patterns will be less slippery. Some tile has a special nonskid surface made by adding an abrasive grit to the tile face or glaze. The downside is that nonglossy surfaces are somewhat harder to keep clean.

Many tile manufacturers use a coefficient of friction (COF) to rate the traction a tile provides. While there are no national standards that specify a required COF, the Americans with Disabilities Act (ADA) recommends a minimum COF of .60 on accessible walking routes. Experts also recommend a minimum COF of .60 for shower stalls, wet bathroom floors, and other wet areas.

Ceramic Tile Types

A wide variety of man-made synthetic and natural tiles are available (Table 6-6). Most fall into one of the categories described below:

Glazed Ceramic Tile. The most common type of tile, glazed ceramic tile, is usually marketed as either a wall or floor tile, depending on the hardness and water-absorption of the underlying clay body of the tile or bisque. Wall tiles are typically $\frac{1}{4}$ inch thick and range in size from 4x4 to 12x12 inches. Floor tiles are generally thicker and are available in sizes up to 13x13 inches and larger.

Some floor tiles may also have a slip-resistant surface, which is advisable in a bathroom. A tile sold for use on floors will not necessarily have a hard scratch-resistant glaze, however. Where scratching is a concern, such as in kitchens and entryways, look for a PEI rating of 3 or higher, or test the scratch resistance of the tile yourself with a cooking utensil or other abrasives the tile might face in service.

Porcelain Tile. Porcelain tile is fired at high temperatures, creating a dense and strong material that is impervious to water absorption. The tile has a smooth texture and sharply formed face. Small imperfections in size due to the high firing temperatures give it a less formal look than standard tiles. Porcelain tile is available unglazed or glazed with a matte or high-gloss finish, and comes in a wide variety of sizes, colors, and shapes. Small 1x1–inch porcelain tiles are commonly used on shower and bathroom floors. These are usually mounted in sheets to simplify installation. Because porcelain is impervious to water absorption and because the mounting sheet can interfere with the bond, it should be installed with a polymer-modified thinset mortar made for use with porcelain.

Mosaic Tile. Mosaic refers to any hard, dense tile, such as porcelain, vitreous clay, or glass, that is typically one inch square. Porcelain mosaics are usually unglazed and are colored by adding pigment directly to the clay. Because it is tough and durable, mosaic tile can work well on just about any application, including floors, shower stalls, and counters. Generally mosaics are mounted in sheets and should be installed with latex-modified thinset mortar (Figure 6-30).

Paver Tiles. Machine-made pavers are $\frac{1}{2}$ to $\frac{3}{4}$ inch thick and are available glazed or unglazed. They are usually made from semivitreous or vitreous clay or impervious porcelain, making them suitable for outdoor use. The glazes used on machine-made paver tiles tend to be very hard and well suited to floors and countertops.

Handmade pavers, on the other hand, tend to be relatively soft, unglazed, nonvitreous tiles, so they are not suitable for wet interior applications or outdoors in areas subject to freeze/thaw cycles. Handmade pavers are generally made in Mexico from yellow or brown terra cotta and

TABLE 6-6 | **Tile Characteristics**

Ceramic Tiles				
Tile Type	Physical Properties	Pros	Cons	Best Uses
Glazed wall tile	Usually nonvitreous tile made by dust press method from talc or other clays. Water absorption not to exceed 20%. Typically $\frac{5}{16}$ in. thick and available in many sizes and shapes.	Glazes impervious to water. Tiles are stain-resistant, nonfading, easy to clean.	Limited ability to withstand abrasion and impact. Bright glazes tend to scratch more easily than matte.	Residential walls
Porcelain tile	Fired at high temp. to create a strong, dense, impervious tile with fine-grain and sharply formed face. Available unglazed, or with a matte or high gloss glaze. Unglazed have pigment throughout tile. 1x1–in. and 2x2–in. mosaics, but larger sizes available.	Very hard and durable, impervious to water absorption. Unglazed type have through color and moderate slip resistance. Available with granular or abrasive nonskid finish.	Due to low absorption, install with polymer-modified thinset or epoxy formulated for porcelain. Requires diamond wet-saw to cut.	Residential walls, floors, wet areas. 1x1–in. mosaics popular on shower and bathroom floors.
Machine-made pavers	Semivitreous or vitreous clay or porcelain tiles. Available unglazed or with a hard, durable glaze. Typically $\frac{3}{8}$ to $\frac{1}{2}$ in. thick and from 3x3 in. to 12x12 in.	Strong, durable tiles with hard, abrasion-resistant glazes.	May require back-buttering when installed.	Indoor or outdoor floors, countertops, wet areas
Handmade pavers (terra-cotta)	Relatively soft unglazed nonvitreous tiles made from yellow or brown terra cotta, mostly in Mexico. May come prefinished or require sealing to prevent staining and to provide a wear surface. One-half to 2-in. thick and from 4x4 in. to 24x24 in.	Suitable for dry interior floors and exteriors in nonfreezing climate. Variations in color, texture, and appearance provide rustic appearance.	Requires sealing before grouting and periodic sealing to prevent staining. Too absorbent for wet interior applications where hygiene is a concern. Must back-butter tiles when installed.	Floors not subject to frequent wetting.
Quarry tile	Glazed or unglazed tile made by extrusion of natural clay or shale. Usually dark red but also brown and gray with through color. Semivitreous or vitreous tiles $\frac{3}{8}$ to $\frac{3}{4}$ inch thick by 4x4 in. to 12x12 in.	The dense, thick tiles have through-color, making them durable and suitable for heavy traffic, as well as wet or exterior applications.	Unglazed tiles require sealing with surface coatings, which are not suitable for food prep areas such as counters. Tile too dense for penetrating oils.	Bath, kitchen floors.
Stone Tiles				
Marble	Natural stone capable of taking a polish. Primarily one or more of calcium carbonate, dolomite, or serpentine. Wide range of colors, abrasive hardness, and quality. Lower quality fabricated with fillers, adhesives, and sometimes fiberglass backing for stability.	Polished finish suitable for walls and bathroom floors. Honed marble suitable for floors or countertops if sealed. Worn floors can be cleaned and repolished.	Must be sealed to prevent stains. Polished finish easily etched by mild acids or worn away on kitchen floors. Polished floors slippery when wet. Serpentine prone to warp when wet.	Polished finish: walls, bathroom floors. Honed finish (sealed): floors, countertops

(*continued*)

TABLE 6-6 *(Continued)*

Stone Tiles

Tile Type	Physical Properties	Pros	Cons	Best Uses
Granite	Hard, granular igneous rock containing mostly quartz and feldspars. Ranges in color from pink and red to light or dark gray or a mix of these. Common finishes include polished, honed, or thermal.	High compressive strength and abrasion resistance. Unaffected by mild acids. Holds polish well. Very low absorption. Resists stains well and most can be removed with effort.	Porous types, such as gneiss, prone to stains. All types subject to stains from oil and grease if not properly sealed. Polished floors slippery when wet. Difficult and expensive to repolish.	Walls, floors, countertops
Agglomerate	Synthetic, from graded marble or granite chips, stone dust, and cement or resin binder. Usually less expensive than natural stone tiles. Available with polished or honed finish.	Similar properties to natural granite and marble tiles.	Limitations similar to natural granite and marble tiles, though typically less strong.	Walls, floors, countertops.

FIGURE 6-30 **Mosaic Tiles.**

Mosaic tiles are tough and durable and can work well on floors, walls, shower stalls, and counters. Mosaics are generally mounted in sheets and should be installed with latex-modified thinset mortar.
SOURCE: Photo courtesy of Dal Tile.

FIGURE 6-31 **Machine-Made Pavers.**

The hard glazes on these machine-made paver tiles are well suited to floors and countertops. Handmade pavers, on the other hand, are relatively soft and unglazed, and require sealing. Their uneven color and texture create a rustic appearance.
SOURCE: Photo courtesy of Dal Tile.

come in various shapes and sizes. Because they are handmade, they have uneven faces and vary in color and texture from tile to tile, giving a floor a rustic appearance (Figure 6-31).

Handmade pavers may come prefinished or require sealing by the installer with either a surface coating or penetrating sealer to provide a wear surface. Penetrating sealers have the advantage of easy refinishing of worn spots, whereas surface coatings need to be stripped before being reapplied. Apply any sealers before grouting.

All pavers should be set with a thinset adhesive. Because of their irregular shapes, the individual tiles may need to be "back-buttered" to provide full support and contact with the substrate material.

Quarry Tile. Originally made from quarried stone, quarry tiles are typically unglazed, semivitreous or vitreous tiles made from extruded slabs of clay or shale. They are $\frac{3}{8}$ to $\frac{3}{4}$ inch thick and come in a variety of square and rectangular shapes. They range in color from gray to browns and reds, depending on the type of clay and firing temperature. These dense, thick tiles have through-color, making them very durable and suitable for heavy traffic as well as wet or exterior applications. The tiles will stain, however, so they should be sealed with surface sealers where that is a concern (the material is too dense for penetrating oil sealers). Since most surface sealers are not suitable for food contact, quarry tile is not a good choice for counters where staining is a concern. Installation should be with thinset adhesive.

Natural Stone Tiles

Stone and agglomerate stone tiles are popular choices in kitchens and baths due to their durability and natural beauty. Stone tiles typically measure from $\frac{3}{8}$ to over 1 inch thick, making some 12-inch square tiles weigh over 10 pounds. The most common stones are marble and granite, since they both can take a high polish. Of the two, granite is much more durable and stain-resistant, and is unharmed by mild acids that will etch marble and limestone. Granite is stained by oil and grease, however. All natural stones are subject to staining if they are not sealed when installed and resealed whenever water does not bead up on the surface.

Most customers prefer a highly polished finish on natural stone in the kitchen or bath. A polished finish offers some protection against stains but can also make floors slippery and requires maintenance to preserve the sheen. On kitchens floors, polished stone will eventually lose its sheen in high-traffic areas, unless protective coatings and sealers are applied regularly. Softer and more porous stones, such as limestone and sandstones, are prone to excess wear and staining and are rarely used in kitchens and baths.

On any stone floor, sand, dirt, and grit do the most damage due to their abrasiveness. Commonsense approaches, such as entry mats to clean shoes and frequent dusting with a dry mop, will go a long way toward preserving the stone surface.

Installation. Installation of stone tile is essentially the same as with ceramic tile except for the following:

- *Deflection of substrate.* Some of the softer stones, such as marble and limestone, require a stiffer floor than the L/360 required for ceramic tile. To meet the stiffer load requirements of L/720 or L/1020, floor joists must be upsized or located closer together. A thicker or double-layer subfloor may also be required.

- *Compatibility with setting materials.* Porous and light-colored stone tiles may be stained or discolored by certain grouts, adhesives, or sealants. With light-colored or translucent stone tiles, it is best to use a white thinset rather than the more common gray type. Also, colored grouts, plumber's putty, and some caulks will stain lighter-colored stones and porous stones. It is best to test a sample for compatibility before installation.

Marble and Limestone. Both marble and limestone are carbonates, made from ancient shells, sand, and mud, although marbles have been additionally heated and squeezed until crystallized. Marbles tend to be denser and less porous than limestones and can take a high polish, but both materials are relatively soft and will etch with mild acids (lemon juice, vinegar), making them unsuitable for kitchen counters unless a rustic appearance is acceptable. Darker stones will etch more noticeably. Limestones and softer marbles are not suitable for floors either.

Marble comes in a wide variety of colors and levels of quality, ranging from Grade A with few to no flaws or voids to Grade D, which has a large proportion of flaws, voids, veins, and lines of separation that need to be repaired with fillers and adhesives during fabrication of the tiles. A fiberglass mesh may also be laminated to the back to provide stability. Some of the most highly prized colored marbles are the least stable.

Tiles are available either polished or honed. Polished marble is suitable for walls or bathroom floors, but the polish will generally not hold up well on kitchen floors or countertops due to wear and tear and mild acid spills. Although worn marble can be cleaned and repolished, a non-glossy honed finish is a better choice for kitchen floors and counters. The honed finish must be sealed to prevent permanent staining of the porous marble.

Dark green marble, called serpentine, is actually a much harder silicate. It will not etch but has a tendency to warp when wet, so it should not be installed in wet areas or exposed to standing water when cleaned. When installing serpentine, epoxy-based mortars and adhesives are preferable to water- or latex-based products. Presealers, applied before grouting, help prevent stains and simplify cleanup, particularly with nonpolished finishes.

Granite. Granite is a hard, granular igneous rock that contains mostly quartz and feldspars and ranges in color from pink and red to light or dark gray or a mix of these. It is generally uniform in color and has high compressive strength and abrasion resistance. It has very low absorption, but some types, especially the popular swirl types (technically called gneiss), are subject to staining if not sealed. Common finishes include polished, honed, or thermal.

Polished granite tiles are suitable for kitchen and bath walls, floors, or countertops. While granite holds its polish longer than marble and is not bothered by mild acids, such as orange juice or vinegar, over time the traffic on a kitchen floor will dull the finish. Use of doormats and frequent sweeping to remove abrasives from the floor will help prolong the finish. Repolishing the granite tends to be expensive due to the hardness of the material. Waxing and special coatings may help protect the polished finish, but frequent stripping and reapplication can, by itself, cause excessive wear.

Another option for floors are honed or thermal finished granite tiles, which are more commonly used outdoors. These are less slippery when wet than polished granite. However, a sealer is required with this type of finish to prevent staining during grouting and to protect from oil or grease stains. These finishes are easy to apply and do not need stripping for touch-up reapplication.

Agglomerates. Agglomerate tiles consist of graded chips of marble or granite mixed with a resin binder and stone dust. Typical thicknesses vary from about $\frac{1}{4}$ inch to $\frac{7}{8}$ inch. Agglomerates usually cost less than natural stone, but have many of the same virtues and limitations as their natural counterparts. For example, granite agglomerates are harder than marble agglomerates, and the polish on granite can withstand greater abrasion and last longer. Also, granite agglomerates will resist mild household acids, such as citrus juice and vinegar, while these will tend to etch marble agglomerates.

Sealers. A wide variety of proprietary sealers are available to protect natural stone against staining from grout, dirt, foods, and household products. In addition, some sealers help conceal minor scratches and increase slip resistance. Some products require regular reapplication and may cause a surface buildup unless stripped. Penetrating oil-type sealers may change the color of some types of stone and can even trap dirt in the finish. For best results, follow recommendations of the stone supplier and use products with an established track record.

Framing under Tile

Tiles, stones, and grout joints crack easily from stresses imposed by movement. For a successful installation, the structure underneath must be very stiff. On walls, 16-inch on-center framing with 2x4s or steel studs is usually adequate. Floors must be level and subject to minimal deflection under uniform or point loads.

Deflection. An insufficiently stiff floor will crack ceramic or stone tiles. The Tile Council of America (TCA) specifies a maximum deflection for floors of L/360 under a 300-pound concentrated load. While building codes limit deflection in living spaces to L/360 under uniform loads, code-compliant floors may still have too much flexing between joists under point loads. Many natural stone tiles require stiffer conditions, ranging from L/480 to as stiff as L/720.

Subflooring. To meet TCA stiffness requirements, floor framing should be no more than 16 inches on-center with minimum $\frac{19}{32}$-inch plywood subflooring. Upgrading to $\frac{23}{32}$-inch plywood will stiffen the subfloor by almost 80% and provide a more solid feeling floor. The subflooring should be level to $\frac{1}{8}$ inch in 10 feet. (TCA specs now permit $\frac{1}{4}$ inch in 10 feet, but this can be problematic for the large tiles popular today.)

To avoid tile cracks caused by tight-fitting plywood joints, it is best to use square-edged subflooring under tile and leave an $\frac{1}{8}$-inch gap between sheets (unless the setting material specifications require tight joints). Lay the plywood with its long dimension across the joists and use solid blocking at all open joints.

Two-Layer Subflooring. To meet the stiffness requirements for natural stone floors may require two layers of subflooring screwed and glued together, with the upper layer serving as the underlayment. Two layers of $\frac{19}{32}$-inch plywood glued and screwed together on 6-inch centers is several times stiffer than a single layer (and over four times as stiff as a single layer of $\frac{23}{32}$-inch plywood). Offset the upper layer so the joints do not line up with the joints in the lower layer or the joists. Also, screws in the upper layer, which serves as underlayment for the tile, should penetrate the subfloor only and not the joists. Use underlayment-grade plywood or plywood rated C-C Plugged or Plugged Crossbands, with a smooth face and no voids.

Substrates for Tile

Ceramic tile can be installed over clean and sound concrete, plywood, cement backerboard, drywall, or plaster. Most substrates can be used with either organic mastic or thinset mortar, but the installer should always check the adhesive label for compatibility with the substrate.

Plywood Underlayment. Because of its stiffness and durability, exterior plywood makes an excellent substrate for tile in relatively dry applications.

- *Plywood type.* The plywood should be free of internal voids such as underlayment grade, CC-Plugged, or Plugged Crossbands , and if tiling directly to plywood, it should have a sanded face free of voids, surface resin, or other surface defects. While TCA specs allow $\frac{15}{32}$-inch plywood underlayment over a $\frac{19}{32}$-inch subfloor in residential work, upgrading to $\frac{19}{32}$-inch underlayment is recommended for a trouble-free floor (Figure 6-32).

- *Plywood installation.* Leave a $\frac{1}{8}$-inch gap between sheets and make sure the edges of adjacent plywood sheets are no more than $\frac{1}{32}$ inch out of plane. Overall, the surface should be level to $\frac{1}{8}$ inch over 10 feet ($\frac{1}{4}$ inch is allowed by the tile industry but is not suitable for large tiles). Fasten with ring-shank nails or screws at 6 inches on-center.

- *Glue and screw.* Where greater stiffness is required in a floor, it is worth the effort to also glue the underlayment to the subfloor and upgrade from nails to screws.

- *Isolation membrane.* On large spans or where significant movement is expected in the floor due to wide moisture or temperature swings, a crack-isolation membrane is recommended as a precaution against cracking. Membranes, if sealed at seams, can also protect the plywood from moisture that seeps through the tile system. This is required in wet applications and recommended in applications subject to occasional wetting, such as bathroom floors.

FIGURE 6-32 Plywood Underlayment for Tile.

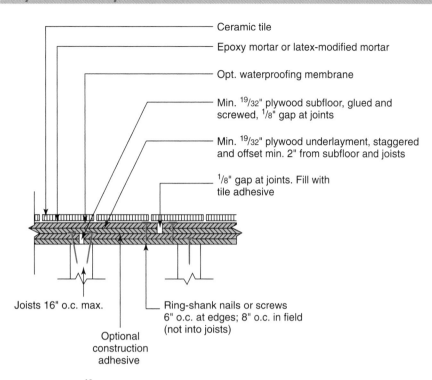

Ceramic tile

Epoxy mortar or latex-modified mortar

Opt. waterproofing membrane

Min. $^{19}/_{32}$" plywood subfloor, glued and screwed, $^1/_8$" gap at joints

Min. $^{19}/_{32}$" plywood underlayment, staggered and offset min. 2" from subfloor and joists

$^1/_8$" gap at joints. Fill with tile adhesive

Joists 16" o.c. max.

Optional construction adhesive

Ring-shank nails or screws 6" o.c. at edges; 8" o.c. in field (not into joists)

Two layers of $\frac{19}{32}$-inch plywood makes a sturdy substrate for tile floors. Where greater stiffness is required, use thicker plywood or glue and screw the underlayment to the subfloor. A waterproofing and isolation membrane will help prevent cracking from structural movement and is essential in wet applications.

- **Countertops.** Plywood also works well as a tile substrate on walls and countertops not subject to regular wetting (Figure 6-33). In applications subject to high humidity or regular wetting, cement backerboard or glass-mat gypsum are better choices.

Cement Backerboard. Developed specifically for use under ceramic tile, cementitious backer units (CBUs) are made of sand and cement and are reinforced by fiberglass facings or chopped fibers within the material itself. Cement backerboard is impervious to moisture, but may let moisture pass through, so it should always be backed by a waterproof barrier, such as polyethylene sheeting or asphalt-impregnated felt. Because the material has little inherent strength, it relies on the structure beneath for stiffness when used on floors and other applications subject to significant loads. A variation on cement backerboard, Hardibacker (James Hardie Building Products), is made of fiber-cement. Because it is only $\frac{1}{4}$ inch thick, it is usually installed over $\frac{1}{4}$-inch plywood, both for stiffness and to bring it flush with the surrounding drywall.

Cement backerboard should be installed with special backerboard screws (not drywall nails) that have an anti-corrosive coating and oversized heads with ridges underneath designed to self-countersink and pull tight. Roofing nails are also acceptable with most backerboards.

Closely follow manufacturers instructions, which vary a little from product to product. Typical installations include:

- **Backerboard on walls:** CBUs can go directly over studs at 16 inches on-center or over drywall, and it is fastened with $1\frac{1}{2}$-inch galvanized roofing nails or special self-countersinking galvanized screws at 6 to 8 inches on-center.

- **Backerboard on floors and countertops:** CBUs are bonded to the plywood with a layer of latex- or acrylic-modified thinset (Figures 6-34 and 6-35). Apply the thinset with the flat side of a $\frac{1}{4}$-inch notched trowel, then rake with the notched side to provide a continuous level setting bed. Seat the backerboard evenly with a beating block, then secure with roofing nails or special backerboard screws. Because CBUs provide little added stiffness, it is best to upgrade to a $\frac{23}{32}$-inch subfloor.

- **Joint details:** Leave a $\frac{1}{8}$- to $\frac{3}{16}$-inch gap between adjoining sheets of backerboard and at edges. Just before tiling, fill the joints with thinset and cover with 2-inch-wide fiber mesh tape (unless the adhesive manufacturer recommends otherwise). Then embed the tape in a thin skim coat of thinset. Reinforce inside and outside corners with three pieces of 2-inch tape or one piece of 4-inch tape.

- **Obstructions:** Leave a minimum $\frac{1}{4}$-inch gap where the CBU meets a tub lip, plumbing fixture, pipe, or any restraining surface and fill the joint with a flexible, waterproof sealant.

FIGURE 6-33 **Light-Duty Tile Counter.**

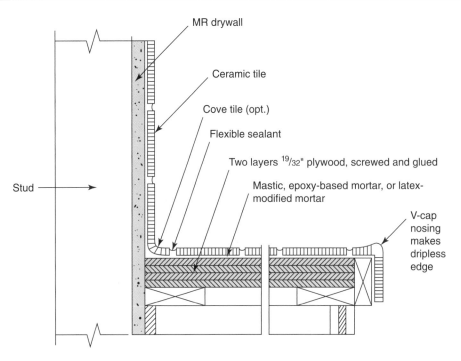

For tile counters not subjected to heavy loads or regular wetting, two layers of $\frac{19}{32}$ plywood make a suitable substrate. Choose plywood with no voids and a smooth face, free of defects.

FIGURE 6-34 **Plywood Subfloor with Cement Backer Board.**

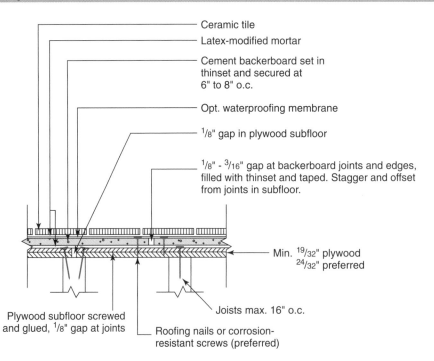

Cement backerboard makes an excellent substrate for floors in wet areas, but it does not provide any structural support. Make sure the joists and subflooring are stiff enough to prevent cracking. Also, if more than occasional wetting is expected, install a waterproof membrane under the tile.

FIGURE 6-35 Wet Counter with Cement Backerboard.

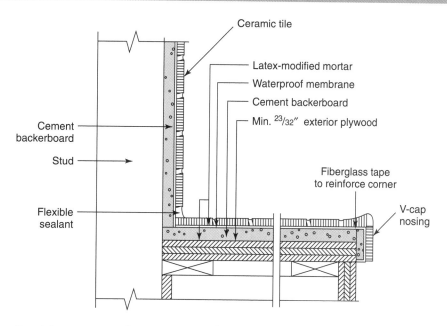

For sink counters or other countertops subject to frequent wetting, cement backer-board will provide the best service. A trowel-on waterproofing membrane between the backerboard and tile will protect the wood underneath from any water intrusion.

Drywall. Regular $\frac{1}{2}$-inch drywall over 2x4 framing or steel studs 16 inches on-center is a suitable substrate for dry installations. Using $\frac{5}{8}$-inch drywall or a second layer of $\frac{1}{2}$-inch will provide a stiffer wall. Joints should be taped and one coat of joint compound applied to joints and nails. If tiles will be set in thinset mortar, use thinset with mesh tape at the drywall joints.

Moisture-Resistant Drywall. Commonly called "greenboard," this offers moderate protection in moist conditions, but it is not recommended in wet areas, such as tub surrounds, unless protected by a waterproof membrane.

Coated Glass-Mat Gypsum Backerboard. Competing with cement backerboard, gypsum products such as Georgia Pacific's Dens-Shield® are designed as tile substrates for damp areas. The gypsum core is treated with silicone to make it water-resistant, and the fiberglass and acrylic facing acts as a surface vapor barrier. While not as strong as cement backerboard, gypsum-based backerboard installs faster since it cuts and installs like drywall. Installation is with roofing nails or galvanized bugle-head drywall screws, which should be driven flush with the surface but should not break the waterproof skin. Panels are butted tightly and the joints are taped and embedded with latex Portland cement mortar, also used to set the tile.

In use since 1987, Dens-Shield® is well-suited to tub surrounds and other light-duty wall applications, but it is not suitable for saunas, steam rooms, or other applications

facing extreme heat and humidity. As with other tile substrates, follow manufacturer's instructions closely regarding framing, installation, and tile application.

Concrete Slabs. Tiles can be applied directly to a clean, smooth concrete slab with a latex Portland cement mortar. The concrete should be properly cured, be level to $\frac{1}{4}$ inch in 10 feet, and have a steel trowel or fine broom finish. Curing compounds or old adhesives used for carpeting or resilient flooring will act as bond breakers, so they need to be removed before installing the tile. Either a power-blast cleaning machine or a rotary sander with a coarse carborundum sanding sheet can be used. If the slab has cracks or control joints, a crack isolation membrane should be used (see Isolation Membranes, page 237).

Other Tile Substrates. In remodeling, tiles can be successfully installed over existing ceramic tile, plastic laminate, or resilient flooring (except the cushioned type) that is well bonded. These must be stripped of any sealers or waxes and scarified with a coarse sander before applying the new adhesive. Wood floors must first be covered with plywood or a cementitious backerboard.

Unsuitable substrates include oriented-strand board (OSB), lower grades of plywood with voids, and interior-grade plywoods such as lauan.

Leveling Compounds. If a wood floor or slab needs leveling, you can use special cement leveling compounds formulated for use as a substrate for ceramic tile or stone tile. These are either site-mixed from a bag or installed by a specialty subcontractor. Some are self-leveling and others require screeding. Gypsum-based underlayments are generally not suitable for tile. Most self-leveling compounds can be installed up to about an inch thick. For thicker applications, let the first layer dry before adding more, or fill lower areas with plywood shims before adding the compound. Use exterior-grade plywood with no voids.

Moisture and Water Barriers

Although glazed ceramic tile is waterproof, neither the grout joints nor the cement backerboard behind the tile are water barriers. To prevent moisture from passing through the tile and substrate to the plywood or wood framing, a moisture barrier is required in areas subject to high moisture levels or occasional wetting, such as tub surrounds and kitchen counters. In areas subject to heavy wetting, such as shower pans and some bathroom floors and counters, a sheet membrane or trowel-on membrane should be used to provide full waterproofing.

Moisture Barriers. On tiled walls, protect the wood framing from water intrusion, using either 6-mil poly or 15-pound asphalt-impregnated felt lapped to shed water. The barrier should go between the tile substrate and the framing. On outside walls, this material can also served as the air and vapor barrier if the joints are sealed with tape or a compatible sealant.

Membranes. Full waterproofing is required in construction that must retain water, such as shower pans and tiled tubs. It is also recommended in areas subject to frequent wetting, such as raised tub surrounds, bathroom floors, and counters with sinks. There are two types of membranes: thermoplastic sheet materials and trowel-on membranes. Sheet membranes can be applied to most tile substrates with either thinset mortar or a proprietary adhesive, and tiles are bonded directly to the membrane. With any membrane, check the label for compatibility with the substrate and adhesives. Most waterproofing membranes also serve as isolation membranes.

- *Sheet membranes:* These single-layer elastomerics, such as chlorinated polyethylene (CPE), are bonded to the substrate—typically cement backerboard, plywood, or a mortar bed—with a compatible thinset mortar or proprietary adhesive (Figure 6-36). After the adhesive cures, tiles are set onto the membrane with a second layer of thinset. For watertight performance, joints in the membrane are caulked with a proprietary sealant or are solvent-welded. In addition to stopping water, CPE membranes act as an *isolation membrane,* protecting the tile and grout from movement in the substrate (see "Isolation Membranes," p. 237). Examples are Noble-Seal TS (Noble Company) and Dal-Seal TS (Dal-Tile Corp.), both 30-mil thick CPE. Special heavier membranes are designed for shower pans, tubs, and other details that must contain standing water.

FIGURE 6-36 Sheet Membranes.

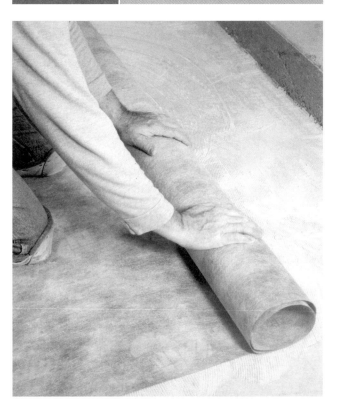

Elastomeric sheet membranes are bonded to the tile substrate with either a latex-modified thinset mortar or a proprietary adhesive. The membrane provides both waterproofing and crack isolation. A second layer of thinset is used to bond the ceramic tile to the membrane.
SOURCE: Photo courtesy of Noble Company.

- *Trowel-on membranes:* These one-part or two-part liquids are troweled or brushed on, some with a reinforcing fiberglass fabric. These are good for less critical areas that are only occasionally wetted, such as countertops, tub surrounds, and some floors. They also provide limited protection against cracking due to movement in the substrate when used in conjunction with a reinforcing fabric. Examples of one-part membranes include Laticrete 9235 (Laticrete International, Inc.) and Mapelastic HPG (Mapei Corp.).

Expansion Joints

Tile and grout are fairly unforgiving of movement in the substrate. To protect against cracking of the tile or grout joints, expansion joints are recommended by the Tile Council of America in the following places:

- Every 24 to 36 feet each direction; or every 8 to 12 feet in each direction if the tiled surface is exposed to moisture or direct sunlight.

FIGURE 6-37 Tile Perimeter Joints and Expansion Joints.

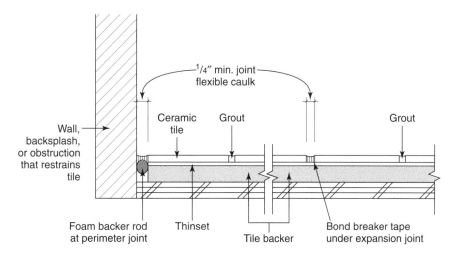

Wherever ceramic tile meets a wall, backsplash, tub lip, or other obstruction, a caulked joint is necessary to absorb any movement. Large areas of tile also need expansion joints to absorb the normal movement in the substrate.

- Around the perimeter of a room or wherever tile meets a different material or abuts restraining surfaces, such as curbs, columns, or pipes.
- At inside corners, such as where a countertop meets the backsplash.
- Wherever changes occur in the backing material, for example, from wood to masonry.
- Directly over any expansion, control, construction, or cold joint

Expansion joints should be at least $\frac{1}{4}$ inch wide and free of grout, backerboard, or tile adhesive. Use an elastomeric caulk with a backer rod or bond-breaker tape to prevent three-sided bonding of the caulk (Figure 6-37). Most tile suppliers now carry colored caulks designed to match standard grout colors, or colored caulks can be ordered from Color Caulk (see Resources, page 264).

Isolation Membranes. Polymer-modified thinset can absorb small amounts of movement, but where more significant movement is expected, a crack isolation membrane is the safest approach. Examples include tiling over concrete with control joints or shrinkage cracks, over radiant slabs, or over existing tile or other problematic remodeling surfaces. Wood-framed floors with long joist spans of 16 feet or more are also good candidates for isolation membranes.

Tile contractors typically use the same type of elastomeric membrane used for waterproofing, such as Noble-Seal TS (Noble Company) or Dal-Seal TS (Dal Tile). It is applied as described above under "Membranes" (previous page), although joints between sheets do not need to be solvent welded unless waterproofing is also required.

Crack Isolation. In repair work or other jobs where covering an entire floor with isolation membrane is not feasible, a strip of membrane can be installed over just a crack, change of materials, or control joint where minor movement is expected. In general, this will work where cracks are less than $\frac{1}{8}$ inch wide, and there is no vertical movement. Also, with some types of membrane, the contractor can offset the sealant-filled "soft joints" in the tile, so they do not have to fall directly over the crack or control joint in a concrete slab. In this case, create soft joints on both sides of the crack or joint (Figure 6-38), using a flexible sealant instead of grout. If a crack in the concrete substrate runs diagonal to the grout joints, the soft joint must run in a zigzag pattern on each side of the crack.

According to tile expert Michael Byrne, crack-isolation membranes used this way should be three times the width of the tile, but he cautions that the preferred approach is to cover the entire floor with membrane. This provides better protection against cracking and eliminates the soft joints and the slight bump in the tile surface, which may be unacceptable visually.

Setting Compounds

Manufacturers offer a wide range of setting compounds formulated for different setting beds and conditions. Choosing the wrong one can result in a failed tile job. For example, applications subject to moisture, temperature extremes, and heavy loading will need a higher quality setting compound than a kitchen backsplash, where organic mastic may serve perfectly well. If applying mortar to a thickness of over $\frac{1}{4}$ inch to even out low spots in the floor or irregularities in the tile, use a medium-set mortar or multipurpose thinset suited to the task. With any product,

FIGURE 6-38   Crack Isolation Membranes.

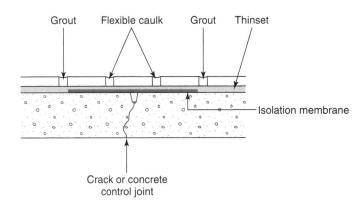

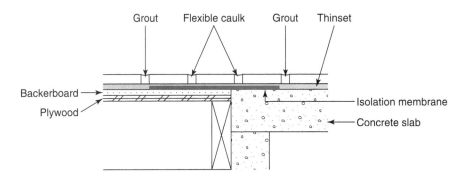

A strip of elastomeric sheet membrane can sometimes be used to prevent cracking over control joints or concrete cracks (top), or changes in the substrate material (bottom). Cut the strip of membrane at least as wide as three tiles and bond it to both the substrate and tiles in accordance with the manufacturer's instructions. Where possible, it is always best to cover the entire surface with membrane.
SOURCE: Tile consultant Michael Byrne.

it is important to follow the manufacturer's instructions regarding application and compatibility with the particular substrate (see Table 6-7).

Organic Mastics. These are ready-to-use adhesives primarily used with drywall and plywood substrates, although some are also approved for concrete and cement backerboards. Most now use a water-soluble formula, although some still use petroleum-based solvents.

In general, mastics are the least expensive setting material and provide the least strength and flexibility. The substrate must be very flat, since mastic is troweled on thin and cannot bridge low spots or uneven joints. Mastic should be avoided in applications subject to high temperatures. Type II mastic is rated for use on walls not subject to wetting. Type I is approved for use on floors and wet areas. However, thinset mortars are better suited for floors and wet applications, for a modest increase in cost.

Dry-Set Mortar. This is a factory blend of sand, cement, retarders, and other additives that is mixed with water on the job site. It provides a strong bond and high compressive strength, making it more suitable on floors

than organic mastic. It is commonly used over concrete slabs, mortar beds, or cement backerboards. It cleans easily with water; but once dried, it is unaffected by water.

Polymer-Modified Mortar. This is essentially dry-set mortar with latex or acrylic polymers added to increase the bond strength, compressive strength, and flexibility of the compound. One formulation uses a latex or acrylic liquid added to the dry mix at the job site. The other has dry polymers blended into the dry mix at the factory so only water is added on site. Polymer-modified mortar bonds well to most substrates, including waterproofing membranes, to provide a premium waterproof system. Some kinds are not recommended for use with plywood, however (always check the label for compatibility with a substrate). Cleanup should be done quickly with a damp sponge while the material is wet, or it is difficult to remove from skin and materials.

Modified-Epoxy Mortar and Grout. This is generally used for floors and countertops that require moderate chemical resistance, or where higher bond strength is needed to withstand greater loading, impacts, and flexing.

TABLE 6-7 Setting Compounds				
Type	Composition	Best Uses	Pros	Cons
Organic Mastic	Ready-mixed adhesive with latex or solvent-based carrier.	Walls in dry areas over wood or drywall. (Type I approved for wet areas, floors, countertops.)	Least expensive. Easy to use. Latex type easy to clean up. Grabs and holds tiles quickly.	Relatively low bond strength and compressive strength. Not flexible. Avoid installation in areas subject to wetting or high heat.
Dry-Set Mortar	Premixed sand, cement, retarders, and additives. Mix with clean water on site.	Wet or dry walls, floors over concrete slabs, masonry, mortar beds, or cement backerboards.	High bond and compressive strength. Easy to mix, use, and clean up. Water-resistant.	Not recommended over plywood or drywall or with large porcelain tiles. Limited flexibility.
Polymer-Modified Mortar (Latex or Acrylic)	Dry-set mortar with polymer additives either dry in mix or added as a liquid. The dry type is mixed with clean water.	Wet or dry walls, floors, and countertops; and showers and tub surrounds. Over most tile substrates, including backerboard, masonry, and waterproof membranes. (Check label for compatibility with plywood.)	Excellent bond and compressive strength and flexibility. Bonds well to large porcelain tiles. Water-resistant. Cleans with water until dry.	More expensive than mastic or dry-set. Cleanup difficult once material dries. Some products not approved for plywood or drywall.
Modified Epoxy Mortar (emulsion epoxy adhesive)	Premixed cement and silica sand plus two-part emulsified epoxy resin and hardener.	Adhesive or grout in wet or dry, interior or exterior installations that must resist moderate stresses and mild chemicals. Over most sound, clean surfaces, including plywood, waterproof membrane, steel, resilient tile, plastic laminate, and plywood.	Very high bond and compressive strength, and flexibility. Little shrinkage. Water-resistant. Moderate chemical resistance. Gains strength quickly. Some can be cleaned with water until dry.	Expensive. Precise measuring required. Respirator and gloves recommended to protect against chemicals, fumes, and silica dust. Cleanup difficult once material dries.
Epoxy Adhesive (100% solids epoxy)	Two-part epoxy (resin and hardener) and silica sand filler.	Adhesive or grout in heavy-duty wet or dry applications needing chemical-resistance. Over most sound clean substrates (see above). Use with green, white, and other moisture sensitive marbles and marble agglomerates.	Very high bond strength and flexibility. Little shrinkage. Water and chemical resistant. Some can be cleaned with water until cured.	Expensive. Precise measuring required. Respirator and gloves recommended to protect against chemicals, fumes, silica dust. May require special solvents to clean once cured.

It is also useful with questionable substrates such as existing tile, sheet vinyl, or plastic laminate, all of which are suitable for epoxy as long as they are sound and the surfaces are roughed up. In addition to their higher cost, these products have precise mixing requirements, a short pot life, and more difficult cleanup than standard thinsets. High-chemical-resistance formulations are also available. Cleanup of dried material requires special solvents, and the silica sand presents a respiratory hazard. Because of the mixing requirements and faster setup, skilled applicators are required.

Epoxy Adhesive and Grout. This is often used to install green (serpentine), white, and other moisture-sensitive marbles and marble agglomerates that may stain or warp with water-based products. It provides high-bond strength and impact-resistance over most sound substrates, including plywood. It also provides water-resistance and moderate chemical-resistance. It requires precise mixing and skilled application, and special solvents are required

to remove any material that dries on the surface of the tile (or the installer's hands). Also, the silica sand presents a respiratory hazard.

Installation with Thinset

Thinset Application. All mastics and thinset mortars are applied in the same way. First apply a thin layer of the adhesive using the flat edge of the trowel for continuous coverage, then comb with the notched edge of the trowel to create a uniform, flat setting bed. Hold the trowel at about 30 degrees from the surface for the continuous layer and at about 45 degrees when combing with the notched edge (Figure 6-36).

Notches range from about $\frac{3}{16}$ inch for thin tile to $\frac{3}{8}$ inch for tiles 12x12 inches and larger or irregular tiles such as handmade pavers. A $\frac{1}{4}$-inch notch works for most other tiles. Always follow the directions on the can. Coverage varies depending on the type of substrate and tile, as well as the heat and humidity in the environment.

It is a good habit to trowel in one direction only, but this is particularly important with tile 12x12 inches or larger. Also apply thinset mortar to only one small section at a time—no larger than the area that can be tiled before the thinset skins over. Any thinset that skins over will have to be discarded.

Medium-Set Application.

Where irregularities in the substrate or tile shape, such as handmade pavers, require a setting bead thicker than $\frac{1}{4}$ inch, choose a medium-set mortar or an all-purpose thinset approved for medium-set usage. Other thinset mortars are not strong enough when built up to that depth. Irregular-shaped tiles will need some adhesive "back buttered" directly to the tile to obtain proper coverage.

Tile Installation.

Next press and twist in a sample tile to check the adhesive coverage. ANSI standards require that dry interior tiles be evenly covered over at least 80% of their surface area (Figure 6-39). Wet or exterior applications require 95% coverage. When set back in place and beat in with a rubber mallet (larger tiles) or beating block (smaller tiles such as mosaics), the setting bed should be about $\frac{3}{32}$ inch but not larger than $\frac{1}{4}$ inch, unless medium-set mortar is used. Irregular-shaped handmade tiles need some adhesive "back buttered" directly to the tile to obtain proper coverage.

FIGURE 6-39 | **Thinset Application.**

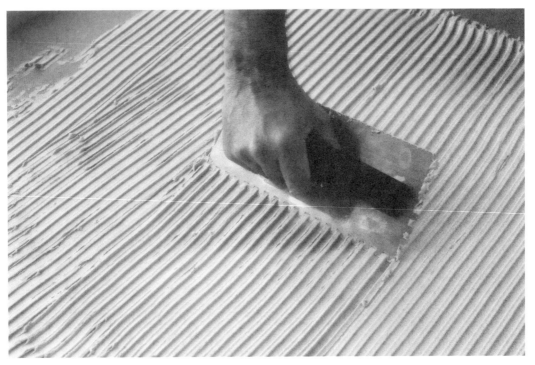

After applying a thin layer of mastic or thinset mortar with the flat edge of the trowel, comb it with the notched edge to create a uniform, flat setting bed (top). Next press and twist in a sample tile to check the adhesive coverage (bottom)—at least 80% for dry interior tiles and 95% for wet applications.
SOURCE: Photos by Francine Cohen.

Grout

A high-quality grout job makes a tile installation stronger, more attractive, and easier to clean. Properly formed joints are generally flat for square-edge tiles and slightly concave for rounded-edge tiles. Standard grout is a mixture of Portland cement and additives to control its texture and cure time, along with pigments if the grout is colored. In general, floors do best with a basic gray or other dark grout. White and light colors show stains the most. Most grouts also contain latex or acrylic additives to improve their performance.

Sanded vs. Unsanded. Sand is added to strengthen the grout where joints are wider than $\frac{1}{8}$ inch. For joints wider than $\frac{1}{2}$ inch, special grout with coarser sand is recommended.

Polymer Additives. Most grouts now have latex or acrylic compounds added either to the dry mix at the factory or as liquid on the job site. These polymer additives make the grout more water-resistant, flexible, and stain-resistant, and, with colored grouts, better able to maintain a consistent color. Although cured grout, like other masonry products, is unaffected by water, it cannot be relied on as a waterproof material.

Epoxy Grouts. For a higher degree of stain-resistance as well as moderate resistance to some chemicals, epoxy grout is a good option for applications such as showers, bathroom floors, or heavily used kitchen counters. One type, called epoxy-emulsion grout, mixes a two-part epoxy with Portland cement and sand. Another, called 100%-solids epoxy, mixes two-part epoxy with silica sand filler. Both types require precise mixing and installation. If joint cleaning is rushed, the grout pulls out and smears the tile; too much delay causes it to harden before you can shape the joints. If installed correctly, however, epoxy grout offers excellent protection against staining and does not require sealing. Epoxy grout will yellow slightly over time, however, particularly if exposed to direct sunlight.

Preparation. Allow the adhesive to dry at least overnight, or longer if recommended by the manufacturer, before grouting. Clean any adhesive or foreign matter from the grout joints. If the tiles are nonvitreous and unglazed, as with handmade pavers, they may need to be sealed prior to grouting to prevent staining. Also, light-colored glazed tiles may need to be sealed to prevent stains from dark-colored grouts.

With nonvitreous tiles, which soak up moisture, it is best to mist or sponge the tiles just before grouting so the grout will cure properly. For best results, maintain the room temperature between 50°F and 80°F during grouting and curing.

Installation. Mix the grout by hand or with a paddle bit run at slow speed (air bubbles from fast mixing will weaker the mix). After letting the grout "slake" for 10 minutes, which allows the ingredients to fully react, the material is remixed and ready to apply as follows, working one small area of several square feet at time (see Figure 6-40):

1. Using a rubber grout float held at about 45 degrees to the tiles, force the grout into the joints from several directions.

2. Scrape away the excess grout with the float held at about 90 degrees to the surface.

3. Once the grout is firm, typically in 15 to 30 minutes, clean the surface of the tiles with a sweeping motion, using a large clean round-edged sponge squeezed as dry as possible (water splashed on the joint lines can weaken the grout or cause splotchy coloring).

4. First remove the excess grout from the face of the tiles, then concentrate on shaping the grout joints. Rinse the sponge frequently, always keeping it as dry as possible.

5. Clean until just a light haze remains, which can be wiped off with a damp rag when the grout is dry.

Now is also the time to pack fresh grout into any voids you discover and clean the grout out of any "soft joints" that are to be filled with sealant. Use the tip of a margin trowel to clean up any corners or problem joints. With sanded grout, it is also a good idea to strike the joints with a curved metal implement, such as the back of a spoon, a steel chisel handle, or the side of a nail set to help force any exposed sand below the surface. This make the joint smoother and easier to clean.

Damp Curing. Traditional Portland cement grout requires several days of damp curing to reach its full strength. This was typically done by covering the freshly grouted tile with Kraft paper and periodically misting or sponging the tile. Most modern grouts with latex or acrylic additives, however, do not require wet curing except in very hot, dry weather. As with all tile products, check the label for instructions.

Grout Sealers. While latex or acrylic additives help protect the grout from staining, sealing the grout after it cures provides the best protection. There are a wide variety of products on the market. Consult the directions regarding when and how often to apply. Many require reapplication annually or more often, depending on the specific use. Regardless, to keep grout from darkening and staining, it will need regular cleaning with a grout cleaner or mild detergent. Avoid oil-based soaps as they tend to darken grout.

Shower Pans

A shower pan must be completely watertight and able to hold pooled water should the drain get clogged. Most are built in place over a mortar bed with a waterproof membrane liner. However prefabricated setting beds have also become available recently.

FIGURE 6-40 | **Grout Installation.**

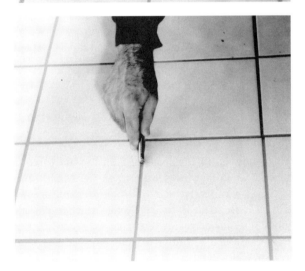

Using a rubber grout float held at about 45 degrees to the tiles, force the grout into the joints from several directions (top). Scrape away the excess grout with the float held at 90 degrees to the surface. Once the grout is firm, clean with a sweeping motion using a clean, round-edged sponge squeezed dry (middle). With sanded grout, it is a good idea to strike the joints to force any exposed sand below the surface (bottom).
SOURCE: Photos by Francine Cohen.

Prefab Cement Pans. For standard shapes and sizes, one option is to install a prefabricated pan made from 40-lb polystyrene foam coated with a reinforced cementitious coating that is ready to receive tile. These range in size from 36x36 inches to 36x60 inches and come presloped with a built-in drain. One unit, called Pro-Form (Bonsal American) is bonded to the subfloor with latex-modified thinset, coated with a liquid waterproofing membrane, and is then ready to tile. As long as the subfloor is sufficiently stiff to prevent flexing, these units should perform well with 4x4-inch tiles. Smaller tiles may exert too much of a point load for the underlying foam, while larger tiles can cause problems with the slope angle.

Prefab Plastic Pans. A less expensive option is to combine tiled walls with a one-piece fiberglass or acrylic shower pan (not to be tiled). These are the same materials used in one-piece shower or shower-tub units. A few companies also sell stand-alone solid-surface shower pans. Like other plastic units, these must be properly supported underneath to prevent flexing and cracking. Some require setting in sand, wet mortar, or plaster. In general, acrylic units cost more than fiberglass but are stronger and less prone to flexing and cracking.

Mortar-Bed Pans. Built-in-place shower floors using a mortar bed and modern waterproofing membranes can provide many years of trouble-free service. For best performance use a heavy-duty sheet membrane, such as 40-mil Chloraloy (Noble Company), which is designed for use in shower pans and similar applications.

To guarantee that any trapped water will drain properly, the membrane is placed on a layer of latex-modified mortar sloped $\frac{1}{4}$ inch per foot toward the drain. Two-piece clamping-type drains are designed to seal to the membrane by compression between the upper and lower flanges (see Figure 6-41).

A layer of sealant applied between the membrane and lower flange serves as backup waterproofing. Weep holes around the bottom of the drain, surrounded by pea gravel or pieces of broken tile, allow any water that accumulates to drain away (see Figure 6-42).

The membrane should run up all sides of the shower, at least 3 inches above the height of the finished curb. Secure the membrane to the framing with galvanized staples or roofing nails along the top edge, being careful to make no holes in the membrane any lower than 2 inches above the finished curb. At the inside wall corners, the extra membrane material is not cut, but pleated and folded over to lay flat against the framing. Avoid making wrinkles here or along the bottom of the pan. At the curb, the membrane must be cut so it can fold over the top of the curb. Seal these corner cuts with either prefabricated "dam corners" or patches of membrane caulked in place with Nobleseal 150, a high-performance Kraton-based sealant.

FIGURE 6-41   Shower-Pan Drain.

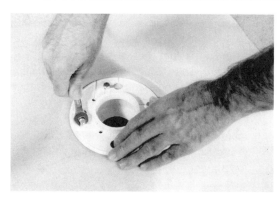

To install a two-piece clamping-type drain, first cut a hole in the CPE sheet membrane using the drain's bottom flange as a guide (top, left). Then bolt the upper and lower flanges together, forming a compression seal to the membrane (top, right). A layer of sealant applied between the membrane and lower flange serves as backup waterproofing. The finish drain threads into the upper flange (bottom), which has weep holes to drain away any trapped water. Use pea gravel or tile chips to keep mortar from blocking the weep holes.
SOURCE: Photos by Francine Cohen.

FIGURE 6-42   Mortar-Bed Shower Pans.

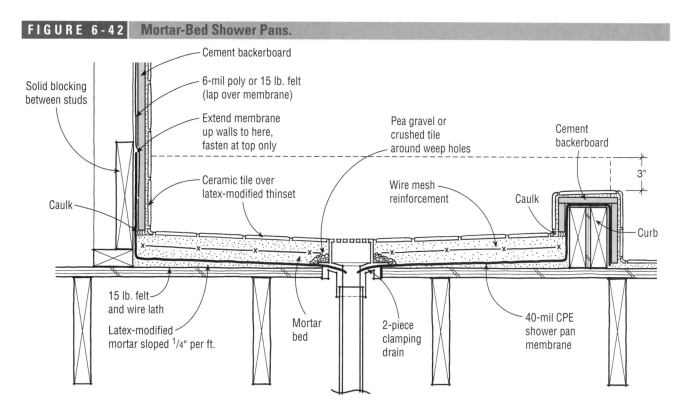

Custom shower floors using a mortar bed and modern waterproofing membranes can provide years of trouble-free service. The membrane clamps to the drain and runs up the walls, creating a watertight pan. Any trapped water escapes away through the weep holes around the drain.

KITCHEN CABINETS

Kitchen cabinets range widely in styles, materials, and levels of quality. Well-made cabinets feature sturdy cabinet boxes and drawers, stiff shelves that resist sagging, and solid hardware that operates smoothly. Higher-end cabinets make greater use of veneer-core plywood rather than medium-density fiberboard (MDF), particleboard, or other composites. Doors, drawer fronts, and visible end panels in premium cabinets make use of solid wood, real wood veneer, or high-pressure plastic laminate. At a glance, it is not always easy to discern quality levels since the best wood-grain vinyl facings do a surprisingly good imitation of real wood, at least until someone dents or scratches a corner.

With so many variables, it is not surprising to find that a set of cabinets for a midsize kitchen could range in price from as little as $3,000 for builder-grade cabinets picked up at a home center to as much as $20,000 for a custom high-end job.

Cabinet Grades

The industry generally divides cabinets into three main grades: stock, semicustom, and custom.

- *Stock cabinets* are mass-produced in factories in standard sizes, typically in increments of 3 inches, although all sizes may not be available for certain items. Each line comes in a limited number of materials, styles, and options. These are the least expensive option with the quickest delivery time, but usually not the best quality. Fillers are used to fit the cabinets into place.

- *Semicustom cabinets* may be similar to stock cabinets in quality level or may be significantly better. Since these are made to order, however, the buyer has many more choices for wood species and finishes, laminates, door styles, and storage options and accessories. More sizes, including special heights and depths, are often available. Like stock cabinets, these are built in 3-inch increments, requiring filler strips for installation.

- *Custom cabinets* are made to order by smaller shops for an individual job. Each shop has its own preferred materials, styles, options, and details; but for enough money, most shops will create whatever is requested. With a custom fit, filler blocks are not needed.

Cabinet Materials

Panel Products. Cabinets use a wide variety of substrate materials for panels. The main panel products, typically covered with a wood, melamine, or vinyl veneer, are listed below:

- *Hardboard,* sometimes referred to by the brand Masonite®, is made from compressed wood fibers and lignin or phenolic resin. It swells and degrades when wet and is used for drawer bottoms, backs, and bottoms of lower-end cabinets.

- *Particleboard* is made from small wood particles that are resin-bonded under pressure and heat. Type I uses urea-formaldehyde resin and Type II uses water-resistant phenol-formaldehyde resin. Density classes are L (low), M (medium), and H (high). Better cabinets use medium-density stock (40 to 50 pounds/cu ft). Some also use Type II, which is water-resistant and has little off-gassing of formaldehyde. Particleboard tends to swell when wet, and it is used widely for cabinet panels, shelves, and doors. Formaldehyde off-gassing may be a concern.

- *Medium-density fiberboard,* or MDF, is a high-quality substrate made from fine fibers and urea-formaldehyde resin. It is more stable than plywood, stiffer than particleboard, and less affected by water. Its surface is smoother than particleboard and can be routed, shaped, and painted. It is widely used for all cabinet panels and shelves. A 36- to 38-lb/cu ft density is adequate for most applications, although some use 42- to 48-pound material. The high formaldehyde content is a concern to individuals concerned about off-gassing.

- *Plywood* is made from thin wood sheets laminated to each other with the grain running at right angles in alternate plies for strength. Interior grades, typically used in cabinets, use urea-formaldehyde resin. Better quality cabinets use plywood for cabinet panels, shelves, and drawer bottoms. Plywood resists water damage.

Finishes. These are the most common finish materials used for cabinet sides, interiors, and door and drawer fronts.

- *Wood veneer.* Wood veneer is a thin layer of wood bonded to particleboard, MDF, or plywood to give the appearance of solid wood. Used in both flat and raised panels, veneer provides good grain matching. Veneered panels rarely have problems, although scratches or dents are easier to repair on solid wood. Very high heat or humidity can cause cracking or delamination. The finish may be a simple solvent-based varnish or a more advanced and expensive multicoat system.

- *High-pressure laminate.* Often called "plastic laminate," high-pressure laminate is composed of layers of resin-saturated kraft paper with a clear melamine finish. High-pressure laminate is widely used on countertops because it is inexpensive, durable, and easy to clean. Scratches and damage are difficult to repair, however. It is used on door and drawer fronts and occasionally on side panels. Color-through laminates are also available at a higher cost. These hide chips and scratches better and do not leave a telltale dark edge at corner seams.

- *Melamine.* Also known as low-pressure laminate, melamine is thinner and less durable than high-pressure laminate. It comes applied to particleboard or MDF with a paper layer under the melamine that provides the color or wood grain. Low-pressure laminate can chip or crack and may discolor over time. It is used widely on cabinet boxes and door backs, and on door and drawer fronts on low-end cabinets.

- *Vinyl.* Vinyl is a plastic sheet material that comes applied to a particleboard or MDF substrate, and is printed with a wood-grain or other pattern. It is typically 2 to 4 mils thick and is not as durable as melamine, although the heavier 4-mil material resists scratches fairly well. Exposed, unfinished edges are prone to damage, and scratches or dents are difficult to repair.

- *Rigid thermofoil (RTF).* RTF is a rigid PVC sheet that is heated, vacuum-formed, and glued onto MDF doors and moldings, creating a seamless face. Most RTF doors are shaped to simulate a raised panel door. Thermofoil is available in many colors and wood-grain patterns, although white is the most common. High-quality RTF is durable, scratch-resistant, and resists yellowing—a problem with some of the early formulations. Better products carry warranties of five or more years. When using wood-grain thermofoil, it is best to use full-overlay doors and matching thermofoil moldings, since real wood finishes will age differently than the thermofoil. Many thermofoil doors have matching melamine backs.

- *Paint.* There are a number of high-quality painted finishes that are durable, lustrous, and resistant to crazing, chipping, or yellowing. *Polyester paint*, also used on cars and appliances, is a very expensive option that requires many coats that are oven-cured and wet-sanded by hand. The finish can be gloss or matte and fills the pores of the wood, giving it a solid appearance. *Catalyzed enamel paint* is a less expensive option that uses a two-part formula to achieve a similar lacquerlike finish. Although these paints resist chipping, nicks are difficult to touch up and blend in. Also hairline cracks will typically appear at the joints in solid wood doors due to expansion and contraction—not a problem with a dimensionally stable substrate such as MDF.

Assessing Quality

Assessing a cabinet's quality is not always easy due to the large number of components involved and the fact that much of the material and joinery is concealed. One good indication of overall durability is certification by the Kitchen Cabinet Manufacturers Association, which has a rigorous testing and certification program that measures such things as structural integrity, shelf strength, hardware durability, and quality of finishes.

Also, many manufacturers offer two or three grades of quality. Comparing the specifications of each line can provide a good idea of what the upgrades are and whether the

FIGURE 6-43   **Face-Frame Cabinets.**

In traditional face-frame cabinets, like the one shown above, doors are mounted on solid wood frames, which are usually visible with the doors closed. Frameless cabinets typically use full-overlay doors, which attach directly to the sides of the cabinet box.
SOURCE: Photo by author.

added expense is worthwhile. The main components to evaluate are covered below.

Boxes. The cabinet box, or carcase, makes up the body of the cabinet and gives it structural integrity. Typically, the only visible parts are end panels, portions of the interior, and the front edges in frameless cabinets or face-frames in framed cabinets (Figure 6-43).

- *Framed cabinets.* These are the traditional style of construction with a frame of $\frac{3}{4}$-inch-thick solid wood rails and stiles fastened to the front of the box. Hinges attach to these face-frames, which are usually partly or fully visible with the doors closed.

- *Frameless cabinets.* Also called 32-mm cabinets or Euro cabinets, these have no face-frames in front. The doors attach directly to the inside face of the cabinet sides with cup hinges. Doors and drawers are usually full overlay, concealing the front edges of the boxes. However, some frameless cabinets now offer more traditional wooden doors that leave a narrow band of the cabinet fronts exposed. The front edge is typically finished in vinyl, melamine, or wood veneer. High-end cabinets may have solid wood banding.

With either type of cabinet, the price is driven by the materials, hardware, and assembly techniques. The cheapest cabinets typically use $\frac{1}{2}$- to $\frac{5}{8}$-inch particleboard with a vinyl or melamine face. Better quality cabinets use medium-density fiberboard (MDF), and the best generally use $\frac{1}{2}$- to $\frac{5}{8}$-inch plywood. The facings on better cabinets are usually high-pressure laminate or real wood veneer. With some cabinet lines, it is possible to order plywood sides only where needed, for example, on the sink base and wherever there is an exposed end panel, which might be subject to wetting or the occasional dent or nick.

Backs and floors range from flimsy $\frac{1}{8}$- or $\frac{1}{4}$-inch hardboard to thicker particleboard, MDF, or plywood (in order of stiffness). The finish inside the cabinet is typically vinyl or melamine. Cabinet bottoms should be rigid enough not to flex under the weight of pots and pans.

Finally, look for securely installed corner blocks or stretchers across the top of the cabinet to hold it square during shipping and installation Some high-end cabinets have a full-size top panel to reinforce the top of the box.

Shelves. Shelves range from $\frac{1}{2}$-inch particleboard, which will bow under the weight of dishes, to $\frac{3}{4}$-inch plywood. Plywood is the strongest shelving material, followed by MDF, then particleboard. In base cabinets, look for full-depth shelves or roll-out shelving. Wall cabinet shelves should be adjustable and have solid clips, preferably metal. In general, shelves are designed to support a uniform load of 15 pounds per square foot (psf) for kitchens, 25 psf for closets, and 40 psf for bookshelves. If loads are likely to exceed these, upgrade to a stronger shelf (see Table 5-11, page 186).

Drawers. It is important to have solidly built drawers, since they get a lot of use and abuse. High-quality drawers typically have solid hardwood or poplar sides and backs, with minimum $\frac{1}{4}$-inch plywood bottoms glued into dados. In the best cabinets, drawer sides are dovetailed or dowelled to the front and back and the drawer front is a separate piece screwed to the box. Respectable drawers are also built with sides of $\frac{1}{2}$-inch or thicker plywood or melamine stock dowelled together. In lower-end cabinets, drawer sides are often particleboard or MDF wrapped in vinyl and nailed or stapled and glued together, with a particleboard or hardboard bottom. Also many lower-quality cabinets use the drawer front as the front of the drawer box, a weaker detail.

Drawer Slides. All drawer slides, even with the same rating, are not alike. Look for heavy-duty epoxy-coated components with ball-bearing rollers that operate smoothly and quietly. At a minimum, use three-quarter extension drawer slides rated to carry 75 pounds. Consider upgrading to full-extension slides rated for 100 pounds, particularly for any large, deep drawers. Undermount slides have the advantage of helping to support the drawer while remaining out of sight. Side-mounted slides that wrap around the drawer bottom also provide good support.

Doors and Drawer Fronts. Doors and drawer fronts are the most visible part of a kitchen and take a lot of abuse. Many cabinet manufacturers buy doors and drawer fronts from large specialty door manufacturers, so they may not reflect the overall quality of the cabinets. When selecting a material and finish, consider durability and ease-of-cleaning as well as appearance. Frame-and-panel wood doors are typically more expensive than laminate or thermofoil doors.

- *Frame and panel* doors have either a raised or flat panel of solid wood or veneer. Veneered panels are more stable but more difficult to repair if nicked or scratched. Avoid frames with mitered corners, as they may open with seasonal changes in humidity.

- *High-pressure plastic laminate* is more durable than melamine (low-pressure laminate). Melamine is fine, however, for the backs of doors.

- *Painted wood* doors may show small gaps (over time) at joints due to seasonal movement of the wood. Also the center panel in a frame-and-panel door may show a paint line if the panel shrinks during the heating season.

- *Thermofoil* (RTF) doors can provide a frame-and-panel look with the convenience of a durable PVC plastic facing. It is best to use matching RTF moldings as well, since colors may change over time.

Frameless cabinets typically have full overlay doors, while framed cabinets may have doors that are inset, rabbeted, or overlaid partially or fully (Figure 6-44). Rabbeted

FIGURE 6-44 **Cabinet Door Styles.**

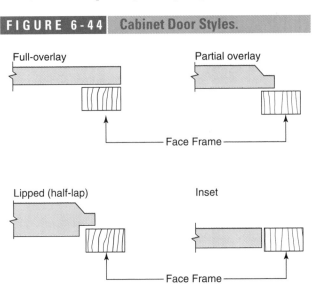

Full-overlay doors with concealed hinges are found on nearly all frameless cabinets today and many framed cabinets as well. A partial-overlay or lipped door leaves more of the face frame exposed, making door adjustments simpler. Inset doors give a more furniturelike appearance but require the most precise setting and adjusting to prevent misalignment.

or partial-overlay doors are the easiest to fit and adjust since they have considerable play. Inset doors provide an attractive furniturelike appearance, but they are also a common cause of callbacks, since the slightest movement in cabinets or hinges can cause the doors or drawers to rub. Full-overlay doors also need careful adjustment, since they have only about an $\frac{1}{8}$-inch gap to the next door. Fortunately, most are hung with easy-to-adjust cup hinges.

Hinges. Good quality hinges are sturdy, smooth to operate, and easy to adjust. Undersized or poor quality hinges, on the other hand, can lead to sagging or rubbing doors and are a common problem with low-end cabinets.

- *Cup hinges.* Originally designed for full-overlay doors on frameless cabinets, concealed cup hinges are now available for most types of doors and cabinets (Figure 6-45). Most can be adjusted in three directions, which makes it easy to align the doors and compensate for minor irregularities in the cabinets. Plus, many cup hinges have a convenient "snap-on" feature that allows removal of the door without tools. Swing angles range from 95 to 175 degrees, with typical doors opening from 105 to 110 degrees. Special hinges are available for nearly every door configuration and self-closing types eliminate the need for magnetic or mechanical catches.

- *Barrel and knife hinges.* These are partially concealed when closed and are usually adjustable in two directions by loosening the hinge-mounting screws. They are common on lower-end cabinets.

COUNTERTOPS

After the cabinets, the countertop is the most visible element in most kitchen and bath designs. In addition to providing a visual focal point, durable and easy-to-clean counters are critical for both hygiene and efficiency. These workhorse surfaces should resist scratching, knife cuts, and burns, be impervious to water and stains, and wipe clean with a sweep of the sponge. Nonporous surfaces like plastic laminate, solid surface, and engineered stone fit the bill well. While plastic laminate still dominates the market based on its combination of good performance and low cost, both natural and manmade stone products now account for a substantial and growing share of the market (Figures 6-46 and 6-47).

Since no one product can meet every need in a home, it often makes sense to mix and match materials,

FIGURE 6-45 **Cup Hinges.**

Most concealed cup hinges can be easily adjusted in three directions, and many conveniently snap on and off without tools. Swing angles range from 95 to 175 degrees, with special hinges available for nearly any door configuration.
SOURCE: Photo by author.

FIGURE 6-46 **Kitchen Countertops in New Homes.**

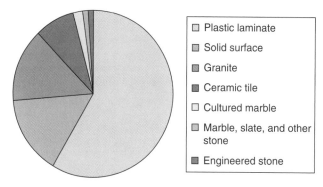

- Plastic laminate
- Solid surface
- Granite
- Ceramic tile
- Cultured marble
- Marble, slate, and other stone
- Engineered stone

Plastic laminate	58%
Solid surface	15%
Granite	14%
Ceramic tile	8%
Cultured marble	2%
Marble, slate, and other stone	1%
Engineered stone	1%

SOURCE: Adapted from *Assessment of K&B Materials in Residential Construction,* 2003, courtesy of NAHB Research Center.

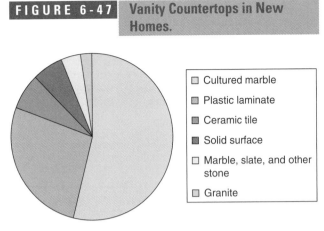

FIGURE 6-47 Vanity Countertops in New Homes.

Cultured marble	54%
Plastic laminate	27%
Ceramic tile	7%
Solid surface	6%
Marble, slate, and other stone	4%
Granite	2%

SOURCE: Adapted from *Assessment of K&B Materials in Residential Construction*, 2003, courtesy of NAHB Research Center.

particularly in the kitchen. The main counter areas might be plastic laminate or solid surfacing with special sections using, for example, stone for baking, wood for a cutting board, and tile for a place to set hot pots. The most common countertop materials and their characteristics are shown in Table 6-8.

Design Issues. For ease of use and maintenance, a countertop will function best if it has the following characteristics:

- A smooth surface that is nonporous, stain-resistant, and has minimal joints and seams.
- A flush or undermount sink that lets the homeowner easily sweep water and dirt into the sink.
- A waterproof joint where the counter meets the backsplash and sidesplashes. Sidesplashes should go wherever a counter end meets a wall or tall cabinet.
- A coved corner at the backsplash makes this joint easier to keep clean.
- In the kitchen: a no-burn area to set down hot pots.
- In the bathroom: a raised dripless counter edge helps keep water contained near the sink.

Cleanup. For most customers, easy cleanup of countertops is a top priority. While most of the materials discussed below are at least moderately stain-resistant, the actual

performance will vary, depending on the specific color, pattern, finish texture (gloss vs. matte), and porosity of the material. If possible, obtain samples of the materials being considered, with sealers applied if planned. Mark each sample with a few stubborn stains: indelible marker, grape juice, salad oil. Let them dry for an hour and then see how easily they clean up with normal household cleansers and nonabrasive cleaning pads.

High-Pressure Laminates

Plastic laminate still dominates the market for kitchen counters, because it provides an attractive, durable surface at a cost of $5 to $10 per square foot versus $50 to $100 per square foot for solid surfacing or stone. The range of colors and textures continues to expand with new printing technologies that have produced realistic looking wood and stone surfaces.

Made from a sandwich of resin-impregnated kraft paper, a decorative paper layer, and a top layer of clear melamine fused under heat and pressure, plastic laminate is impervious to moisture and resists scratches, dents, and chips. It can be scratched with a knife, however, or scorched with a hot pot. Small chips can be repaired with special sealers, but scratches and burns are permanent.

Before selecting a color, it is a good idea to test a few sample squares with an indelible marker to see how easily the marks clean off. Matte finishes stain much more readily than gloss, and there may be considerable variation from one pattern and finish to another. More expensive color-through plastic laminates show wear less than the standard type, and they eliminate the dark band at exposed edges.

If using a postformed counter with a miter joint, it is essential to mechanically draw the joint tight and to seal that joint to prevent water intrusion, which will degrade the particleboard and open the joint. Silicone or special laminate seam sealer can be used.

Edge Treatments. There are now many alternatives to the traditional square edge that exposes the dark edge of the horizontal sheet of laminate. The least expensive is a postformed counter with either a seamless square edge or raised dripless edge. A wide variety of upgrades are available to dress up the counter. Beveled laminate, solid surfacing, and hardwood edging are common details. Site-applied wood edging, however, can lead to problems over time if the joint is not watertight.

Cast Polymers

Cultured marble, solid surfacing, engineered stone, and most other composites used in kitchen and bath counters, sinks, and wall panels are different types of cast polymer. Cast polymer products consist of a plastic resin, either acrylic or polyester, and a mineral filler. The type and amount of filler largely determines the hardness, stain-resistance, and overall durability of the product. Cultured marble, for example, which uses crushed limestone as a

TABLE 6-8	Countertop Materials			
Type	Cost	Pros	Cons	Recommendations
Plastic laminate	Low	Inexpensive, versatile, and durable. Many colors and patterns. Moderate stain and heat resistance. Nonporous surface is easy to clean and maintain. Color-through option more durable and can form clean bevels.	Can be scratched by knives, scorched by hot pots, stained by coffee, tea, ink, etc., particularly matte finishes. Can be etched or discolored by strong acids or alkalis in household cleansers. Joints vulnerable to water damage. Difficult to repair.	Test samples before installation for staining and ease of cleaning. Textured pattern can help hide minor damage. Do not cut or place hot items on surface. To remove stains, soak in household cleanser and use nonabrasive scrubbing pad. If necessary use alcohol or bleach sparingly.
Solid surfacing	High	Seamless and nonporous with wide range of color-through patterns. Seamless undermount sinks available. Moderate stain and heat resistance. Easy to clean and fix minor damage. Larger problems repairable by trained technician. Workable with woodworking tools.	Can be scratched by knives. Surface can crack or scorch if heated to over 360°F by hot pot, burning cigarette, etc.	Avoid high gloss finish, particularly with dark colors (shows scratches). Do not cut or place hot pots on surface. Clean with nonabrasive pad and mild cleansers. For stubborn stains, bleach is OK or use abrasive cleansers on matte finish only. Buff out minor scratches, stains, and burns with very fine wet sandpaper.
Ceramic tile	Medium	Very durable, scratch-resistant, heat-proof, and nonstaining (glazed tiles). Damaged tiles can be replaced.	Grout must be periodically sealed to minimize stains. Tiles can chip. Hard, uneven surface tough on dishes.	Choose a tile rated for floor or counter use. Set with epoxy grout for best protection against stains. Avoid white grout. Avoid unglazed, porous tiles and tiles with uneven surfaces.
Natural stone	High	Extremely durable and impervious to heat and water. Granite very scratch resistant. Easy to clean. Good work surface for bakers.	Most types require periodic sealing to prevent stains. Acids etch marble. Oil stains granite and marble. Hard, cold surface tough on dishes. Not repairable.	Apply sealers as recommended by fabricator. Avoid abrasive pads and cleansers on marble. Remove minor scratches from slate with steel wool.
Engineered stone (Quartz, granite)	Medium to high	Extremely durable, nonporous surface very resistant to scratching, stains, and heat. No sealing required. Easy to clean and maintain. Seamless undermount sinks available with some products.	Less natural appearance than stone. Prolonged contact with hot pots not recommended. Strong solvents, drain cleansers, and other harsh chemicals can mar finish. Edges can chip from sharp blow.	Wipe up spills and stains quickly. Do not cut or place hot pots on surface. Remove tough stains with ScotchBrite™ pad and nonabrasive household cleanser. Avoid bleach, strong solvents, and harsh chemicals, such as paint stripper or nail polish remover.
Wood butcher block	Medium	Warm appearance. Can be used as cutting board. Surfaces can be restored by sanding and oiling.	Scratches and scorches easily. Needs periodic treatment with oil. Can absorb food odors. Standing water can blacken wood.	Sand surfaces and reseal to repair scratches. Periodically apply mineral oil to keep in good condition.
Cultured Marble	Low	Look of marble with nonporous surface that is stain-resistant and easy to clean and maintain. Lavatory tops can have integral bowls. Small scratches, chips repairable.	Less durable than solid surfacing or engineered stone. Thin gel coat can scratch, chip, or burn.	Primarily used in bathrooms. Use only ANSI-labeled products. Avoid abrasive cleansers and pads. OK to use thinner or alcohol on tough stains. Restore sheen by using Gel Gloss or light wax.

filler, is relatively soft and porous and needs to be protected by a gel coat. The new granite and quartz "engineered stone" composites, on the other hand, are nearly indestructible. These materials are described in greater detail below.

Cultured Marble. Cultured marble has been produced since the 1960s. It combines ground-up marble dust with polyester resins and pigments to make a sheet material for use in kitchens and bathrooms, as well as fireplace surrounds and other ornamental applications. Nearly all cultured marble products are finished with a thin clear or colored gel coat, which provides the color and pattern and creates a relatively hard and stain-resistant skin. Quality varies widely, as there are hundreds of small-volume manufacturers around the country producing the material. Some of the lower-quality products have had problems with crazing (small cracks) in the gel coat around drains and other areas subjected to thermal shocks. High-quality products should carry a label certifying compliance with ANSI standard Z-124.

While cultured marble is stronger and less brittle than natural marble, it is less impact-resistant and scratch-resistant than other cast polymers such as solid surfacing.

The nonporous surface resists mildew and most stains and is easily cleaned with nonabrasive cleaning agents. Cultured marble is commonly used for vanity tops with integral sinks, as well as shower and tub enclosures, but it is rarely used for kitchen counters. Scratches and small chips can be repaired using a special gel-coat compound available from the supplier, but cracks or breaks cannot be repaired. Since the gel coat provides the color and pattern, significant damage to the gel-coat cannot be repaired. The sheen, however, can be restored with a thin coat of auto wax or Gel Gloss (TR Industries).

Because many of the light-colored products are translucent, they should always be installed on a light-colored background material with clear silicone adhesive.

Solid Surfacing. Initially introduced as Corian® by Dupont almost 40 years ago, solid surfacing products consist of mineral fillers, usually alumina trihydrate (a product of bauxite), and acrylic or polyester resins. Solid surfacing is a hard and durable homogenous material with color throughout and is easily machinable with woodworking tools. It can be seamlessly welded to itself at joints and to undermount sinks made of the same material. This creates an attractive, continuous work surface that is easy to wipe clean. An economical veneer version, laminated over a particleboard substrate, is available from some manufacturers.

Because solid surfacing is nonporous, stains do not penetrate and it will not support mold or bacterial growth. Most stains can be wiped away with a nonabrasive pad and mild cleanser, although bleach is OK if needed. It also resists mild chemicals, but it should not be exposed for long to harsh chemicals, such as acetone or paint thinner. The alumina trihydrate filler also makes solid surfacing fire-resistant, although it is possible to scorch the surface.

While relatively easy to scratch or stain, solid surfacing is easily repaired. Tough stains as well as minor scratches or burns can be buffed out of a matte finish with an abrasive cleanser and Scotch-Brite pad, or wet sanded with very fine sandpaper (start with 1000-grit and use coarser grits as needed). Most solid surfacing used on countertops has a matte-satin finish, which is the easiest to maintain. For a gloss finish, follow the recommendations of the manufacturer.

Engineered Stone. The newest class of cast polymer, sometimes called composite or engineered stone, uses a high percentage of quartz, quartz silica, and granite to produce a material with the hardness of natural stone and the easy maintenance of solid surfacing. Engineered stone typically has over 90% stone aggregate with just enough acrylic resin and binders to hold it together. Combined under heat and pressure, the resulting material is uniform throughout and has greater flexural strength than stone.

The nonporous surface is virtually stain-proof and very scratch-resistant, although it should not be used as a cutting board. Unlike natural stone, it does not require any sealing or waxing. Although it will not burn or scorch, placing hot pots directly on the surface can cause surface damage from the thermal shock. Some manufacturers are able to add a seamless undermount solid-surface sink to the engineered stone slab, adding to the appeal of the material. Products include Silestone® (Cosentino USA), Zodiaq®, (Dupont), Cambria®, (Cambria), Technistone® (Technistone USA), Ceasarstone® (U.S. Quartz Products), and Granyte® (Halstead International).

Ceramic Tile Countertops

Ceramic tile is a popular countertop material in the West and Southwest. Its main advantages are high durability and imperviousness to heat and water. Glazed tiles will not stain, and tiles rated for use on floors and counters are very scratch resistant. Softer tiles are prone to chipping, but damaged tiles can be removed and replaced if necessary. Installing ceramic tile on one section of counter near the range can provide a handy place to set down hot pots and pans.

The main problem with tile counters are the grout joints, which tend to discolor over time. Using a latex-modified grout and sealing the grout will help but will not prevent stains altogether. The best solution is to use an epoxy grout and to choose a dark or neutral grout color such as gray. Lighter colors are generally OK on backsplashes and walls. Tile is also very hard and prone to breaking fragile glasses or dishes that strike it.

Two details that will enhance a tile counter are a V-cap nosing tile, which will create a clean-looking dripless edge, and a coved corner at the backsplash, which will make it easier to keep the corner clean. Tile counter details are shown in Figures 6-33 and 6-35, pages 234–235.

Natural Stone Countertops

Natural stone has become as popular as solid surfacing in the past few years as stone prices have dropped and finished stone slabs have become more widely available. In general, natural stone is hard, heavy, and cold, and is unaffected by heat and water. Bakers like the cool, smooth surface for handling dough. Stain and scratch resistance varies with stone type, but all stones need some type of sealer to prevent staining. The most common choices are discussed below:

- *Granite:* Very hard and resistant to scratching or chipping. Takes high polish. Oil and grease may stain granite if not sealed.

- *Marble:* Acidic foods etch or dull surface. Porous surface absorbs oils, stains, and some odors if not sealed. Avoid abrasive pads or cleansers.

- *Soapstone:* Very dense and stain-resistant but relatively soft. Usually treated with mineral oil. Scratches can be sanded out.

- *Slate:* Very stain-resistant. May need sealers depending on quality of stone. Minor scratches can be buffed out with steel wool.

PLUMBING FIXTURES

Sinks

Because of its durability and economy, stainless steel accounts for over 60% of kitchen sinks in both new homes and remodels. In new construction bathrooms, cultured marble is the leading material for lavatory sinks, while the more expensive vitreous china leads in remodels.

Both stainless steel and vitreous china are extremely durable, easy to clean, and impervious to rust, stains, and heat. Cultured marble quality varies, based largely on the thickness and quality of the thin gel coat that provides the color and wear surface. Well-made cultured marble, however, treated with reasonable care, can provide years of satisfactory service.

With any sink, choose a bowl that is large enough for the intended use. A kitchen sink should be deep enough to handle large pots and have vertical sides and tight-radius corners to increase the usable space. A relatively flat bottom allows dishes to sit without sliding toward the drain, and an offset drain also increases the usable space. Models with multiple bowls and built-in cutting boards, draining racks, and other accessories can simplify both food prep at the sink and cleanup. A raised or gooseneck-type faucet allows large pots to be easily rinsed or filled (Figure 6-48).

Many "builders' grade" lavatory sinks are undersized for basic grooming tasks, such as tooth brushing and face washing without splashing water across the vanity top. If the client plans to wash hair, water plants, or perform other household chores at the lavatory, an oversized bowl is recommended. A sink with the faucet offset to one side, with a pivoting spout, provides still more usable space.

FIGURE 6-48 | Kitchen Sink Features.

A mix of large and small basins allows sink space for both food prep and large pots. High-arch or gooseneck faucets should be high enough to accommodate large pots, and able to swing out of the way when extra work space is needed.
SOURCE: Photo courtesy of Kohler.

FIGURE 6-49 | Sink Installation.

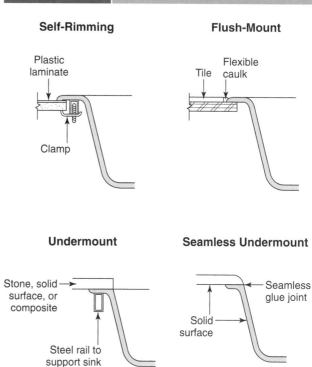

Self-rimming sinks (top left) are the easiest to install but create a barrier on the counter surface that traps dirt and grime. Flush-mount sinks are easier to keep clean (top right). While used primarily with tile counters, some can also be adapted for use with laminate counters. Undermount sinks (bottom left) are streamlined and easy to clean, but costly to install. Solid-surface sinks glue in place with no visible seam (bottom right).

Mounting. Most kitchen and bath sinks are self-rimming and sit on top of the counter surface. However, demand has been growing for flush-set and undermount sinks, which cost more to buy and install, but have the benefits of a more streamlined appearance and easy cleaning (Figure 6-49). With the sink set level with or beneath the counter, food debris can be easily swept into the sink and grime does not collect at the joint of the sink rim and counter.

- *Self-rimming.* Standard cast-iron and stainless-steel sinks have a metal flange that hides any rough cuts in the countertop but tends to collect grime where it joins the counter. The rim should be sealed to the counter with silicone sealant.

- *Flush-set.* This type of sink, sometimes called a "tile-edge" sink, is designed to sit level with the countertop and is often used with tile counters. With tile counters, a flexible sealant such as Color Caulk (Color Caulk, Colton, CA) should be used to seal between the sink and tile. Flush-set stainless-steel or cast-iron sinks can also be set into laminate counters by carefully routing a hole for the sink and using special trim that is virtually flush with the counter. European-based sink manufacturer Karran USA makes a reinforced solid surface sink specially designed to flush set in laminate tops.

TABLE 6-9	Sink Materials			
Material	Cost	Pros	Cons	Recommendations
Stainless steel	Low to medium	Impervious to odors, heat and most chemicals. Scratch and stain resistant. Relatively easy to clean. Abrasive pads and cleansers OK except on polished deck.	Polished steel finish difficult to maintain. Light-gauge steel may flex, dent, and be noisy with a disposal. Steel wool or iron pots can leave rust stains.	Brush finish easier to maintain than polished. 18-gauge or thicker steel will help avoid flexing, denting, and noise. Undercoatings also dampen noise. Avoid steel wool pads, concentrated bleach solutions, and strong acids.
Enameled cast iron	Medium	Durable, glossy, nonporous finish impervious to stains, odors, chemicals, heat. Very scratch-resistant. Cleans easily. Solid feel and quiet with disposal.	Sharp blow can chip enamel, leading to rust. Abrasive cleansers will dull finish over time. Drain cleaner may stain. Hard surface can chip delicate dishes.	Avoid abrasive cleansers or pads.
Enameled steel	Low	Same characteristics as enameled cast iron and lighter weight.	Less solid and more noisy than cast iron. Also more likely to chip if struck with heavy object.	Do not use abrasive cleansers or pads. Consider Americast steel composite material as an upgrade.
Solid surfacing	Medium	Custom sizes and seamless undermounting available. Nonporous surface with moderate stain- and scratch-resistance. Scratches, stains, burns can be sanded out.	Limited colors for stock sinks. Less scratch- and stain-resistant than porcelain. Can discolor from harsh chemicals, such as oven cleansers and acetone. Dropped heavy object can shatter solid surface.	Clean with soapy water or ammonia-based cleaner. For tough stains or small scratches, scour with ScotchBrite™ pad. Sand bigger scratches with fine sandpaper and finish with ScotchBrite™ pad. Run cold water in sink before pouring in boiling water.
Composite	Medium to high	Very durable nonporous material with excellent scratch, stain, and heat resistance.	Limited colors. Sharp blow can chip. Can be stained by rust from cast iron or discolored by concentrated bleach, drain cleaner, or solvents.	Clean with gentle cleansers and nonabrasive nylon pads. ScotchBrite™ and mild abrasives cleansers OK on some products (consult manufacturer).
Cultured marble	Low	Inexpensive. Integral sink vanity tops easy to clean and maintain. Nonporous gel coat has good stain resistance. Small scratches, chips repairable.	Thin gel coat subject to chipping, scratching, burning and other damage. Not recommended for kitchen sinks. Low-quality products may craze around drains and other stress points.	Use only ANSI-labeled products. Avoid abrasive cleansers and abrasive scouring pads. OK to use thinner or alcohol on tough stains. Can restore sheen by using Gel Gloss or light wax.
Vitreous china	Low to medium	Durable, nonporous surface easy to clean. Very scratch resistant.	Tends to chip from a sharp blow.	Used mainly in bathrooms. Not recommended for high-usage kitchen sinks.
Acrylic	Medium	Durable, nonporous, very stain-resistant finish with through-color. Small scratches can be sanded out and the surface restored with polishing compounds.	Relatively soft material easy to scratch. Vulnerable to excessive heat or petroleum-based chemicals. Abrasive pads and cleansers will mar surface. Steel wool will stain.	Only use nonabrasive pads and cleansers. OK to bleach to remove stains. Surface can be restored with auto polishing compound or acrylic polish from supplier.

• *Undermount.* This type of sink provides the most streamlined appearance and easiest cleanup, but it is also the most costly to buy and install. Used with solid surfacing, stone, or composite tops, the cutout must be perfect, and most sinks must be independently supported from underneath the counter. Solid-surfacing sinks are typically glued to the countertop from underneath, creating a seamless transition.

The main material choices for kitchen and bathroom sinks are outlined below (Table 6-9).

Enameled Steel. This uses the same process as enameled cast iron, but with a substrate of stamped 14-gauge steel. These sinks weigh half or less than a comparable cast-iron model, making them easier on the installer. But the lighter, less rigid substrate does not dampen noise as well and is more likely to chip if a heavy object is dropped. One alternative is a hybrid from American Standard called Americast, introduced in the late 1980s, which uses enameled steel on the inside and a cast-polymer composite on the outside to create a sturdy tub with half the weight of cast iron.

Solid Surfacing. The main advantage of solid-surface sinks is that they can be installed seamlessly to the underside of a solid-surface counters. This provides a very streamlined appearance, easy cleanup, and no joints to collect dirt. Solid surfacing also offers great design flexibility, as most fabricators can build nearly any configuration desired. Most solid surface sinks have a matte finish, which is easier to maintain than a semigloss or high-gloss. While relatively easy to scratch or scorch, this type of damage is superficial and is easy to sand away with fine sandpaper or a ScotchBrite™ pad.

Composite. Similar to solid surfacing, composite sinks are a cast polymer using crushed quartz or granite as the filler. High-quality composites have similar characteristics to engineered stone counters, and in some cases are seamlessly cast from the same material (Figure 6-50).

FIGURE 6-50 | **Composite Sink.**

Composite vanity tops with integral or undermount sinks provide a streamlined appearance with easy cleanup and no rim to catch dirt and grime.
SOURCE: Photo by author.

In general, they provide excellent resistance against stains, scratches, chips, and fading. They also tolerate heat well. For example, the sinks made from Moenstone (Moen) and Kindred Granite (FHP Kindred) can tolerate temperatures up to 530°F for short periods. Finishes range from matte to a satin semigloss. Cleaning instructions vary from manufacturer to manufacturer, but most recommend mild nonabrasive cleansers and nylon scrub pads for everyday cleaning and ScotchBrite™ pads and abrasive cleansers as needed. Scratches or stubborn stains can be sanded out as with solid surfacing, although matching a glossy finish may be difficult (one solution is to sand the whole sink).

Metal scouring pads or cast-iron cookware can leave rust stains on composite sinks. Concentrated bleaches, paint strippers, or products containing formic acid (drain cleaner) can discolor the surface. An application of Gel-Glos™ (T.R. Industries) or Invisible Shield® (Unelko Corp.) is recommended by some manufacturers to maintain the sheen and ease of cleaning.

Cultured Marble. See page 249 and Table 6-9 for a discussion of the performance and care of cultured marble.

Vitreous China. Also called porcelain, vitreous china is a heavy ceramic product also used to make toilets. It is highly scratch-resistant and less affected by abrasive cleansers than enameled cast iron or cast-polymer materials (solid surfacing, stone composites). It is widely used in vanity sinks, but because the china substrate will chip more easily than other materials, it is not commonly used in kitchens. Some high-end ornamental china sinks are available for kitchens, but may be too fragile for a kitchen work center.

TABLE 6-10 | **Tub and Shower Materials**

Material	Cost	Pros	Cons	Recommendations
Enameled cast iron	Medium	Durable, glossy, nonporous finish is easy to clean and impervious to stains and odors. Very scratch- and chip-resistant. Feels solid and quiet.	Sharp blow with hard object can chip enamel. Very heavy for installers. Abrasive cleansers will dull finish over time. Cold surface to touch.	Avoid using abrasive cleansers or pads.
Enameled steel	Low	Same finish as enameled cast iron and lighter weight.	Less solid feel. More noisy and prone to chip.	Avoid using abrasive cleansers or pads. Consider Americast, a steel-composite material, as an upgrade.
Fiberglass	Low	Lightweight, contoured shapes. Relatively easy to repair.	Not as hard or as stain-resistant as acrylic or enameled steel. Protected by thin gel coat, which is easy to scratch or chip. Vulnerable to caustics such as lye-based drain cleansers.	To maintain a glossy shine, apply an auto wax or fiberglass bath wax. Clean with nonabrasives only.
Acrylic	Medium	Lightweight, contoured shapes. Durable, stain- and scratch-resistant, high-gloss finish. Through-color. Small scratches can be sanded out and the surface restored with polishing compounds.	Relatively soft material. Abrasive pads and cleansers will mar surface. Vulnerable to excessive heat or petroleum-based chemicals. Difficult to repair.	Only use nonabrasive pads and cleansers. OK to bleach to remove stains. Surface can be restored with special acrylic polish or auto polishing compound.

Acrylic. A relative newcomer, acrylic sinks are made of the same materials as acrylic tubs or showers. Made from heat-molded $\frac{1}{8}$-inch thick acrylic plastic sheets, the fixtures are molded into a wide variety of shapes, then reinforced on the back side with fiberglass and resin. The surface is nonporous and very stain-resistant, but it is relatively soft and easy to scratch. It is also vulnerable to petroleum-based chemicals and heat, for example from a hot skillet. Burns are not repairable.

On the plus side, acrylic has good noise dampening characteristics and can tolerate bleach when needed for a difficult stain. The color goes all the way through the material, so it is possible to sand or buff out small scratches with auto polishing compound or special acrylic polish. For larger scratches, use 400- to 600-grit sandpaper and buff with baking soda.

Stainless Steel. A basic stainless-steel sink is one of least expensive options, and one of the most durable. It is unaffected by heat and most chemicals and the surface will not absorb stains, odors, or oils. It is relatively easy to clean and can be scrubbed with abrasive cleansers and pads when needed. Avoid the cheapest sinks, which use lightweight steel (20 to 23 gauge), as they can flex or dent; also avoid low-nickel alloys, such as 18-8, which can tarnish. Lightweight steel sinks also tend to be noisy with a waste disposer. Good quality sinks are typically 18-gauge or thicker and use high-quality alloys, such as 18-10.

Also avoid steel sinks with a polished finish, which is difficult to maintain. A brushed (matte) finish hides scratches from normal use and cleaning. And although good quality stainless is tough to damage, it is not indestructible. It can develop rust stains from steel wool residue or prolonged contact with cast iron cookware. Also, prolonged contact with concentrated bleach solutions, strong acids, or salty materials can cause pitting. Still, for function and economy, steel is hard to beat.

Enameled Cast Iron. The hard, glossy finish on a cast-iron sink is made by fusing a porcelain enamel glaze to a heavy cast-iron substrate under high heat. This creates a solid fixture with a tough, lustrous finish that is impervious to stains, chemicals, odors, and heat, and cleans easily with a sponge. The heavy mass of the cast-iron base dampens any disposal sounds, but the same rigid mass will readily crack a dropped dish. The porcelain finish is durable but not indestructible. Harsh, abrasive cleansers and abrasive pads will dull the surface and a hard enough blow with a sharp object can chip the coating exposing the iron beneath to rust.

Tubs and Tub/Shower Units

The classic tub for many homeowners is still the rock-solid, cast-iron model with a porcelain enamel finish. Driven by cost savings on both materials and installation, manufactures have introduced lighter-weight alternatives using molded plastics (Table 6-10). While not as hard as the

FIGURE 6-51 **One-piece Tubs and Showers.**

Economical one-piece fiberglass and acrylic tubs and showers provide a seamless, leak-free enclosure that is easy to clean and maintain. Quality improvements and a vast array of choices have made these a popular alternative to custom surrounds. SOURCE: Photo by author.

original iron tubs, the plastic alternatives have improved over the years and provide great flexibility for designers. Most units include integral soap dishes, ledges, grab bars, and other molded features that help bathers and also provide rigidity to the unit. For new construction, most units are one-piece, creating a seamless, leak-free bathing enclosure (see Figure 6-51). Retrofit tub/shower units typically have four or five pieces that fit together with sealant. The fewer pieces, the better.

Porcelain-Enameled Cast Iron. The classic tub material is very solid, very quiet, and resists scratches, chips, and stains. On the other hand it is very heavy, cold to the touch, and available only in pretty basic shapes. It does not easily mold into the body-hugging contours found on many of the plastic units. Although the surface is the most durable for molded tubs, it is not indestructible. If homeowners use abrasive cleansers and pads, over time the finish will dull.

Porcelain-Enameled Steel. This weighs less than cast-iron and feels less substantial, as the material has some give under the weight of water and bathers. It is also more prone to chip than cast iron. A hybrid alternative is American Standard's Americast, which is a porcelain-enameled steel reinforced on the back with a cast polymer. The material is stiffer and quieter than regular enameled steel, but it is half the weight of cast iron. The company literature claims that the surface is more durable and slip resistant than standard porcelain enamel and that the composite layer helps retain heat.

Fiberglass. Also called FRP (fiberglass-reinforced plastic), or gel coat, this material is manufactured by spraying a thin coat of gel coat into a mold followed by several layers of polyester resin mixed with chopped fiberglass. Between coats, fabricators typically reinforce tub and shower walls and floors with blocks of wood or corrugated cardboard. One of the premium manufacturers of fiberglass tubs, Aqua Glass, uses a layer of rigid polyurethane foam between fiberglass coats, creating a stiffer assembly that helps the tub retain heat and deadens sound. The company also adds an antimicrobial compound to the gel coat to inhibit the growth of bacteria and mold. With any FRP tub, the thickness of the gel coat and thickness of the overall lay-up affect its strength and durability. One way to assess quality is to look for tubs that conform to the voluntary standard ANSI Z-124, which requires that a random sampling of acrylic and gel-coat tubs undergo a variety of durability tests.

The thin gel coat can be damaged, but it is relatively easy to repair with gel-coat repair kits sold in marine and automotive stores. Color matching, however, can be difficult. Also if a chip is not repaired quickly, water can penetrate to the backing and cause the damage to spread. Overall, a gel-coat surface is less scratch- and stain-resistant than acrylic. Abrasive cleansers must not be used. Many manufacturers recommend an autowax or special fiberglass bath wax available from tub suppliers.

A proprietary FRP composite called Vikrell™ (Sterling/Kohler), makes tubs from a mix of resin, color, and chopped fiberglass molded under compression. The material has no gel coat or layers to chip or crack and has color all the way through. The manufacturer claims that the high gloss finish has the durability of acrylic and the ease of repair of fiberglass at roughly 40% less than the cost of acrylic.

Acrylic. See description (previous page), under sink materials. One interesting high-end product called Armacryl (Kallista/Kohler) uses a thick clear layer of acrylic over a second color layer. Like other acrylic fixtures, it is reinforced with a fiberglass backing. The result is a lustrous and thick acrylic finish with excellent durability.

Soaking Tubs and Jetted Tubs

Material choices for jetted tubs and other oversized "soaking" tubs are the same as described above. However, with the added stress of higher water levels, multiple bathers, and the vibration of jets, the choice of material is more critical. The most durable type of tub, enameled cast iron, has limited offerings and are very heavy so they are not often used. Acrylic is an excellent choice, as are some of the proprietary materials, such as Americast (American Standard), Vikrell (Sterling/Kohler), and Armacryl (Kallista/Kohler). A tub with a nonslip surface is recommended for safety.

Tub Design. Jetted tubs come in a wide range of shapes, sizes, and depths and are designed for either a drop-in or niche installation with a tile flange. Typical rectangular or oval units range from 5 to 6 feet long, 32 to 45 inches

FIGURE 6-52 Jetted Tub Surround.

Most jetted tubs are designed to either drop into a custom surround or fit into in a niche with a skirt to finish the front. Drop-ins, like the one shown, use more floor space and require an access panel to service the motor or plumbing.
SOURCE: Photo by author.

wide, and 18 to 24 inches high, and many offer an optional skirt to finish the front and conceal the motor and plumbing. A drop-in requires a custom-built surround and requires more floor space (see Figure 6-52).

- **Tub-showers.** Some models are available as tub-shower combos and others can be adapted for use as a tub/shower. However, most designs do not lend themselves to showering due to placement of controls, impractical shapes for doors or curtains, high walls, and rounded interiors. A hand-held shower is always an option.

- **Primary bath.** If this is the home's only tub, be aware that deep tubs with a wide top edge or tile ledge can be awkward for bathing small children. Also pay attention to the number of gallons required to take a bath, often double the volume of a standard tub.

- **Two-person.** For two people, look for a tub with a center drain, or a triangular corner model. Taller people will generally need larger tubs to stretch out.

- **Tight spaces or retrofits.** Many manufacturers make a tapered model designed to replace a standard 30x60–inch tub. Corner models can also save space in a tight layout.

Noise. Jetted tubs can be rather noisy. If this is a concern, look for a unit with flexible hoses connected to the pump, which helps reduce the vibration and noise transmitted to the tub. Some manufacturers also offer quieter jet mechanisms as an upgrade.

Jets. Some manufacturers offer different types of jets with different flow characteristics. Some circulate air, some water, and some both, producing the most vigorous massage. It is best to try out a system at a showroom if possible. Also pay attention to jet locations. Jets high and on the sides will provide swirling water, while jets placed where bathers will sit and directed toward the back and other stress points will provide a more effective massage. Some manufacturers will customize jet locations. Other options to consider are:

- Recessed jets that do not protrude
- Individually adjustable jets, including on-off controls
- An in-line heater if long soaks are desired
- A built-in cleaning system to keep the internal plumbing free from mold and other residues.

Mechanical. Pumps are usually located at the opposite end from the drain and fill area. Some manufacturers allow the buyer to specify the pump location for better access. Look for a pump with automatic protection against burn-out if accidentally run dry. Pumps range from $\frac{1}{2}$ to 3 horsepower, and many manufacturers offer a horsepower upgrade for a modest price increase, often a worthwhile expense.

The pump comes with a cord and plug that requires a GFCI-protected receptacle. Since the receptacle is inaccessible, it should be either downstream of an accessible GFCI-receptacle or on a circuit with a GFCI breaker. Locate the receptacle where it does not block access to the pump. To ensure safety, make sure that the unit purchased is UL-listed for the entire system, not just the individual components.

Heaters. If the occupants expect to take long soaks in the tub, an accessory heater is advisable. Energy-efficient models capture waste heat from the motor and pump to provide a modest degree of supplemental heat, while others use a conventional electrical heating element that can maintain the water temperature indefinitely. For installations where the volume of the tub exceeds the capacity of the home's water heater, some manufacturers offer an instantaneous water heater that will heat the water as it enters the tub.

Installation. The simplest and least expensive installation is into a two-wall or three-wall cove with factory-supplied skirts to conceal the mechanicals. However, many customers prefer the look of a drop-in model with a custom tile surround (see Figure 6-53). With either type of installation, it is critical that the weight of the tub be supported by the base, not the lip. With larger tubs, often holding 50 to 80 gallons of water (at 8 pounds per gallon), the framing and subfloor may need to be reinforced to support the total weight of the tub, water, and bathers. A larger water heater may also be needed.

The most common method to support the tub bottom and dampen vibration is to set the tub onto several small

FIGURE 6-53 | **Jetted Tub Installation.**

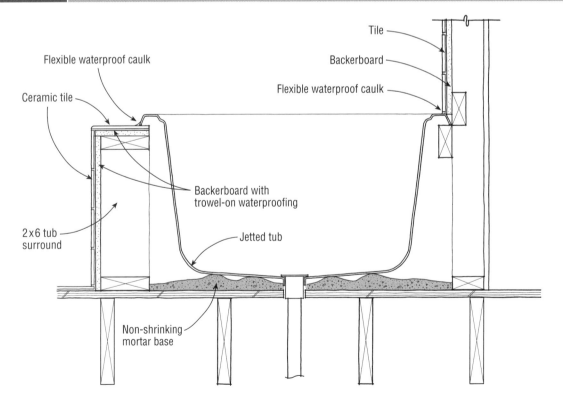

A drop-in tub requires a sturdy wood-framed surround. The weight of the tub, however, should rest on a mortar bed below the base, not on the tub lip. Caulk the joint with a long-lasting flexible sealant where the tub rim meets the tile or other wall finish. Large tubs may require reinforcing of the floor joists for added support.

mounds of wet plaster or mortar mix that compress and conform to the shape of the tub bottom as it is installed. A nonshrinking grout such as Sonogrout 10K (Sonneborn) is best since it will not leave gaps as it dries. Partially filling the tub with water until the mortar sets will ensure a good fit. Many contractors recommend using mortar even with units including "self-leveling" bases, which are designed to sit on a level subfloor with no additional support.

A few tubs are designed with individual bearing points molded into the bottom of the tub. With this type, it is important to make sure that the bearing points fall on joists or solid blocking. Solid bridging should also be added to help distribute the load to the surrounding joists.

Access Panel. With a custom-built surround, it is necessary to leave an access panel. Although access to the mechanicals is rarely needed, it should not require tearing apart the tile surround. A simple approach is to tile the access panel, hang it with magnets, and caulk the joint between the access panel and surrounding tiles. If access is necessary, the caulk can be slit with a knife and the panel removed.

Where the tub rim meets tile at a wall or the top of the surround, use a flexible caulk rather than grout, which will crack as the tub flexes, and expands and contracts with temperature changes.

Faucets

Manufacturers offer a vast array of faucet styles and finishes. Most homeowners choose single-level models for the convenience of one-handed control, and chrome remains the most popular finish because of its durability and easy cleaning. New finishing technologies permit nearly any color or metallic finish with similar durability. Most faucets also have improved valve systems that make leaky faucets a thing of the past.

Finishes. Most faucet finishes are guaranteed for life. Chrome, the most popular choice, is easy to clean and very resistant to scratching. For a colored finish, the best choice is an epoxy coating. These come in a wide range of colors and are also very durable and easy to maintain, but they can be scratched with an abrasive scouring pad. Also solvents, such as nail polish remover (acetone), can soften the epoxy coating.

Many higher-end faucets now use a new high-tech plating technology called physical-vapor deposition or PVD, which can imitate almost any metallic finish from brass to pewter to gold. PVD finishes are as hard as chrome and easy to maintain, although they can be stained by drain cleansers containing lye. Also, abrasive pads, such as steel wool or ScotchBrite®, can scratch these or any finish with enough effort. Where scratching is a concern, a brushed or satin finish is preferable, since it will help conceal scratches.

Solid brass faucets usually come with a lacquer or similar coating to protect against tarnishing. The metal will tarnish, however, when the coating eventually wears off or where it gets scratched. As an alternative, many manufacturers now offer "tarnish-free brass" finishes using PVD technology. Marketed under a variety of names, these finishes are actually applied over a chrome-plated faucet and have the durability of chrome.

Styles. Most homeowners find single-handle faucets more convenient, and experts consider them safer for kids than separate hot and cold levers, which are more likely to cause a scalding injury. Whatever style is chosen, levers are easier to manipulate than knobs, particularly for anyone with restricted mobility. Lavatory faucets should be low enough that they do not hit people in the head if they lean down to wash their face. An offset faucet or one that swings out of the way can provide greater useful space.

The spout in a kitchen should be high enough to accommodate large pots and long enough to reach all the basins in a multibasin model. One that swings out of the way can increase the useful space. High-arch or *gooseneck* designs are often a good solution (see Figure 6-48, page 251). Pullout spray spouts are also very convenient, but avoid those with flimsy plastic hoses. Look for a durable double-interlock hose that will not kink.

Faucet Materials. Underneath the finish, the base material for most faucet bodies is brass, zinc, steel, or cheap alloys called "pot metal." Some economy models use plastic. The highest-quality faucets use solid forged or machined brass, which will last the longest and require the least care, particularly in areas with hard water. Cast brass is also used, but it is not quite as durable. Midrange faucets use brass or chrome plating on zinc alloys, which provides good durability, but must be replaced when the plating wears through. Plastic, steel, and pot-metal bodies are usually the least durable.

Valve Types. The faucet valve is the mechanism that regulates the flow of water when you turn the lever or knob. There are four basic types, each with pros and cons.

- *Compression valves* are the old standby used in traditional two-handle faucets. These have a rubber washer on the bottom of a threaded stem, which opens and closes as the knob is twisted. The washers wear out over time, more quickly with hard water or overtightening. New washers cost less than a dollar and are generally easy to replace. The metal valve seats also get worn over time and need to be ground with fine sandpaper to prevent leaks.

- *Sleeve-cartridges* have a hollow stem that rotates inside a metal or plastic sleeve to control the water flow in both one- and two-handle faucets. Usually the whole cartridge is replaced when it leaks, although replacing the O-ring at the top or bottom of the cartridge may work. Most cartridges are easy to replace, but some faucets require a special tool to pry out the cartridge. Choose brass cartridges, rather than plastic, if available. Replacements cost $15 to $20.

- ***Ball valves.*** Developed originally by Delta, these are still widely used in single-lever faucets. They use a hollow brass or plastic ball with spring-loaded seals to control water flow. Ball valves last a long time, and replacement parts are inexpensive (under $10), but the small parts can be tricky to assemble. Brass replacement parts will outlast plastic.
- ***Ceramic-disk cartridges.*** Considered the most reliable approach, this new type of valve uses hard ceramic disks that rotate in a cartridge until the holes line up to release water. The polished disks are nearly indestructible, but eventually the grease between the disks washes out or the washers or O-rings wear out. Typically, the whole cartridge is replaced. This is a simple operation, which costs about $10 to $20 for parts.

Scald Protection

There are many styles of mixer valves in tubs and showers, but all should have some sort of protection against scalding. The young and elderly are at greatest risk due to thinner skin and slower reaction times. More than 35,000 children, most under age six, are treated each year in emergency rooms for tap-water scald burns, according to the National Safe Kids Campaign. A child exposed to 140°F water for as little as three seconds can sustain a third-degree burn requiring hospitalization and skin grafts. Although most scald burns occur in the kitchen, according to the Consumer Product Safety Commission, the most severe scald burns are caused by water flowing into the tub or shower. In response, many state and local codes now require antiscald protection in residential tubs and showers.

Many organizations advocate setting water heaters to no more than 120°F, which gives bathers significantly more time to move away or adjust the temperature before a burn occurs. While this strategy is helpful, it is not always reliable and can cause other problems:

- Noncompliance: many homeowners turn the thermostats up to increase supply.
- Water heater thermostats are often inaccurate. The ANSI standard for gas water heaters allows the temperature to vary by plus-or-minus 10°F.
- Stacking effect: water at the top of a gas water heater can exceed the set point by as much as 30°F.

Pressure-Balancing Valves. This is the most commonly used point-of-supply approach. These automatically adjust the water pressure to maintain the mix of hot and cold water to within 2°F to 3°F of where the user set the shower control. If cold water is diverted to a flushing toilet or other appliance and the pressure drops, the valve automatically reduces the hot water flow proportionately to maintain the temperature. If the cold water pressure plummets or stops altogether, the flow is reduced to a trickle. To guard against someone accidentally turning the shower valve to scalding temperatures, these valves typically use a temperature limit stop that prevents the user from turning the shower control past a set point—typically set at or below 120°F.

Thermostatic Mixing Valves. These are more expensive than pressure-balancing valves and not widely used in the United States. They can be installed either inline near the water heater or as part of the shower control. The inline type of valve, also called a tempering valve, adds cold water to the hot water as it leaves the water heater to maintain a constant temperature, set by the installer. These are commonly used with solar water heaters but can also be tied to a conventional water heater. A check valve is required on the cold water side to prevent backflow, and a hot-water expansion tank is recommended to prevent excessive pressure on the hot water side.

The other type of thermostatic valve is built into some high-end showers (Figure 6-54). These allow the user to set the temperature on a dial when showering. The unit will compensate for changes in either pressure or temperature to maintain a constant delivery temperature and flow rate.

FIGURE 6-54 | **Dial a Shower Temperature.**

High-end thermostatic shower valves allow the user to set the temperature on a dial when showering. The valve will compensate for changes in either pressure or temperature to maintain a constant delivery temperature and flow rate.
SOURCE: Photo courtesy of Kohler.

If the cold water fails or the tempered water is still too hot for any reason, the unit will shut off the flow. As with the pressure-balancing valve, the installer sets a temperature limit stop to prevent the user from turning the shower control to scalding temperatures.

Retrofits. In retrofits, point-of-use devices can be installed by a plumber or homeowner to limit water temperatures to 120°F. These include antiscald showerheads, as well as point-of-use devices that fit into individual plumbing fixtures, such as showerheads and bath and sink faucets. For example, MemrySafe and ScaldShield (Antiscald Inc.) are inexpensive retrofit devices that reduce the water flow to less than $\frac{1}{4}$ gallon per minute when the water temperature at the faucet or showerhead exceeds 120°F. These devices do not regulate temperature or pressure, but do offer protection against serious burns.

Toilets

When water-saving toilets were first introduced in the 1980s, they reduced water usage from 5 to 6 gallons per flush (gpf) to 3.5 with little effect on performance. However, when Congress mandated in 1992 that manufacturers had two years to reduce water usage to 1.6 gpf, the engineering challenges were much greater. Most early models were essentially 3.5 gallon designs hastily modified with smaller tanks and narrower trapways intended to increase the flow rate. Most did not work well and required two and sometimes three flushes, negating much of the benefit to water conservation.

Better design and new technologies have led to greatly improved performance in some models. In addition to traditional gravity designs, two types of pressurized designs have been introduced: pressure-assisted and vacuum-assisted. Both types benefit from the additional flushing power. However, with the pressure-assisted, the increased flush comes with increased noise and complexity (Table 6-11).

Because there is no standardized testing, figuring out which models perform well is difficult. Test results on individual models performed by the National Association of Home Builders, Consumer Reports, and other groups are a useful guide, but results are not consistent due to differing test methods. While all toilets sold in the United States must meet the American Society of Mechanical Engineers (ASME) performance standard A112.119.2M, this only guarantees basic functionality.

Gravity-Flush. This is the traditional method, which relies on water dropping from the tank to cleanse the bowl and start a siphon action in the trapway. Although the first low-flow gravity toilets had trouble, many of the newer models have been reengineered with higher tanks and larger trapways to improve the flush. Some store three or more gallons in the tank to create greater pressure, but only let 1.6 gallons drain in each flush. One side benefit to this approach is less sweating in hot weather, since only a portion of the water in the tank is replaced with new cold water. Gravity-flush models are generally easy to maintain, although they are a little more finicky than older 3.5 gallon models. For example, the water level must be accurately set for them to work as designed and worn out flapper valves must be replaced with the proper model rather than a generic one.

Among the gravity flush models that consistently rated well in independent tests are any of the Toto models that use its G-Max flushing system and the Kohler Wellworth and Santa Rosa.

Power-Assisted. Widely used in commercial settings, these toilets have a sealed chamber in the tank that fills after each flush, using the home's water pressure to compress the air trapped inside. With each flush, the compressed air forces the water down into the bowl with a loud whoosh, which is the primary complaint about these models. As long as household water pressure is at least 25 psi, these are very effective at clearing the bowl.

Because of the tank within a tank, there is no sweating on the outside of the porcelain, but condensation and mildew can form on the surface of the power-assist unit. While these toilets are not for everyone, they are

TABLE 6-11	Toilet Performance		
Flush type	Pros	Cons	Recommendations
Gravity	Simple mechanism. Easy to maintain. Least expensive. Units with 3-plus gallon tanks have reduced sweating.	Some models have trouble removing waste. Many require 2 or 3 flushes.	Only buy models with high ratings. Set water level accurately. Replace worn out flapper with proper model only.
Power-assisted	Very effective at removing waste and cleaning bowl. No sweating outside tank.	Noisy flush. Mold and mildew can grow on surface of power-assist unit and in stagnant water at bottom of tank.	Good solution for high-use toilets or where drainage is marginal.
Vacuum-assisted	Very effective at removing waste and cleaning bowl. Quiet as gravity flush. Easy to maintain. Moderate cost.	Relatively new design. Few models available.	Good option where power flush is desired without noise.

particularly useful in retrofits where drainage is a problem, for example, where venting is inadequate or where the toilet sits at the end of a long drainage run with no fixtures upstream to wash the line. When these toilets malfunction, the complete pressure-assisted unit is replaced. Power-assisted toilets cost about $100 to $150 more than comparable gravity models.

Examples of top-rate power-assisted models include the Crane Economiser, American Standard Cadet PA, and Gerber Ultra Flush.

Vacuum-Assisted.
A relatively new design, these also have a sealed chamber. With each flush, a partial vacuum is created in the chamber, which connects to the trapway with tubing. The vacuum boosts the suction during the next flush, achieving a powerful flush but without the noise of a power-assisted unit. Vacuum-assisted toilets use the same flush and fill valves as gravity flush toilets, making them easy to maintain. Only a few models are currently on the market, but they have performed very well in tests and are moderately priced at $50 to $100 more than comparable gravity models.

Among the top-rated models are the Briggs Vacuity 4200 and the Crane VIP Flush 3999.

One-Piece vs. Two-Piece.
One-piece units tend to have low sleek lines and cost substantially more than standard two-piece models where the tank is bolted to the bowl during installation (Figure 6-55). One-piece models are generally easier to clean, since there are fewer crevices

FIGURE 6-55   **One-Piece Toilet.**

One-piece units tend to have low sleek lines and are generally easier to clean, since there are fewer crevices between the bowl and tank to collect dirt. They also cost substantially more than standard two-piece models.
SOURCE: Figure courtesy of Kohler.

between the bowl and tank to collect dirt. With gravity-flush toilets, it is harder for designers to achieve a sufficient head of water without the height of a standard tank, although redesigning other components has overcome this in at least some models.

Sweating.
In humid areas with cold incoming water, sweating on the outside of the porcelain tank can be a significant problem, in some cases rotting the flooring around the toilets. Some of the new flushing strategies alleviate the problem somewhat:

- In gravity toilets that store 3 or more gallons of water in the tank, the incoming 1.6 gallons of cold water is tempered by the retained water in the tank.

- In pressure-assisted toilets, the incoming water resides in a small tank within the larger porcelain tank, so the porcelain tank is not chilled. However, condensation can form around the inner tank, leading to mold and mildew growth.

One approach with older toilets is to add special foam insulation inserts inside the tank. These may not work with low-flow designs, however. Also this does not prevent dripping from the bowl or water supply line. Where the problem persists, consider added an antisweat valve (Beacon Valves) that tempers the incoming cold water with a little bit of hot water to bring it up to room temperature.

Installation.
Before installing a toilet, examine it for manufacturing defects that can cause leaks or prevent it from sealing fully to the closet flange. Occasionally the inlet where the tank connects to the bowl, or the outlet (horn) at the bottom of the bowl that seals to the closet flange, is deformed enough to cause problems. Also make sure the base (foot) of the toilet is flat or it will have a tendency to rock and break the seal, leading to odors and leakage.

Next, make sure the framing is adequate. Ideally the toilet should sit between two joists set no more than 12 inches on-center, with blocking nailed on either side of the drain to reinforce the area around the closet flange (Figure 6-56). This will limit any movement in the fixture that could break the seal at the wax ring. The hole in the subfloor should be just a bit larger than the drain. Also, it is best if the toilet sits on top of the finished flooring rather than having a dirt-trapping joint where the finished floor abuts the fixture.

Next, install the closet flange and secure it to the subfloor with brass or stainless-steel screws. With vinyl flooring, the flange can sit directly on the subfloor. With tile, use a plywood spacer to raise the flange to the height of the tile. To insure longevity, use brass closet bolts, nuts, and washers rather than the plated steel that is often packaged with the toilet. Then place the wax ring in the closet flange and set the bowl in place. Gradually tighten the nuts on the closet bolts, alternating from one side to the other

FIGURE 6-56 | Framing Under Toilet.

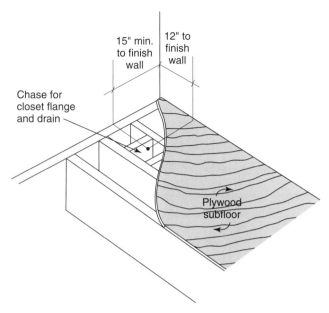

Use closely spaced framing and solid blocking around the toilet flange to create a sturdy floor under the toilet, preventing movement that could break the seal at the wax ring.

until it is snug without overtightening. Finally, apply a bead of silicone caulk around the sides and front of the toilet foot, leaving the back unsealed. This will allow leaks to be spotted before serious damage develops.

KITCHEN AND BATH VENTILATION

Kitchens and bathrooms are key sources of indoor moisture and other pollutants. Kitchens produce particulates and atomized grease from cooking, and with a gas range, they also produce combustion by-products including nitrogen dioxide and carbon monoxide. Bathrooms produce moisture, odors, and VOCs from aerosols and various personal hygiene products. Effective spot ventilation in these areas is critical for maintaining healthy levels of indoor humidity levels and an overall healthy indoor environment (see Chapter 7).

Bathroom Ventilation

Bathroom ventilation should be powerful enough to remove excess moisture before it has time to condense on cold walls and windows, potentially damaging finishes, or escape into wall or ceiling cavities, where it may lead to decay or peeling paint.

Ventilation Rate. The minimum ventilation rate for bathrooms required by the 2003 International Residential Code (IRC) is 50 cfm intermittent or 20 cfm continuous

TABLE 6-12 | Minimum Ventilation Rate per Fixture for Baths Over 100 Square Feet

Toilet	50 CFM
Shower	50 CFM
Bath Tub	50 CFM
Steam shower	50 CFM*
Jetted Tub	100 CFM

*Separate fan recommended.
SOURCE: Guidelines courtesy of the Home Ventilating Institute. © 2004 Penton Media.

(if part of a whole-house ventilation system). While this may be adequate for a small bath, the guidelines of the Home Ventilating Institute (HVI) are more suitable for larger rooms:

- **Small baths:** For bathrooms up to 100 sq ft, HVI recommends using an exhaust fan that provides 1 cfm per square foot of floor space. This will provide approximately 8 air changes per hour. So, for example, an 8x10-foot bathroom would require 80 cfm of ventilation.

- **Large baths:** For bathrooms over 100 sq ft, the HVI recommends a ventilation rate based on the number and type of fixtures as shown in Table 6-12. For example, a bathroom with a toilet, shower, and jetted tub would require 200 cfm (50 + 50 + 100) of ventilation either in a single fan or multiple fans placed over the fixtures being vented.

Noise. The biggest homeowner objection to bath fans, their noise, has been addressed with the introduction of whisper-quiet bath fans from a number of manufacturers. Choose the quietest fan for the job, preferably one rated 1.5 sones or less (one sone roughly equals the sound of a quiet refrigerator motor).

Fan Location. For optimal ventilation, locate the exhaust grilles near the source to be ventilated, typically over the tub or shower. In smaller baths, a single, central exhaust point is usually adequate, while in larger baths, multiple smaller fans (or a remote fan with separate pickups) will be more effective. Bathrooms with enclosed toilet areas or steam showers should have separate exhaust grilles in those areas. Since all exhaust fans require makeup air, the bathroom door needs to be undercut to provide makeup air when bathroom windows are closed.

Controls. The Home Ventilating Institute (HVI) recommends that a fan be left on for 20 minutes after use of a bath or shower to remove all excess moisture from the room and ductwork. A timer switch is the simplest way to accomplish this. Another option is a dehumidistat switch, which will

TABLE 6-13	Equivalent Duct Length for Bath Fans	
	Duct Diameter	
Duct Type or Fitting	4 inch	6 inch
Smooth metal duct	Same as measured length	
Flex duct	2xduct length	2xduct length
Insulated flex duct	2.5xduct length	2.5xduct length
Roof or wall cap	30 ft	40 ft
45° adjustable elbow	4 ft	6 ft
90° adjustable elbow	15 ft	12 ft
90° smooth elbow	4 ft	6 ft

automatically operate the fan whenever the humidity level rises above a preset level. A manual override allows normal operation of the fan for other bathroom uses.

Ductwork. A ventilating fan must overcome the resistance, called "static pressure," created by the ductwork, including transitions, elbows, and the wall or roof cap. The advertised airflow of bath fans is typically based on a static pressure of 0.1 (inches of water gauge) although some also publish the airflow rating at 0.25 inches, which gives a better estimate of actual airflow in most installations. A static pressure of 0.1 is roughly equivalent to 15 feet of straight, smooth 4-inch duct with a 100 cfm fan. A typical installation with about 20 feet of duct, two elbows, and a wall cap has an equivalent duct length closer to 80 feet (see Table 6-13).

How well a particular fan can overcome the duct-work's resistance to airflow is highly variable and is indicated by its fan curve, available from the manufacturer. Inline fans mounted remotely and exterior fans are generally the most powerful. A typical installation with two elbows and 20 to 30 feet of duct reduces the rated airflow of a standard fan by anywhere from 10 to 30%. With long runs, multiple elbows, or corrugated flex duct, airflow may be reduced by 50% or more. To ensure good airflow, follow these guidelines:

- Use smooth ductwork of the same size or slightly larger than the fan outlet. Thin-walled PVC pipe works well and is easy to seal.

- Keep duct runs as short and straight as possible.

- Where possible, use broad sweeps rather than 90-degree angles to change direction.

- Seal all joints in metal duct with foil-backed tape or duct mastic, not cloth duct tape.

- For a standard installation, choose a slightly larger fan size than required. Where duct runs exceed about 25 feet plus two elbows, choose a larger fan size and check the fan curve to determine actual airflow.

Condensation in ductwork is also a concern in cold climates. To avoid problems, insulate the ductwork to at least R-5 or run it below the ceiling insulation. Also keep any metal duct seams facing upward and slope the duct slightly toward the exterior outlet so that any condensation drains to the outside. Avoid any sags in the ductwork, which are potential pooling areas for condensation.

Kitchen Ventilation

Kitchen cooktops produce large amounts of water vapor, atomized grease, particulates, and cooking odors. In addition, gas cooktops produce combustion by-products, including carbon dioxide, nitrogen dioxide, nitrous oxide, and carbon monoxide (see "Combustion Appliances," page 293). The most effective way to remove the moisture and contaminants from a cooktop is with an overhead range hood vented to the outdoors. Unvented range hoods offer no protection against moisture and combustion gases and provide only minimal protection against grease, smoke, and odors trapped by the filter.

Range Hood Location. Research at the University of Minnesota has shown that many range hoods are too small, too high, or not oriented properly to do their job well. According to the study, the most effective range hoods are at least as wide as the cooktop and rectangular rather than angled in front. For best performance, standard hoods should be mounted no more than 24 inches above the cooking surface and project out at least 20 inches from the wall (see Figure 6-57). Some high-capacity ventilators are designed to work from 24 to 30 inches above the cooking

FIGURE 6-57	Range Hood Location.

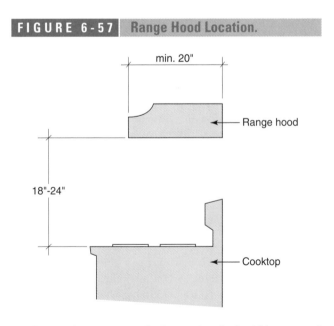

For best performance, standard range hoods should be mounted no more than 24 inches above the cooking surface and should project out at least 20 inches from the wall. Shallow hoods designed to fit under a microwave are not effective at venting the front burners.

surface, but the stronger fans increase the risk of back-drafting other combustion appliances. Hoods integrated with microwaves typically project out only 13 to 15 inches and miss most of the contaminants generated by the front burners.

Ventilation Rate.

The minimum ventilation rate for kitchens required by the 2003 International Residential Code is 100 cfm intermittent or 25 cfm continuous (if part of a whole-house ventilation system). Most industry experts recommend that overhead range hoods have a minimum capacity of 150 cfm and recommend higher capacity fans for open grilling, high-BTU commercial-style ranges, and other cooking styles that generate greater than average steam or smoke (see Table 6-14). A rule of thumb for high-output ranges is 1 cfm of ventilation per 100 BTU's of output. With high-powered fans, multispeed controls are best so the fan speed (and noise) can be lowered when full power is not required.

Island Hoods.

Island hoods are typically installed at least 27 inches above the cook surface so they do not interfere with sightlines. Because they are farther from the cooktop and subject to passing air currents, they require more powerful fans than standard wall-mounted hoods. HVI recommendations for wall-mounted and island hoods are shown in Table 6-14.

Downdraft Fans.

Because they lack a canopy to catch contaminants and must overcome natural convection, downdraft fans are less effective than overhead range hoods. Typically sized at 400 to 600 cfm or more, they do a reasonable job of venting barbecues and grills, but are less effective with pots and pans taller than about 3 inches. Downdraft fans are either flush-mounted at counter level or pop up about 8 inches at the back of the range.

- **Flush-mounted fans** are typically located in the center of the cooktop or to one side. They generally do an adequate job of capturing combustion gases,

as well as water vapor and cooking odors from pans 3 inches or shorter. Vapors from taller pots tend to escape.

- **Pop-up fans** in the rear do an adequate job of capturing combustion gases and moisture, except for tall pots on the front burners.

Noise.

Look for the unit with the lowest noise rating that meets the ventilation requirements. Some hoods are rated as low as 2.5 sones, although most range from 4 to 7 sones at full power (one sone roughly equals the sound of one quiet refrigerator).

Ductwork.

Like bathroom fans, the airflow from kitchen exhaust fans is generally rated at a static pressure of 0.1 in. (see "Ductwork," page 262). This is roughly equivalent to 30 feet of smooth 7-inch round or $3\frac{1}{4}$x10 in. metal duct venting a 200 cfm fan. For good performance, the total equivalent duct length, not counting the wall or roof cap, should not exceed about 30 feet. Equivalent duct lengths for common fittings are shown in Table 6-15.

Backdrafting.

With large-capacity exhaust fans of 200 cfm or greater, there is a risk of causing backdrafting of a fireplace or atmospherically vented boiler, furnace, or water heater. The potential for backdrafting can be tested by a heating system technician and should be conducted with all household exhaust fans running to simulate the worst-case scenario. A simple preliminary test can be done by holding a stick of incense next to the draft diverter or dilution port of each combustion appliance with the furnace fan on and off.

If the smoke spills into the room for more than 30 seconds, then dedicated makeup air is most likely required. Modest amounts of makeup air can be delivered through a passive duct with an automatic damper, but significant depressurization may require an active supply fan. The makeup air supply should be delivered into the kitchen or a nearby room not blocked by a door. (For more information, see "Backdrafting," page 295).

TABLE 6-14	Recommended Ventilation Rates for Overhead Range Hoods	
	Minimum for Regular Cooking	Minimum for High-Intensity Cooking*
Wall-mounted hoods	40 cfm/ft	100 cfm/ft
Island hoods	50 cfm/ft	150/cfm/ft

*Recommended for open grilling, high-BTU commercial ranges, and other cooking styles that generate high levels of steam or smoke.
NOTE: Multiply the cfm rate given by the width of the range hood (in feet) to find the required exhaust airflow in cubic feet per minute.
SOURCE: Courtesy of the Home Ventilating Institute. © 2004 Penton Media.

TABLE 6-15	Equivalent Duct Length for Range Hoods	
	Duct Size	
Duct Type or Fitting	7-inch round	$3\frac{1}{4}$x10 in.
45° elbow*	5 ft	7 ft.
90° elbow*	10 ft	15 ft**
Straight transition 7-inch round to $3\frac{1}{4}$x10 in.	4 ft	4 ft
90° transition: 7-inch round to $3\frac{1}{4}$x10 in.	25 ft	25 ft

*Adjustable-type elbows
**20 ft. for flat elbow

RESOURCES

Manufacturers

Ceramic Tile-Setting Materials

Bonsal American
www.bonsal.com
Setting compounds, grouts, preformed shower pans, curbs, and niches. Also, distributor of backerboards, isolation membranes, and other tile-setting products

Color Caulk, div. of Roanoke Companies Group
www.colorcaulk.com
Color-matched caulking

Custom Building Products
www.custombuildingproducts.com
Elastomeric and liquid-applied membranes, self-leveling underlayments, setting compounds, and grouts

Laticrete International
www.laticrete.com
Trowel-on membranes, self-leveling underlayments, setting compounds, grouts, and sealants

Noble Company
www.noblecompany.com
CPE sheet membranes, trowel-on membranes, clamping ring drains, and preformed slopes, niches, and curbs

Mapei
www.mapei.com
Trowel-on and sheet membranes, self-leveling underlayments, setting compounds, grouts, and color-matched sanded caulks

Ceramic Tile Backerboards

Custom Building Products
www.custombuildingproducts.com
Wonderboard cement backerboard, Easyboard cement and polystyrene lightweight backerboard, and Rhinoboard fiber-cement backerboard

Georgia-Pacific Gypsum
www.gp.com/build
Denshield gypboard backer with glass-matt facing

James Hardie Building Products
www.jameshardie.com
Fiber-cement backerboard

National Gypsum
www.nationalgypsum.com
Permabase lightweight cement and polystyrene backerboard

Schluter Systems
www.schluter.com
Kerdi tile membrane goes directly over drywall or other substrates

T. Clear Corp./Fin Pan Inc.
www.finpan.com
Util-A-Crete lightweight concrete backerboard

U.S. Gypsum
www.usg.com
Durock cement backerboard

W. R. Bonsal
www.bonsal.com
Extruded polystyrene backerboard with fiberglass-reinforced cement facing

Cabinet Hardware

Accuride International
www.accuride.com
Ball-bearing drawer slides

Amerock Corp.
www.amerock.com
Comprehensive catalog of cabinet hinges, pulls, slides, and accessories

Grass America
www.grassusa.com
Cup hinges, slides, shelf supports, and storage accessories

Hafele America
www.hafeleamericas.com
Cup hinges, ball-bearing slides, pulls, and KD connectors

Mepla Inc.
www.mepla-alfit.com
Cup hinges, ball-bearing slides, drawer systems, shelf supports, and KD connectors

Hettich America
www.hettichamerica.com
Cup hinges, ball-bearing drawer slides, shelf supports, and KD connectors

Plastic Laminate

Abet Inc.
www.abetlaminati.com

Arborite
www.arborite.com

Formica Corp.
www.formica.com

Nevamar Decorative Surfaces
www.nevamar.com

Pionite Decorative Surfaces
www.pionite.com

Wilsonart International
www.wilsonart.com

Solid Surface

Avonite Surfaces
www.avonite.com

Centura Solid Surfacing
www.centurasolidsurface.com

Dupont
www.corian.com

Formica Corp.
www.formica.com
Formica solid surfacing, formerly Surell and Fountainhead

Halstead International
www.e-topstone.com
Topstone solid surface distributor

Lippert Corp.
www.lippertcorp.com

Sansung Staron
www.getstaron.com

Swanstone
www.theswancorp.com

TFI
www.tficorp.com

Transolid Inc.
www.transolid.com

Wilsonart International
www.wilsonart.com

Engineered Stone

Cambria
www.cambriausa.com

Cosantino USA
www.silestoneusa.com

Dupont
www.zodiaq.com

Halstead International
www.halsteadintl.com
Granyte distributor

Technistone USA
www.technistoneusa.com

U.S. Quartz Products
www.caesarstoneus.com

Jetted and Soaking Tubs

American Standard
www.americanstandard-us.com

Aqua Glass Corp.
www.aquaglass.com

Aquatic Industries
www.aquaticwhirlpools.com

Bains Oceana Baths
www.bainsoceaniabaths.com

Eljer Plumbingware
www.eljer.com

Hydro Systems
www.hydrosystem.com

Jacuzzi Whirlpool Baths
www.jacuzzi.com

Kohler Co.
www.kohler.com

Lasco Bathware
www.lascobathware.com

Maax
www.maax.com

MTI Whirlpools
www.mtiwhirlpools.com

Bath Fans

American Aldes
www.americanaldes.com
Remote location single- and multi-port exhaust ventilators

Broan-Nutone LLC
www.broan.com
Low-sone Broan bath fans, also single- and multiport remote location exhaust ventilators; Nutone ceiling-mount bath fans

Fan Tech
www.fantech.com
Remote location inline-duct fans

Kanalflakt
www.kanalflakt.com
Remote location inline-duct fans

Marley Engineered Products
www.marleymeh.com
Ceiling-mount bath fans and general kitchen and room exhaust fans

Panasonic
www.panasonic.ca/English/ventilationfans
Low-sone, Energy-Star-compliant ceiling-mount, inline, and wall bath fans

Kitchen Exhaust Fans

Amana
www.amana.com
Pop-up and traditional range hoods

Bosch
www.boschappliances.com
Chimney style, downdraft, and traditional range hoods, stainless steel and colors

Kitchenaid
www.kitchenaid.com
Retractable downdraft, slide out, island, and traditional range hoods

Broan-Nutone LLC
www.nutone.com
Nutone downdraft, island, and traditional range hoods

Dacor
www.dacorappl.com
Commercial-style, stainless-steel pop-up and overhead range hoods; hood liners for custom canopies

Gaggenau
www.gaggenau.com/us
Chimney-style, pop-up, slide-out, and island range hoods in stainless-steel, aluminum, and glass

Jenn Aire/Maytag Corp
www.jennaire.com
Pop-up, under-cabinet, wall, soffit, island, and fans and hood liners for custom canopies

Thermador
www.thermador.com
Updraft, downdraft, and island-range hoods

Wolf Appliance Co.
www.wolfappliance.com
Pop-up, chimney-style, wall, and island-range hoods in stainless steel

For More Information

Association of Home Appliance Manufacturers (AHAM)
www.aham.org

National Kitchen and Bath Association (NKBA)
www.nkba.org

Ceramic Tile Institute of America
www.ctioa.org

Home Ventilation Institute (HVI)
www.hvi.org

Marble Institute of America
www.marble-institute.com

Porcelain Enamel Institute (PEI)
www.porcelainenamel.com

Tile Council of America (TCA)
www.tileusa.com

| # Indoor Air Quality

OVERVIEW

Two trends have conspired to place significant stresses on the indoor environment over the past two decades. First, houses are being built much tighter today than they were a generation ago, either deliberately by energy-minded builders or simply as a by-product of using modern building materials, such as plywood, drywall, insulation, and tight-fitting doors and windows. Second, the number of synthetic building materials has rapidly expanded to include synthetic carpeting, a wide variety of plastics, wood composites, adhesives, sealants, and finishes. These, along with the wide variety of cleaning, personal care, and hobby products stored and used indoors provide most homes with an ample source of airborne chemicals, many of which have not been well studied, either alone or in combination with others. Some leading indoor-air-quality advocates have referred to this unknown mix of airborne compounds as "chemical soup."

Individuals with allergies, asthma, or strong chemical sensitivities were, like the proverbial canary in the coal mine, the first to call attention to the higher concentrations of chemicals that were building up in our new, tighter homes. While scientists had thoroughly studied the outdoor air in cities and indoor air in occupational settings, little was known about air quality in homes.

What's in the Air? A growing body of scientific evidence has demonstrated that the air inside homes is typically more polluted than outdoor air, even in polluted urban areas. For example, the U.S. EPA TEAM study of over 600 residents in seven cities in the 1980s found that exposure to toxic chemicals was much greater at home or at work than outdoors. Levels of about a dozen common organic pollutants were found to be two to five times higher inside homes than outside, regardless of whether homes were in rural or industrial areas. And since the average person spends far more time indoors than outside, the study concluded that health risks from the indoor environment pose a greater risk to most people than outdoor air pollution.

Fortunately, as builders, designers, and homeowners, we potentially have much greater control over our indoor environment than out of doors. Public health professionals and researchers both in the private sector and in state and federal agencies have identified the most significant threats posed by indoor air pollution, as well as a number of straightforward strategies that enable us to minimize or eliminate the health risks.

Acceptable Risk. Remember, there is no environment—indoor or outdoor—that is 100% free of hazardous materials, many of which (like radon, asbestos, and airborne particulates) occur naturally in the environment. And while many of these substances have been studied extensively in the workplace, the effects of long-term exposure to the lower levels found in most homes are not well understood. For some pollutants, like radon, scientists have a fairly precise understanding of the health effects and recognize that that no exposure level is safe. However, the cost of reducing the indoor radon level to zero (below outdoor levels) would be prohibitive for most people, so homeowners, health professionals, and regulatory agencies do their best to find a "cost-effective" goal that balances costs against perceived health risks.

In the absence of clear indoor air guidelines, and taking into account that all building projects have budget limitations, the goal of this chapter is to identify reasonable steps that builders and designers, and, in some cases, homeowners can take to produce a healthy indoor environment by eliminating or substantially reducing known hazards. The emphasis will be on getting the greatest benefit for the least cost, starting with the most significant hazards. How much an individual invests in clean indoor air is a matter of personal choice. Fortunately, with good planning, a great deal can be accomplished for a modest investment.

For individuals with special sensitivities to chemicals, dust, or biological materials, the measures described here may not be adequate. A more comprehensive approach under the guidance of environmental health specialists is advisable. Information resources are listed at the end of the chapter for those seeking additional assistance or a deeper understanding of the issues.

Health Effects

Indoor air pollutants at high levels can cause acute illness, while lower levels may lead to health problems only after years of exposure. In the case of certain carcinogens, such as radon, health professionals believe that a single exposure could lead to health problems many years later (although the greater the total exposure over time, the greater the risk). While the effects of some pollutants are well understood, for others further research is needed to determine what concentrations and types of exposure will impair health. Also, it is important to bear in mind that different people react very differently to indoor pollutants.

Even in the absence of definitive studies on every pollutant, there is little disagreement that reducing exposure to volatile organic compounds, combustion gases, radon, common allergens, and other indoor pollutants is a worthwhile goal for all homeowners and particularly vital for the very young or for those with allergies or respiratory problems.

Short-Term Health Effects.
High levels of indoor pollutants can cause immediate symptoms after one or more exposures. The symptoms may look like those of a cold or virus, including irritation of the eyes, nose, and throat, headaches, dizziness, and fatigue. These effects are usually short-term and reverse quickly once the person leaves the building or the pollutant is identified and eliminated. Short-term exposures can also trigger asthma episodes and lead to other serious allergic responses, including hypersensitivity pneumonitis and humidifier fever, both of which may first appear as flu-like symptoms.

For many pollutants, the exposure level at which symptoms first appear is highly variable. Key factors include a person's age, preexisting medical conditions, and his or her individual sensitivity to the chemical or biological compound in question. For example, mold, pollen, and animal protein (dander, etc.), elicit a range of allergic reactions in some, while others are unaffected. Also, the level at which formaldehyde elicits symptoms ranges from as little as .04 ppm to as much as 5.0 ppm (parts per million), depending on an individual's sensitivity. To complicate matters, people can develop sensitivities to both biological and chemical pollutants at any point in their lives, possibly from repeated exposures to low levels of the substance.

Long-Term Health Effects.
Some of the most toxic substances in our homes, such as lead, asbestos, and radon, can under some circumstances cause long-term irreversible damage to health. Many types of air pollutants increase the frequency and severity of asthma attacks. Combustion by-products have been linked to reduced lung function in developing children. Some health problems, including certain cancers, have long latency periods and may show up years after exposure to a pollutant such as tobacco smoke or radon. There is also ample evidence that some materials, such as formaldehyde, are "sensitizers," which can cause a person to become hypersensitive after years of low-level exposure. Whether indoor air quality contributes to other chronic health problems, such as heart disease, respiratory diseases, and cancers (other than lung cancer from radon and secondhand smoke), is unclear; but there is evidence that all major internal systems can be strained and become symptomatic as a result of poor indoor air quality.

Key Strategies

The key principles for creating and maintaining a clean indoor environment are straightforward and can be summarized in order of effectiveness as follows:

1. *Source control:* Keep it out
2. *Spot ventilation:* Vent it out
3. *Whole-house ventilation:* Dilute it
4. *Air cleaning:* Filter it out

Source Control.
The most effective way to avoid a household hazard is not to bring it into the house in the first place. In the case of building materials, this typically requires a material substitution. For example, one of the most common indoor air pollutants is formaldehyde, widely used in wood composites such as particleboard, hardwood plywood, and medium-density fiberboard (MDF). If acceptable substitutions can be found at an affordable price, the problem is solved. Another example is fiberglass duct board, which releases small amounts of fiberglass, a lung irritant, into the air stream. Use rigid metal ducts or flexible metal-lined duct instead.

In cases where there is no acceptable alternative, look for ways to seal the chemicals in. For example, medium-density fiberboard (MDF) that is sealed on all six sides by plastic laminate, as is the case on some laminated cabinets, emits only low levels of formaldehyde. In general, a material that is impervious to water vapor can effectively block formaldehyde emissions.

Combustion devices are another major source of both gases and particulates. To keep emissions to a minimum, avoid the use of fireplaces, woodstoves, and unvented combustion appliances, including gas stoves and heaters. If gas cooking is desired, select a unit with a pilot-less ignition. Also, substitute sealed-combustion appliances for atmospherically vented heating, ventilating, and air-conditioning (hvac) equipment. This eliminates the possibility of flue-gas spillage and usually has higher efficiency ratings as well.

Other steps that can have a big impact on indoor air involve lifestyle changes that are decidedly low-tech, including the following:

- Close windows when outdoor air is full of pollen or other particulates.

- Remove footwear at the entry to prevent spreading pesticides, lead, biological materials, and a wide range of pollutants around the house.

- Avoid bringing strong chemicals into the home for cleaning or hobbies.

- Keep smoking, pets, workshops, and other sources of pollutants and allergens out of the main living area.

- Minimize the use of dust-collecting surfaces such as carpeting, open shelves, and upholstered furniture.

- Vacuum frequently with an efficient vacuum cleaner. If allergens are a concern, use a HEPA vacuum or a central vacuum that vents to the exterior.

Spot Ventilation. Some pollutants are created by our daily living patterns. It is far more effective to exhaust these directly at the source than to try to remove them after they are distributed throughout the household air.

The most common examples are kitchens and bathrooms. Both produce large amounts of water vapor, not a pollutant in itself, but a contributor to other problems. Too much moisture in the air significantly increases formaldehyde emissions and can lead to mold and mildew growth. An effective range hood also removes atomized grease, particulates, and, in the case of gas ranges and cooktops, combustion by-products. For details, see "Kitchen and Bath Ventilation," page 260.

Spot ventilation is also important for darkrooms and other hobby areas that can produce high concentrations of chemical fumes. Home offices with high-capacity laser printers or photocopiers can also generate enough pollutants to justify spot ventilation.

Whole-House Ventilation. Whole-house ventilation is designed to provide a low level of fresh air to all habitable spaces, particularly bedrooms and main living areas, and to help flush out the low levels of pollutants generated by occupants, pets, and building materials. Occupants and pets produce moisture, carbon dioxide, and odors. In addition, most homes have a certain amount of chemical and biological pollutants from pets, cleaning, and hobbies and from outgassing from paints, plastics, pressed wood products, fabrics, and other household materials.

Whole-house ventilation is not meant to take the place of spot ventilation, which is still required to exhaust concentrated pollutants from cooking, bathing, and hobby areas. Although not yet required in most current building codes, whole-house ventilation is being incorporated into more and more new homes, and is recommended by model energy codes and standards organizations, such as the American Society of Heating, Refrigerating, and Air-Conditioning Engineers (ASHRAE) and the National Fire Protection Association (NFPA). Different approaches to whole-house ventilation are discussed under "Whole-House Ventilation Strategies," page 270.

- *Operating costs.* It is important to note that mechanical ventilation costs money both to operate the fans and to heat or cool the incoming fresh air. To provide the recommended ventilation levels, using an efficient fan in either an exhaust or supply system, the annual cost for a 1,500-square-foot house ranges from about $150 to $200 per year, depending on climate and fuel costs. Heat-recovery ventilators (HRVs) have higher electrical costs with dual fans but save money through heat reclamation, so annual energy costs are similar.

- *New homes.* Plan to run the ventilation system at high speed for at least the first few months of occupancy, since paints, plastics, pressed wood products, and many other materials will outgas at their greatest rate during this period. If the house still smells of fresh paint or new carpet, volatile organic compound (VOC) levels are still too high.

Air Cleaning. Air cleaning is the least effective strategy for maintaining a healthy indoor environment, but it can play a role along with source control and ventilation. There are many different types and sizes of air filters on the market, both portable units and filters integrated into the home's hvac system. Situations that may call for air cleaning equipment are:

- Where the outside air is polluted or full of pollen and needs to be filtered before bringing the air into the house.

- Homes with high pollutant sources, such as tobacco smoke or certain hobbies.

- Individuals with asthma, allergies, or chemical sensitivities.

Different approaches to filtering pollutants from indoor air are discussed under "Air Cleaning Strategies," page 277.

WHOLE-HOUSE VENTILATION STRATEGIES

There are a number of strategies for providing whole-house ventilation, which vary in cost, complexity, and effectiveness. All strategies, however, can be categorized as either *exhaust-only, supply-only,* or *balanced* (see Table 7-1).

Sizing

ASHRAE Standard 62-1989 recommends a minimum ventilation rate in houses of 15 cfm per person, or .35 air changes per hour (ACH), whichever is greater. Based on the ACH method, a three-bedroom house of 1,500 sq ft with 8 ft ceilings would require:

$$(1{,}500 \times 8) \times .85 \times .35 \div 60 = 60 \text{ cfm.}$$

Multiplying the volume by .85 accounts for partitions and exterior wall thickness.

Using the per person method and assuming two people in the master bedroom and one in each other bedroom, the rate is also 60 cfm.

The revised ASHRAE standard 62.2, released in 2003, uses the formula of 7.5 cfm per person (based on the number of bedrooms plus one) plus an factor of .01 cfm for

| TABLE 7-1 | Whole-House Ventilation Strategies | | | | | |

Type	Pros	Cons	Controls	Cost (1—low 4—high)		Recommendations
				Installation	Operating	
Single-port exhaust (page 271)	Simplest system. Low energy use, low maintenance. Can also serve as bath fan.	Little control of incoming air. Doors must be undercut. Slight negative pressure.	Programmable timer with manual override	1	1	Choose a quiet, efficient fan. Works best in small, tight homes with few pollutant sources. Consider upgrading to a dedicated fan in central location.
Multiport exhaust (page 271)	Simple design. Improved air distribution. Heat-recovery option with heat-pump water heater. Typically exhausts all bathrooms.	Negative pressure can cause backdrafting, increase radon levels, and cause condensation problems in hot, humid climates.	Programmable timer or continuous with high-speed switches in bathrooms	3	1	Use in tight homes with electric or sealed-combustion appliances. Add passive air inlets for better distribution. Avoid in hot, humid climates. Use separate range hood.
Multiport supply (page 274)	Simple design. Good distribution. Incoming air can be filtered or dehumidified. Positive pressure protects against backdrafting and radon.	Positive pressure can cause condensation problems in cold climates. Discomfort from nontempered air.	Programmable timer with two or more speeds	2	1	Good choice in temperate climates. In cold climates, couple with a bathroom or central exhaust fan to avoid depressurization.
Multiport forced-air supply (page 273)	Uses existing ductwork and air handler. Incoming air can be filtered. Good distribution. Positive pressure protects against backdrafting and radon.	Most air-handler fans noisy and inefficient. Discomfort from nontempered air. Positive pressure can cause condensation problems in cold climates.	Special programmable controller required for air handler and damper.	1	4	Same recommendations as for multiport supply, above. Also consider upgrading to a quiet, efficient, variable-speed hvac fan.
Heat-recovery ventilators (page 275)	Excellent distribution. Tempers incoming air. Heat recovery.	Most expensive and complicated systems to install and maintain.	Continuous operation with manual switching for high-speed spot ventilation. Optional dehumidistat.	4	1	Good choice for very tight homes in cold climates. Use separate kitchen-range hood. Should include a maintenance plan.
Energy-recovery ventilators (page 275)	Excellent distribution. Tempers, dehumidifies incoming air. Heat recovery.	Most expensive and complicated systems to install and maintain. Cannot be used to ventilate high humidity areas such as baths.	Continuous operation with manual switching for higher speeds.	4	1	Good choice for tight homes in hot, humid climates. Use separate kitchen and bath fans. Should include a maintenance plan.

each square foot of house area. For example, based on the new ASHRAE standard, the same three-bedroom, 1,500-square-foot house would require:

$$(7.5 \times 4) + (1,500 \times .01) = 45 \text{ cfm.}$$

As these calculations show, a low ventilation rate is adequate if run on a continuous basis. A higher continuous rate would be advisable for a home with higher-than-average moisture levels or pollutant sources such as smoking. Intermittent ventilation can also work as long as the total daily ventilation rate is equivalent, but is most effective when the system is timed to operate when people are home breathing air and generating pollutants. A two-speed or variable-speed fan provides flexibility, allowing the ventilation rate to be raised when needed, for example when painting a room or during a party. More important than the precise number of cubic feet per minute, however, is a well-designed system that is quiet, reliable, and low-maintenance, ensuring it will actually be used.

Installation

Whole-house ventilation systems should be installed by people familiar with the equipment. Since they normally operate at 100 to 200 cfm rather than the much larger fans found in air handlers, they are less forgiving of errors. Numerous field studies have found heat-recovery ventilators performing poorly due to installation errors and poor maintenance. For good performance with whole-house ventilation systems, follow these general guidelines:

- Size the system correctly. Oversizing will increase heating and cooling costs.

- Choose quiet, efficient fans.

- Keep duct runs as short and straight as possible.

- Locate fresh air intakes away from pollution sources such as cars, pesticides, and outlets from hvac equipment or exhaust fans.

- Seal all ducts and insulate where required. Examples: Insulate intake ducts that run though a hot attic or exhaust ducts that pass through a cold, unheated space.

- Integrate spot ventilation in bathrooms or provide separately.

- Use separate spot ventilation in kitchens due to grease.

- Place supply registers high on walls and away from beds, sofas, chairs, and other places likely to cause occupant discomfort.

- Keep controls as simple and automatic as possible.

- Educate homeowners about the system and maintenance requirements.

Exhaust-Only Ventilation

Exhaust-only ventilation is the most common approach, due to its simplicity and use of familiar components such as bathroom fans. However, unless houses are built very tight, there is little control over where fresh air enters the building. Also, depressurization can be a problem, particularly with high-capacity fans. In addition to the increased potential for backdrafting, a depressurized house tends to draw more soil gases, including radon if it is present. And in hot, humid climates, moist air infiltrating through exterior walls can condense on interior finishes such as the back face of vinyl wallpaper that is chilled by air conditioning.

Single-Port Exhaust. The simplest and least expensive central ventilation system consists of an automatic timer wired to one centrally located bathroom or laundry fan so it cycles on and off for a portion of every hour or for the 8 to 12 hours per day when most people are home, typically mornings and evenings (see Figure 7-1).

Since the fan is doing double duty as a bath or laundry fan, it must have a manual override switch for intermittent use. In larger homes, two fans at separate locations can be used. Another upgrade is to use a dedicated fan in a central location, such as a hallway ceiling, which will provide better distribution of both exhaust and supply air.

For the system to work well, it is important to use a quiet fan of one sone or less and choose a central location. Also, the door to the bathroom with the exhaust fan must be undercut by $\frac{3}{4}$ to 1 inch, along with doors to all bedrooms and other rooms that require ventilation. An alternative is to connect the rooms with through-the-wall transfer grilles.

The biggest drawback to exhaust-only ventilation is that there is little control over distribution of the incoming air. Makeup air will come via the path of least resistance. In a leaky house, this might be a window or drop ceiling in the bathroom with the exhaust fan, leaving the rest of the house unserved by the ventilation system. For this reason, single-port exhaust-only ventilation works well only in relatively small, tight houses.

- *Passive air inlets.* Some contractors install passive air inlets in an effort to direct makeup air into bedrooms and main living areas. For these to work properly, however, the house must be extremely tight and doors must be left open or be cut at least an inch above the carpet. If a house is too leaky or rooms are cut off from household airflows, the inlets will function like other random holes in the building shell, leaking air inward or outward, depending on the wind, stack effect, and imbalances in the hvac system. The inlets typically require at least 10 Pascals of negative pressure to operate. They do not eliminate depressurization as sometimes thought. In fact, they require it to work properly.

Multiport Exhaust. This type of system uses a more powerful exhaust fan that is remotely mounted, typically in the attic or basement (Figure 7-2).

The system is ducted to exhaust grilles in bathrooms, laundries, and other wet areas, and sometimes to a centrally located pickup point in the main living space. A room with no outside walls would also benefit from a pickup point.

FIGURE 7-1   **Single-Port Exhaust Ventilation.**

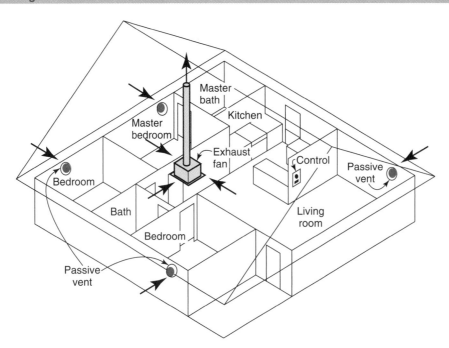

Key Components
- Quiet, efficient exhaust fan
- Programmable timer and speed switch

Operation
- Runs on timer or continuously at low speed.
- Occupants can temporarily boost ventilation rate.
- Spot fans required in kitchen and bathrooms.

The simplest ventilation system uses a single, centrally located exhaust fan that runs on a timer or continuously at a low speed. The fan may also serve as a bathroom or laundry fan, but a dedicated fan is optimal. Passive air inlets are sometimes installed but will only work properly in very tight homes.

SOURCE: *Recommended Ventilation Strategies for Energy-Efficient Production Homes,* 1998, by Judy A. Roberson, et al., Lawrence Berkeley National Laboratory.

Systems typically run on a low background speed with timer switches in bathrooms for higher-powered spot ventilation. If installed correctly, these systems are very quiet and provide good distribution of ventilation.

Multiport exhaust systems may incorporate passive air inlets (see description above) that install either in windows or through the wall, providing some control over supply air. The inlets, typically three or four for a small house, go in bedrooms, main living areas, and other occupied rooms, such as dens or home offices. Inlets should be placed high on the wall away from beds, chairs, or other places where drafts might cause discomfort. Placement near a window is preferred.

Because these systems use more powerful fans that depressurize the house, they should not be used in houses with fireplaces or atmospherically vented combustion appliances. They are also not recommended in hot climates, since hot, moist exterior air may be drawn into walls and condense behind interior surfaces chilled from air conditioning. Packaged systems (Figure 7-3) are available from American Aldes, Fantech, and a few other few manufacturers (see Resources, page 297).

Ventilating Heat-Pump Water Heater. This variation on exhaust-only ventilation passes the exhaust air through a heat-pump water heater, reclaiming heat from the outgoing air stream. Some systems can be reversed in summer, functioning as a supply ventilation system while cooling and dehumidifying the incoming air. A packaged heat-pump ventilating system is available from Therma-Stor.

Supply-Only Ventilation

While not widely used, supply-only systems have distinct advantages over exhaust systems. The incoming air is easily filtered and can be directed to bedrooms and main living areas. The slight positive pressure helps guard against radon, backdrafting, and other problems associated with negative pressures.

In cold climates, however, delivering nontempered air can lead to uncomfortable drafts. Also, forcing moist, interior air out through gaps in the building shell could contribute to condensation problems in building cavities and between prime and storm windows. In airtight homes in

FIGURE 7-2 **Multiport Exhaust Ventilation.**

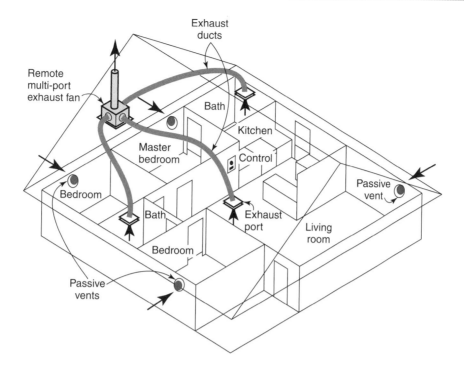

Key Components
- Quiet, efficient multi-port exhaust fan
- Programmable timer and speed switch
- Several passive wall or window vents
- 3″ to 4″ diameter ventilation ductwork and grilles

Operation
- Runs continuously at low speed.
- Occupants can temporarily boost exhaust rate.
- Bathrooms vented through exhaust ports.

A multiport exhaust system improves air distribution by picking up exhaust air from bathrooms and main living areas. These are often used in conjunction with passive air inlets. Exhaust-only systems are best used in homes with electric heating or sealed-combustion appliances where backdrafting is not a concern.
SOURCE: *Recommended Ventilation Strategies for Energy-Efficient Production Homes,* 1998, by Judy A. Roberson, et al., Lawrence Berkeley National Laboratory.

very cold climates, supply-only systems have reportedly iced up keyholes to entry doors as exfiltrating warm, moist air came in contact with the cold metal. Adding a single exhaust fan wired to operate whenever the supply fan switches on can alleviate these problems.

Forced-Air Supply.

This system piggybacks on the ductwork of a central heating or cooling system by running an intake duct from outside into the return ductwork. The screened intake has a motorized damper that is timed to open when ventilation is needed, blending fresh air into the hvac system and slightly pressurizing the house. A special controller is needed to control the damper and fan, activating the damper whenever ventilation is needed and activating the fan whenever the air handler has not run for a set period of time, typically 20 minutes. Several manufacturers, including Lipidex Corporation, Tjernlund, and Honeywell, make controllers for this application. One manufacturer, American Aldes, offers a packaged system for warm, humid climates: the DHV-100V,

which dehumidifies and filters incoming air and integrates with the home's central air-conditioning system.

This approach is relatively inexpensive since it uses existing ductwork, and it provides good distribution of fresh air. The chief drawback is that ventilation is required at regular intervals, often when the hvac system is not operating. At those times, the controller will switch on the air handler, which is typically noisy and inefficient, making this the most expensive system to operate. Also, delivering untempered outside air through the duct system can cause discomfort in very hot or cold weather. And if the return ducts are not well sealed, they can pull contaminants from attic or basement spaces into the ventilation system.

The operating costs can be cut in half by replacing the standard air-handler blower with an efficient, variable-speed fan with an integrated control motor (ICM). The fan would work on high speed for heating and cooling and continuous low speed for ventilation-only, cutting operating costs in half. However, the damper adjustment

FIGURE 7-3 Packaged Multiport System.

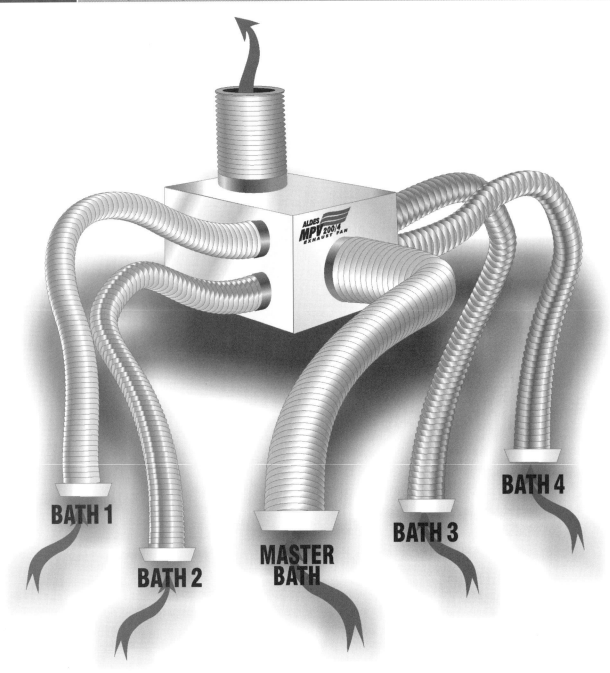

This packaged multiport exhaust system can provide spot ventilation for all bathrooms as well as continuous whole-house ventilation at low airflow rates. Optional engineered intake grilles are available to regulate airflow at each pickup point.
SOURCE: Courtesy of American Aldes.

that provides the right amount of ventilation air at 500 cfm will produce too little when run at 100 cfm. Either the setting has to be changed seasonally or a compromise level found.

Multiport Supply. Although the least common type of ventilation system, the multiport supply system was identified in a recent study by Lawrence Berkeley National

Laboratory (LBNL) as the optimal system for tract homes based on first cost, operating costs, air distribution, and the potential health and safety benefits of positive pressures. It is also easy to filter and, if necessary, to dehumidify the incoming air (Figure 7-4).

While few packaged systems are currently available, one manufacturer, Tamarack Technologies, offers a unit with a replaceable filter and an efficient variable-speed fan

FIGURE 7-4 Multiport Supply Ventilation.

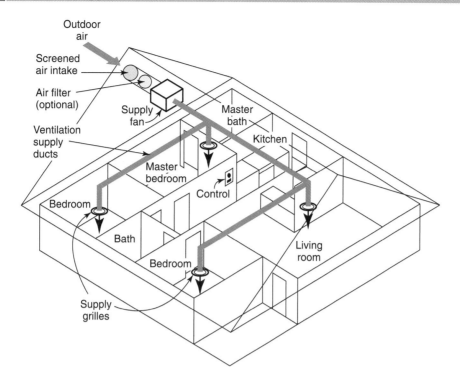

Key Components
- Quiet, efficient supply fan with air filter
- Programmable timer and speed switch
- Ventilation ductwork and supply grilles

Operation
- Fan operates continuously at low speed.
- Spot fans exhaust air from kitchen and bathrooms.
- Residents can boost ventilation rate as needed.

Though not widely used, supply-only ventilation has several advantages: Incoming air can be easily filtered and distributed to the rooms where it is needed, and positive pressures help guard against backdrafting and radon entry.
SOURCE: *Recommended Ventilation Strategies for Energy-Efficient Production Homes*, 1998, by Judy A. Roberson, et al., Lawrence Berkeley National Laboratory.

that provides 90 cfm of ventilation through one 3-inch and three 2-inch ducts.

Since these systems pressurize the house, the LBNL study recommends that, in cold climates, the supply fan be balanced by a single-port exhaust fan, which could also serve as a bathroom fan. In this type of system, a central fan, typically in the attic or basement, draws in outside air through a filter and delivers it through ducts to bedrooms and main living areas. The supply ductwork should be in conditioned space or insulated and sealed airtight. Supply grilles should be placed high on the wall away from beds, chairs, or other places where drafts could cause discomfort.

Balanced Ventilation

Balanced ventilation uses both a supply and exhaust fan to provide fresh air while keeping house pressures neutral. Linking a multiport supply system (described above) with a bathroom exhaust fan on the same switch is a form of balanced ventilation. Most balanced ventilation systems, however, use a heat-exchanger to transfer heat and, with energy recovery ventilators (ERVs), humidity between the two air streams. These systems, sometimes called air-to-air heat exchangers, are the most expensive option for whole-house ventilation; but, if installed properly and well-maintained, provide optimal comfort and ventilation (Figure 7-5). Depending on the type of heat exchanger, balanced ventilators are referred to as either heat-recovery ventilators (HRVs) or ERVs.

Controls. HRVs and ERVs are typically run continuously, but they also may be set to run 8 to 12 hours per day when people are at home. In addition, most have a high-speed mode that can be manually activated for spot ventilation of kitchens and bathroom. Some also use dehumidistats to automatically turn on or increase ventilation when the air reaches a preset humidity level.

Ducting. While a dedicated ductwork system is the best approach for HRVs and ERVs, to save money they are often

FIGURE 7-5 **Balanced Ventilation.**

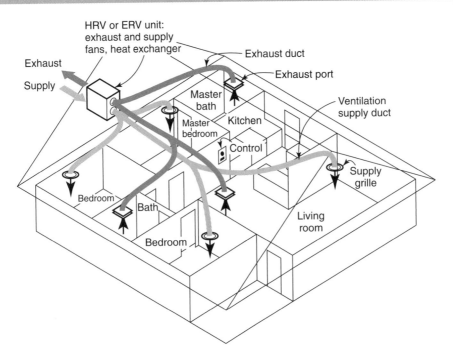

Key Components
- HRV or ERV unit with exhaust and supply fans, heat exchanger
- Programmable timer with speed switch
- Exhaust and supply ducts and grilles

Operation
- Runs continuously at low speed.
- Air is exhausted from bathrooms, supplied to bedrooms.
- Residents can temporarily boost ventilation rate.

Well suited to extreme climates, balanced ventilation provides optimal distribution of incoming fresh air and tempers it for comfort. HRVs reclaim heat from the exhaust air while ERVs, recommended for hot, humid climates, dehumidify and cool incoming air. For good performance, systems must be properly installed, balanced, and maintained.
SOURCE: *Recommended Ventilation Strategies for Energy-Efficient Production Homes,* 1998, by Judy A. Roberson, et al., Lawrence Berkeley National Laboratory.

piggybacked onto the home's hvac ductwork. In one approach, the HRV or ERV draws exhaust air from the return ductwork and feeds fresh air into the furnace's return plenum. In a slightly better arrangement, the hvac ductwork is used only for supply, while the exhaust side picks up stale air in bathrooms, laundry, and kitchen. Neither approach distributes fresh air as well as a dedicated duct system.

Also, since HRV/ERV fans operate at 100 to 200 cfm while air-handler fans are often sized at over 600 cfm, trying to integrate the controls, balance airflows, and provide the correct amount of ventilation air is challenging and rarely works well. The most common approach is to run the ventilation system only when the thermostat calls for heating or air-conditioning, providing too little ventilation. Heat-recovery efficiencies are also compromised, typically due to unbalanced airflows.

Maintenance. One drawback of HRVs and ERVs is that they require more maintenance than other ventilation systems. Numerous studies have found that many of these systems significantly underperform in the field due to both installation errors and poor maintenance. In addition to cleaning or changing intake, outtake, and internal filters, the homeowner or service person needs to clean the core once a year or more to prevent mold and bacteria growth. It is important to follow closely the manufacturer's recommendations. Unless the homeowner enjoys the responsibility of hvac maintenance, the work is best handled by a professional service company.

Cost-Effectiveness. For HRVs and ERVs to work properly and achieve the rated efficiencies, they must be installed correctly and balanced well, and the house must be very tight. Even so, the added cost over a basic ventilation system will be recouped only in the most extreme United States climates with the highest energy costs. However, in very cold or hot climates where mechanical ventilation is needed during most of the year, the added cost may be justified by the comfort of tempered, filtered ventilation air, the effectiveness of the distribution, and the lack of pressure-related problems.

FIGURE 7-6   Heat Recovery Ventilator.

Used primarily in cold climates, heat-recovery ventilators (HRVs) pass two air streams by one another within a plastic or aluminum heat exchanger, recapturing 60 to 75% of the heat from the outgoing air stream. This unit, installed in an attic, is hung from chains to reduce noise and vibrations.
SOURCE: Photo courtesy of David Hanson, Memphremagog Heat Exchangers.

HRVs. Used primarily in cold climates, HRVs have two air streams that pass over one another in a plastic or aluminum heat exchanger (Figure 7-6).

Recovery of heat from the exhaust air typically ranges from 60 to 75%, if properly installed and balanced. During the summer, if air-conditioning is used, the heat transfer reverses, cooling the incoming hot air. Systems generally have exhaust ports in rooms that generate moisture or pollutants, including bathrooms, laundry, and kitchen and supply ports in bedrooms, living rooms, and other main living spaces. Because they have both supply and return ducting, HRVs provide the best distribution, exhausting air from bathrooms and other wet areas and providing fresh air to primary living space. The kitchen typically has its own range hood, so grease does not get into the HRV system.

Defrost. In cold climate applications, a defrost cycle is required. It usually switches on at about 20°F to keep frost from building up in the core as condensation from the exhaust stream begins to freeze. Systems either recirculate indoor air or preheat incoming air to prevent freeze-ups.

ERVs. Energy-recovery ventilators are primarily used in air-conditioned homes in hot, humid climates. They are generally recommended for climates where the cooling load exceeds the heating load and where sustained freezing temperatures are rare. Sustained temperatures below 10°F can damage the permeable core material used in many ERVs.

ERVs either use a dessicant-coated plastic wheel or a special "enthalpic" core material to move moisture (latent heat), as well as sensible heat, between the two air streams. In summer, incoming air is cooled and dehumidified. Since dehumidification is the biggest component of air-conditioning costs in humid climates, it is important to find a unit with a high TRE (total recovery efficiency) rating, indicating that it can transfer large amounts of moisture. To achieve the rated efficiencies, the units must be run at the recommended airflows.

In cold weather, an ERV will tend to humidify the incoming air, since the moisture transfer is always toward the less humid air stream. This is rarely a problem, however, since the cold incoming air holds so little moisture to begin with that the net effect of the air exchange is to remove humidity from the house.

AIR CLEANING STRATEGIES

There are many types and sizes of air cleaners and filters on the market, both stand-alone units and those integrated with hvac equipment. Different types of air cleaners work on different types of pollutants and none handles everything. The effectiveness of a device depends on a number of factors including the type and efficiency of the filter, how much air flows through it, how well the polluted air reaches the filter, and how effectively the clean air is delivered to occupied areas. (Some small units tend to draw in the same air they just exhausted, creating a short circuit with little impact on the larger space). Also, with electronic air cleaners, performance drops off rapidly if the filters are not kept clean.

Another limiting factor is that many allergies are linked to larger particles, such as pollen, house dust, animal dander, and some molds, that are more likely found settled on surfaces than suspended in the air. A high-efficiency vacuum is needed for these, not an air cleaner.

Particles vs. Gases

Some filters are effective with particles, such as dust and pollen, and others are effective with gases, such as combustion fumes and formaldehyde. Certain pollutants such as tobacco smoke contain both gases and particles, so they require two types of filters for effective removal.

Particles. Sometimes called "particulates," these are small solid or liquid particles suspended in the air. They can be captured in mechanical or electrostatic filter elements. How many get captured depends on the size of the particle along with the type, size, and efficiency of the filter and the rate of airflow (Figure 7-7).

- **Respirable particles.** These are small, invisible particles, typically ranging in size from 0.5 to 2.5 microns (millionths of a meter) that can penetrate deep into the lungs and cause acute or chronic illnesses. Examples include asbestos, viruses, bacteria, and the particles in

FIGURE 7-7 Particle Size and Filter Efficiency.

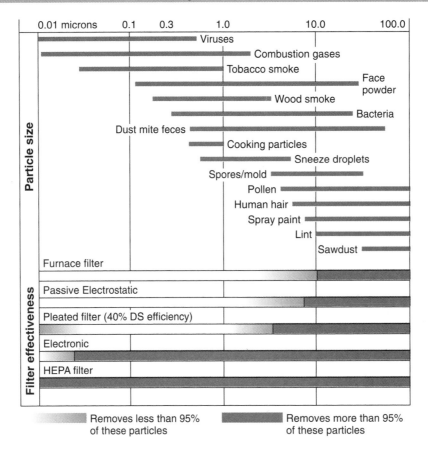

Tiny, invisible respirable particles pose the greatest risk to health, while larger inhalable particles may be irritants and allergens. Electronic air cleaners and HEPA filters provide the best filtering performance across all particle sizes.

SOURCE: Reprinted with permission from Oikois.com. © 1994 Iris Communications, Inc.

tobacco smoke. Other sources include unvented kerosene and gas space heaters, woodstoves, fireplaces, poorly adjusted furnace flues, and cracked heat exchangers. Health effects vary with the type of particle, degree of exposure, and individual sensitivity, and range from eye and respiratory irritation to chronic diseases, such as cancer.

- *Inspirable particles.* Particles ranging in size from about 2.5 to 10 microns include dust, pollen, animal dander, and some mold spores. These can be inhaled, but they generally do not penetrate deeply into the lungs. They may cause allergic responses and other health problems in some individuals.

- *Visible dust.* Most particles over 10 microns get trapped in the nose and upper airways and do not generally cause health problems.

Gases. Gaseous pollutants include combustion gases and a huge array of organic chemicals that have been detected in homes. Gaseous organic compounds can originate indoors from combustion appliances, cigarette smoking,

cleaning and personal hygiene products, or hobby materials, or can outgas from building materials, such as pressed wood products, paints, adhesives, and caulks. Others, such as auto emissions and pesticides, originate out of doors and are drawn into the home with outside air.

Health effects vary with type of pollutant, level of exposure, and individual sensitivity, and range from eye and respiratory irritation and allergic responses to cancer and other serious diseases affecting the respiratory, liver, cardiovascular, and nervous systems.

Gaseous pollutants can be removed from the air by passing them through special adsorbents, such as activated carbon, that adhere to the gas molecules.

- *Radon.* Radon is a radioactive gas that may enter a building from soil or groundwater. The gas breaks down into short-lived particles, which can get trapped in the lungs and cause cancer. Although some adsorbents can reduce radon gas levels and some high-efficiency filters can trap radon progeny, this has not been adequately tested and is not currently endorsed by the EPA as a radon mitigation method (details below).

Air-Cleaner Types

There are three main types of particulate air cleaners on the market: mechanical filters, electrostatic filters, and ion generators. In addition, there are filters with adsorbents, such as activated carbon, that are capable of removing certain gases (Table 7-2). Most filters are available as stand-alone units or as add-ons to the home's hvac or ventilation system. Some hybrid systems on the market combine two or more strategies, such as a filter to remove particles plus activated carbon to remove odors and organic gases.

Mechanical Filters. These use a matrix of fiberglass or synthetic fibers and resin to filter particles out of air passing through. Some are coated with an oil or adhesive to help trap particles, and others have a static electrical charge. Some types of mechanical filters can be cleaned, but most need to be replaced when full. As mechanical filters clog, they become more efficient at trapping particles, but airflow is reduced. They can either work in stand-alone units or be incorporated into the hvac or ventilation system.

- *Flat filters* are the standard fiberglass elements found in furnaces and air handlers. They are designed to catch large dust particles and have little effect on the smaller respirable particles that affect health.

- *Pleated filters,* or "extended media" filters, have smaller pores designed to capture small- and medium-sized particles. They are pleated like automobile air filters to provide greater surface area for improved airflow. Some will fit into a standard furnace filter slot, but the higher-efficiency types are generally too thick and require ductwork modifications. For good performance, they need to be replaced regularly.

- *Electrostatic filters* use a plastic element that is permanently charged with static electricity or captures an electric charge from the passing air. They are a little more effective than a standard furnace filter at capturing larger particles such as pollen and mold spores, and can be washed and reused when full.

- *HEPA* stands for "high-efficiency particulate accumulator." These filters range from 95% to over 99% efficient for particles over 0.3 microns, including mold spores, mites, pet dander, and some viruses. Because of their high resistance to airflow, HEPA filters typically require a separate fan and housing. Replacement elements last a year or longer but cost well over $100 versus $10 to $20 for a medium-efficiency pleated filter. Most have a prefilter to catch large particles that would prematurely clog the filter. Prefilters need to be changed regularly.

Electronic Air Cleaners. Electronic air cleaners (EACs) use a series of electrically charged metal plates or media filters to pull particles out of the air stream. They are either portable units designed to clean the air in one or two rooms or central systems connected to the return ductwork of the hvac system. EAC systems create little resistance to airflow but require a separate fan, which along with the electrical charging device use about 30 to 50 watts of electricity.

Electronic air cleaners are highly efficient at removing both small and large particles, but require more maintenance than many homeowners would like. To remain effective, the filters must be removed regularly and hosed down in a large sink or tub. Some are designed to fit in a dishwasher.

TABLE 7-2	**Particulate Air Cleaner Comparison**				
	Large Particle Efficiency*	Small Particle Efficiency**	Maintenance	Cost	Recommendations
Flat filters	75%	<5%	Replace every 1–3 months	Very low	Little effect on air quality.
Pleated filters	90–99%	20–50%	Replace every 3–12 months	Low to Medium	Can provide some benefit at a moderate cost. Look for 25% or higher "dust spot efficiency."
Electrostatic	75–95%	10–15%	Clean regularly	Medium (one-time expense)	A little better than a standard furnace filter, but must be cleaned frequently.
Electronic	>95%	70–95%	Remove and clean plates regularly.	High (one-time expense)	Good choice for people with allergies, but must be cleaned regularly to be effective.
HEPA	>95%	>95%	Change every 1–3 years. Change pre-filter 2–3 times a year	Very high to install plus high annual costs	Good choice for people with allergies or chemical sensitivities.

*Weight Arrestance as measured by ASHRAE Standard 52-76. Measures the ability to remove large, heavy particles that have relatively little impact on health.
**Atmospheric Dust-Spot Efficiency as measured by ASHRAE Standard 52-76. Measures ability to capture particles between 0.3 and 6.0 microns.

In charging the particles with high voltage, EACs also produce small amounts of ozone, which can be an eye or respiratory irritant at high levels. Most people are not bothered by the amount produced. If this a concern, however, look for a unit with an activated carbon filter to remove the ozone.

Negative Ion Generators. These work by releasing electrically charged ions, which attach to dust particles in the air causing them to settle on walls, ceilings, furniture, and draperies. Placed too near a wall, they might leave a smudge of particles. Some units contain an optional collector to trap the charged particles in the unit, functioning similarly to an EAC. Over time, however, the particles can lose their charge and reenter the air. Like EACs, they produce small amounts of ozone. There is little scientific evidence supporting the effectiveness of these units.

Turbulent Flow Precipitators. The turbulent flow precipitator (TFP) is a new proprietary technology from Canadian-based Nutech Energy System. The device, which attaches to the return ductwork of an hvac or ventilation system, contains a fan and a labyrinthine core made of aluminum plates and synthetic fibers. Turbulence in the air flings the suspended particles out of the airstream where they are trapped by a primary and secondary core, which need replacement in one and three years, respectively. Nutech claims that TFPs will capture 99% of particles larger than 5 microns, 97% from 2 to 3 microns, and 90% from 0.5 to 0.9 microns. A TFP unit with HEPA filtration is also available.

Gas Removal. To remove gases, such as formaldehyde, combustion fumes, or volatile organic compounds, from the air requires the use of special adsorption media. These media contain materials, such as activated carbon or aluminum oxide, which trap the gases in tiny pores. Different chemical adsorbents are effective with different gases, and none is effective with every gas found in the typical home. Relatively small quantities of activated charcoal can be very effective at reducing odors, but how well they filter out the low levels of multiple chemical compounds typically found in household air is unclear.

In general the rate of adsorption of a gas is reduced as more of the target gas is captured in the filter media. Researchers have also found that, in many cases, some of the gas is reemitted from the filter back into the air. Scientific evidence about the real-life usefulness of these filters in homes is very limited.

Measuring Filter Effectiveness

Different types of air cleaners use different rating systems. It is important to know what is being measured, since a "90% efficient filter" may actually capture 0% of respirable particles.

In-Duct Filters. Under ASHRAE Standard 52-89, low-efficiency hvac filters are evaluated for "arrestance" efficiency and medium- and high-efficiency filters are rated for "dust spot" efficiency. HEPA filters have their

own measure as follows:

- *Arrestance efficiency:* This measures how well a filter captures large, heavy particles and is generally used only for the low-efficiency filters typically found in residential hvac systems.

- *Dust-spot efficiency:* This measures how well a filter captures a mixture of fine particles ranging from 0.3 to 6.0 microns, and it is a good guide to how well a filter handles many respirable particles. However, it does not accurately predict the filtering of particles less than 1.0 micron. Also, two filters with the same dust-spot efficiency may perform very differently depending on the size of the particles being captured (see Table 7-3).

- *DOP:* Based on Military Standard 2823, this uses a fine aerosol of the chemical dioctylphthalate (DOP) to rate high-efficiency filters and requires removal of 99.97% of 0.3-micron particles. Filters meeting this standard are often called "true HEPA" filters. High-efficiency filters that do not meet the standard may still have very good performance and may be called "near HEPA" or similar terms.

- *MERV:* The Minimum Efficiency Reporting Value is based on ASHRAE Standard 52.2 1999. This new

TABLE 7-3	Filter Efficiency and Particle Removal	
Atmospheric Dust-Spot Efficiency	Typical Applications	Particles Removed
10%	Heating systems and window air conditioners	Useful for lint and somewhat useful for ragweed. Not good for smoke or staining particles.
20%	Air conditioners; central heating and air-conditioning systems	Fairly useful on ragweed pollen. Not good for smoke and staining particles.
40%	Heating and air-conditioning systems; prefilters to high-efficiency cleaners	Good for fine airborne dust and pollen. Reduces smudge and staining particles. Slightly useful for nontobacco smoke, but not tobacco smoke.
60%	Heating and air-conditioning systems; prefilters to high-efficiency cleaners	Good for pollen, coal and oil smoke, and smudge and staining particles. Partially useful on tobacco smoke.
80%	Hospitals and controlled areas	Very good for coal and oil smoke and smudge and staining particles. Quite useful for tobacco smoke.
90%	Hospitals and controlled areas	Excellent against all smoke particles.

SOURCE: Adapted from Exhibit 1 in U.S. EPA document 400/1-90-002.

rating system evaluates a filter at a particular air velocity over 12 particle sizes ranging from 0.3 to 10 microns. MERV values range from 1 to 16. A higher MERV indicates that a filter is more efficient and more effective against smaller particles. Filter manufacturers are beginning to report MERV values.

Portable Units. While tabletop units are generally ineffective, larger console-style filters can be very effective for one or two rooms. These units may contain one or more of the filter types described above. Many include HEPA filters and some also use adsorbents to capture odors and gases. The Association of Home Appliance Manufacturers (AHAM) has developed an ANSI-approved standard for portable air cleaners called "Clean Air Delivery Rate" (CADR), which measures how quickly the filter removes dust, smoke, and pollen particles. The CADR is a function of both filter efficiency and airflow rate and assumes the filters are new and clean. To calculate how large a space a unit can effectively handle, AHAM recommends using the formula:

Square footage of room = Smoke CADR × 1.55

So, for example, a unit with a CADR of 100 can service a 155-square-foot room. The formula is based on the requirement that the filter remove 80% of the smoke particles on a continuous basis.

Real-World Effectiveness. While air cleaners work efficiently in laboratory tests, their effectiveness in the typical household is less clear. Research conducted by the Canada Mortgage and Housing Corporation (CMHC) found that upgrading the hvac systems to use medium and high-efficiency filters had only a modest effect on personal exposures to particulates in homes. During peak activity periods, the best performing filter in the study, an electronic air cleaner, reduced particulate levels in the ductwork by 95%, but cut household levels by only 31%. In nonactive times, like the middle of the night, the filter reduced particulate levels by over 70%, but levels were already so low that the reduction had little impact on the occupants' personal exposure or health (see Table 7-4).

Researchers attributed the modest effect of the filters to two key facts: First, the filters only worked while the furnace fans were operating, about 20% of the time on average. Second, personal exposures to high levels of particulates were almost always caused by the occupants themselves, who in their daily tasks of cooking, vacuuming, or just walking on the carpet stirred up small clouds of surrounding dust. Once the activity ended, the particles tended to rapidly settle out on their own before the filter, far away down a duct, could have much of an impact. If furnace fans were run continuously or the filters were part of a continuously operating ventilation system, their impact might have been greater.

Also, many of the larger, heavier particles linked to allergies, such as pollen, house dust, animal dander, and some molds, are more likely to be found settled on surfaces than in the air. A high-efficiency vacuum is needed for these, not an air cleaner.

TABLE 7-4	**Real World Performance of Upgraded Furnace Filters**	
Filter Type	Particulate Reduction While Occupants Were Active	Particulate Reduction While Occupants Were Nonactive
25-mm premium pleated	21%	57%
100-mm pleated	9%	13%
Electrostatic charged pad	9%	29%
HEPA bypass*	23%	38%
Electronic Air Cleaner**	31%	71%

*The HEPA filter handled only 51% of total airflow through hvac system, lowering overall performance.
**Electronic plate and wire type.
NOTE: Data based on sample of five houses during winter in Canada.
SOURCE: Canada Mortgage and Housing Corporation (CMHC). *Your Furnace Filter. About the House, CE22,* 2001. All rights reserved. Reproduced with the consent of CMHC. All other uses expressly prohibited.

Conclusions. The conclusion drawn by most air-quality experts is that filtering the air is costly and the effects are modest unless a high-efficiency filter is used, maintained well, and run frequently. A properly sized console unit can be effective at limiting personal exposure to pollutants, at least for the time spent in the room with the device. First costs for whole-house systems, typically in the hundreds and sometimes thousands of dollars, plus maintenance costs, may be a justifiable expense for individuals with special health problems—but probably not for the average homeowner.

Anyone intent on keeping their household air clean should focus first on keeping pollutants out of the house in the first place (see "Source Control," page 268), along with regular vacuuming with a HEPA vacuum or central vacuum exhausted to the exterior. That, along with filtering any outdoor air brought into the home's ventilation system, will go a long way toward providing wholesome indoor air.

COMMON INDOOR POLLUTANTS AND SOURCES

Public health officials have focused their efforts on outside air pollution for decades, establishing laws and guidelines for what is acceptable air to breath. As it turns out, however, people spend a lot more time indoors than out and, on average, the air is far dirtier indoors. Pollutants range from the merely unpleasant to the potentially deadly. The health effects of low exposures to many of these substances over long time periods is not well understood. Yet, it seems prudent to have a working knowledge of what's in the air you're breathing and to take commonsense steps to eliminate or minimize materials that, in larger doses, make most people sick. A summary of common indoor pollutants is shown in Tables 7-5 and 7-9 (pages 282 and 294).

TABLE 7-5   **Common Indoor Air Hazards**

	Description	Primary Household Sources	Health Effects	Typical Exposure Levels	Steps to Reduce Exposure
Radon (page 284)	Colorless, odorless, radioactive gas released during natural decay of uranium	Soil and rock beneath home; well water.	Estimated 20,000 lung cancer deaths per year. Smoking increases risk by factor of 8.	1.3 picocuries per liter of air (pCi/L) indoors; 0.4 pCi/L outdoors.	• Test home for radon • If levels are above 4 pCi/L, install subslab ventilation or other EPA-recommended techniques.
Formaldehyde (page 287)	A volatile organic compound (VOC) widely used in building materials and household products as an adhesive or preservative. Has a pungent odor at high concentrations.	Pressed-wood products, such as particleboard, MDF, and interior plywood; permanent-press drapes and fabrics; unvented heaters and stoves, tobacco smoke. (Outgassing from materials decreases over time.)	Sensitivity varies widely. Above 0.1 ppm, may cause watery eyes or burning sensation in eyes, nose and throat. Also nausea, coughing, chest tightness, wheezing, skin rashes, and allergic reactions. Possible carcinogen.	Typically below 0.01 ppm in older homes. In homes with significant amounts of new pressed wood products, may be over 0.3 ppm.	• Use exterior-grade pressed wood products with phenol-formaldehyde or MDI resin rather than urea-formaldehyde (UF). • Any UF-based wood products should be sealed with plastic laminate, vinyl, or a heavy paint or clear finish. • Reduce humidity levels. • Increase ventilation rates during first few months in a new home.
Biologicals (page 289)	Bacteria, molds, mildew, viruses, animal dander, dust mites, pollen.	Wet or moist walls, ceilings, carpets, furniture. Poorly maintained humidifiers, dehumidifiers, and air conditioners; bedding; pets and pests (roaches, mice, rats).	Eye, nose, and throat irritation. Shortness of breath, digestive problems, dizziness, lethargy, fever. Can trigger asthma or cause humidifier fever or influenza.	Indoor levels of pollen and spores are lower than outdoors unless molds are present. Dust mite levels are higher indoors. Levels are typically higher in warm, moist climates and in homes with damp conditions.	• Maintain interior humidity level at 30–50% • Vent kitchens, baths, dryers to outdoors • Clean carpets regularly with HEPA vacuum or central vacuum with outdoor exhaust. • Remove or clean and dry water-damaged materials within 24 hours. • Regularly empty water trays in air conditioners, dehumidifiers, and refrigerators. • Properly maintain cool-mist and ultrasonic humidifiers.

(continued)

TABLE 7-5 *(Continued)*

	Description	Primary Household Sources	Health Effects	Typical Exposure Levels	Steps to Reduce Exposure
Volatile Organic Compounds (VOCs) (page 289)	Volatile organic chemicals, found in many household products as well as paints, caulks, and adhesives, are released when products are used and, to some degree, when stored.	Paints, paint strippers, other solvents. Wood preservatives, aerosol sprays, cleaners and disinfectants, moth repellants, and air fresheners. Stored fuels, hobby supplies, dry-cleaned clothes.	Range from highly toxic to no known health effect. Safe levels in homes unknown. Effects may include eye, nose, and throat irritation, headaches, loss of coordination, nausea, and damage to liver, kidney, and central nervous system. Some may cause cancer.	Often 2–5 times higher indoors than outdoors. May be 1,000 times outdoor levels during and for several hours after activities such as paint stripping.	• Choose low-VOC products when possible. • Provide ample ventilation when using these products. • Buy small quantities and safely discard unused containers. • Closely follow manufacturers directions. • Avoid exposure to methylene chloride, benzene, and percholorethylene, all known or suspected carcinogens.
Pesticides (page 290)	A wide range of products used to kill household pests as well as fungi and microbes. Both the active and inert ingredients can be volatile organic compounds with similar health effects.	Insecticides, termiticides, rodenticides, fungicides, and disinfectants. Includes sprays, liquids, sticks, powers, crystals, and foggers.	See "VOCs" above for health effects. Highly variable and not well studied at low exposures. High levels of exposure to aldrin, dieldrin, heptachlor, or chlordane (now banned) cause headaches, weakness, dizziness, muscle twitching, nausea, and possible damage to liver and central nervous system. Possible carcinogens.	Research shows widespread presence of pesticides in homes. However, effects of long-term low-level exposures not well understood.	• Control pests with nonchemical methods. • Keep indoor spaces clean and well-ventilated. • Do not store pesticides in home. • Strictly follow instructions and apply only recommended quantities. • Mix or dilute outdoors. • Increase ventilation when using indoors. • Keep clothes with moth repellants in separate ventilated area.

(continued)

TABLE 7-5   **Common Indoor Air Hazards (*Continued*)**

	Description	Primary Household Sources	Health Effects	Typical Exposure Levels	Steps to Reduce Exposure
Lead (page 291)	A metal once commonly used in paints, gasoline, and water pipes. Considered the leading environmental hazard for children in the United States.	Deteriorating lead-based paint, dust from scraping or sanding paint, and water carried through lead pipes in older home. Contaminated soil close to homes with lead-based paint. Hobby activities, such as soldering and stained-glass making.	Low blood levels affect the central nervous system, kidneys, and blood cells and can impair children's mental and physical development. High-blood levels can cause convulsions, coma, and death.	Most homes built before 1960 contain heavily leaded paint. Homes built until 1978 may contain lead paint. Paint dust may occur from abrasion at window and door frames. High dust levels can be caused by remodeling activities. Lead dust can be inhaled or ingested.	In homes with lead-based paint: • Keep areas where children play clean and dust-free. • Leave paint undisturbed, if in good condition. Do not sand or burn off paint. • Use only a trained, qualified contractor to remove lead paint. • If occupation involves lead, change clothes and use doormats before entering home. • Feed children a diet rich in calcium and iron, so they will absorb less lead.
Asbestos (page 292)	A mineral fiber once widely used as a fire-retardant and in a wide range of building materials. The most dangerous fibers are too small to be visible.	Deteriorating or damaged pipe or furnace insulation, fireproofing, acoustical insulation, floor tiles, textured paints. Also roofing and siding.	Lung cancer, mesothelioma (chest and abdominal cancer), and asbestosis (irreversible lung scarring that can be fatal). Most cases occur many years after workplace exposure. Risk greatest to smokers.	Elevated levels may occur in home where asbestos-containing materials are damaged or disturbed by remodeling or other activities.	• Leave undamaged asbestos material alone if it is not likely to be disturbed. • Use licensed, qualified contractors for asbestos encapsulation, removal, or cleanup. • Follow proper procedures when replacing woodstove door gaskets that may contain asbestos.

SOURCE: Table adapted in part from *The Inside Story: A Guide to Indoor Air Quality,* 1995, by the U.S. EPA and Consumer Products Safety Commission.

Radon

Radon is a colorless, odorless gas released from the breakdown of uranium and radium, which is found in rocks and soil and sometimes in water. The gas enters the house primarily through cracks and gaps in the foundation, floor drains, and sumps, and concentrations build up indoors. Radon can also enter the home through well water and be released during showering or other uses. In rare cases, it is found in masonry building materials. Radon is thought to be the second leading cause of lung cancer in the United States, after smoking (Table 7-6).

Radon is drawn into buildings by the stack effect and by depressurization from mechanical equipment. During warm weather when the stack effect is reduced and buildings are often well-ventilated, indoor radon levels are usually one-third or more lower. Also, levels in the basement are typically over twice the level on the first floor.

Health Effects. Radon gas breaks down into short-lived decay products that can be inhaled either unattached or attached to other particles in the air and penetrate deeply into the lungs. According to its *2003 Assessment of Risks from Radon in Homes,* the EPA estimates that radon causes about 20,000 lung cancer deaths annually in the U.S. This makes radon the second leading cause of lung cancer in the United States, where an estimated 1 out of 15 homes has elevated levels. The cancer typically occurs 5 to 25 years after exposure, and the risk goes up dramatically if the person is also a smoker (see Table 7-6).

- *Drinking water.* While much less of a problem than airborne radon, radon in water is also a concern. If indoor radon levels are high and the household uses well water, the water should also be tested. In general, every 10,000 pCi/L of radon in household water contributes about 1 pCi/L (picocuries per liter) of radon to indoor air level. The radon gas is released from the water when it is aerated during showering, washing dishes, or laundering. There also may be an increased risk of stomach cancer from swallowing the water. According to the Centers for Disease Control, ventilating bathrooms, kitchens, and laundry rooms is usually adequate to reduce risks from radon in water. However, where water levels are high, the radon can be removed by aeration treatment or carbon filtering.

Measuring Radon. While some regions of the country have more homes with elevated radon levels, high indoor levels can occur anywhere. For that reason, the EPA recommends that all homes be tested. Although soil testing is possible prior to building, there is currently no reliable way to predict what household levels will be until a home is completed. Testing indoor levels is straightforward, using inexpensive test kits available from hardware stores or by mail order. Select a kit that is nationally or state-certified. The "action level" established by the EPA for remediation in homes is 4 pCi/L, although it recommends that people consider taking action at 2 pCi/L or above—since no exposure level is without risk. The average indoor level in the U.S. is 1.3 pCi/L, while outdoor levels average 0.4 pCi/L.

- *Testing methods.* Tests should be conducted away from drafts, high heat, and high humidity in a regularly used room on the lowest level in the home that is used as living space. Short-term tests last for 2 to 90 days, and long-term tests run for up to a year. Because radon levels vary daily and seasonally, longer test periods are better indicators of the average level. However, if two short-term tests yield an average result greater than 4 pCi/L, the EPA recommends taking steps to lower the level to 2 pCi/L or lower. Since 1985, millions of homes have been tested for radon, and an estimated 800,000 homes have been mitigated.

Radon Remediation. The EPA and the U.S. Geological Survey have rated every county in the United States as

TABLE 7-6	Lung Cancer Risk from Radon Exposure			
Radon Level (pCi/L: picocuries per liter of air)	Nonsmokers Lung Cancers per 1000 People*	Smokers Lung Cancers per 1000 People*	Cancer Risk for Smokers Compares to:	What To Do: If You Smoke, Stop, and
20	36	260	250 times the risk of drowning	Lower radon levels
10	18	150	200 times the risk of dying in a home fire	Lower radon levels
8	15	120	30 times the risk of dying from a fall	Lower radon levels
4	7	62	5 times the risk of dying in a car crash	Lower radon levels
2	4	32	6 times the risk of dying from poison	Consider lowering radon levels
1.3	2	20	(Average indoor radon level)	Reducing radon levels below 2 pCi/L is difficult.
0.4	—	—	(Average outdoor radon level)	

*Based on a person's average radon exposure over a lifetime at the given level.
SOURCE: Adapted from *A Citizen's Guide to Radon,* U.S. EPA 402-K02-006.

FIGURE 7-8 Radon Mitigation System.

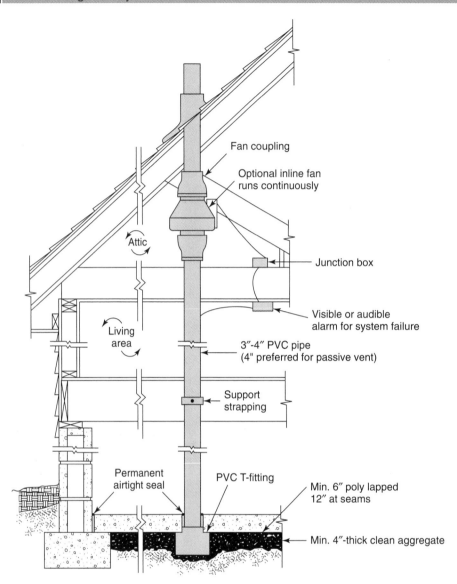

The EPA recommends that builders add a passive subslab depressurization system to all new homes built in areas prone to elevated radon levels (www.epa.gov/radon/zonemap.html). The key components are a gas-permeable layer below the slab, airtight sealing of the slab, and an exhaust duct to vent soil gases. If the passive system does not sufficiently lower radon levels, adding an inline blower will usually solve the problem.
SOURCE: Illustration adapted from *Building Radon Out, U.S. EPA/402-K-01-022,* 2001.

Zone 1 to 3 for radon risk. Links to state maps with county-by-county risk levels can be found at *www.epa.gov/radon/zonemap.html.* The EPA recommends that all homes in Zone 1 counties be built with radon-resistant features, which can be easily upgraded to a radon remediation system if needed. Since homes in Zones 2 and 3 can also have high levels, it is best to check with your state radon office to see if they are aware of any local "hot spots."

The techniques for radon-resistant building vary for different foundation types and site conditions, but all contain the six basic elements described below. Following these steps creates a *passive soil depressurization system,*

which sufficiently lowers radon levels in about 50% of homes requiring mitigation. If radon levels need to be lowered further, the system can be easily converted to an active system by adding an inline fan, which can meet the target levels in nearly all cases (see Figure 7-8). The goal of remediation is to lower the average indoor level to less than 4 pCi/L, and preferably 2 pCi/L.

• **Gas permeable layer.** This is usually a 4-inch layer of clean, course gravel installed beneath the slab for drainage, but which also allows the gas to move freely beneath the house. In areas where native soils are sufficiently permeable to build on, a loop of perforated pipe inside the footings is an option, and may also serve as a drain tile. The perforated pipe should be about 12 inches in from the foundation wall and 1 inch below the slab, with a minimum diameter of 3 inches for slabs under 2,000 sq ft and 4 inches for slabs up to 4,000 sq ft. Where subgrade soils are compacted or frozen, another option is to use geotextile drainage mats to create a gas-permeable layer on top of the subgrade but beneath the slab.

- *Plastic sheeting.* Lay minimum 6-mil polyethylene sheeting (or 3-mil cross-laminated sheeting) on top of the gas permeable layer. This helps keep soil gases from entering the home and also keeps concrete from clogging the gravel layer. Overlap seams by at least 12 inches, and repair any punctures or tears with tape or a patch of sheeting material.

- *Vent pipe.* Run a 3- or 4-inch PVC pipe from the gas-permeable layer up through the house to the roof to vent soil gases above the house. Where better suction is needed, connect the subslab tee to a minimum 10-foot length of perforated, corrugated pipe run horizontally in the gravel layer. The vertical pipe should be as straight as possible and should be located inside the insulated shell of the building to keep it warm, inducing the stack effect.

  Field data has indicated that 4-inch vent pipes work better than 3-inch vent pipes for passive systems. Some builders cap the stub just above the basement slab and connect the riser to the roof only if the house tests high for radon. If so, clearly label the capped pipe so no one mistakes it for a plumbing drain in the future.

- *Sealing and caulking.* Seal all cracks, perimeter joints, control joints, and other openings in the foundation floor with long-lasting materials to reduce soil gas entry. Seal large openings with expanding foam or nonshrink mortar or grout. Seal smaller holes with a high-grade elastomeric sealant conforming to ASTM C920-87. If the home has a sump, it should have an airtight cover and, if needed, can have a floor drain with a trap (filled with oil so it will not evaporate). If the sump is not connected to the drain tile loop, it can be vented into the radon system with a 3-inch pipe connected to a special sump cover available from suppliers of radon mitigation products. Also seal and caulk the rest of building envelop to reduce the stack effect in the home. The tighter the home, the less the building will draw radon out of the soil. Also tightly seal any return air ducts that pass through basements or crawlspaces.

- *Seal ducts and air-handling units.* Placing any return-air ductwork under the concrete slab is not recommended, since this will tend to draw radon into the ductwork and distribute it around the house. If supply ductwork must pass through a subslab space, it should be seamless or sealed airtight with durable aluminum tape or duct mastic. Also tightly seal any air-handling units or ductwork passing through basements, crawlspaces, or any areas in contact with the slab. In addition to saving energy, this will prevent the hvac system from drawing radon out of the soil.

- *Junction box.* Install an unswitched junction box in the attic or attached garage within about 6 feet of the vent pipe. A dedicated circuit is not needed. In the event that the passive system is not enough to keep radon levels below 6 Pci/L, then an inline fan will need to be added and run continuously. The fan should be located so that all positively pressurized sections of the system (from the fan to roof outlet) are located outside of habitable space. An active vent system should also have a visible or audible alarm to alert the occupants in the event of a loss of pressure or airflow in the vent pipe.

A postmitigation radon test of 2 to 7 days should be done within 30 days of system installation. For an accurate reading, all windows and doors must be closed 12 hours before and during the test, except for normal use for entry and exit.

Formaldehyde

Formaldehyde is a ubiquitous volatile organic compound (VOC) that occurs in nature and is widely used in building products, finishes, and furnishings because of its desirable properties and low cost. Nearly all products made with formaldehyde outgas to some extent, but only a few contribute significantly to indoor air problems.

Sources. Formaldehyde is used to add permanent press qualities to clothing and drapes, as a preservative in many paints and coatings, and as the adhesive resin in some carpeting, fiberglass insulations, and pressed wood products. It is also a product of combustion found in tobacco smoke and the fumes from gas stoves and other unvented combustion.

During the 1970s, it was used in urea-formaldehyde foam insulation (UFFI), which was blown into the walls of many homes in the U.S. and Canada and later banned after elevated levels of formaldehyde were found in a small number of homes. Testing has since shown that, in most cases, any excess formaldehyde was released within a few days of installation. Nonetheless, the material was removed from a large number of homes and banned for several years in the United States and permanently banned in Canada.

By far, the most significant source of formaldehyde in homes today is pressed wood products made with urea-formaldehyde resins. These include particleboard, interior hardwood paneling, and medium-density fiberboard (MDF), which has the highest concentration of urea-formaldehyde of any pressed wood product. The relative contributions of new materials to a single room are shown in Table 7-7.

Formaldehyde is normally present at low levels, usually below 0.03 ppm both indoors and outdoors. However,

TABLE 7-7	Contributions of Formaldehyde to Room Air	
Product	**Area or Use Pattern**	**Concentration (ppm)***
Particleboard (uncovered)	108 sq ft	0.16
Decorative paneling with print overlay	108 sq ft	0.11
Particleboard underlayment with cushion and carpet	172 sq ft	0.8
Furniture (MDF)	11 sq ft	0.06
Kerosene heater	8 hrs per day	0.05
Decorative paneling	108 sq ft	0.05
Gas oven	$\frac{3}{4}$ hours per day	0.02
Cigarettes	10 per day	0.02
Furniture (particleboard)	11 ft sq ft	0.01
Fiberglass insulation (under drywall)		<0.01
Exterior plywood subfloor	172 sq ft	<0.01
Particleboard underlayment with tile over	108 sq ft	<0.01
Resilient flooring	108 sq ft	<0.01
Carpeting	108 sq ft	<0.01

*Parts per million (ppm)

NOTE: Based on published emissions levels and their own tests, researchers at Oak Ridge National Labs modeled the formaldehyde contributions of common building materials to the air of a 11x16-ft. room with 0.5 air changes per hour. When there are several concurrent sources, the emission rates cannot be added; rather, the strongest source dominates.

TABLE 7-8	Formaldehyde Exposure Limits
0.03 parts per million	Average outdoor level
0.10 ppm	Recommended upper limit for residences by ASHRAE, ANSI, EPA.
0.10 ppm	NIOSH recommended upper limit for 15-minute exposure in workplace.
0.016	NIOSH recommended upper limit for 10-day-average exposure in workplace
0.40	Recommended upper limit for manufactured homes.
0.50 ppm	OSHA workplace limit.
0.75 ppm	OSHA 8-hour exposure limit.
0.80 ppm	Level at which most people first detect odor.
2.00	OSHA 15-minute exposure limit.

buildings with high levels of pressed wood products can have higher indoor levels. For example, many manufactured homes have levels well above 0.03 ppm, due to their relatively small volume and large surface area of formaldehyde emitting materials. HUD standards that limit the formaldehyde emissions of materials used in manufactured housing are designed to bring the ambient level to below 0.40 ppm, still over four times the 0.10 ppm limit recommended by most health and standards organizations, including ASHRAE and ANSI. No standard exists for site-built homes.

Health Effects. Sensitivity to formaldehyde varies widely. At levels between 0.40 and 3.0 ppm, most people experience watery eyes, burning sensations in the nose or throat, nausea, and difficulty breathing. Most people detect the chemical's pungent odor at about 0.80 ppm, but many can smell it at concentrations as low as .05 ppm. High concentrations may trigger asthma attacks, and there is strong evidence that some people can develop a sensitivity to formaldehyde from exposure. Formaldehyde has been conclusively linked to nasal cancer in rats, while human studies have suggested a link to nose and throat cancer in humans, but are not conclusive. Based on the current evidence the EPA and the International Agency for Research in Cancer have consider formaldehyde a probable carcinogen prompting the lower workplace limits suggested by the National Institute for Occupational Safety and Health (NIOSH) (see Table 7-8).

Reducing Exposure. The best way to limit exposure to formaldehyde is to avoid the use of bare pressed wood products made with urea-formaldehyde resins. Also avoid cabinets, flooring, and furniture finished with acid-catalyzed urea formaldehyde coatings, which emit a very high level of formaldehyde when new. Individuals with formaldehyde sensitivity should take further steps to avoid permanent press draperies, wallpaper, and conventional paints, many of which use formaldehyde as a preservative.

The highest emitting products are typically medium-density fiberboard, particleboard, interior hardwood plywoods, such as lauan, and prefinished interior plywood paneling. In addition to underlayments and decorative panels, these product are widely used in cabinets, countertops, shelving, and furniture.

Where possible, substitute solid wood, softwood plywood, or products certified as low emitters of formaldehyde. All exterior-grade plywood and pressed-wood products and all APA-stamped plywood panels use phenol-formaldehyde resins, which are more chemically stable than urea-formaldehyde and have negligible emissions. Homasote products are also free of formaldehyde and can be used as underlayment and sound insulation.

For cabinetwork, look for low-emitting substrates using phenol-formaldehyde or methylene diisocyanate (MDI) resin, such as SierraPine's formaldehyde-free medium-density-fiberboard called Medite II. There are also many "low-formaldehyde" panel products developed to comply with U.S. Department of Housing and Urban Development (HUD) requirements for manufactured housing. While these generally have lower formaldehyde emissions than noncertified products, they may still have over three times the emissions of products made with phenolic or MDI resin.

Where panel products with urea-formaldehyde resins must be used, they should be covered or coated on as many surfaces as possible. Panels covered with an impermeable facing, such a vinyl or plastic laminate, have low emissions. Another option is to coat the panels with two or more coats of a water-resistant finish, such as polyurethane, lacquer, or alkyd paint. In general, unless a finish is visibly thick and an effective vapor barrier, it probably has little effect on formaldehyde emissions.

Controlling heat and humidity is also important, since hot, humid conditions significantly raise the level of formaldehyde emissions. Sensitive individuals should also launder permanent-press draperies before using and should avoid newly painted rooms for several days. Prior to use, any new furnishings or surfaces with formaldehyde-based materials should be allowed to air out for several days to several weeks in a well-ventilated space. Generally, formaldehyde levels will drop off rapidly at first and eventually level off at very low levels. Monitoring of 40 new houses by Oak Ridge National Laboratory found that, after five years, nearly all houses, including those insulated with urea-formaldehyde (UF) foam insulation, had formaldehyde levels below 0.1 ppm.

Biological Pollutants

Biological pollutants are or were living organisms. At least some of these are found in every home. Common examples include molds, pollen, dust mites, animal dander, and cockroaches, as well as viruses and bacteria. Whether these are brought in inadvertently by humans or pets or ride along on houseplants or their soil, living organisms tend to stay longer and breed more successfully in warm, damp conditions. In fact, given temperatures of 50°F to 90°F and a material that stays wet for more than 48 hours, a colony of mold or other fungi will rapidly develop from their invisible spores, which are everywhere in our environment just waiting for the right conditions to spring to life.

Health Effects. Allergic reactions are among the most common health problems associated with indoor air quality. They are often connected with molds, pollen, animal dander (mostly from cats and dogs), and dust mites, which are microscopic animals living in carpets, bedding, and furnishings. Allergic reactions can range from annoying to life-threatening, as in a severe asthma attack. Common symptoms include watery, itchy eyes, runny nose, sneezing, nasal congestion, coughing and breathing difficulties, headaches, and fatigue.

- *Asthma.* Children and adults with asthma are particularly at risk. Asthmatics have very sensitive airways that react to irritants by narrowing, making breathing

difficult. Between 1980 and 1994, asthma rates in the United States rose by 75%, affecting over 20 million people today, including over 6 million children, according to the Centers for Disease Control and Prevention. Asthma "triggers" vary from person to person, but some of the most common indoor triggers are biological allergens, such as dust mites; molds; animal dander, urine, and saliva; and cockroach body parts, secretions, and droppings.

Reducing Exposure. Since most biological pollutants thrive in a moist environment, the key to reducing them is good moisture control. This starts with building houses correctly with good foundation drainage and waterproofing, proper flashings, continuous air and vapor barriers, and adequate ventilation. Household humidity levels should be maintained between 30% and 50%. Basements should only be finished if they are dry year-round and detailed so finish materials and carpeting are not wetted by capillary action or condensation. If a building sustains water damage for any reason, it is essential that the wet materials be dried or removed within 24 to 48 hours or mold will grow. While wood, concrete, and other solid materials can be cleaned and disinfected, porous materials should be removed and replaced.

Hvac equipment and appliances that come in contact with water are other breeding grounds for biological contaminants. Have all such equipment serviced regularly and keep filters clean. Air conditioners can help filter out pollen, but dirty coils and drain pans can also become a source of biological pollutants. If using humidifiers, clean them according to manufacturers' instructions and refill with fresh water daily. Evaporation trays in air conditioners, dehumidifiers, and refrigerators should also be cleaned frequently. Duct cleaning may also be justified if an occupant is suffering from allergies and a visual inspection reveals that the air ducts are contaminated with large deposits of dust or mold. If so, choose a reputable company that follows the standards of the National Air Duct Cleaners Association (NADCA).

Good housekeeping is also an important part of the strategy for controlling household allergens. Bedding should be washed at 130°F. Using a HEPA vacuum or central vacuum with an exterior exhaust is recommended. Minimizing the use of carpeting, upholstered furniture, and dust-collecting shelving and furnishings can also help by eliminating hiding places for dust and contaminants (see "Carpeting," page 292). In some case, portable or central air filtration may also play a role, but these are not a panacea for removing allergens (see "Air Cleaning Strategies," page 277).

Volatile Organic Compounds

Many organic compounds are used during construction. Others are used daily in cleaning fluids, cosmetics, and hobby materials. These include the solvents in paints, caulk, and adhesives, as well as the ingredients in hair sprays, carpet and oven cleaners, floor and furniture polishes, and pesticides. In its TEAM study, the Environmental Protection Agency found that the average level of 12 common organic pollutants was two to five times higher in houses than outdoors, although still 1,000 times less than short-term occupational limits. The health effects of high concentrations of VOCs vary from the highly toxic and carcinogenic to no known effect. The impact of long-term exposure at the levels found in households, however, is less well understood.

Health Effects. As with most pollutants, the health effect depends on individual sensitivities as well as the level and duration of the exposure. Common acute symptoms from moderate levels of exposure include eye and respiratory irritation, headaches, dizziness, visual disorders, and memory impairment. Effects on the nervous system are similar to those from alcohol consumption. Common chemicals that should be avoided include:

- *Benzene.* Benzene is a known human carcinogen. The main indoor sources are tobacco smoke, stored gasoline, and auto emissions from attached garages. It is also found in some adhesives, paints, furniture waxes, and detergents. Acute inhalation exposure may cause drowsiness, dizziness, and headaches, as well as eye, skin, and respiratory tract irritation, and, at high levels, unconsciousness.

- *Methylene chloride.* Found in paint strippers, adhesive removers, and aerosol spray paints, methylene chloride is known to cause cancer in animals and is considered by the EPA to be a probable human carcinogen. Also, it is converted to carbon monoxide in the body and can cause symptoms associated with CO (carbon monoxide) poisoning including decreased visual, auditory, and motor functions. Avoid use if possible or use outdoors.

- *Perchloroethylene.* This is the most widely used dry-cleaning chemical. The most common effects of moderate overexposure to perchloroethylene are irritation of the eyes, nose, throat, or skin, and nervous system effects, such as dizziness, headaches, and nausea. If dry-cleaned clothes have a strong odor, do not accept them until they have been properly dried.

Reducing Exposure. Paints and coatings, adhesives, sealants, and a variety of other building products and materials produce high concentrations of VOCs when they are first applied or installed. At these levels, even nonsensitive individuals might experience symptoms such as eye and respiratory irritation. To avoid problems, new homes should be allowed to air out for at least a couple of weeks before being occupied, particularly if the weather is too cold to leave windows open. In cold weather, the home should be heated with ventilation systems run at full speed to help drive off the volatile compounds.

To limit exposure to household VOCs, the best strategy is to find alternative products. When that is not possible, carefully follow directions, use in well-ventilated areas, and do not store partially used containers in living spaces.

Pesticides

Pesticides are a special class of organic chemicals designed to kill living organisms. In addition to the compounds used in the home and garden, the class of chemicals regulated as pesticides also include kitchen and bath disinfectants, flea and tick products, and swimming pool chemicals. In most cases, both the active ingredient targeted to one or more pests and the "inert" carriers are organic chemicals that are toxic to humans.

Studies indicate that up to 80% of most people's exposure to pesticides occurs indoors and that measurable levels of up to a dozen pesticides have been found in the air inside homes. Because of its widespread use for over 30 years, more than 80 percent of Americans already have traces of Dursban in their bodies, according to the Centers for Disease Control. Another study found Dursban in the carpet dust of 67 percent of homes surveyed.

Also, remember that a pesticide found "safe" to use today may be determined to be unsafe tomorrow. Chlordane, the most widely use termiticide for decades, was banned in 1988 because of its toxicity to humans and its persistence in the environment. It was largely replaced by Dursban (chlorpyrifos), an organophosphate. Dursban became the most widely used pesticide in the United States until it was phased out starting in 2000, along with the popular pesticide diazinon, because of the risks they posed to humans, especially to the growth and nervous system development of children.

Health Effects. There are nearly 900 pesticides registered for use in the United States. Nearly all are at least moderately toxic to humans and pets and many are highly toxic. Symptoms of overexposure to pesticides include irritation to the eyes, nose, and throat, headaches, blurred vision, nausea, loss of coordination, muscular weakness, and damage to the central nervous system, liver, and kidneys. Every registered pesticide has a "signal" word on the label, ranking the level of toxicity to humans, as follows:

- *Danger—Poison:* highly poisonous
- *Danger:* poisonous or corrosive
- *Warning:* moderately hazardous
- *Caution:* least hazardous

Some of the more problematic pesticides used in and around households include:

- *Organophosphates and carbonates.* These two classes of chemicals, including Dursban and Lorsban, kill insects by disrupting their nervous systems. Studies indicate that they affect the birth weight and neural development of infants. From 1993 to 1996, nearly 63,000 reports were made to U.S. poison control centers about residential exposures to organophosphates, according to the U.S. EPA. Almost 25,000 of these incidents involved children under 6, who are particularly vulnerable to organophosphate poisoning and at least 482 resulted in hospitalization.

- *Mothballs.* Mothballs contain either of the chemicals paradichlorobenzene or naphthalene. Paradichlorobenzene is classified as a possible human carcinogen by the EPA, and its vapors can irritate skin, eyes, and the respiratory tract. Large doses can damage the liver. Exposure to naphthalene promotes hemolytic anemia, associated with fatigue in mild cases and acute kidney failure in severe cases. Poisonings of infants have been reported after dressing the children in clothing stored in naphthalene mothballs.

Reducing Exposure. When possible, the best approach is to find nonchemical approaches to pests. When chemicals must be used, choose the least toxic option, and use it outdoors, if possible, and away from areas used by pets and children who will track it back into the house.

- *Use insect-resistant construction materials and techniques.* The use of termiticides can be reduced or eliminated by careful detailing of entry points, and by using alternative building materials, such as steel, masonry, concrete, insulating concrete forms (ICFs), or treated lumber. Borate-treated lumber is nontoxic to humans and very effective against termites and carpenter ants as long as it is not exposed to regular wetting.

- *Use nonchemical methods of pest and weed control.* Since outdoor pesticides and herbicides invariably end up indoors on carpets and in the air, it is prudent to reduce the use of chemicals indoors and out. Options include integrated pest management, biological pesticides, and planting disease-resistant plants.

- *If using chemicals, choose the least toxic.* Look for products with the signal word "warning" or "caution" rather than "danger." Baits and traps are better than sprays or "bug bombs."

- *Read the label and closely follow instructions.* If you must handle pesticides, wear gloves and long sleeves and avoid breathing the vapors. Always keep these chemicals away from children. Carefully follow directions with regard to concentration, protective gear, and restricting access to treated areas. Always ventilate the area well after use, and mix or dilute chemicals outdoors if possible.

- *Dispose of unwanted pesticides safely.* Most of these chemicals contain VOCs that will vaporize and get into the household air. If you cannot dispose of partially used containers, store outside the living space.

- *Minimize exposure to moth repellants.* When used, place in a well-sealed trunk or other container that can be stored in ventilated areas outside of the main living space, such as attics or attached garages. Paradichlorobenzene is also the active ingredient in many air fresheners and should be avoided.

Lead

In 1991, the Secretary of the Department of Health and Human services called lead "the number-one environmental threat to the health of children in the United States." The leading source of lead exposure today is old lead-based paint in homes built before 1960, although homes built until 1978 may also contain lead paint. Other sources include contaminated soil and drinking water that runs through old lead piping. Hobby activities, such as soldering and stained-glass making, can also introduce lead into the home.

Where two painted surfaces abrade, such as door and window frames, lead dust can be released and later ingested by children. High-level exposures leading to acute illness can be created when lead-based paint is removed by sanding, scraping, or open-flame burning. The soil around old houses can also contain high levels of lead from paint scrapings over the years, and the soil around highways can have high levels from leaded gasoline. Playing in contaminated soil can be a threat to children, and contaminated soil can also be tracked into homes, contributing significantly to indoor levels.

Health Effects. Lead affects most systems of the body. Even at low levels, harm to fetuses and young children can be significant. Blood lead levels as low as 10 micrograms per deciliter can impair mental and physical development, leading to lower IQ levels, shortened attention spans, and increased behavioral problems. Lead is more easily absorbed into the bodies of fetuses, infants, and children, and they are more sensitive to the damaging effects. Also, children often have higher exposures, since they are more likely to get lead dust on their hands and then put their fingers or other lead-contaminated objects into their mouths.

Acute exposures to high levels of lead generated from remodeling activities can cause adverse health effects on the central nervous system, kidney, and blood cells. At very high levels (above 80 micrograms per deciliter of blood), lead can cause convulsions, coma, and even death.

Reducing Exposure. Since lead paint is the leading cause of exposure, preventive measures focus on keeping paint in good condition and cleaning up any lead-containing dust before children are exposed. In older homes with lead paint, experts recommend mopping floors and wiping window ledges and other smooth flat areas with damp cloths frequently, keeping children away from areas where paint is chipped, peeling, or chalking and preventing children from chewing on window sills and other painted areas. Also, ensure that toys are cleaned frequently and hands are washed before meals. If the paint is in poor condition, it should be removed by a licensed lead-abatement professional. Recommendations include:

- Keep areas where children play as dust-free and clean as possible.

- Leave lead-based paint undisturbed if it is in good condition; do not sand, scrape, or burn off paint that may contain lead.

- If the paint needs to be removed, hire a licensed professional with training in lead abatement.

- Do not bring contaminated soil or lead dust into the home.

- If your work or hobby exposes you to lead, change clothes and use doormats before entering your home. Demolition and work along roads and highways are examples.

- Eat a balanced diet, rich in calcium and iron. A child who eats enough iron and calcium will absorb less lead.

Asbestos

Asbestos is found mainly in older homes in pipe and boiler insulation, asbestos shingles, textured paint, and floor tiles. It becomes a health hazard only if it is disturbed by cutting, sanding, or other remodeling activities. Loose, "friable" pipe insulation is a problem since it can be easily damaged and may spread fibers into the air.

Health Effects. The most dangerous asbestos fibers are too small to see. If inhaled, they can penetrate deep into the lungs and accumulate there. Asbestos can cause lung cancer, mesothelioma (cancer of the chest and abdominal linings), and asbestosis (irreversible lung scarring). Most people with asbestos-related diseases were exposed to high levels on the job or were exposed to asbestos fibers brought home on the clothes and equipment of workers.

Reducing Exposure. Undamaged asbestos is best left alone. If asbestos materials are more than slightly damaged, or if they need to be altered or removed due to a remodeling project or equipment replacement, hire a state-licensed asbestos-abatement professional to evaluate and then encapsulate or remove the asbestos.

Carpeting

Concerns about the health effects of carpeting first gained national attention in 1988 when new carpeting installed at the EPA headquarters in Washington, D.C., was linked to a rash of health complaints among EPA staff. While a definitive cause never was identified, experts focused on two main compounds: (1) the solvent-based adhesive used to install the carpeting and (2) the chemical 4-PC (4-phenylcyclohexene), a compound found in the synthetic latex backing used in 95% of all U.S. carpets. The compound 4-PC gives carpeting its distinctive "new carpet" odor and is detectable by most people at very low levels. Styrene, a known health hazard and suspected carcinogen, is also found in the latex backing.

Since 1988, over 500 people have made complaints to the Consumer Products Safety Commission (CPSC) about new carpeting. The most frequently reported symptoms have been watery eyes, runny nose, burning sensation in the eyes, nose, and throat, headaches, rashes, and fatigue. In response, the CPSC commissioned a study of off-gassing from new carpeting and identified 31 compounds, but none approached airborne levels known to be hazardous for short-term exposure. Long-term effects were not studied. While some suspected formaldehyde, a common respiratory irritant, it has not been used in the manufacture of U.S. carpeting since the late 1980s (with the exception of some vinyl-backed carpet tiles used in commercial installations).

Labeling Program. The Carpet and Rug Institute (CRI), an industry association representing carpeting manufacturers, also took action by launching its "Green Tag" program in 1992. The voluntary program tests new carpeting for four categories of emissions: total VOCs, styrene, 4-PC, and formaldehyde. Since national standards do not exist for carpet emissions, the industry established its own acceptable levels. While these might not be as stringent as some health advocates would like, they have led to a lowering of emissions by manufacturers eager to display the Green Tag label.

Labeling of Pads and Adhesives. Since 1992, the CRI program has expanded to include carpet pads and adhesives, suspected by some to be a greater source of volatile compounds than the carpeting itself. Also, while no chemical stands out as the source of most complaints, the synergistic effect of multiple compounds is not well understood. Also, the sensitivity to chemical emissions varies among individuals, making the effects of new carpeting on individual occupants difficult to predict.

Air Out Carpet Before Installation or Occupancy. Both CRI and independent health advocates agree that new carpet emissions drop off rapidly in the first 24 to 72 hours after being unrolled and exposed to ventilation air. By increasing ventilation during that time, or if possible, airing out the carpet for several hours to several days before installation, most of the chemical emissions can be avoided. In glue-down installations, seek out low-VOC adhesives rated at less than 50 grams of VOC content per liter of adhesive.

Carpeting Alternatives for Sensitive Individuals. Once installed, carpets can act as reservoirs for contaminants filtered from the air or tracked in on shoes, including hydrocarbons, pesticides, and other particulates. Also, in high-humidity conditions, dust mites, a powerful allergen, can thrive in carpets. In homes with small children, people with allergic conditions, or high-sensitivity individuals, consider alternatives to carpeting. Area rugs that can be washed periodically in 130°F water are an option. Where carpeting is installed, health experts recommend frequent vacuuming with a HEPA-type vacuum or central vacuum with an outside exhaust, and periodic deep cleaning using a hot-water extraction system.

Combustion Appliances

A combustion appliance is any device that burns fuel for heating, cooking, or decorative purposes. This includes central-heating systems, space heaters, water heaters, ovens and cooktops, woodstoves, and fireplaces. The major pollutants associated with combustion are carbon monoxide, nitrogen dioxide, sulfur dioxide, and particles (Table 7-9).

Unvented space heaters and gas stoves without range hoods dump combustion products directly into the living space and have no place in the modern home. Vented appliances, such as boilers, water heaters, and fireplaces, are designed to exhaust combustion products to the outdoors, but they are vulnerable to backdrafting in today's tightly built houses. When appliances are malfunctioning or out of adjustment, they produce more pollutants, including carbon monoxide. The combination of backdrafting and the high production of carbon monoxide can be deadly.

Health Effects of Combustion Products. Possible health effects from combustion products include eye and respiratory irritation, persistent coughing, headaches, fatigue, and dizziness. In the case of carbon monoxide, symptoms can include nausea and confusion, and, at very high levels, loss of consciousness and death. Effects associated with specific pollutants are discussed below:

- *Carbon monoxide.* CO is a colorless, odorless gas produced by incomplete combustion. Common sources include blocked chimneys or vents, cracked or rusted heat exchangers, poorly adjusted appliances, smoldering fireplaces, and auto exhaust from an attached garage. CO interferes with the blood's ability to deliver oxygen to the body. Low concentrations may increase chest pain in people with heart disease. Sustained concentrations above 70 ppm can cause fatigue, headache, weakness, and nausea, and may be confused with the flu or food poisoning. Fetuses, infants, the elderly, and people with anemia or heart disease are especially vulnerable. At very high levels, CO causes confusion, loss of consciousness, and death. CO alarms are programmed to sound before levels reach 100 ppm for 90 minutes, 200 ppm over 35 minutes, or 400 ppm over 15 minutes.
- *Nitrogen dioxide.* NO_2 is a colorless gas with an acrid odor at high levels. The primary source of NO_2 in homes is unvented gas and kerosene space heaters, gas stoves without a range hood, and stoves with continuously burning pilot lights. Studies have shown that homes with unvented gas appliances have elevated NO_2 levels, and there is some evidence linking this with impaired lung function and increased respiratory infections in children. At high levels, NO_2 is an eye, nose, and respiratory irritant. Children and people with asthma and other respiratory problem are more susceptible to exposure.

- *Sulfur dioxide.* SO_2 is a colorless gas with a pungent odor and is primarily associated with oil- and coal-burning appliances. At low levels of exposure, SO_2 can cause eye, nose, and respiratory tract irritation. At high exposures, it can cause the airways to narrow, leading to chest tightness and breathing problems. People with asthma are particularly susceptible to SO_2 exposure.
- *Particles.* The health effects of breathing particles depend on several factors, including the size and chemical makeup of the particles. In general, suspended particles can cause eye, nose, and throat irritation, and increased respiratory symptoms for people with chronic lung or heart disease. In addition, a number of pollutants, including the carcinogens radon and benzo(a)pyrene, attach themselves to small particles and are then inhaled and carried deep into the lungs.

Reducing Exposure. The three main sources of combustion products in household air are unvented appliances, appliances or flues that are broken or poorly adjusted, and backdrafting. To minimize exposure, follow these general guidelines:

- *Unvented space heaters.* Do not use unvented space heaters in living spaces. If required for temporary use, closely follow manufacturer's directions, open a window, and open doors to adjoining rooms. A persistent yellow-tipped flame is generally an indicator of poor adjustment and increased pollutants.
- *Cooking.* With gas ranges and cooktops, always use a range hood vented to the exterior. Choose appliances with electronic ignition rather than a continuously burning pilot light. Or replace with electric appliances.
- *Sealed combustion.* In new construction, avoid the use of atmospherically vented boilers, furnaces, or water heaters. Instead, use power-vented appliances, preferably with sealed combustion.
- *Inspections and maintenance.* Have central heating systems and water heaters inspected and adjusted annually. Inspect all flues and chimneys for blockages or damage and promptly repair any problems. Blocked or leaking chimneys or flues can result in serious illness or death from carbon monoxide poisoning.
- *Woodstoves.* Make sure doors in older woodstoves are tight fitting with intact gaskets. New stoves should meet EPA emissions standards. Burn only seasoned wood. Make sure there is adequate air for combustion and that the house is not depressurized by exhaust fans (see "Backdrafting," page 295).
- *Fireplaces.* Fireplaces should have inserts with tight-fitting doors and a dedicated outside air supply. Fireplaces should not be used if the house is depressurized by exhaust fans (see "Backdrafting," page 295).

TABLE 7-9   Combustion Products and Indoor Air Quality

	Description	Primary Sources	Health Effects	Typical Indoor Levels	Steps to Reduce Exposure
Carbon Monoxide	Colorless, odorless gas is a by-product of incomplete combustion, often produced by inadequate air supply or poorly adjusted burners. May be indicated by a persistent yellow-tipped flame.	Unvented kerosene and gas space heaters, gas stoves; leaking chimneys and furnaces; cracked heat exchangers; back-drafting from furnaces, gas water heaters, woodstoves, and fire-places; auto exhaust from attached garages; tobacco smoke.	• Low levels (70–100 ppm): headaches, eye and nose irritation, flu-like symptoms; chest pain in people with heart disease. • Medium levels (150–300 ppm): dizziness, drowsiness, impaired vision, nausea. • High levels (over 400 ppm): unconsciousness, brain damage, death.	In homes without gas ranges: 0.5–5 ppm. Near properly adjusted gas stoves: 5–15 ppm. Near poorly adjusted gas stoves: 300 ppm or higher.	• Keep gas appliances adjusted. • Do not use unvented heaters. • Vent gas stoves to the outdoors. • Supply outside air to fireplace and leave flue open when in use. • Inspect, clean, and tune heating system annually. • Do not idle car in garage.
Nitrogen Dioxide	Colorless gas with acrid odor is a by-product of combustion.	Kerosene heaters. Unvented gas heaters and gas stoves. Tobacco smoke. Auto exhaust from attached garages.	Eye, nose, throat irritation. Impaired lung function and increased respiratory infections in young children. Common asthma trigger.	About half outdoor levels in homes without combustion appliances. Often exceeds outdoor levels in homes with gas stoves, kerosene heaters, or unvented space heaters.	See steps under carbon monoxide.
Sulfur Dioxide	Colorless gas with a pungent odor is a by-product of combustion.	Primarily associated with oil and coal burning appliances. Also, unvented or malfunctioning kerosene, gas, wood, or coal burning appliances.	• Low levels: eye, nose, and respiratory tract irritation. • High levels: chest tightness and breathing problems. People with asthma are particularly susceptible.	May exceed outdoor levels in homes with unvented combustion appliances.	• Vent all combustion equipment to outdoors • Inspect, clean, and tune heating system annually
Respirable Particles	Particles released when fuels are incompletely burned. A number of pollutants, including radon, attach to small particles and get carried deep into the lungs.	Fireplaces, woodstoves, and kerosene heaters; tobacco smoke.	Eye, nose, and throat irritation; respiratory infection and bronchitis. Increased respiratory symptoms for people with chronic lung or heart disease.	The same as or lower than outdoors in homes without smoking or other strong particle sources.	• Vent all combustion equipment to outdoors. • Use only properly sized woodstoves with tight-fitting doors and EPA certification. • Inspect, clean, and tune heating system annually.
Second-Hand Tobacco Smoke	A complex mixture of over 4,000 compounds. At least 40 are known to cause cancer.	Cigarette, pipe, and cigar smoking	Eye, nose, and throat irritation; headaches; lung cancer (3,000 deaths per year). Especially in children: bronchitis and pneumonia; ear infections; decreased lung function; and increased frequency and severity of asthma episodes.	Homes with one or more smokers have particle levels several times outdoor levels.	Do not permit smoking in home, particularly if infants or toddlers are present.

NOTE: Table adapted in part from *What You Should Know About Combustion Appliances and Indoor Air Pollution,* 2005, by the Consumer Products Safety Commission, and *The Inside Story: A Guide to Indoor Air Quality,* 1995, by U.S. EPA and Consumer Products Safety Commission.

Backdrafting

While the trend is toward power-vented appliances, most furnaces, boilers, and water heaters still use atmospheric or "natural" venting. Atmospheric venting relies on the natural buoyancy of warm air in the flue or chimney to carry exhaust gases from the home. The strength of the draft depends on the temperature difference between the flue gases and outside air, the height of the chimney, and the indoor air pressure.

- *Leaky vs. tight homes.* While brief spillage of flue gases has always occurred in homes, natural-draft appliances had little trouble establishing and sustaining an adequate draft in older, leaky homes. Not only was there a ready supply of combustion and dilution air, but the flue gases were also hotter than they are in today's more efficient appliances.

In newer, tighter houses, significant negative pressures can be generated by kitchen and bath exhaust fans, gas dryers, and unbalanced air flows in the home's air distribution system. Unbalanced pressures can also be caused by leaks in return ductwork, by the use of building cavities as ducts, or by the simple closing of bedroom doors in homes with a central return register. Leaky return ductwork in a basement may be enough to backdraft a water heater or furnace.

- *Spillage vs. backdrafting.* If a naturally vented appliance lies in an area of the house with strong enough depressurization, the flue gases will spill into the home. When the flow reversal lasts for 30 seconds to a minute, it is called *spillage;* longer sustained spills are called *backdrafting,* a far more serious condition. Once backdrafting begins and the flue gets cold, it may be sustained for a long time. Research has shown that negative pressure of as little as 5 Pascals (Pa) creates a risk of backdrafting with naturally vented boilers and furnaces. Numerous studies have documented the prevalence of high negative pressures and frequent spillage in new homes built to current codes but not intentionally built airtight.

If the heating equipment is well adjusted and has adequate combustion air, the flue gases will contain primarily water vapor and carbon dioxide, along with nitrous and sulfur oxides, and particulates. If the burner is malfunctioning for any reason, it may put out large quantities of carbon monoxide and turn a backdrafting situation deadly. Fireplaces and poorly sealed woodstoves are most likely to reverse flow late at night when the fire is smoldering, producing a weak draft and high levels of CO.

Preventing Backdrafting. There are three key elements to preventing backdrafting:

1. *Equipment maintenance:* In existing equipment, make sure all burners are properly adjusted and that flues are properly sized and free of cracks or blockages. Inspect annually.

2. *Venting:* In new construction, eliminate all atmospherically vented appliances, including woodstoves and fireplaces.

3. *Depressurization:* Minimize depressurization by reducing exhaust fan sizes and balancing airflows in heating, cooling, and ventilation systems.

Chimney Problems. A chimney or flue that is too large, too small, or blocked by a bird's nest or loose brick will not draw properly and will be prone to spillage problems. Uninsulated chimneys on outside walls are also prone to poor draft and to condensation problems that can deteriorate flue materials. These problems should be fixed first before addressing problems inside the house.

Mechanical-Draft Appliances. Heating systems with fan-powered exhaust systems can withstand higher negative pressures than natural-draft appliances. Some types of fan-powered systems are much better than others, however. In order of effectiveness, the choices are:

- *Sealed-combustion.* Also called "direct vent," these appliances draw all combustion and dilution air from outside. These can typically tolerate negative pressures in the range from 25 to 50 Pa.

- *Power-vented.* These draw their makeup air from indoors and are also called fan-assisted, forced-draft, or mechanical-draft. These can typically tolerate up to 15 to 20 Pa of negative pressure.

- *Induced-draft.* These have a small fan added for energy performance, not to overcome house depressurization. These can typically tolerate 5 to 15 Pa of negative pressure.

By comparison, an atmospherically vented furnace can backdraft with as little as 5 Pa of negative pressure, and a gas water heater will have spillage at 2 or 3 Pa. Fireplaces can start having problems at about 3 Pa. Canadian codes limit negative pressures in homes with atmospherically vented equipment to 5 Pa. U.S. codes do not currently address the issue.

Reducing Depressurization. To keep indoor depressurization to a minimum, do not oversize bathroom and kitchen fans (see "Kitchen and Bath Ventilation," page 260), and avoid the use of downdraft and island fans, which can draw 600 cfm or more. If large fans must be used, they should be interlocked with a supply fan to provide makeup air. Canada's 1995 National Building Code requires that in homes with fuel-burning appliances vented through a chimney, any exhaust fan with a net capacity greater than 160 cfm must have fan-supplied makeup air. The makeup air fan should be sized to reduce the net exhaust rate to no more than 160 cfm and can be delivered to an adjacent room or through the forced-air distribution system. For example, a 300 cfm exhaust fan should have at least 140 cfm (300 minus 160) of makeup air.

How much an exhaust fan will depressurize a house depends on the tightness of the house. A 1993 study of several newly built energy-efficient homes in Minnesota found that exhaust airflows of 300 to 550 cfm depressurized the homes to 5 Pa, the level at which natural-draft appliances start having spillage problems. Other studies indicate that a 600-cfm exhaust fan can produce negative pressures from 3 to over 20 Pa, depending on house tightness. Without an adequate source of makeup air, a fan this size (or a combination of exhaust fans running at the same time) will pull air from the path of least resistance—often a nearby chimney or flue. Unless makeup air is provided, exhaust fans of this strength should not be used in homes with chimneys.

Worst-Case Test.

In homes with the potential for backdrafting, a simple test can be conducted to determine the likelihood of problems:

1. Close all interior doors except those leading to the furnace room and rooms where exhaust fans are located.
2. Switch on all exhaust fans, dryers, and other exhaust equipment, including the air handler if the home has forced-air heating.
3. Turn up the thermostat to turn on the boiler or furnace, and run hot water to turn on the water heater burner.
4. Hold a smoke indicator, such as an incense stick, about 3 inches from the draft hood of a gas furnace or water hater or near the barometric damper of an oil furnace. Test a fireplace near the top center of the firebox opening, and a woodstove near the doors or where the stovepipe connects to the stove.

Perform the test with the air handler both on and off, since unbalanced airflows can be a significant factor. If smoke spills into the room for more than 30 seconds at any combustion appliance, the home has a potential backdrafting problem that requires attention. A more scientific procedure for determining backdrafting potential, using a pressure gauge, can be found in Step 7 of the "Recommended Procedures for Safety Inspection" in Appendix H of the National Fuel Gas Code.

Fireplaces and Woodstoves.

Traditional open fireplaces and older leaky woodstoves burn very inefficiently and produce hundreds of chemical compounds, including carbon monoxide, organic gases, particulates, and some of the same cancer-causing agents found in tobacco smoke. Minor spillage of these pollutants occurs regularly, primarily when starting or stoking the fire. However, the larger concern is when the fire smolders late at night, producing high levels of CO and a weak draft. Backdrafting at this time can be dangerous or even fatal.

Another problem, particularly with fireplaces, is created when the fire is roaring and drawing up to 400 cfm of combustion air. At this point, its voracious appetite for air can cause backdrafting in other combustion appliances such as a gas water heater. Also, the need to reheat all the makeup air drags down the fireplace's heating efficiency to less than 15% and, if the fireplace is allowed to smolder all night, it becomes a net heat loser.

Woodstove efficiency has improved dramatically in response to EPA emissions standards (begun in 1988 and updated in 1990), which apply to most freestanding wood stoves and to fireplace inserts with air-supply controls and tight-fitting doors. To meet these standards, manufacturers use either a catalytic converter, similar to the ones used in cars, or a reengineered firebox. The new fireboxes have primary and secondary combustion zones capable of reaching system efficiencies of 60% or more and reducing combustion air intake to as little as 10 cfm. If installed with an outdoor air supply, these can be successfully decoupled from household air pressures.

While many fireplaces are fitted with glass doors, and some have outside air intakes, nearly all of the glass doors leak air. Even with low levels of depressurization, these fireplaces can still backdraft, and the fireplace's outdoor air supply might become the makeup air for the kitchen range hood or other exhaust fans, drawing fireplace fumes along with it. The best solution is an airtight fireplace insert.

To minimize pollution, indoors and outside, from wood-burning appliances:

- Choose a properly sized stove or insert certified as meeting EPA emissions standards.
- Make sure the door gaskets are in good shape, the doors fit tightly, and the stove is free of air leaks.
- Make sure the flue is the correct diameter and height, and have it inspected and cleaned annually.
- Use wood that has been split and dried for at least six months. Try to use small pieces, and do not overload the firebox. Leave enough room for air to circulate freely around the wood.
- For safety purposes, install a smoke alarm and carbon monoxide detector in the same room as the woodstove or fireplace.

RESOURCES

Manufacturers

Whole-House Ventilation Systems

American Aldes Ventilation Corp.
www.americanaldes.com
HRVs, ERVs, forced-air supply system, and multiport exhaust systems

Aprilaire (A division of Research Products Corp.)
www.aprilaire.com
ERVs, fresh-air hvac intake systems, electronic air cleaners, and media air filters

Broan-Nutone
www.broan.com
HRVs, ERVs, and remote inline-exhaust fans

Des Champs Laboratories
www.deschamps.com
Heavy-duty HRVs for residential and commercial applications

Fan Tech
www.fantech.com
HRVs, ERVs, whole-house HEPA filters, multiport exhaust fans, radon fans

Honeywell
www.honeywell.com
HRVs, ERVs, fresh-air hvac intake systems, electronic air cleaners, media air filters, and whole-house HEPA filters

Lifebreath (A divison of Nutech Brands Inc.)
www.lifebreath.com
HRVs, clean-air furnaces, and turbulent-flow precipitators

Memphremagog Heat Exchangers
www.mhetv.com
HRVs

Nu-Air Ventilation Systems
www.nu-airventilation.com
HRVs and integrated HRV and air handlers

RenewAire
www.renewaire.com
Static plate-core ERVs

Stirling Technology
www.ultimateair.com
ERVs with MERV 8 filtration and window-mounted ERV

Summeraire Manufacturing (A division of Trent Metals Ltd.)
www.summeraire.com
HRVs

Tamarack Technologies
www.tamtech.com
Multiport supply systems, fan timers, humidistats, and exhaust fans

Venmar Ventilation Inc.
www.venmar-ventilation.com
HRVs, ERVs, and whole-house HEPA filtration

Ventilation System Controllers

Aprilaire (A division of Research Products Corp.)
www.aprilaire.com
Fresh-air intake damper controls

Honeywell
www.honeywell.com
Complete line of controllers

Lipidex Corp.
www.lipidex.com
Air Cycler hvac ventilation timer

Tamarack Technologies
www.tamtech.com
Fan timers, humidistats, and exhaust fans

Tjernlund
www.tjernlund.com
Complete line of controls

For More Information

American Academy of Environmental Medicine (AAEM)
www.aaem.com

American Lung Association Health House
Indoor air quality information
www.healthhouse.org

Association of Home Appliance Manufacturers (AHAM)
Standards for air cleaners
www.aham.org

Carpet and Rug Institute (CRI)
www.carpet-health.org

Consumer Products Safety Commission (CPSC)
www.cpsc.gov

Healthy House Institute

Help with improving the indoor environment

www.hhinst.com

Home Ventilating Institute

www.hvi.org

National Center for Environmental Health

Air pollution and respiratory health information

http://www.cdc.gov/nceh/airpollution

Occupational Safety and Health Administration (OSHA)

www.osha.gov

U.S. Environmental Protection Agency (EPA)

Indoor air quality information

www.epa.gov/iaq

Index